SYDNEY OPERA HOUSE GLASS WALLS – 50 YEARS ON

J. A. Hemsley

Published by the Society of Glass Technology

United Kingdom Government Registered Charity, Number 237438

The objects of the Society are to encourage and advance the study of the history, art, science, design, manufacture, after treatment, distribution and use of glass of any and every kind.

These aims are furthered by meetings, publications, the maintenance of a library and the promotion of association with other interested persons and organisations.

Society of Glass Technology
9 Churchill Way
Chapeltown
Sheffield S35 2PY
England

Tel: +44(0)114 263 4455
E-mail: info@sgt.org
URL: http://www.sgt.org

First published 2023

A catalogue record for this book is available from the British Library

ISBN 978-0-900682-95-7
EAN 9780900682957

Copyright on text and new illustrations
© J. A. Hemsley, 2023
Front cover photograph by courtesy of N.S.W. Government

Typeset by Riverside Publishing Solutions Ltd, Salisbury, UK
Printed and bound by Bookvault, Peterborough, UK

CONTENTS

Preface vii

Glass walls, Sydney Opera House 1
1. Introduction 1
2. Site location: historical backdrop 5
3. Design of main glass walls 7
   - 3.1. Roof structure geometry and terminology 7
   - 3.2. Evolution of glass wall schemes 14
   - 3.3. Wind-tunnel tests 22
   - 3.4. Glazing steel support structure 28
   - 3.5. Glass panel support system 30
   - 3.6. Choice of glass 36
   - 3.7. Design of remaining glass walls 40
4. Architect's proposed patch-support glazing 42
5. Research on laminated glass 47
   - 5.1. Basic rationale 47
   - 5.2. Preliminary loading tests on patch fixing system 48
   - 5.3. Bending resistance of laminated glass 51
   - 5.4. Laminated glass under sustained loading 54
   - 5.5. Dynamic fatigue loading of laminated glass 55
   - 5.6. Full-scale loading tests 57
   - 5.7. Calculated design bending stresses: primary loading 62
   - 5.8. Calculated design bending stresses: secondary loading 65
6. Glass colour measurement and control 67
   - 6.1. Standard glass specimens 67
   - 6.2. Chromaticity definitions 67
   - 6.3. Measured results 70
   - 6.4. Interpretation of measured data 72
7. Construction of main glass walls 73
   - 7.1. Erection of support structure 73
   - 7.2. Manufacture of laminated glass panels 75
   - 7.3. Shaping of laminated glass panels 80
   - 7.4. Glazing operations 83
   - 7.5. Construction of remaining glass walls 90

8.	Post-construction assessment	90
9.	References	107

A1. Elastic flexure of simply supported laminated glass beam — **115**
A1.1. Introduction — 115
A1.2. Governing equations — 117
A1.3. Beam under symmetric partial uniform loading — 118
    A1.3.1. General case — 118
    A1.3.2. Limiting case of complete uniform loading — 126
    A1.3.3. Limiting case of central line load — 129
    A1.3.4. Other load cases — 134
A1.4. Beam without overhang under symmetric four-point line loading — 134
A1.5. Beam with overhang under symmetric four-point line loading — 139
A1.6. References — 148

A2. Full-scale loading tests on glass panels — **151**
A2.1. Introduction — 151
A2.2. Equations of flexure — 152
A2.3. Test arrangement — 154
A2.4. Test results — 156
A2.5. Reference — 169

A3. Elastic flexure of uniformly loaded thin monolithic rectangular plate simply supported on opposite edges — **171**
A3.1. Introduction — 171
A3.2. General solution — 171
A3.3. Uniformly loaded plate — 175
A3.4. Numerical results — 182
A3.5. Supplementary computed results — 194
A3.6. References — 220

A4. Hailstone impact: illustrated historical prologue, with literary allusions in prose and verse — **223**
A4.1. Introduction — 223
A4.2. Ancient history (pre 500 A.D.) — 224
A4.3. Early Middle Ages (500–1100) — 269
A4.4. Late Middle Ages (1100–1500) — 275
A4.5. Early modern period (1500–1750) — 294
A4.6. Late modern period (1750–1900) — 316
A4.7. Near-contemporary history (post 1900) — 346
A4.8. Hail prevention measures since antiquity — 353
A4.9. References — 370

A5. Hailstone impact: simplified flexural analysis of laminated glass beam-strip   **393**
A5.1. Introduction   393
A5.2. Hailstone characteristics   397
A5.3. Basic hailstone dynamics   404
A5.4. Elastic beam-strip modelling   421
A5.5. References   427

Principal notation   **441**

Subject index   **445**

PREFACE

The glass walls that envelope the three separate buildings of the Sydney Opera House complex were constructed some fifty years ago (1970–1972), with the official opening taking place in 1973, and the present monograph celebrates this notable golden anniversary. Its primary purpose is to describe and explain the glass science and technology aspects of the project, for which the writer – then based in London and Sydney – was largely responsible. This inclusive first-hand account represents a non-profit enterprise created without external funding or sponsorship, free of tribalism and of specific constraints or hidden agendas, and written without fear or favour, or any trace of *quid pro quo*. The interpretation of events and the views expressed herein are solely those of the writer, and are completely independent of those held by any commercial company or non-commercial organisation, government agency, public or professional institution or corporation, or any private individuals past or present. The intended readership ranges widely, from architectural and construction professionals to inquisitive members of the general public.

The glass walls constituted a pivotal strand in the latter stages of the protracted Opera House saga, and substantial technical challenges were encountered in their design, to a marked degree epitomising the earlier convoluted conundrums associated with the shell structures. Moreover, following a prolonged cessation in on-site construction in the 1960s, it was essential to build the external glass walls with minimal further delay so that the internal fit-out could proceed apace. In stark contrast to the present era, there were very few practitioners in the field of advanced architectural glazing, with little formal published guidance on several key topics: since when an entire sub-culture of '*façade* engineering' has developed exponentially, along with aficionado companies and all manner of international conferences, symposia, colloquia and similar jamborees, some with hundreds of participants. This increased activity has been accompanied by a veritable avalanche of published material to satisfy a voracious worldwide appetite. Indeed, it is now possible to be unaware not only of an individual scientific paper, but even of the journal in which it appears.

What follows in this narrative is essentially a coherent amalgam of several published papers and unpublished reports by the writer, augmented by selected extracts from other sources and encompassing the architectural evolution of various glass wall schemes over almost a decade, together with a broad range of hitherto unpublished material. For although most detailed records of the original theoretical and experimental work are no longer available, further particulars have been recovered from personal archives to give a more embracive contribution, including additional colour photographs taken by the writer during both full-scale loading tests and on-site construction. Moreover, typical distributions of calculated bending stress across the laminated glass sections are shown graphically, and new theoretical results for beam and plate flexure have been incorporated into self-contained appendices.

Further technological discussion on possible hailstone impact damage is preceded by a liberal historical prologue on hailstorms comprising a modest excursion into classical art and literature, referred to privately as the *Coronavirus Chronicles*, and written primarily for individual recreation and cultural edification during the recent global pandemic; a period of comparative isolation that otherwise might have been squandered away in a circle of worthless follies. The inclusion of such extra-curricular material is not intended merely as an idiosyncratic indulgence or an elitist distraction, but hopefully will impart some tangible benefit to both specialist and general readers, commensurate in principle with the public entertainment bestowed by the customary performing arts staged within the opera and concert halls. Indeed, it might be said that the international scope of the quoted artistic sources reflects the world-wide participation of architects, engineers and others in the compendious design and construction of the Opera House. For despite the continuing trend towards specialisation in a post-industrial world, there can surely be an imaginative affinity between disparate activities in art and science, to the extent that one can enlighten the other.

The foremost design objective for the glass walls was to produce an innovative three-dimensional glazed surface that would remain safe and durable over a period of many decades, and that would complement the original graceful edifice created by a visionary architect. Following a demonstrably promising start, underscored by the absence thus far of any serious shortcomings, it is just possible that such worthy aims will continue to be achieved long into the future, as also envisaged for the main building structure. Upon completing the first half-century, perhaps '*we may allow ourselves a brief period of rejoicing*', to quote a Churchillian phrase celebrating victory on an earlier momentous occasion. Yet, as ever, 'the show must go on', continuing the Opera House tradition with the same optimistic and energising spirit in all its multifarious guises. Few original participants in this wider Antipodean adventure remain alive, so in part this modest offering is an enduring tribute to their steadfast endeavours.

The present tome is the fifth and final volume of the writer's 'memoirs' covering various aspects of engineering science derived mostly from personal experience, prepared partially as a valedictory assignment yet content at the future prospect of a less demanding professional schedule. Although ill-health and decrepitude have prevented more rigorous checks on the theoretical results, unwavering efforts have been directed towards providing full and accurate reportage, whilst any remaining mistakes or blemishes are entirely the responsibility of the writer. Although focussed predominately on the singular engineering needs of an iconic building structure, and furnishing a reasonably comprehensive account of project analysis and design, the narrative may also be of interest to a wider audience in view of the abundant tranche of generic material. Perhaps readers might stumble across a few golden nuggets among a profusion of harsh technicalities, whilst finding enjoyment and inspiration in the cited examples of art and literature, albeit presented in a somewhat unconventional setting. That is all.

John A. Hemsley
Christchurch, Dorset

GLASS WALLS, SYDNEY OPERA HOUSE

1. Introduction

Just as the structure of the Sydney Opera House is unique in its shape and complexity, so are the 'glass walls' that enclose the openings between the roof shells and the podium. Many of the problems encountered in the design and construction of these walls were entirely without precedent, and had to be solved from first principles. Moreover, such solutions were needed urgently in order to achieve the earliest possible building completion, following prolonged delays of an unusual nature. The objective of this narrative is firstly to provide a brief general description of the project, and then to summarise those aspects directly concerned with glass and glazing, which formed the core of the writer's original work during the building design stage. The present expanded text follows similar lines to that of an earlier publication – namely the first chapter in a monograph [1] covering wider issues of glass in engineering science – with minor corrections and modifications, augmented by substantial additional data and illustrations. Further detailed material on specialised topics is consigned to several new appendices.

Safety considerations resulted in the deployment of 3-ply laminated glass panels, where both intact and fractured layers of annealed glass are held together by means of strong adhesion to a tough plastic interlayer, thereby preventing the hazardous detachment of sharp glass shards. But because little was then known on the structural behaviour of architectural laminated glass, with a dearth of published technical literature and no formal codes of practice, a robust and reliable design method had to be developed without undue delay. In a combined approach, theoretical studies and laboratory tests were undertaken to meet this objective, extending well beyond the conventional roles of architect and engineer.

The original theoretical results derived by the writer covering the flexure of laminated glass beams – abstracted from possibly the first such publication to appear in the technical literature – are reproduced in appendix A1, together with a more general analytical continuation formulated principally for checking purposes and including some new results. Archive material recovered from personal records is presented in appendix A2 to provide further details of the full-scale pressure loading tests on glass panels. A recent comprehensive analysis of the elastic flexure of a monolithic rectangular plate supported only on two opposite edges is summarised in appendix A3, and includes benchmark results that hitherto have not been published. This latter work assists in understanding the anticlastic deflection mode of the laminated glass panels under self-weight and wind loading, and may also be of wider use in applications with different materials. The possible effects of hailstone impact also receive attention, with a historical literary prologue of novel composition presented in appendix A4, and a simplified dynamic flexural analysis given in appendix A5.

The establishment of a modern opera house for the State of New South Wales in Australia was first officially promoted in 1954, two of the leading advocates being the State Premier John Joseph Cahill (1891–1959), and the English musician Eugene Aynsley Goossens (1893–1962), resident conductor of the Sydney Symphony Orchestra, who felt strongly that Sydney Town Hall was utterly inadequate as a venue for concerts and that a new purpose-built hall was sorely needed. The site chosen was Bennelong Point, a small peninsula extending into Sydney Harbour that previously had been occupied by the military Fort Macquarie and later by the city tram depot.

Early in 1957, an international design competition for a new opera house was won by Jørn Utzon (1918–2008), then a little-known Danish architect, whose last-minute entry in December 1956, outlined in only 12 drawings, was selected from over two hundred submissions [2]–[4]; not least due to the enthusiastic support of the Finnish-American architect Eero Saarinen (1910–1961), who extolled the merits of Utzon's design proposals and recognised their marked contrast to the quotidian. Figure 1(a) shows one of the competition entry drawings (east elevation) submitted in 1956, while in Figure 1(b), Utzon is seen with a Danish 1:200 scale model of the basic structure at Sydney Town Hall in August 1957. Further preliminary designs were prepared by Utzon [5] in March 1958, and work on site started in March 1959 [6]. Construction was divided into three stages: stage 1 (foundations and part podium); stage 2 (remaining podium plus roof shells); stage 3 (glass and louvre walls, auditoria, podium cladding, services and finishes).

Early in 1966, Utzon withdrew from the project in unfortunate circumstances – his design contract having effectively been terminated by the Client – and returned to Denmark shortly thereafter, alas never to return to Sydney. Advocates of a return to the *status quo ante* did not prevail, and Utzon was replaced by the local architectural consortium of Hall, Todd and Littlemore, whose principal designer was Peter Hall (1931–1995). By this time, construction

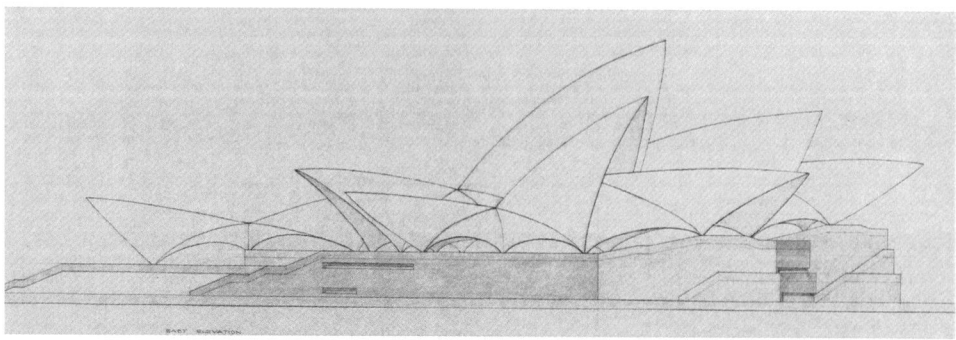

(a)

FIGURE 1. Winning design of architect Jørn Utzon for new Sydney Opera House: (a) competition entry drawing (east elevation), 1956 (by courtesy of *New South Wales State Archives, Sydney*)

(b)

FIGURE 1. (continued) Winning design of architect Jørn Utzon for new Sydney Opera House: (b) Utzon (left) with Danish 1:200 scale model of Opera House at Sydney Town Hall, 1957 (by courtesy of *National Museum of Australia, Canberra*)

of the shells was almost complete, but no acceptable design for the glass walls had yet been produced; a key issue in the evolution of the entire project, and resulting in moribund construction site activity for several years. A fresh perspective on this challenging issue was urgently required.

The basic concept finally selected, namely that of a continuous glass surface supported by a steel structure, dates from 1967, although detailed design did not commence until the following year. Construction of the glass walls took place during the period 1970–72, and the Sydney Opera House was officially opened by Her Majesty Queen Elizabeth II in October 1973; see also the contemporary views in Figure 2. As one of the great buildings of the modern era, this civic monument has since held an international reputation as a centre of excellence for the performing arts, and is now a U.N.E.S.C.O. World Heritage Site.

Papers and reports describing various technical issues that were written during or shortly after construction are to be found in the literature [7]–[35]. The Sydney Opera House rapidly acquired an iconic status, even before completion, attracting the attention of numerous authors and resulting in a mixed assortment of books and other documents covering most aspects of the project [36]–[101]. Countless articles have also been published worldwide in newspapers, magazines and journals; the overwhelming majority, it must be admitted, written by those not actively engaged in its design and construction, leading inevitably to many oversights and omissions. Moreover, the advent of the ubiquitous computer internet has increased the scope for unregulated and inaccurate commentaries.

(a)

(b)

FIGURE 2. Sydney Opera House in September 1973 from harbour viewpoints (photographs by the writer): (a) glass walls A4 (right) and B4 (left); (b) side elevation looking eastward

In what follows, the writer can affirm the veracity of information concerning glass technology and glazing, to the best of his recollection, while endeavouring to base the remaining text covering other aspects (*e.g.* architectural form, main concrete and steel structures) upon patently reliable sources. The various experimental studies supervised and mostly conducted by the writer took place in Sydney; the corresponding numerical stress analyses were carried out by the writer in London, in conjunction with visiting glass and related manufacturing facilities worldwide. The principal objective of these investigations was to ensure the long-term structural integrity of the laminated glass panels and support fixings, whilst also seeking where possible to enhance the aesthetic quality of the glass walls.

2. Site location: historical backdrop

The fascinating history of the Opera House site can best be encapsulated pictorially. Its geographical location is named in recognition of Woollarawarre Bennelong (*circa* 1764–1813), a gifted member of an indigenous Aboriginal clan who, following his capture in 1789, learnt to speak English and served as an interlocutor between the Eora people and the British settlers. For a while he also lived with his family in a small dwelling on the headland at Tubowgule (now Bennelong Point), built by order of Captain Arthur Phillip who had landed nearby in 1788 as commander of the First Fleet. The early development of the site is conveyed by the 19th-century watercolour paintings shown in Figure 3. An anonymous painting, *circa* 1804 and entitled *Bennelong Point from Dawes Point*, is shown in Figure 3(a). A wider perspective (*circa* 1806) is depicted in *Eastern view of Sydney Cove* by the pardoned English convict John Eyre (1771–) in Figure 3(b), with Bennelong Point – originally a small island at high tide – located centre-left. Figure 3(c) portrays a painting entitled *View of Sydney Cove*

(a)

FIGURE 3. Historical views of Bennelong Point, Sydney, in 19th-century watercolour paintings: (a) anonymous painting *Bennelong Point from Dawes Point*, *circa* 1804 (by courtesy of *Mitchell Library, State Library of New South Wales, Sydney*)

(b)

(c)

FIGURE 3. (continued) Historical views of Bennelong Point, Sydney, in 19th-century watercolour paintings: (b) *Eastern view of Sydney Cove* (*circa* 1806) by John Eyre (by courtesy of *Mitchell Library, State Library of New South Wales, Sydney*); (c) *View of Sydney Cove from Bunker's Hill* (*circa* 1836) by Conrad Martens (by courtesy of *National Gallery of Victoria, Melbourne*)

(d)

FIGURE 3. (continued) Historical views of Bennelong Point, Sydney, in 19th-century watercolour paintings: (d) *Fort Macquarrie, Sydney Harbour, N.S.Wales, December 13th, 1878* by Thomas Glover (by courtesy of *National Library of Australia, Canberra*)

from Bunker's Hill (*circa* 1836) by the English artist Conrad Martens (1801–1878), which features Fort Macquarie (1821–1901) located on Bennelong Point. This fort was built as a military installation for defensive purposes, holding several naval cannon, and also appears in the much later watercolour shown in Figure 3(d), painted in 1878 by the talented amateur Thomas George Glover (1826–1881), formerly of the Bengal Engineers during service in the Indian Army.

Eventually Fort Macquarie was demolished and replaced by a tram depot (1902–1958) – whose crenulated construction resembled that of the fort – including a loop line and sidings, as part of the city transport system. The albumen photoprint (*circa* 1906) in Figure 4(a) shows a paddle-wheel tug boat in the foreground, with the tram sheds in the far distance. Figure 4(b) is a photograph taken from a Harbour Bridge pylon in the 1940s, with Government House on the far right and the Garden Island naval base in the distance; while also noting the passenger ship moored alongside. The 1958 aerial photograph in Figure 4(c) also features the Fort Macquarie tram depot, which closed in 1955 and was demolished in 1958, while the aerial view in Figure 4(d) displays an Opera House setting in joyous celebration mode in May 2010, complete with flags, bunting and boats.

3. Design of main glass walls

3.1. Roof structure geometry and terminology

Over a period of years following the competition entry (Figure 1), in which the main shells were geometrically undefined, several structural roof schemes were developed based on parabolic, elliptical and circular profiles; which notable results – although not always

(a)

(b)

FIGURE 4. Photographic records of Bennelong Point, Sydney: (a) paddle-wheel tug boat (*circa* 1906) against background of Fort Macquarie tram sheds (by courtesy of *State Library of New South Wales, Sydney*); (b) Fort Macquarie tram depot, 1940s (by courtesy of *City of Sydney Archives*)

(c)

(d)

FIGURE 4. (continued) Photographic records of Bennelong Point, Sydney: (c) aerial view, Fort Macquarie tram depot, 1958 (by courtesy of *Max Dupain Photography & Mitchell Library, State Library of New South Wales, Sydney*); (d) aerial view, Sydney Opera House, May 2010 (by courtesy of *Pavel Sigarteu*)

infallible – stemmed in no small measure from the outstanding cognitive abilities of Ronald Jenkins (1907–1975) and John Blanchard (1928–1997) in London, especially in spearheading novel methods of structural analysis [102]–[104], including the use of tensor matrices. The final external surface geometry took the form of a single sphere of 75 m radius. The derivation of this elegant solution is indicated by the 1961 drawing attributed to Utzon shown in Figure 5(a), where each half-shell appears in elevation as a curvilinear triangle standing on one vertex, whilst the sectioned wooden model in Figure 5(b) more clearly illustrates the three-dimensional nature of the segmented roof formation. Besides simplifying the structural analysis, while preserving much of the sculptural quality of the original design, the resulting

(a)

(b)

FIGURE 5. Final surface geometry of Sydney Opera House roof structure derived from single sphere: (a) Utzon drawn version, 1961; (b) sectioned wooden model

geometric discipline of a *constant* shell curvature greatly facilitated the production of concrete roof segments, especially as the elaborate formwork moulds could be put to multiple use.

Upon reflection, however, by encompassing the enormous advances in computer technology over the past sixty years, particularly in mathematical and structural modelling, together with developments in materials and construction technology, it would be of consummate interest to carry out an in-depth retrospective analysis based on the *original* architectural form, to determine whether a workable engineering stratagem would now be possible or even feasible. For although the adopted discipline of spherical geometry was a brilliant contemporary solution derived specifically to overcome formidable technical difficulties, some would claim that the original subtlety and free-flowing artistry of the original shell cluster became somewhat diminished, as well as reducing the volume available for internal modules.

With regard to general terminology, the broad layout of the Opera House is sketched in Figure 6, representing an extensive building complex or 'arts centre' rather than a unitary structure. The main buildings are the concert hall, opera hall and restaurant, which stand on the variable-level podium substructure. The roof shells of the concert hall are prefixed with the letter A, and the top of the largest shell (A2) is 54.6 m above its arched springing points. The length of the concert hall roof structure is 121 m, measured between the tips of the end shells, and the widths of the main shells range from 22 m to 57 m. The opera hall is geometrically similar to the concert hall, but smaller, and the shells of the roof structure are numbered in the same way but prefixed with the letter B. The glass walls are referred to by the names of the shells they enclose. Each shell has a glass wall except A2, A3, B2 and B3, which are connected to the shells below by bronze louvre walls.

These so-called 'shells' are not strictly shells in the structural sense, as support from intrinsic membrane action is comparatively small. Referring to the concert hall, for example, the main shells A1, A2, A3 and A4 – which might better be described as 'fan vaults' – are made up of contiguous ribs to form ogival arches, which spring from reinforced concrete pedestals on the podium, east and west symmetric half-shells meeting at a ridge beam at the

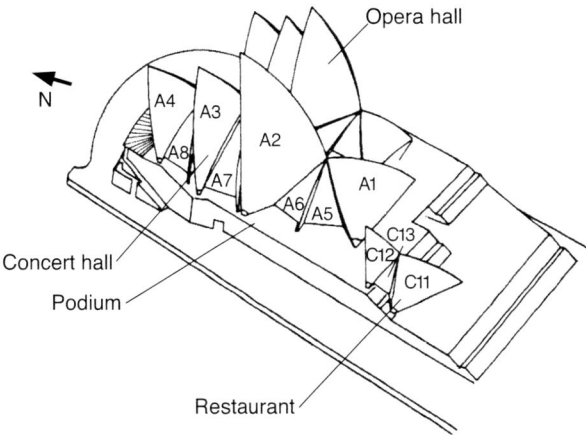

FIGURE 6. Simplified sketch of general layout of Sydney Opera House main structure

top along the central axis of the building. These ribs (arc-lengths up to 64 m) comprise a series of precast concrete segments that are post-tensioned. The side shells A5, A6, A7 and A8, whose outer geometric surfaces (spherical triangles) are based on the same reference sphere, span between the main shells on each side of the building, with narrow transitional warped surfaces connecting adjacent shells, defined by straight lines joining points equidistant from the origin along each circular boundary. Again, the opera hall layout is similar, whereas the smaller restaurant building has only two main shells C11 and C12, together with a pair of side shells C13 (east and west). Figure 7 illustrates the general state of construction in May 1969.

The colour and texture of the cladding to the concrete roof structure was of major significance to Utzon, who chose to develop special ceramic tiles in conjunction with the Höganäs company in Sweden [10]. After many prolonged trials, a total of around one million off-white and cream-coloured tiles were manufactured, typically 120 mm square and 10 mm thick, in a subtle combination of matt and glazed surface finishes to give the desired vitreous sheen or lustre; as partially illustrated in Figure 8(a) for a location near the base of one of the concrete pedestals. The unique visual qualities of this tile pairing under various conditions of natural lighting naturally had a strong influence on the eventual selection of glass colour. For the main shells, the tiles were affixed to curved precast concrete panels (usually referred to as 'tile lids'), chevron-shaped (2.3 m long, maximum weight 2540 kg) and encompassing a contrasting surface finish between the inner glossy tiles and the matt tiles on the periphery. These panels were bolted on to the shell to form a decorative zig-zag pattern, with the lateral

(a)

FIGURE 7. General state of construction of Sydney Opera House in May 1969 from harbour viewpoints (photographs by the writer): (a) elevation looking eastward

(b)

(c)

FIGURE 7. (continued) General state of construction of Sydney Opera House in May 1969 from harbour viewpoints (photographs by the writer): (b) opera hall (left) and concert hall (right); (c) view from western side of Circular Quay

(a)   (b)

FIGURE 8. Roof tile cladding: (a) combination of matt and glazed off-white and cream coloured ceramic tiles; (b) typical layout of chevron-shaped 'tile lids' on main shell roof, September 1966 (by courtesy of *Harry Sowden Photography*)

inter-panel joints coinciding with those between the shell ribs; see Figure 8(b). Modified versions of this construction format were adopted for the side-shell and warped-surface panels, while the general regularity of the tile pattern greatly benefited from the imposed discipline of spherical roof geometry. The visual qualities of this tiled roof cladding turned out to be spectacularly successful, and without parallel in modern architecture.

3.2. Evolution of glass wall schemes

Glass wall representations presented by Utzon in the original 1956 design competition entry were rudimentary. From a longitudinal section of the major hall, given in Figure 9, it appears that each end-enclosure essentially comprises an opaque partial vertical screen suspended from the shell (well back from the tip), together with near-horizontal glazing extending outwards from the base of the screen. Part of the covering note to the submission states:

> 'Ceiling and walls of wooden acoustic panels continue in 'overhead' doors closing toward foyers and entrance-areas, which in their turn are closing toward open-air-areas with 'overhead' glass-doors. This construction implies the possibility of a complete opening of halls, foyers and public areas towards open-air during intermission whenever weather permitting and presents to the audience the full sensation of the suspended shells while moving through the foyers commanding the beautiful view of the harbour.'

which far-fetched musing is entirely understandable at such an early concept stage. More realistic proposals of a radically different nature emerged in Utzon's 'Red Book' [5]

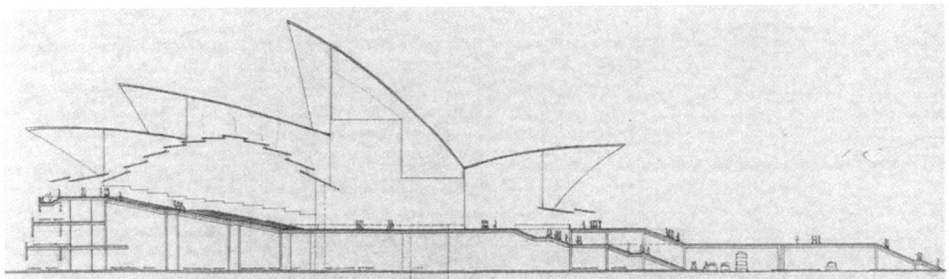

FIGURE 9. Drawing (longitudinal section, major hall) from Utzon's competition entry, 1956 (by courtesy of *New South Wales State Archives, Sydney*)

produced in 1958. Whence the building was to be enclosed between shells and podium by vertical full-height glass walls, zig-zag in plan (except for the side shells); see Figure 10, showing a schematic arrangement of rectangular glass panels and staggered horizontal joints or transoms.

Thereafter, various alternative geometrical forms and materials were investigated during several design studies, including the use of concrete, bronze and reinforced laminated plywood for the mullions. Among the most prominent of these inchoate schemes is that represented by Figure 11 for the north-facing (A4) glass wall, taken from Utzon's 'Yellow Book' [7] published in 1962. Figure 11(a) indicates the proposed layout of deep plywood mullions (50 mm thick), uniformly spaced 0.91 m apart and covered by hot-bonded bronze sheeting; for upon drawing inspiration from nature, Utzon [10] envisaged these unprecedented mullions to be 'hanging as the folds of a bird's wing'. Figure 11(b) gives a more complete picture of the same wall, while noting that its lateral boundaries stop well short of the edges of the podium structure.

Subsequent modifications and refinements were made over a period of several years, as described by Utzon [12] and later by Compagnoni [24], [97], Fromonot [61], Prip-Buus [66], Compagnoni and Buhrich [76] and others, yet without reaching the stage of producing a finalised design. Among several architectural models constructed during the Utzon era, those illustrated in Figures 12–14 give some indication of the proposed external appearance of the major hall glass walls at various stages of development, the last of which (October 1965) surprisingly shows the plywood mullions in the northern foyer of the concert hall extending down to podium floor level. Figure 15 shows Utzon's drawings of the restaurant glass walls, dated October 1964. Further project details can be obtained from the large archive collection of Utzon's architectural drawings held by the State Library of New South Wales in Sydney.

In a 1965 journal article, Utzon [12] commented as follows on what was intended to be the final design configuration for the glass walls:

> '*As in the corridor panelling, I am applying the idea of using flexible connection points between elements also to the glass walls.*
>
> *A vertical glass wall kills the unsupported shell effect as seen in many existing examples, because by its verticality and because of the reflections in the glass, it will appear as a load-carrying wall.*

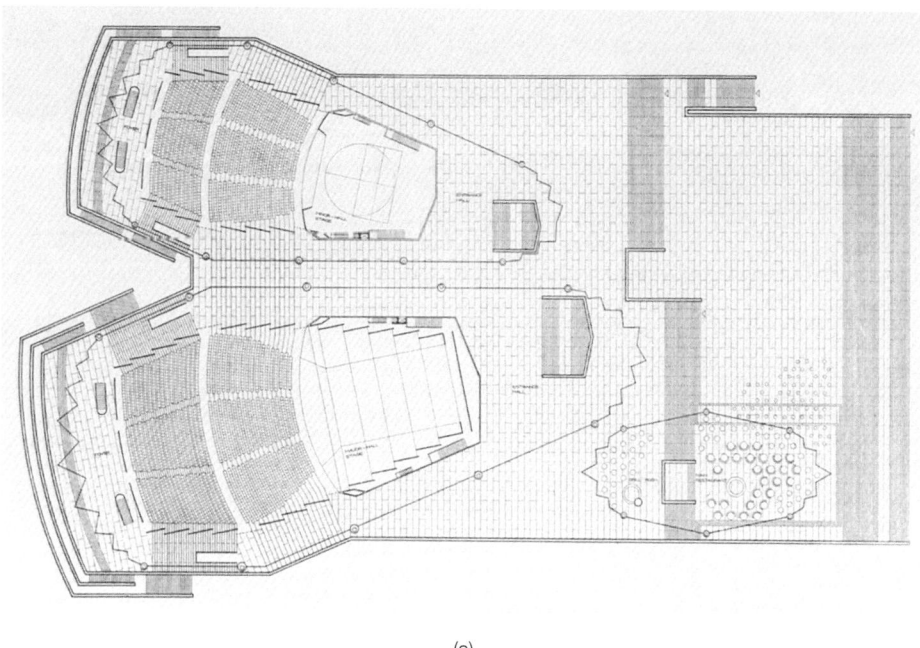

(a)

(b)

FIGURE 10. Drawings from Utzon's 'Red Book', 1958 (by courtesy of *Atelier Elektra, Copenhagen*): (a) plan, indicating zig-zag pattern of main glass walls; (b) west elevation, indicating layout of vertical glazing

> *The problems were therefore:*
> - *to create a hanging feeling, and also*
> - *to form the glass wall as a connecting surface between two different geometrical systems – the spherical geometry of the shell ribs from which the glass wall is hung, and the rectangular geometry of the pavement of the base where the glass wall ends, and*
> - *in spite of an extremely complex form world, to create the glass wall of a relatively small number of interfitting elements to be mass-produced.*

FIGURE 11. Proposed layout of A4 glass wall given in Utzon's 1962 'Yellow Book' (by courtesy of *State Government of New South Wales*): (a) plywood mullions (elevation and half-plan); (b) complete wall (cross-section and half-elevation)

(a)

(b)

FIGURE 12. Sydney Opera House model showing concert hall glass walls, *circa* 1963 (by courtesy of *Mitchell Library, State Library of New South Wales, Sydney*): (a) northern wall; (b) southern wall

(a)

(b)

FIGURE 13. Sydney Opera House model, November 1964 (by courtesy of *Max Dupain Photography & Mitchell Library, State Library of New South Wales, Sydney*)

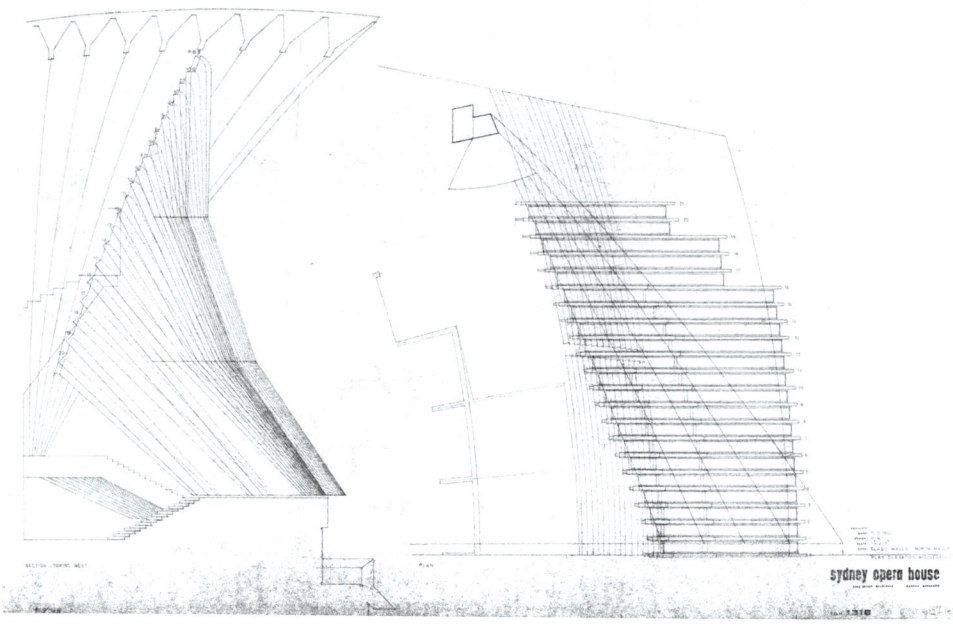

(a)

(b)

FIGURE 14. Northern glass wall, major hall (by courtesy of *Mitchell Library, State Library of New South Wales, Sydney*): (a) Utzon drawing (1:48 scale), October 1965 (PXD 492/SOH 1318). Left side, sectional elevation looking west. Right side, half-plan; (b) part Sydney Opera House 1:128 scale model, *circa* 1965 (*Finecraft Scale Models Pty Ltd, Sydney*)

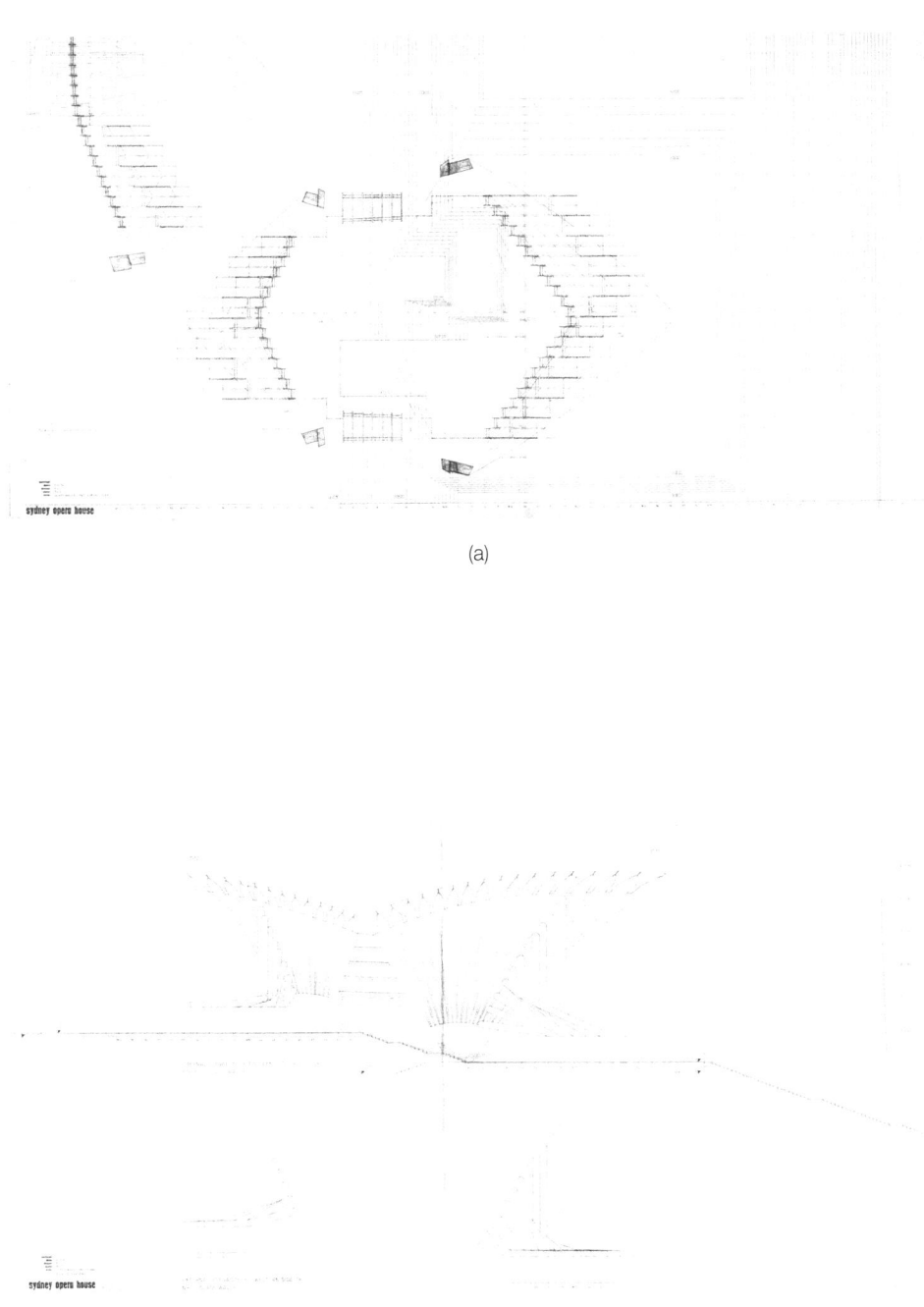

FIGURE 15. Utzon drawings (1:48 scale), October 1964 (by courtesy of *Mitchell Library, State Library of New South Wales, Sydney*). Glass walls, restaurant: (a) plan (PXD 492/SOH 1153); (b) east and west elevations, north and south glass walls (PXD 492/SOH 1154). Upper, internal elevation of eastern side; lower, external elevation of western side

A glimpse of my final solution is shown in photos. Four stages are being shown:
  1. *flexibility;*
  2. *the necessary elements;*
  3. *one example of inter-connection;*
  4. *one example of a mullion.*

The mullions are built up of plywood with a hot-bonded skin of bronze on the exterior, glass sheets are attached in overlapping positions like a hot-house.'

The only sectional drawing proffered by Utzon [12] is that for glass wall A1 reproduced in Figure 16(a), while a companion photograph of architectural models illustrating the composite construction of the prefabricated mullions is shown in Figure 16(b), together with two notional cross-sections outlining a staggered wooden laminate bonded to external bronze cover pieces.

Eventually, despite much effort and expenditure, all the earlier schemes were discarded for multifarious reasons, although some of the basic design features were retained. Figure 17(a) shows Peter Hall with an office model of glass wall A4 in May 1968, while Figure 17(b) illustrates a more elaborate architectural model produced later the same year, both of which more closely resemble the final shape. But here the corner detail is relatively benign compared to the as-built version, as discussed briefly in section 8.

The final detailed design of the glass walls was carried out in parallel with their construction. This was necessary because of time constraints, as weatherproofing the buildings was critical to the overall construction programme. Initial design focussed on the largest glass wall A4 (maximum height above podium 24 m, maximum width 55 m), whose final layout is given in Figure 18. Most of the basic glazing details developed for this primary structure were adopted for the other glass walls. Having regard to the special nature of the building, glazing and ancillary materials of the highest quality were selected to maximise service life, and additional funding was made available by the Client for any research and development necessary to complete the project. Indeed, the able assistance provided by resident staff at several N.S.W. State laboratories was indispensable in achieving a successful outcome.

3.3. Wind-tunnel tests

To assist with early design estimates of wind loading applicable to the entire building, wind-tunnel tests were carried out on a 1/96 scale wooden model of the upper part of the concert hall structure, prior to the adoption of the final spherical shell geometry. Initial tests were conducted in 1960 at Southampton University, where the limited tunnel cross-section (1.22 m square) could barely accommodate the model in particular cross-wind modes, thereby introducing substantial errors from boundary-wall interference, especially in overestimating the leeward suctions. However, useful results were obtained for other wind directions, where the windward surface pressures were little affected by partial tunnel blockage.

Subsequent tests at the National Physical Laboratory, Teddington, utilised a much larger tunnel (2.74 × 2.13 m), where side effects were reckoned to be negligible. The modelling of the glass walls was rudimentary as final details of their shape had yet to be established, but the original model was modified to represent a more likely shape of the lower part of the A4 glass wall, as illustrated in Figure 19. No attempt was made to reproduce variations of

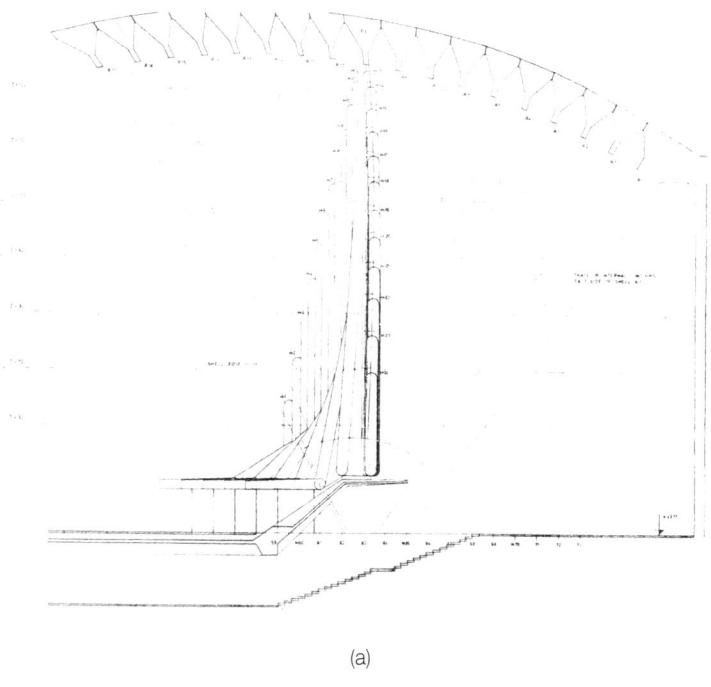

(a)

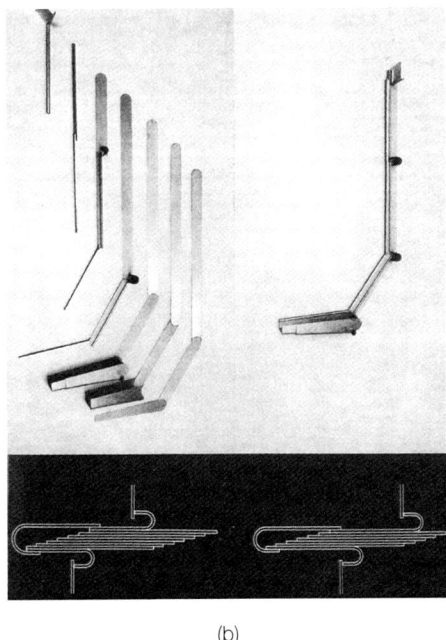

(b)

FIGURE 16. Images by Utzon published in 1965 illustrating proposed final design of glass walls (by courtesy of *Edizioni di Comunità, Milano*): (a) glass wall A1, sectional elevation looking west; (b) articulated models of mullion, with cross-sections

(a)

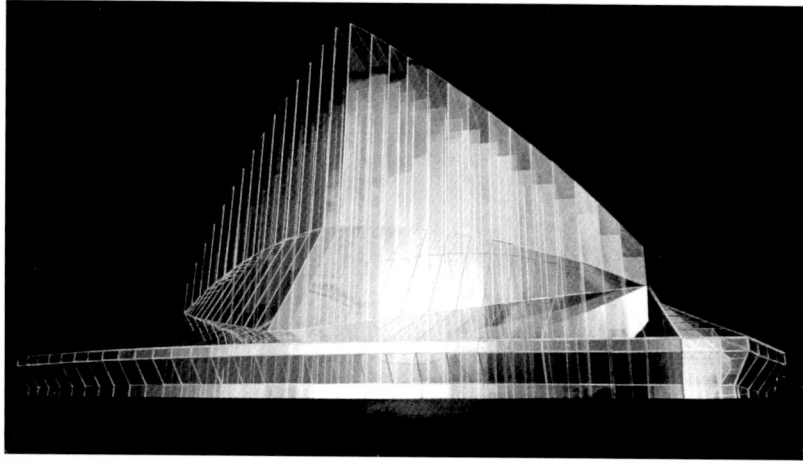

(b)

FIGURE 17. Architectural models of A4 glass wall, 1968: (a) Peter Hall with office model (by courtesy of *R. L. Stewart & Fairfax Media*); (b) complete model (by courtesy of *Pat Purcell Photography*)

wind speed with height, and the initial air velocity profile was sensibly uniform; a condition deemed appropriate for design purposes, by neglecting any wind-speed reductions at lower levels. Likewise, it was considered unlikely that the surface pressure distributions would be adversely affected by the presence of the adjacent opera hall building; notwithstanding localised perturbations caused by the funnelling of a northerly wind, for example, blowing between the two buildings.

In the terminology of fluid dynamics, the model was regarded as a bluff-shaped solid with sufficiently sharp edges to create flow separation, thereby appearing to meet the guidelines required for aerodynamic similarity between the model and the full-scale structure over a wide range in the flow scaling parameter (Reynolds number). Indeed, during preliminary tests

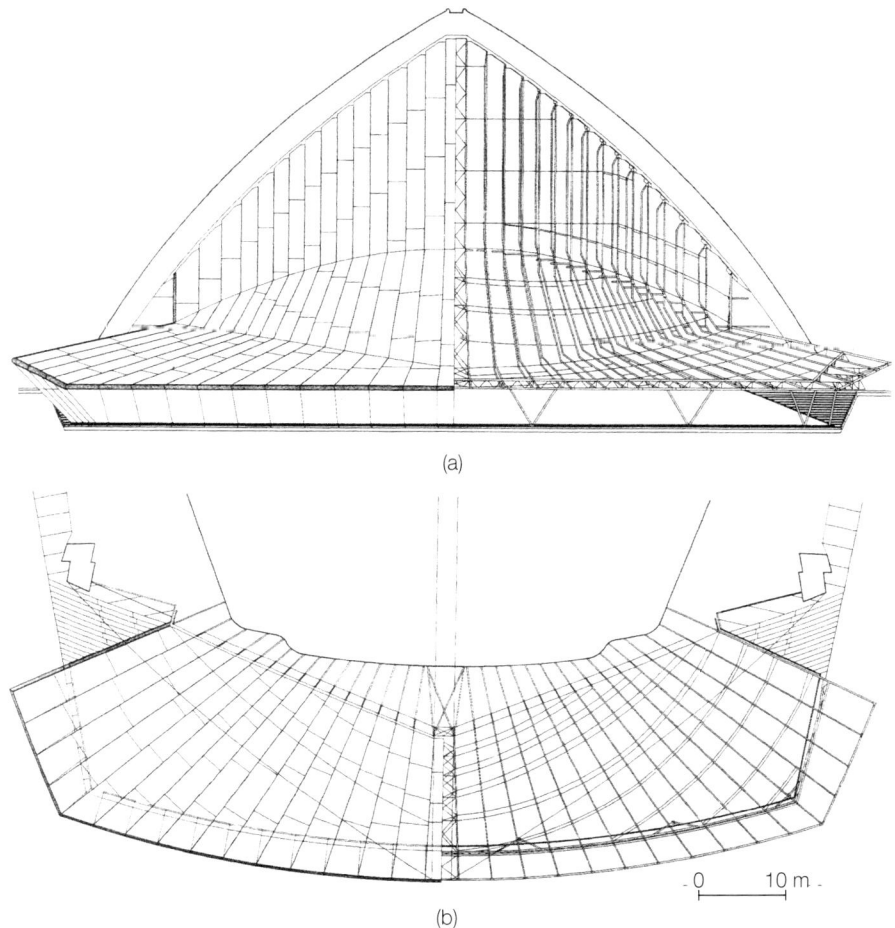

FIGURE 18. Final architectural configuration of glass wall A4, showing glass panel layout (left side) and supporting steel structure (right side) (*Hall, Todd and Littlemore*, drawn by Romano Consigli): (a) front elevation; (b) plan

to examine the airflow, tufts attached to the shell surfaces demonstrated that flow separations were confined to the sharp edges of the model, indicating that the test measurements would give reliable predictions of the full-scale pressure distribution.

The model was mounted on a central turntable, with pressures measured by means of multitube manometers at a steady wind speed of 18.3 m/s (60 ft/s). These measurements were confined to shell A2, together with the glass wall and cantilever section to shell A4. The effect of wind direction was obtained by rotating the model incrementally through 180°. Some of the results reported in late 1962 by Whitbread and Packer [9] for part of the prototype A4 glass wall are given in Figure 20, while noting that maximum recorded pressures and suctions steadily decreased with $\theta > 90°$. These results for the surface pressure (p) and suction ($-p$) are expressed in terms of the non-dimensional pressure coefficient $c_p = 2p/\rho_a v^2$, where p is the local surface wind pressure in excess of the free-stream static

FIGURE 19. Modified wooden model of Sydney Opera House (1:96 scale) used in 1962 wind-tunnel tests (by courtesy of *National Physical Laboratory, Teddington*)

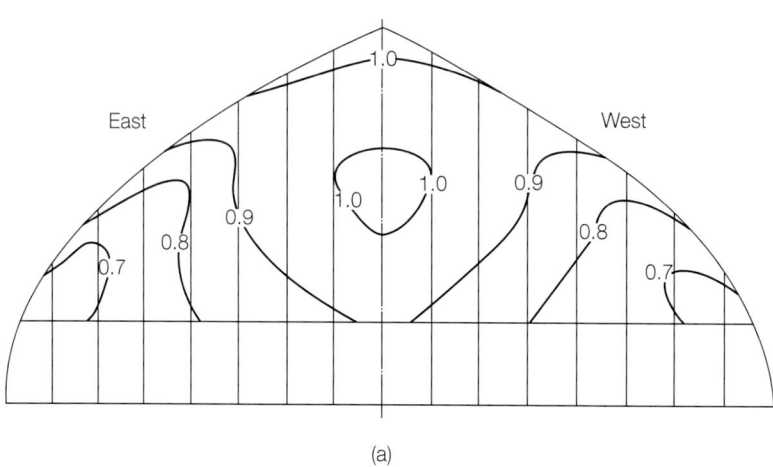

(a)

FIGURE 20. Non-dimensional coefficient $c_p = 2p/\rho_a v^2$ representing distribution of pressure and suction ($\pm p$) under steady stream of air (density ρ_a) at velocity (v) measured on prototype A4 glass wall of scale model at various orientations (θ) east of north (by courtesy of *National Physical Laboratory, Teddington*): (a) $\theta = 0$

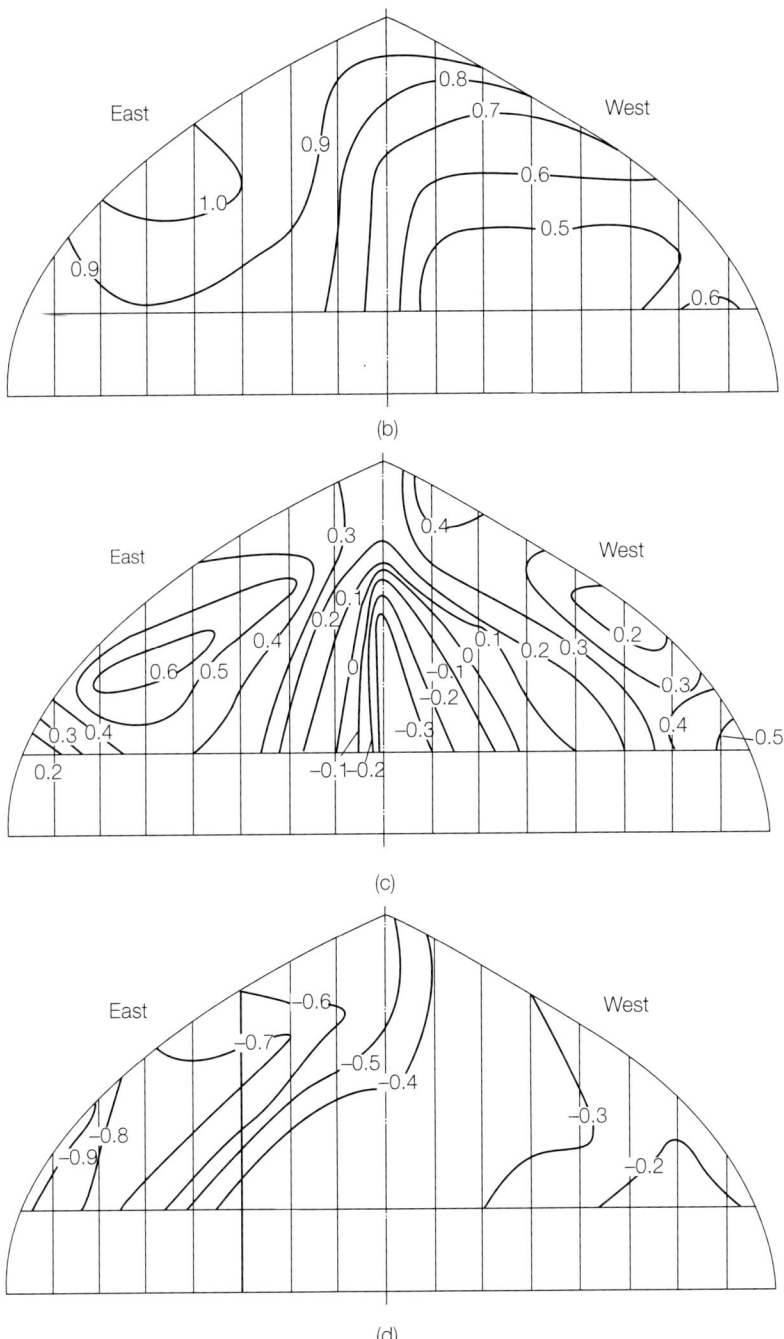

FIGURE 20. (continued) Non-dimensional coefficient $c_p = 2p/\rho_a v^2$ representing distribution of pressure and suction ($\pm p$) under steady stream of air (density ρ_a) at velocity (v) measured on prototype A4 glass wall of scale model at various orientations (θ) east of north (by courtesy of *National Physical Laboratory, Teddington*): (b) $\theta = 30°$; (c) $\theta = 60°$; (d) $\theta = 90°$

pressure (*i.e.* remote from any disturbance), and where ρ_a and v denote the free-stream air density and velocity, respectively. With regard to orientation, the terrestrial north–south meridian was assumed to coincide with the longitudinal axis of the building; see also Figure 6. Other results gave maximum values for shell A2 as $c_p = 1.0$ (east face, $\theta = 105°$) and $c_p = -1.85$ (east face, $\theta = 150°$), noting that a coefficient of unity represents a specific condition whereby the wind momentarily is brought to rest against the windward face of the model, generating a so-called 'stagnation pressure'.

In view of the sparsity of local wind-speed records and of uncertainties associated with the much later revised size and shape of the glass walls (*e.g.* the inclined upper and lower cone segments), together with the intended long life-span of the building, the maximum 3-second gust pressure for design purposes was taken as ±1.44 kPa (±30 psf); which value corresponds, for example, to the notional values of $c_p = \pm 1.45$ for a wind speed of 145 km/h (90 mph), $c_p = \pm 1.17$ for a wind speed of 161 km/h (100 mph), or $c_p = \pm 0.97$ for a wind speed of 177 km/h (110 mph), taking $\rho_a = 1.225$ kg/m^3 for the standard air density (strictly at sea level and 15 °C). By comparison, meteorological records for Sydney summarised by Whittingham [105] and others cite measured peak wind gusts of 137 km/h (85 mph) during a tropical cyclone in January 1911, 153 km/h (95 mph) during a tornado in October 1940, and 113 km/h (70 mph) during a 'southerly buster' in December 1948.

3.4. Glazing steel support structure

From an architectural perspective, it was imperative that glass wall A4 should encompass a lightweight aspect and be visually subservient to the shell. In other words, following Utzon's earlier concept, the glass wall should appear to hang from the shell, rather than propping it up. The structural support system for the glazing evolved concurrently with the development of the geometry [106]–[109], which is represented diagrammatically in Figure 21. The main glass wall surface takes the shape of a cylinder and two cones. Starting from the top, the surface is a vertical elliptic cylinder; next comes an upper cone, which forms a transition surface between the cylinder and a lower cone. So-called 'view windows' (not shown) of reverse slope (inclined outwards from the base) span between the bottom of the lower cone and the podium floor, partly to emphasise the minimal supporting structure; see Figure 18(a).

These glass surfaces conform to an independent geometry that satisfies the boundary conditions of a given shell rib along the upper edge, and the periphery of the horizontal podium at the base, while noting that substantial regions of the glazed podium are not covered by the shell roof. The vertical axis joining the apices of both upper and lower cones is termed the mullion origin, and through it pass the vertical planes containing the mullions. This arrangement allowed each glass panel to be defined in space by two straight-line generators that are coplanar; an essential prerequisite in practical terms, to avoid the costly manufacture of curved or warped glass panels. The use of flat glass naturally means that the glazed surface representing the idealised vertical cylinder is faceted rather than curved.

Mild steel was chosen as the main structural material on account of its strength and stiffness, and also because it can readily be welded. The standard mullions comprise two

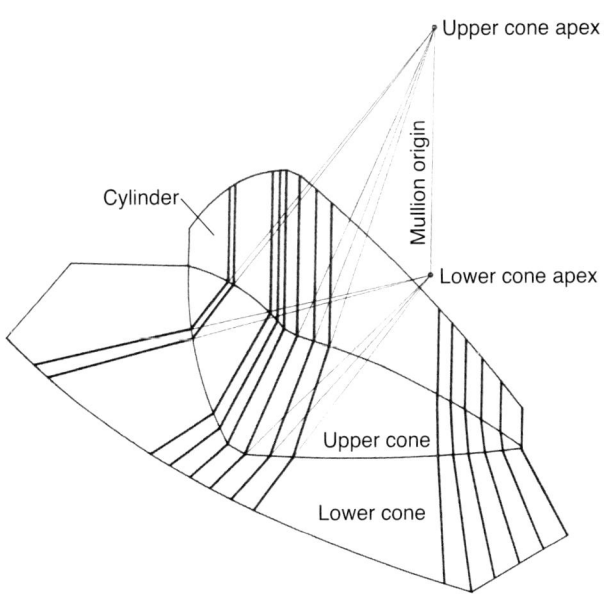

FIGURE 21. Basic geometry of glass wall A4

parallel 90 mm diameter tubes at 530 mm centres joined by a 6 mm thick plate web, and are painted to give a rust-coloured appearance. An early suggestion by the writer to lighten the blank visual appearance of these large solid webs by means of perforations – possibly a combination of circular and oval cut-outs – apparently was not pursued by others. The mullions were fabricated in two sections; one for the cylinder, the other for the upper and lower cones. The upper sections are bolted to a cast *in-situ* reinforced concrete corbelled strip fixed to the strengthened pair of shell ribs, and are tied back at the base by struts to the rear wall of the auditorium. The lower sections are bolted to the upper sections, and are supported *via* horizontal trusses by light V-columns or '*pilotis*' fixed to the podium.

The structural analysis of this statically determinate steel structure was straightforward. Components of wind load inclined to the mullion planes are transmitted through the ties between mullions to the centre truss, which is formed by the two central mullions braced diagonally together. These ties also restrain the mullion chords in compression against lateral buckling. Generally, the main design criterion for the support structure was stiffness rather than strength, in order to minimise deflections both during and after glazing operations. Similar details apply to the B4 glass wall, but for the smaller glass walls in the restaurant structure (C11 and C12) it was possible to dispense with the '*piloti*' supports, thereby giving virtually uninterrupted visions through the 'view windows'. The substantial task of computing the shapes and sizes of most of the various components for each glass wall (and also the bronze inter-shell louvre walls), together with their geometrical location, was carried out over a prolonged period by the writer's erstwhile colleague David Croft, mostly based in the Architect's site office. As for the main shells, the various subcontractors played a vital and often understated role in achieving a high standard of construction.

3.5. Glass panel support system

From an aesthetic viewpoint, the glazing system initially preferred by the Architect was one based on corner patch supports, with silicone rubber butt-joints on all four sides of each flat glass panel; the use of curved glass having been proscribed on practical grounds. Various preliminary designs were produced by the Architect during 1967–68, and full-scale loading tests were carried out on annealed glass panels (see section 4). This type of so-called 'structural glazing' had begun to appear worldwide in numerous 1960s buildings, but was used in conjunction with monolithic thermally toughened glass in order to accommodate the relatively high localised tensile stresses generated in the patch support regions, particularly at the surface of each transverse hole. However, laminated toughened glass was not generally available for use in architectural glazing, partly because the tolerances on out-of-plane glass distortion were considered too large to achieve adequate interlayer adhesion during manufacture. Moreover, the structural safety of such a product was regarded as distinctly suspect, mainly because it seemed likely that typical compound fractures in the toughened glass would leave the laminated panel with little residual strength, possibly resulting in complete detachment from the supporting structure. In any event, the specific need for on-site shaping of the glass panels effectively precluded the use of toughened glass.

At this late stage, some three years after the appointment of the new Architects, it was clear that there had been little real progress in resolving many of the issues concerning glass technology and related matters. Accordingly, full responsibility for the technical aspects of the glazing was transferred by the Client to the Engineer. Yet the broad objective remained the same: namely, to produce a world-class solution commensurate with an existing structural masterpiece. In this endeavour, having recently returned from a four-year leave of absence to undertake university research in civil engineering field instrumentation utilising photoelastic glass transducers, the writer was asked by Arup London partner Jack Zunz to visit Sydney early in 1969 to assess the prevalent issues of glass technology and to explore practical methods of glazing the various walls. It was clear from the outset that such investigations would involve detailed considerations of theoretical and experimental stress analysis not normally encountered by architects and their engineers, and hence require specialist technical input.

In a spirited attempt to meet the Architect's initial requirement of minimising the visual structural support to the large glass panels, substantial efforts were made by the Engineer during the first half of 1969 to develop a corner patch-fixing detail suitable for use with annealed laminated glass. Laboratory loading tests (section 5.2) were conducted in parallel with extensive finite element computations to assess and minimise stress concentrations in the glass, at a time when applications of this form of numerical analysis within the construction industry were the exception rather than the rule. Upon the strong recommendation of the writer, the corner-support option was abandoned when it became evident that the patch fixings, together with the associated adjustment mechanism needed for their accurate location in three dimensions, were likely to be too large and cumbersome for practical purposes, thereby compromising the original objective. Moreover, the additional development work necessary to progressively refine the design would have taken far too long in the prevailing circumstances, notwithstanding the unpredictable final outcome of such research.

Accordingly, a simpler yet elegant glazing system was developed. Over the main elevated surfaces, each glass panel is supported along its two 'vertical' edges by narrow glazing bars, as indicated in Figure 22, whereas the top and bottom butt-joints are filled with silicone rubber sealant. Architecturally, this scheme was designed to impart a vertical emphasis to the glass walls whilst minimising visual obstruction, and in particular to avoid the bland appearance of conventional 'four-sided' framed glazing. For the inclined 'view windows' at podium level, large laminated panels are supported along top and bottom horizontal edges, with vertical butt-joints in silicone rubber. Because most of the glazing is located directly above areas accessible to the public, it was also important that the glass walls should remain completely watertight over a prolonged period, while recognising that intense rainfalls are not uncommon in Sydney; see, for instance, Pilgrim *et al.* [110].

It is as well to recall that although the deployment of silicone rubber as a sealant has now become commonplace, the material was not widely used in the 1960s. In particular, its long-term performance was uncertain, and could not formally be guaranteed for external glazing. Accordingly, the writer visited manufacturers in Europe and the U.S.A. during 1969 to discuss their latest research, mostly centred on accelerated weatherometer laboratory tests on small specimens, together with outdoor trials on larger specimens exposed to natural weathering. The findings were encouraging, and indicated that the service life of external silicone rubber would be in excess of 20 years.

However, it was recognised that positive results from such laboratory testing are rarely conclusive, so that the long-term performance of a sealant in 'real time' might differ

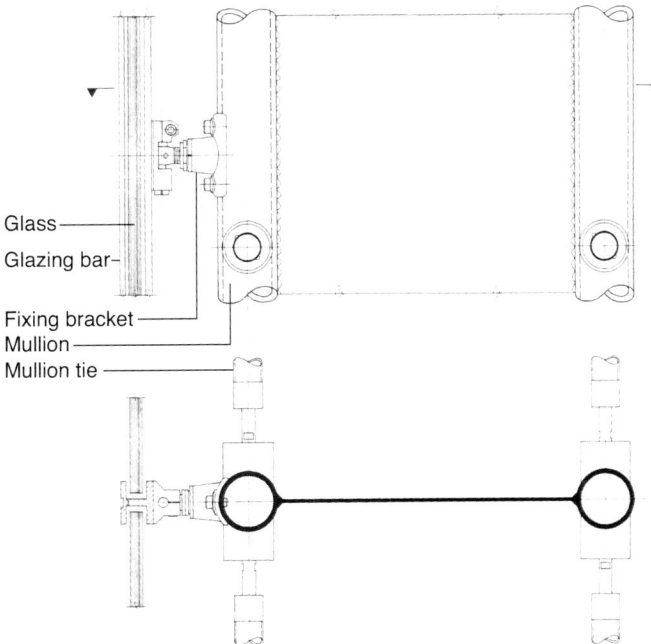

FIGURE 22. Basic structural glazing support system for main glass walls

significantly. Even less certain was the chemical compatibility of the silicone rubber with substrates other than pristine glass, and the durability of the resulting bond. Here the detailed nature of the chemical interaction becomes more complex, and frequently there is little option in practice but to rely upon specialised technical data supplied by the sealant manufacturer. Moreover, while the estimated service life is adequate for some structures, it is comparatively short compared to the anticipated life of the Opera House, so that future joint maintenance and repair on a substantial scale were regarded as unavoidable. But these potential risks and difficulties were deemed acceptable in order to achieve the desired aesthetic effect sought by the Architect.

An early architectural preference was for broadly 'silver-coloured' glazing bars, but stainless steel and aluminium alloys were rejected; the former on grounds of excessive fabrication costs, and the latter due to insufficient resistance to surface corrosion in a marine environment. Whence the glazing bars were made from manganese bronze (Austral 412 alloy) to give the required strength and durability, although technical difficulties initially were encountered in the extrusion and machining processes. The basic form consists of a T-section (width 63 mm) together with an outer structural cover piece fixed to the web (stainless-steel screws at 75 mm centres) after the final positioning of the glass and subsequent curing of the inner silicone rubber seals. The two bronze components then act together as an I-section (Figure 23). A continuous flat bearing strip of neoprene rubber (with an inner air space) is interposed between the glass and the T-section flanges, and held in place temporarily by double-sided tape. Inert polyethylene foam rod is positioned at the base of the glazing bar joints to reduce the depth of the silicone rubber seal, primarily to enable the seal to deform more readily (*e.g.* where possible, the depth of a seal should not be greater than its width). Thus all four edges of each glass panel are encompassed by silicone rubber, which can accommodate the in-plane displacements resulting from differential thermal expansion and other small movements between the glass and the supporting metal structure.

To complete the glazing, a D-section neoprene rubber extrusion is fitted between the external face of the glass and the bronze cover piece, and an adjacent secondary seal applied using translucent silicone rubber. The outer central rebate of the cover piece is filled with bronze coloured silicone rubber, partly to give weather protection to the screw heads. In principle, it might have been feasible to omit the outer cover piece, and rely entirely upon the shear strength of the silicone rubber joints to resist the transient outward forces generated by wind suction. However, the associated safety risk was deemed unacceptable, especially upon taking account of the less predictable long-term behaviour of the bond between silicone rubber and bronze. Moreover, the presence of a structural cover piece afforded the welcome opportunity of providing an additional outer weather seal.

In the absence of horizontal glazing bars, the writer devised a new and innovative solution for supporting the weight of the glass. Whence each vertical glass panel is held near the bottom corners by two specially designed stainless-steel pins projecting from the flange of the bronze T-section, but hidden from view within the horizontal silicone butt-joint. With the thinner glass layer of the laminate positioned internally, there is a sufficient flat bearing area on the pin to provide direct support to *both* glass layers, thereby preventing vertical creep deformation within the plastic interlayer under in-plane self-weight loading. The same detail

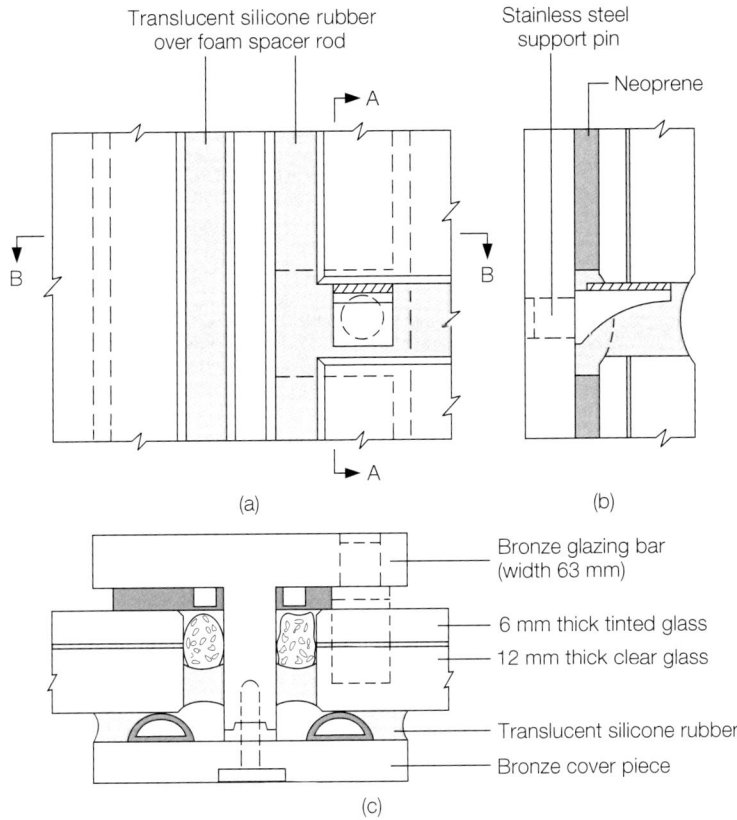

FIGURE 23. Schematic glazing bar details on vertical glass wall: (a) exterior front elevation (outer cover piece removed); (b) section A–A; (c) section B–B (with cover piece)

applies to the inclined glass panels above 'view window' level, where the internal positioning of the thinner glass layer reduces the maximum bending stresses under certain conditions. Also it was surmised – albeit without hard evidence – that the external fire-polished surface of the clear 'float' glass would offer more robust long-term resistance to weathering than the mechanically ground and polished surface of the tinted plate glass.

In the detailed design of the glass panel support pins (flat bearing area 12×14 mm), recourse was made to elastic contact theory to ensure that localised stress concentrations were within acceptable limits for annealed glass. In particular, it was essential to bond a thin layer of suitably compressible dense material (vulcanised fibre) to the flattened surface of the steel pin, to avoid direct glass–metal contact. Here it was necessary to minimise any lateral expansion of the 'platen' material in contact with the glass. It is noted that interfacial shear stresses generated by compressive applied loads can tend to induce tensile stresses in the glass, which in extreme cases might lead to glass fracture; see also Hooper [111] for related work on localised stresses induced in glass by concentrated loading. It is further recognised that, because the self-weight loading imposed on the support pins is permanent, the inherent physical property of static fatigue in annealed glass had to be taken into account. This aspect has added significance when recalling that the glass-bearing region encompasses not only an inner arrised edge, but also two raw (saw-cut) glass edges doubtless exhibiting minute spalling and other flaws that tend to initiate crack propagation.

The selection of silicone rubber (Rhône Poulenc type 3B, translucent, acid cure) as the principal sealant was governed mainly by the presence of the horizontal butt-joints (width 14.3 ± 1.6 mm). The translucent version was chosen to add lightness to the appearance of the glazed surfaces, and to enable any extraneous bubbles or inclusions within the joint to be directly visible upon initial application, thereby facilitating immediate replacement where necessary. These joints are directly exposed to the atmosphere, and silicone rubber – besides having an excellent long-term adhesion to glass – has a high resistance to ultraviolet radiation and other weathering agencies. Furthermore, its elastic properties ensure that it remains permanently in place on the inclined glass surfaces; whereas most other sealant materials exhibit time-dependent flow properties at high ambient temperatures that would cause them to sag and possibly become detached from the joint altogether. However, there was a discernible risk in the adoption of such a new material as a fundamental element in the glazing system. Moreover, the relatively deep cross-section of these solid silicone rubber joints is far from ideal, especially regarding tolerance to imposed deformations.

Essentially, the narrow width of the glass-to-glass butt joints was chosen for aesthetic effect, while amounting to around one-half of that required to limit tensile stresses at the glass surface to more acceptable levels in the presence of in-plane thermal displacements within the glass and the supporting structure. By way of a possible remedy, the insertion of an oval-shaped foam spacer as part of a two-stage joint formation procedure was suggested by the writer, both to improve the sectional geometry and reduce curing times, with the added benefit of obtaining a double seal; see Figure 24 for a schematic layout of a horizontal joint between vertical laminated glass panels. Whence in Figure 24(a), a temporary backing strip (or plastic tube) coated with a release agent is attached to the internal glass surface, and the first silicone bead is applied, followed shortly thereafter by a foam spacer rod (polyethylene or similar inert material of suitable stiffness and porosity) with an oval cross-section (oversized along the major axis) lightly pressed into the rebate. In Figure 24(b), following an initial curing period, the backing strip is removed and the second silicone bead applied, to be flush with the exterior glass surface but with the final application of a curved finishing tool. A similar approach would be followed for vertical joints between the 'view window' glass panels, but instead starting with the exterior seal and using a larger foam spacer.

However, the tests necessary to assess the efficacy of this tentative proposal were not carried out, and the design configuration was not developed further. No subsequent action was taken, and the problem was not solved to the writer's satisfaction by the start of glazing operations. In retrospect, insufficient effort at the early design stage was devoted to some of the detailed practical issues concerning the sealing of the glass walls, with an over-reliance placed upon technical data provided by sealant manufacturers. It might have been possible to carry out laboratory and field trials to simulate local conditions, although the results of such testing may well have been inconclusive. In the event, the writer was not present on site during this period to monitor glazing operations on a regular basis, and no formal records

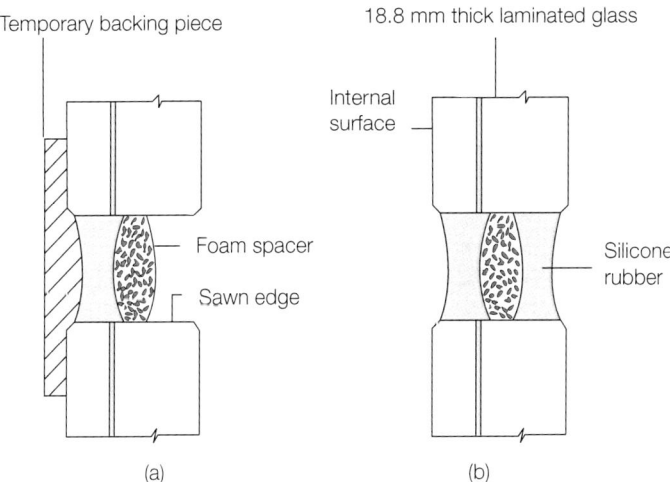

FIGURE 24. Schematic representation of proposed two-stage sealing of horizontal glass-to-glass butt joints between vertical laminated glass panels: (a) stage 1, formation of inner silicone rubber seal, with added foam spacer; (b) stage 2, formation of outer silicone rubber seal following partial cure of inner seal

appear to be available that summarise the detailed sealing methodology finally adopted, although a few anonymous comments are reported by Sowden [106].

In particular areas, the basic glazing system had to be modified, requiring bronze extrusions of various cross-sections (*e.g.* where adjacent glass panels are located in different planes) or of completely different shape (*e.g.* the 'view windows'). Some of the additional bronze extrusions are displayed in Figure 25. There was a late design proposal by the Architect for an outer chromium-plated metal clip-on cover strip to be attached to the bronze glazing bar for decorative purposes, but this encountered strong resistance from several quarters (including the writer) and was soon withdrawn.

The principal physical consequences of replacing horizontal glazing bars with silicone rubber glass-to-glass butt joints were as follows:
(a) the removal of structural support, thereby increasing the bending stresses in the glass panels, and also requiring the provision of separate fixings to accommodate vertical self-weight loading;
(b) the reliance of primary weatherproofing on the long-term integrity of the glass–silicone rubber interface, being fully exposed to natural ultraviolet radiation and in-plane thermal movements of the glass panels;
(c) to render visible to the naked eye any unsightly edge delamination in the glass panels.

Perhaps the only marginal benefit is the elimination of predominantly in-plane thermal stresses at the 'free' edges of a glass panel caused by temperature gradients, although this partial shading effect will still occur in the supported edge regions.

The glazing bars are attached to the outer circular tube of the steel mullion by brackets located at approximately 900 mm centres (see Figure 22). These brackets had to be sufficiently

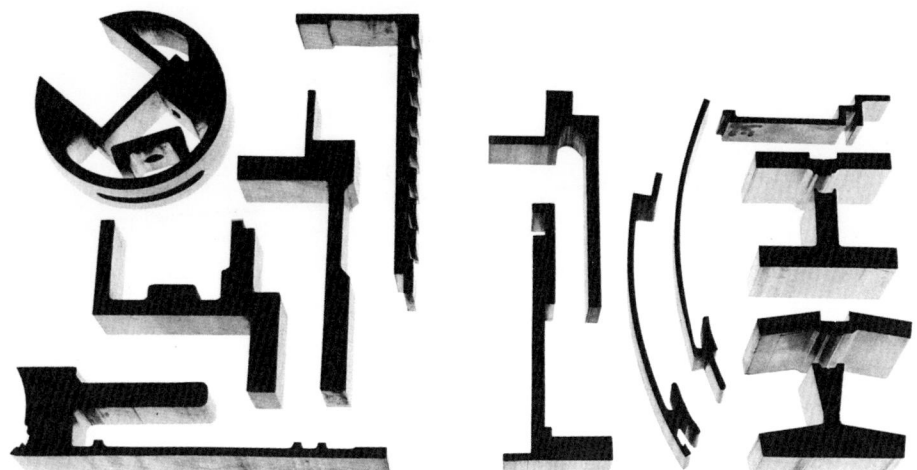

FIGURE 25. Examples of bronze extrusions deployed in glazing system (by courtesy of *Harry Sowden Photography*)

adjustable to accommodate the geometrical variations in angle and distance between structure and glass. They also had to take up the construction tolerances encountered in the fabrication and erection of the main building segments. For design purposes it was assumed that any point on the steel structure might be up to 15 mm from its specified theoretical position, in view of the rather labyrinthine shape of the glass walls. In contrast, the dimensions between adjacent glazing bars needed to be within a tolerance of ±2 mm to ensure satisfactory placement of the glass panels.

The detailed design of the fixing brackets turned out to be more complicated than expected, and extensive assistance was sought from the local aircraft industry (Hawker de Havilland), thereby benefiting from its expertise in producing complex mechanical components, together with experience of the associated alloy metallurgy and long-term behaviour in service. Besides their necessary aesthetic qualities, the brackets had to be adjustable in three dimensions and yet readily locked in their final position; see Figure 26. Approximately 2300 brackets were manufactured in aluminium bronze; a material suitable for casting and machining, while offering high strength together with resistance to stress-corrosion and fatigue.

3.6. Choice of glass

The main structural glazing requirement was for a safety glass that could be cut to shape on site. Hence monolithic panels of ordinary annealed glass were entirely unsuitable as breakage could lead to large shards with sharp edges falling from a considerable height, possibly with lethal effect. Thermally toughened or tempered monolithic glass was rejected because of the great variety of panel shapes and sizes that were required. Specified glass panel dimensions were to be verified on site, just prior to installation. Hence there would have been insufficient time available for the necessary heat treatment of each panel, after the annealed glass had been cut to its final dimensions. Moreover, such monolithic panels would not have remained

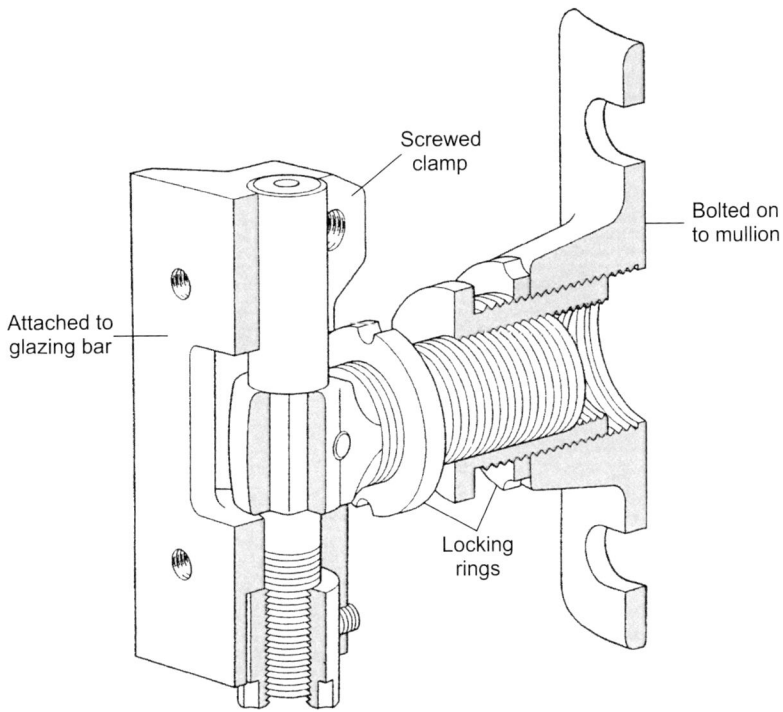

FIGURE 26. Adjustable bronze fixing bracket used for main glass walls

in place in the event of multiple glass fracture, leaving the interior of the building partially open to the elements and, more important, showering the foyer below with a large quantity of glass 'dice' (small fragments with relatively blunt edges); neither contingency being likely to encourage a sense of contentment and well-being amongst the visiting public. The use of laminated toughened glass also was discounted on similar grounds, notwithstanding the additional disadvantages consequent upon thermal distortion of initially plane surfaces generated during the toughening process.

Thus annealed laminated glass was chosen for the glass walls, although at that time there was little standard documentation or readily available guidance covering its use in buildings. More particularly, there was a dearth of technical information on non-standard applications of such laminates, including those encompassing inclined glazing and irregular support conditions. Hence a short but intensive research programme was rapidly carried out to investigate the related glass technology, as summarised in section 5. A further advantage of using annealed glass is the relatively high level of surface flatness of the resulting laminate, thereby minimising any visual optical distortion experienced by occupants of the building; see also section 8.

A further benefit associated with laminated glass is that coloured layers can readily be incorporated within the composite section, either by using a body-tinted glass or plastic interlayer, or by deploying an additional thin film or surface coating (*e.g.* a metalic layer deposited by a vacuum sputtering process). In the present case, the tinted interlayer and coloured film options were rejected because the long-term colour stability could not be guaranteed, and because of the limited range of available colours. It was also preferable on aesthetic grounds to avoid glass having a highly reflective coating (despite assertive recommendations from some glass manufacturers, partly on grounds of reducing solar heat gain), especially as the glass then becomes opaque in daylight when viewed externally. However, it was intended that a distinctive form of tinted laminate should become a primary feature of the external glazing, for two main reasons. Firstly, to complement the external colour and texture of the tiled shells, and to a lesser extent of the podium sub-structure. Secondly, to reduce solar heat gain and glare within the peripheral zones of the enclosed structures, particularly as the two largest glass walls (A4 and B4) are facing northwards (*i.e.* the source direction of maximum solar radiation in the southern hemisphere). Fortunately, these are largely foyer spaces with transient occupancy generally concentrated during evening periods, which circumstances substantially reduce the impact of such heat gain.

Regarding the latter requirement, it may be recalled from a written account by Hunt [112] in 1847 that similar problems were addressed more than a century earlier when designing the glazing for the Palm House located in the Royal Botanic Gardens at Kew in southern England, where the additional expense of installing blinds or other mechanical shading over such a large building (111 m long, 30 m wide, 19 m high) was considered to be prohibitive. Whereupon, to prevent the scorching effect of direct sunlight, and yet maintain sufficient transparency for healthy plant growth, the chosen glass was of a pale yellow-green colour, achieved by the addition of copper oxide and the elimination of manganese oxide in the melt. Preliminary experiments on numerous test specimens were conducted by Messrs Chance of Birmingham, who also manufactured the selected glass, which was then installed in preference to the sheet glass in common usage at that time.

The slight green cast visible in ordinary soda–lime window glass stems from the small amount of iron oxide (Fe_2O_3) contained as an impurity in the natural sand used in the manufacturing process. Thus 'clear' glass (*n.b.* the word 'clear' is familiar glass industry jargon, loosely meaning 'colourless') typically contains about 0.1% Fe_2O_3 by weight, which occurs in both ferrous (Fe^{2+}) and ferric (Fe^{3+}) forms. Because these two ionic forms exhibit markedly different absorption spectra, the resulting glass colour will depend on the relative proportion of each component, as well as on the total Fe_2O_3 content. Glass without the normal green cast can be produced by substantially reducing the iron oxide content, or by means of a 'decolorisation' process, whereby complementary colorants (*e.g.* selenium) are introduced into the molten glass. However, while these and similar methods are appropriate for most optical glasses, manufacturing costs can become prohibitive for large quantities of architectural 'low-iron' glass; although in any case, such 'ultra-clear' glass was not considered to be aesthetically compatible with the shell roof tiles. Conversely, to produce the type of conspicuous body-tinted green glass sometimes deployed for architectural and decorative purposes, and also used for storage containers, the iron oxide content can more readily be increased to around 0.5%.

In searching worldwide for a suitable tinted (*i.e.* coloured) glass for the Opera House glass walls, it soon became evident that all standard products were far too dark; a difficulty also encountered by Hunt [112] in his quest for a tinted glass, whilst recognising the rather different requirements of plant and human species with respect to solar radiation. For the Opera House glazing, it was deemed essential from the aesthetic viewpoint to avoid any green coloration in the glass, and to seek a relatively subdued colour compatible with those of the surrounding surface structures. Preliminary investigations indicated that only one standard product was worthy of closer examination; namely a French bronze-coloured glass, probably too dark but with the unusual and appealing capability of being modified to give a bespoke colour of choice. Small specimens were first shown (by the writer) to Peter Hall early in 1969, much to his delight and relief in having eventually come across a broadly acceptable body-tinted flat glass that could be manufactured in comparatively large quantities.

Following the preparation of several different samples of pale bronze coloration for consideration by the Architect, the chosen product was a *demi-topaze* polished plate glass manufactured to special order by Boussois Souchon Neuvesel (B.S.N.) in France. The approximate percentages (by weight) of the three crucial metallic oxides were given by the manufacturer as 0.05% iron oxide (Fe_2O_3), 0.5% manganese dioxide (MnO_2) and 0.025% nickel oxide (NiO); which quantities amounted to one-half of those used to produce its traditional *topaze normale* glass. However, the manufacturing process utilising antiquated machinery was known to be slow and laborious, which gave added impetus to solving the structural problem of determining the glass thickness at the earliest possible date. It was assumed that the glass colour would remain sensibly constant almost indefinitely, notwithstanding the colour changes observed by Gaffield [113] during the 1860s on numerous test specimens of clear and body-tinted glass exposed to direct sunlight.

This unique tinted glass formed the inner face of the laminate, and gave the required aesthetic effect: namely, to blend well with the internal and external finishes of the building and its surroundings, and to neutralise the green tint associated with clear 'float' glass; see section 6 for a description of the measures taken to monitor glass colour quality during manufacture. It needs to be recognised, however, that the broad visual perception of the glass wall coloration will depend on both the ambient lighting and the line of sight of the observer. These conditions vary hugely (*e.g.* during daylight or at night, and whether viewed from inside or outside the building). Hence it is likely that the glass walls occasionally may appear rather too light or too dark; the latter condition being mentioned by Hall [55] in 1990, notwithstanding the practical difficulties of keeping the glass clean. The tinted glass also provides a welcome degree of shading from solar radiation, but automatically leads to higher glass and interlayer temperatures. However, the resulting integrity of the plastic interlayer was reckoned not to be compromised, while the estimated in-plane thermal stresses in the edge support regions of the glazed panels generated by differential thermal expansion of the glass were of acceptable magnitude; see also section 8.

The 'standard' laminate finally selected comprises a nominal 12 mm thick layer of clear 'float' glass and a 6 mm thick layer of body-tinted *demi-topaze* polished plate glass, together with a 0.76 mm thick clear interlayer of plasticised polyvinyl butyral. This form of plastic interlayer had been used widely in the glass industry for many years, largely on account of its long-term optical clarity and resistance to discoloration, together with its outstanding mechanical properties of toughness and durability over a sufficiently wide temperature range. Moreover, it has excellent adhesion to glass, which highly beneficial characteristic stems from the presence of hydroxyl groups within the polyvinyl butyral, and the resulting chemical and hydrogen bonding with glass silanols (*i.e.* the availability of 'free' hydroxyl groups at intervals on the long-chain synthetic polymer, resulting in a natural affinity to glass upon application of modest heat and pressure). Indeed, the impressive performance of such laminates over many decades is due largely to the perceptive early development of this particular interlayer material.

Because the tinted glass was relatively expensive to manufacture, its thickness is much less than that of the clear 'float' glass. For the 'view windows' at podium floor level, which are inclined outwards from the base at 42° to the vertical in order to eliminate internal optical reflections at eye level, the nominal thickness of the clear glass was increased to 15 mm to accommodate the larger 'free' span of the laminated panels. The crucial physical property of such panels, whose maximum size is around 4 m long by 2.1 m wide with an approximate weight of 450 kg, is that they retain sufficient residual strength and remain temporarily in place should fracture occur in one or both glass layers.

An added advantage of laminated glass is that it has rather better sound attenuation properties than monolithic glass of the same overall thickness, due mainly to the damping effect of the plastic interlayer. These characteristics are further improved if the laminate cross-section is asymmetric, as in the present case. The resulting enhancement in the acoustic performance of the glass walls was judged to be of tangible benefit, particularly in reducing the transmission of harbour traffic noise (*e.g.* from ship sirens) and aircraft noise (*e.g.* from helicopters) into the opera and concert halls. However, the glazing design was not directly influenced by any sound reduction requirements for the buildings, as there was very limited scope for providing additional sound attenuation; hence the internal acoustic design overwhelmingly relied upon independent sound-proofing. Accordingly, although the glass walls form a completely sealed façade of comparatively thick laminated panels, it was assumed that the additional acoustic insulation material needed to achieve acceptably low levels of extraneous sound would largely be incorporated within the outer structural skin of the major and minor halls, together with an inner surface layer having appropriate sound absorption properties.

3.7. Design of remaining glass walls

Although the aforementioned design details specifically relate to glass wall A4, they could be carried through to the smaller but otherwise similar wall B4; see the side elevations in Figure 27. Here it is recalled that the broad concept of upper vertical glazing hanging from the shell roof, with lower inclined glazing projecting outward, is evident in the early Utzon designs, albeit with a substantial part of the deep mullion structure positioned outside the plane of the glass.

Although the other glass walls differ substantially in size and shape, many of the glazing details remain the same. The simpler south-facing glass walls A1 and B1 essentially comprise vertical or near-vertical glazing, extending from an upper pair of shell ribs down to the roof of a concrete entrance canopy or in part to floor level. The lower cone region is omitted in the smaller glass walls C11 and C12, as seen on the architectural layout in Figure 28, where part

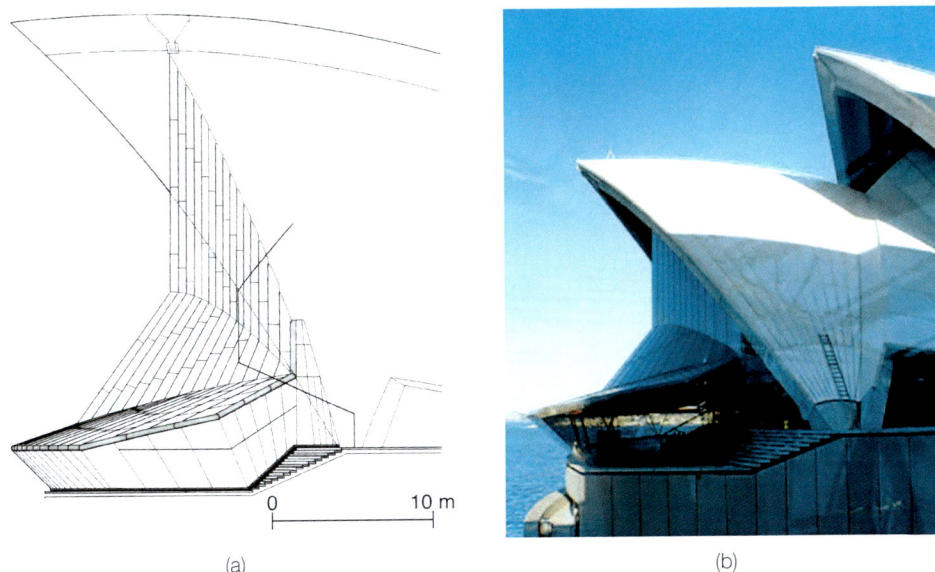

FIGURE 27. Side elevations of similar north-facing glass walls: (a) architectural configuration of wall A4 (*Hall, Todd and Littlemore*, drawn by Romano Consigli); (b) completed wall B4, September 1973 (photograph by the writer)

of the architectural brief was to maximise the seating floor area of the split-level restaurant. Many would contend that it is aesthetically the most pleasing and coherent of the three building structures, especially as the spectacular shell rib structure remains largely visible to the internal occupants. Interesting comparisons between a pair of distant views are illustrated in Figure 29, where the complete void beneath the restaurant shell structure prior to glazing is clearly visible in the earlier photograph. Further simplifications were implemented on the much smaller side shells. Here it was feasible to cantilever the short steel mullions downward from the side shell box-beams to support a bronze nosing at the head of the 'view window', and to eliminate the mullion ties and fixing brackets; see the illustrations in section 7.5. Figure 30 shows the general state of construction of the western side-shell regions of the concert hall building in March 1971.

Design calculations were undertaken to ensure that thermal stresses and displacements within the glass panels induced by solar radiation would be of acceptably low magnitude. In particular, substantial in-plane tensile stresses can be generated by thermal gradients resulting from partial shading, provided either by glazing frame members or by projecting segments of the external building structure. As an aid to this investigation, glass temperatures were measured by thermocouples attached to internal surfaces in the lower cone area of the north-facing A4 glass wall (see section 8). Out-of-plane stresses and deflections are also generated by thermal gradients through the thickness of the glass laminate, especially as the inner body-tinted *demi-topaz* glass layer will tend to absorb more solar energy than the outer clear glass layer.

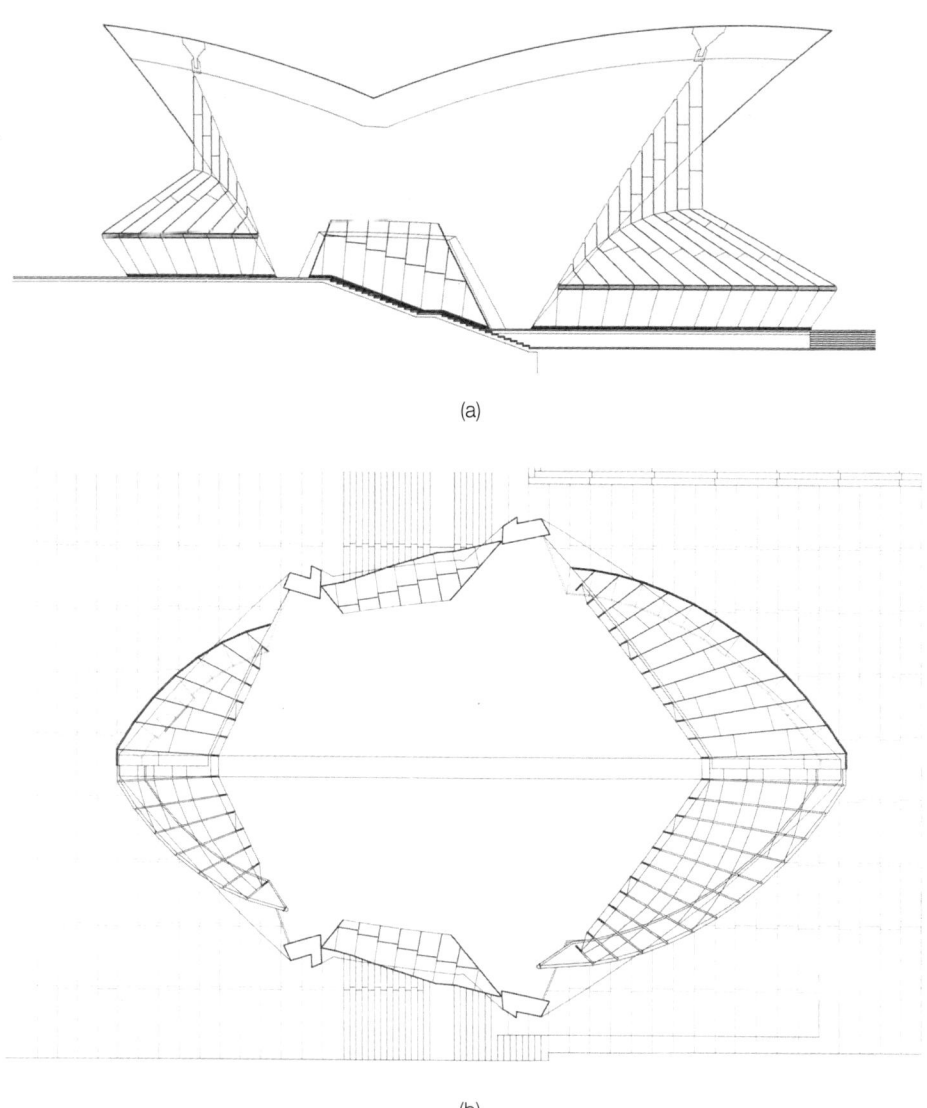

FIGURE 28. Architectural configuration of restaurant glass walls (*Hall, Todd and Littlemore*, drawn by Romano Consigli): (a) side elevation (looking east); (b) plan, glass panel layout (upper), supporting steel structure (lower)

4. Architect's proposed patch-support glazing

During the 1960s, an emerging architectural trend in the 'frameless glazing' of large office buildings comprised small corner patch fixings (typically circular or rectangular) supporting *monolithic* panels of thermally toughened glass, with silicone rubber butt-joints between adjacent panels. The viability of such systems is due mainly to the high tensile strength of this

(a)

(b)

FIGURE 29. Companion views of restaurant and concert hall buildings looking north from similar viewpoints (photographs by the writer): (a) March 1971; (b) September 1973

(a)

(b)

(c)

FIGURE 30. General state of construction of western side-shell regions of concert hall building, March 1971 (photographs by the writer)

form of glass, especially in resisting the concentrated stresses imposed in the vicinity of each transverse hole drilled to accommodate a clamp fixing bolt. It came as no surprise, therefore, that a patch-fixing system initially was proposed by the newly created architectural 'glass walls team' whose ephemeral leader, according to Yeomans [41], had opined that: *'The glass will be about 32 mm thick (it will have to be imported from Italy or Germany) and the biggest panes will be about 3 m high by 1.5 m wide'*.

In late 1968, the Architect independently commissioned full-scale static loading tests on glass panels supported by prototype adjustable patch fixings, and a single complete unit is illustrated in Figure 31(a). This heavy steel fixing could be attached to one of the supporting steel tubes within the test rig, and provision was made in its construction to adjust the position of the fixing bolt passing through a hole in the glass. Figure 31(b) gives a closer view of the bolt components incorporating 38 mm diameter steel clamps, which include flexible annular discs to prevent direct contact between metal and glass.

The general testing arrangement is illustrated in Figure 32(a). In each test, four rectangular *monolithic* panels of clear annealed 'float' glass (approximately 2.2 × 1.4 m) formed one vertical face of a suction box, and were connected to three vertical steel tubes by 38 mm diameter patch fixings located near each corner of each panel (but closer to the vertical edge than the horizontal edge). The vertical and horizontal butt-joints between adjacent panels were formed using silicone rubber. Experimental details are sparse, but Figure 32(b)

(a)   (b)

FIGURE 31. Prototype adjustable patch fixing used in suction-box loading tests commissioned by Architect in late 1968 (photographs by the writer): (a) complete assembled fixing; (b) bolted clamp components (38 mm diameter)

FIGURE 32. Initial full-scale static loading test at ambient temperature on prototype vertical glazed assembly of four panels (2.2 × 1.4 m) of monolithic annealed clear 'float' glass supported near each corner by 38 mm diameter patch fixings, commissioned by Architect in late 1968 (by courtesy of *Pat Purcell Photography*): (a) glazed suction box on completion of test; (b) failed upper right side glass panel; (c) details of adjustable metal fixing and localised glass fracture; (d) glass fracture pattern in edge region of panel

illustrates that failure in one loading test occurred in the upper right-side panel, in the vicinity of the small circular patch fixing. Figures 32(c) and (d) indicate that glass fracture initiated at the transverse hole formed to accommodate the fixing bolt.

It was clear that much additional work along similar lines would be required, and in the absence of a more plausible approach from the Architect, their test programme was terminated by the Client. Suffice to mention that there had been genuine concerns voiced by others from the beginning, set within the context of a somewhat febrile atmosphere. When a preliminary architectural drawing of a circular patch system for the complete A4 glass wall was shown (by the writer) to Ove Arup in London in mid-1969, he was distinctly unimpressed: '*Too many blobs*', he said (his exact words). Moreover, on technical grounds, it was anticipated (by the writer) that much larger fixing clamps would be needed when supporting the comparatively weak annealed glass used in architectural laminates, even after increasing the glass thickness. In view of the technical complexity of the proposed glazing system, all subsequent work on structural aspects of the glazing was assigned to the Engineer. Despite having been in post during the preceding three years, the Architect's team did not supply the writer with any test results or other technical information relating to the glass walls. But on a personal note, the writer was always on the best of terms with Peter Hall; a thoughtful individual with an engaging personality whose abilities – regardless of much *ad hominem* criticism – invariably were underrated and whose wider design contributions have yet to be fully recognised.

5. Research on laminated glass

5.1. Basic rationale

By the late 1960s, laminated glass had long been deployed within the aircraft and automobile industries, but was not in widespread use in architectural glazing. In windshield applications, for instance, it is the impact resistance of the glass laminate that usually is of prime concern, and this aspect had been studied extensively within the industry, almost entirely on an empirical basis. Conversely, it became evident from a preliminary investigation that very little work had been carried out on the response of architectural laminates to normal structural loads, such as those imposed by wind pressure and self-weight. Of particular concern was the likely behaviour of the interlayer under service conditions; for being a thermoplastic material, its physical properties undoubtedly would depend upon plasticiser content, ambient temperature and duration of loading.

Accordingly, a special research programme was undertaken to determine the structural behaviour of architectural laminates, at least to a level sufficient for project design purposes. This work included laboratory and full-scale loading tests, together with theoretical studies. However, the calendar dates for finalising the glass specifications and ordering the glass panels were such that only around nine months were available in which to organise and complete these investigations, and to formulate the basic design principles. This allowable time period included the design and fabrication of test equipment, and the manufacture and importation of laminated glass test specimens.

5.2. Preliminary loading tests on patch fixing system

Despite the aforementioned inauspicious circumstances, the development of a corner patch fixing of sufficient structural capacity was regarded by the writer as essentially straightforward, at least in principle, and following an assessment of the limited experimental results obtained by the Architect, steps were taken to make more positive progress. In particular, preliminary loading tests were conducted by the writer [114] in 1969 to investigate the feasibility of supporting laminated glass panels by means of corner patch fixings. These tests took place at the laboratories of the National Materials Handling Bureau in Sydney, with the assistance of resident staff.

Clear annealed 'float' glass was used throughout, and all panels were 86 cm square with 1.5 mm wide arrises along each edge (ground and polished). Holes of 25 mm diameter were drilled near each corner (centred 15 cm from the plate edges). The prototype bolted patch fixings were 140 mm diameter mild steel discs, 13 mm thick and containing a 13 mm diameter axial hole. Annular fibre gaskets (external diameter 140 mm, internal diameter 51 mm, thickness 3 mm) were used to prevent glass–metal contact. To achieve a state approximating to pure bending with an anticlastic deformed surface, the vertical loads near two diagonally opposite corners were directed upward on the horizontal test specimen, whilst the remaining two corner loads were directed downward; see Figure 33(a). These loads were applied to the circular metal clamps through stiff rubber pads (external diameter 76 mm, internal diameter 25 mm, thickness 51 mm), at a nominal rate of 3 kN per minute.

All test panels were loaded to failure at room temperature, and in no case did fracture initiate at the holes drilled through the glass to accommodate the fixing bolts, potentially the zones of greatest weakness. Indeed, during initial control tests on six 12 mm and 18 mm thick monolithic panels (average total failure loads 6.3 kN and 10.5 kN, respectively), where convulsive fracture took place due to the uninhibited release of strain energy, the glass enveloped by the circular steel clamps was usually the only test material remaining intact.

For the two laminated panels (5.8 mm and 7.5 mm thick outer glass layers) with a 1.02 mm thick 'hard' plastic interlayer (average total failure load 6.7 kN), the general fracture pattern comprised a large number of cracks running parallel to both main diagonals. For the two test panels comprising identical glass layers but with a 1.14 mm thick 'soft' interlayer (average total failure load 4.7 kN), the fracture pattern consisted of several cracks running parallel to only a single diagonal. In one of these tests, rapid load removal shortly after failure occurred (at a total load of 4.3 kN) enabled the initial fracture location to be ascertained with certainty; see Figures 33(b) and (c), where fracture initiated at the edge of the thicker glass layer, 20 cm from the corner.

As part of a proposed extended research programme, a special 'stand-alone' load cell utilising photoelastic glass transducers was designed by the writer to assess the long-term behaviour of laminated glass subjected to sustained clamping forces, possibly in conjunction with moment loading; see Figure 34(a). A simplified version of the original engineering drawing (dated August 1969) is given in Figures 34(b)–(d). In its basic construction, the axial load cell comprises three identical circular glass discs or cylinders compressed diametrically between two flat circular stainless-steel platens. The load carried by each glass transducer

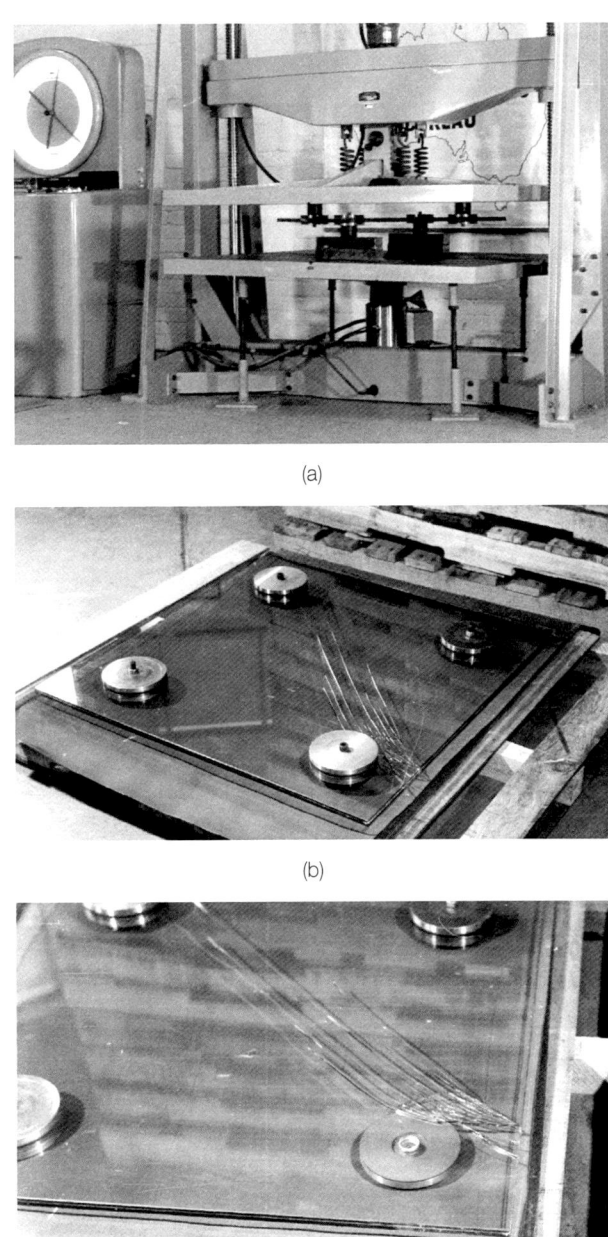

FIGURE 33. Laboratory test (1969) at ambient temperature on square laminated glass panel (side length 86 cm, 5.8 mm and 7.5 mm thick layers of annealed clear 'float' glass, 1.14 mm thick 'soft' plastic interlayer) loaded to failure in anticlastic flexure through prototype 140 mm diameter steel patch fixings (by courtesy of *National Materials Handling Bureau, Sydney*): (a) horizontal test specimen under vertical loading; (b) fractured laminate, with all patch fixings in place; (c) fractured laminate, with one upper fixing plate and gasket removed

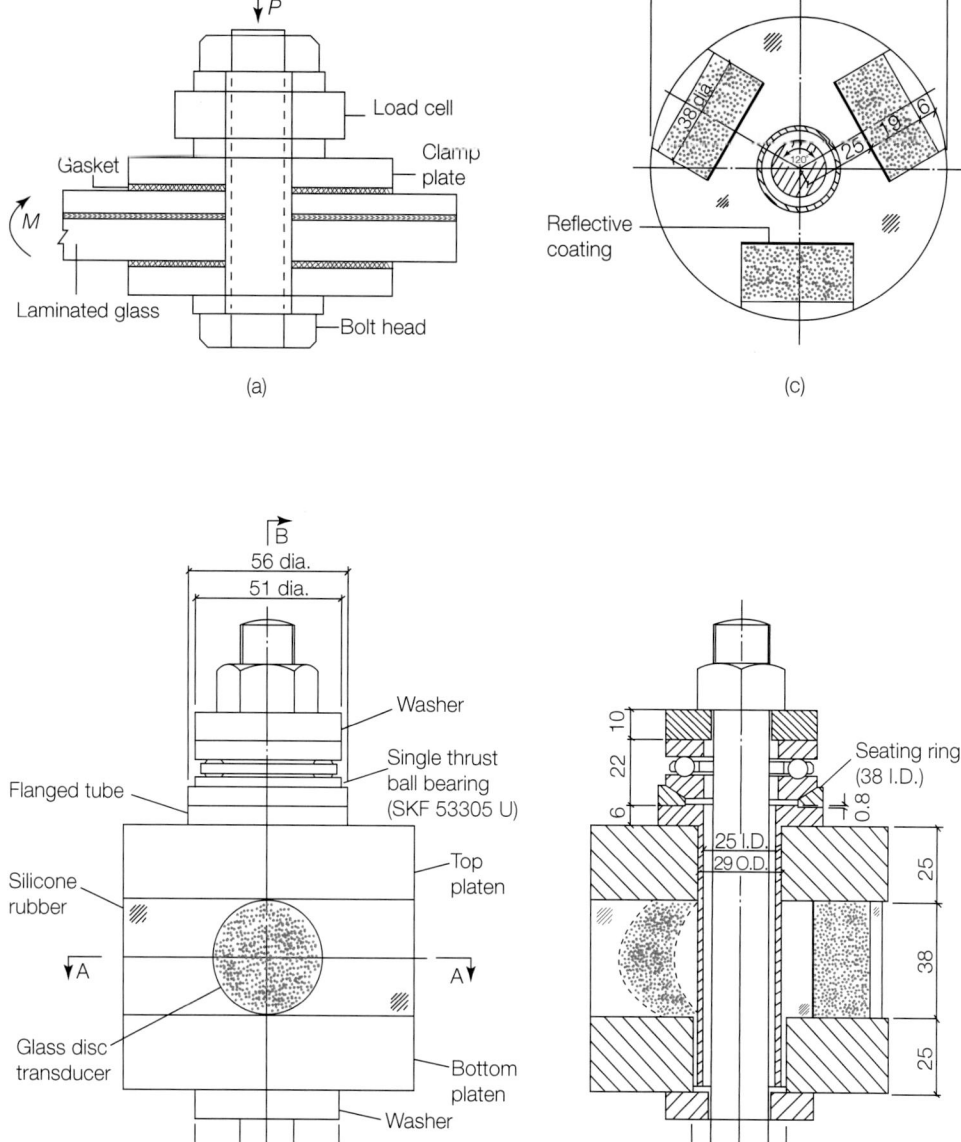

FIGURE 34. Schematic details of 22 kN bolt tension load cell (EN57 stainless steel, dimensions in mm) utilising photoelastic glass transducers designed to investigate long-term behaviour of clamped patch fixing on laminated glass: (a) test arrangement; (b) elevation; (c) section A-A; (d) section B-B

is determined by measuring the central isochromatic fringe order under reflected polarised light, having first ascertained the stress-optical coefficient of the glass at an appropriate wavelength. This robust load cell is essentially a self-contained instrument, suitable to leave unattended, for example, in temperature-controlled environments, to be read manually when required using a small portable combined light source and analyser. Steel-body photoelastic load cells developed along similar lines, but designed to measure much higher loads, had previously been deployed in several practical applications; as explained by Hemsley [115].

However, substantial further investigations would have been required to progressively reduce the patch diameter and to optimise the detailed design, including more refined laboratory tests and numerical stress analysis. Of special concern were the potential long-term effects of patch clamping forces on the laminated glass, such as creep deformation of the interlayer material at elevated ambient temperatures, possibly leading to additional localised stress concentrations in the hole region. Moreover, it seemed likely that the size of the patch fixing deemed necessary to satisfy structural and safety requirements would be too large from a purely visual perspective. The subsequent decision to discard the patch-fixing scheme was made primarily on account of the projected excessive period of time required to fully develop and verify the final product, having regard to the dearth of previous test data or specialised practical experience, and in the wider context of several other crucial aspects of the glazing system needing urgent investigation.

5.3. Bending resistance of laminated glass

To examine the fundamental behaviour of laminated glass in flexure, static load tests were carried out by the writer on a series of small beams that were approximately 560 mm long, 50 mm wide and of various cross-sectional proportions. These test beams comprised two outer layers of annealed plate or 'float' glass, not necessarily of the same thickness, and a thin plastic interlayer joined in adhesive contact. Two types of commonly available interlayer (polyvinyl butyral) were used, one having a much smaller plasticiser content than the other; they were designated as 'hard' and 'soft' interlayers. The 'hard' interlayer (low plasticiser content) was intended mainly for use in aircraft windscreens, whereas the 'soft' interlayer (high plasticiser content) was designed for use in architectural laminates. For selected test beams, values of interlayer moisture content obtained using near-infrared spectroscopy were found to be around 0.4% in all cases, which from industrial experience implied a high degree of adhesion between glass and polyvinyl butyral.

The tests were carried out at the Materials Testing Laboratory of the New South Wales Public Works Department in Sydney, again with the help of staff members. Numerous beam specimens were loaded in four-point bending using a carefully calibrated universal testing machine, and measurements were taken of outer surface longitudinal strain and central deflection; see Figure 35(a). Each horizontal beam was simply supported on transverse cylindrical steel bars fixed to the lower machine crosshead, giving a 'free' span of 508 mm. To ensure accurate alignment, vertical loads were applied through a spherical seating to a rectangular steel bar incorporating two tilting rollers set 254 mm apart. Hence a uniform bending moment was imposed over the central half-span. Dial gauge and strain gauge

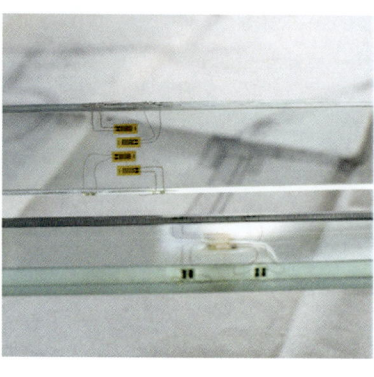

(a)   (b)

FIGURE 35. Static 'short-term' loading of laminated glass test beams (560 mm long, 50 mm wide) in four-point bending at room temperature (photographs by the writer): (a) laboratory test arrangement; (b) beam specimens with internal and external strain gauges attached

readings were taken manually at several load increments, and each test took about three minutes to complete. The ambient temperature was 21 °C, and for beams containing a 'soft' interlayer, creep deformation was negligible and the observed load–deflection curves were sensibly linear. Some creep deformation did occur in beams with a 'hard' interlayer, and subsequent analysis was based on the initial tangent to the curves.

In a few selected cases, resistance foil strain gauges (Micro-Measurements, Michigan, U.S.A: 120 Ω, 6 mm gauge length) were attached to all four glass surfaces near the beam centre, as indicated in Figure 35(b), in order to determine the distribution of bending stress throughout the depth of the laminated section. This procedure was not entirely straightforward as the two inner gauges and associated lead wires had to be affixed prior to the application of heat and pressure that formed part of the laminating process; the latter being carried out with the generous assistance of Pilkington Pty in Geelong, Victoria, in the presence of the writer. Special gauge bonding techniques were required to ensure the complete reliability of subsequent strain measurements.

Detailed test results are given elsewhere [116], while a closed-form analytical solution to the corresponding theoretical problem in plane elasticity is relegated to appendix A1 for reference purposes. In particular, the degree of coupling between the two glass layers was shown to depend mainly on the shear modulus of the plastic interlayer, which key parameter could be deduced by matching experimental data with theoretical results. At the ambient test temperature (21 °C), the shear modulus of the 'soft' interlayer was lower than that of the 'hard' interlayer by around one order of magnitude. In the case of 0.76 mm (0.030 inch) thick interlayers (a common industrial standard thickness), for example, these modulus values were approximately 0.8 MPa and 10 MPa, respectively. Hence the bending stiffness of a composite beam with a 'soft' interlayer was much lower than that for a monolithic glass beam of the same overall thickness, based on a Young's modulus for glass of 72.4 GPa deduced from measured deflections.

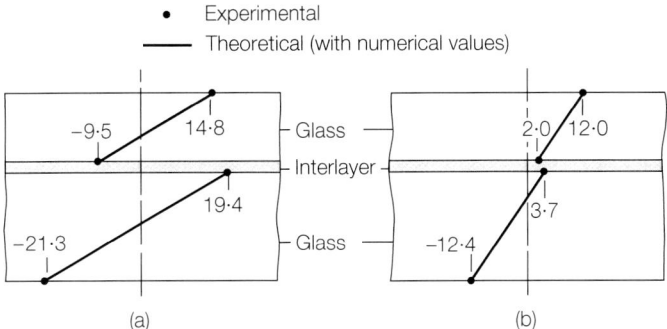

FIGURE 36. Theoretical and experimental values of bending stress (MPa) at mid-span of horizontal laminated glass test beams (length 560 mm, width 50.5 mm) subjected to 'short-term' four-point loading at room temperature (total vertical applied load 400 N, span 508 mm, 5.87 and 9.52 mm thick layers of annealed clear 'float' glass, 0.76 mm thick plastic interlayer, tension negative), after Hooper [116]: (a) 'soft' interlayer; (b) 'hard' interlayer

The effect of this reduction in stiffness on mid-span bending stresses (tension negative) is typified by the results given in Figure 36 for a 50 mm wide beam with nominal 10 mm and 6 mm thick glass layers and a 0.76 mm thick plastic interlayer. The horizontal test beam was positioned with its thinner glass layer uppermost, and the total vertical load was 400 N. There was excellent agreement between theoretical and measured stress distributions, based on the deduced material properties of the glass and plastic. In this example, the maximum tensile stress in the beam with a 'soft' interlayer is 70% greater than the corresponding stress in a similar beam containing a 'hard' interlayer, and 80% greater than the maximum bending stress in a monolithic glass beam of the same overall thickness.

Although the primary objective of the static beam loading tests was to ascertain the flexural stiffness properties of annealed glass laminates with various layer configurations, together with the distribution of bending stress throughout the cross-section, selected tests were carried through to failure. Two broken specimens are illustrated in Figure 37, comprising 10 mm clear 'float' glass and 6 mm *demi-topaz* plate glass joined by a 0.76 mm thick 'soft' interlayer. As in almost all such tests, glass fracture initiated at an edge of the sawn beam, weakened by the machining process when forming the specimens from larger blanks – a condition that, notwithstanding the addition of fine arrises, essentially applied also to the laminated glass panels used in the glass walls. It is also noted that the cracked glass remained securely held together by the plastic interlayer.

During testing, the thicker glass layer was positioned on the 'tension side'. It appears that upon initial failure of this layer, load was transferred rapidly to the thinner glass layer, causing almost instantaneous fracture in a similar manner, sometimes at multiple locations. The observed bifurcation in the crack patterns is characteristic of high-energy fracture. Referring to the upper test specimen in Figure 37, the initial fracture in the 10 mm glass layer occurred close to mid-span, generating a similar single crack pattern directly above it in the thinner glass layer. In the lower test specimen, an initial primary crack in the 10 mm layer

FIGURE 37. Test specimens of laminated glass beams (nominal 10 mm clear 'float' glass, 0.76 mm 'soft' interlayer, 6 mm *demi-topaz* plate glass) loaded to failure in static four-point bending at ambient room temperature (photograph by the writer)

occurred well away from mid-span, followed by a pair of typical edge-crack patterns in the 6 mm layer, one of which was located above the primary crack. Also observed was a series of secondary cracks in the thinner glass layer, probably associated with the propagation of stress waves radiating from the initial fracture location.

5.4. Laminated glass under sustained loading

Early in the project design stage, it was evident that large areas of glass (*e.g.* the lower cone panels of walls A4 and B4) were to be glazed in a near-horizontal position, and therefore would be subjected to sustained self-weight loading of appreciable magnitude. Similar considerations applied to the inclined 'view windows', having particular regard to their larger span and thicker cross-section. It was also beyond doubt that the laminated panels in these prominent locations would be exposed to large variations in temperature, especially as solar radiation can raise glass temperatures to well above that of the surrounding atmosphere. Of particular concern was the effect of interlayer softening on the long-term flexural behaviour of the glass laminates.

In recognition of these pivotal factors, experiments were carried out by the writer to determine the combined effect of temperature and sustained loading on a range of laminated glass test beams, with specimen dimensions and loading spans similar to those adopted in the aforementioned short-term tests. Creep-loading frames designed to accommodate three sets of identical beams were fabricated and installed in three temperature-controlled rooms (maintained constant at 1.4, 25 and 49 °C) that normally were used in studying the influence of storage temperature on the longevity of food products at the C.S.I.R.O. Division of Food Preservation, Sydney. Each steel frame accommodated six laminated glass beams (nominally 560 mm long, 50 mm wide) of different cross-sectional proportions, together with one monolithic glass beam that served as a control specimen (Figure 38). Vertical loads were applied to the simply supported horizontal beams by means of dead-weights positioned at the quarter points. These constant loads were maintained for around 80 days, and vertical deflections at the centre of each beam were measured using dial gauges at intervals throughout this period.

FIGURE 38. Laboratory test arrangement for sustained vertical loading of horizontal laminated glass beams (560 mm long, 52 mm wide) in four-point bending at various ambient temperatures (photograph by the writer)

At the start of the creep tests, the dead-weights were applied manually, with the beams becoming fully loaded after about one minute. The resulting load–deflection curves were linear in all cases, and deduced values of interlayer shear modulus relating to this initial 'short-term' loading are plotted in Figure 39, which also gives curves suggesting the likely continuous variation of shear modulus with temperature. These results reflect the dominating influence of ambient temperature upon beam flexure, with values of interlayer shear modulus differing by two orders of magnitude over a modest temperature range. However, even at the lower temperatures, the calculated maximum interfacial shear stress (appendix A1) is low compared to the anticipated shear strength of the glass–plastic bond, which suggested that interfacial slip or loss of contact is unlikely to occur under service conditions.

Typical creep test results for a set of six similar beams (52 mm wide, nominal 6 mm and 10 mm thick glass layers, 0.76 mm thick interlayer) at the three ambient temperatures are summarised in Figure 40. The deflection measurements clearly demonstrate that, except at comparatively low temperatures, the effect of sustained loading is to cause the laminate to behave as though the glass layers were separated by a material having a shear modulus close to zero. As expected, the deflections of the three monolithic glass beams remained sensibly constant throughout the test period.

5.5. Dynamic fatigue loading of laminated glass

Because the natural effect of wind buffeting against the walls will be to cause individual glass panels periodically to flex and vibrate, dynamic fatigue loading tests specified by the writer

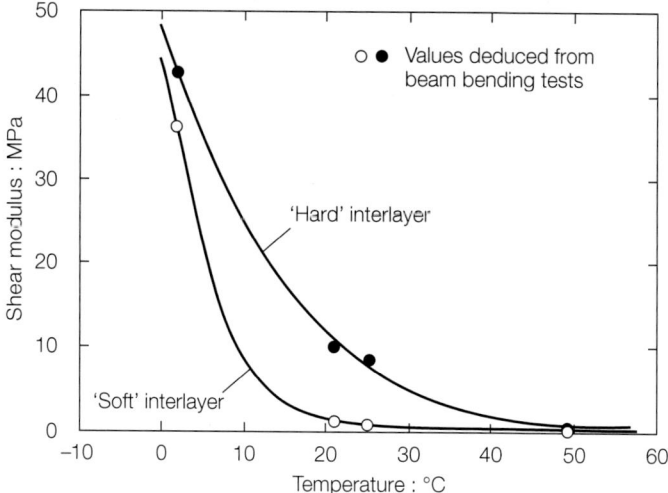

FIGURE 39. Suggested variation of shear modulus with ambient temperature for 'hard' and 'soft' polyvinyl butyral interlayers under 'short-term' loading deduced from four-point bending tests on laminated beams (560 mm long, 52 mm wide, nominal 6 mm and 10 mm thick glass layers, 0.76 mm thick plastic interlayer), after Hooper [116]

were carried out at the Civil Engineering Department, University of Sydney. The beam test specimens were of similar dimensions to those in the previous experiments, and comprised laminates of various cross-sectional proportions. The main objective of these tests was to ascertain whether there was any tendency for widespread delamination to occur.

The laboratory test arrangement is illustrated in Figure 41. Because the minimum load that could be applied by the large hydraulic fatigue machine far exceeded the strength of the laminated test beams, load was transferred to the glass by means of a purpose-built rig mounted on the machine crosshead. The main component of this rig was a tapered steel beam (high strength alloy) with adjustable clamps fitted to the tension face, designed with assistance from the local aircraft industry (Hawker de Havilland). These clamps enabled both upward and downward loads (or, more correctly, imposed vertical displacements) to be applied to the attached glass beam.

As before, the laminated test beams were subjected to four-point bending, and the applied loads were such as to induce stresses in the beams of slightly higher magnitude than those expected in service. Strain levels in the glass were measured using electrical resistance foil gauges bonded to the outer faces of the beams. During testing, where the frequency of loading was maintained at either 240 or 480 cycles per minute, strain levels were checked periodically by channelling the gauge signals into an auxiliary circuit to give an oscilloscope display.

The laminated glass responded well to the dynamic fatigue testing. Each beam was subjected to more than one million cycles of stress reversal, but there was no perceptible deterioration in any of the test specimens. Moreover, static calibration values of bending strain and deflection obtained on completion of the fatigue testing were very similar to the initial measured values.

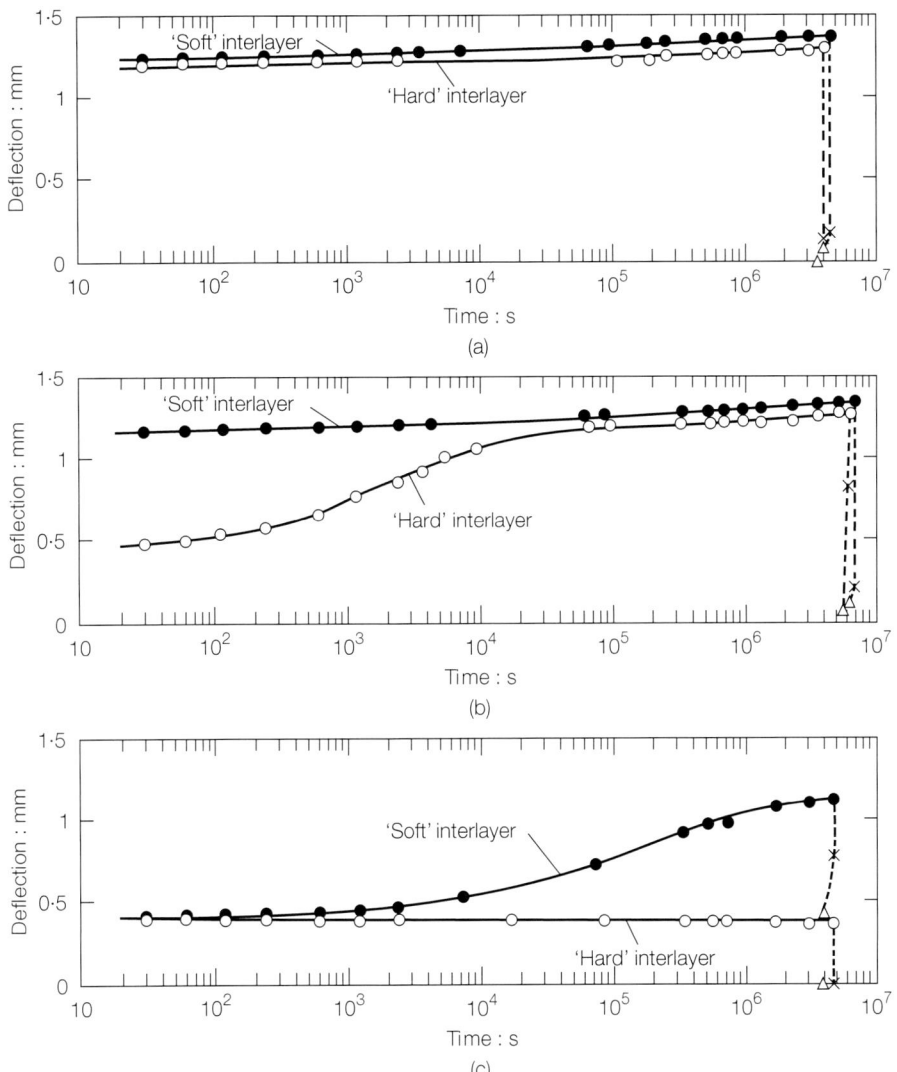

FIGURE 40. Measured vertical mid-span deflections of similar horizontal laminated glass test beams (560 mm long, 52 mm wide, nominal 6 and 10 mm thick layers of annealed clear 'float' glass, 0.76 mm thick plastic interlayer) subjected to sustained vertical loading in four-point bending for 80 days at various ambient temperatures (total applied load 285 N, span 508 mm; points × relate to immediate unloading, points Δ relate to 7 days after unloading), after Hooper [116]: (a) temperature 49 °C; (b) temperature 25 °C; (c) temperature 1.4 °C

5.6. Full-scale loading tests

In terms of 'characteristic' wind loading, the design 3-second gust pressure for the glass walls was taken as ±1.44 kPa (±30 psf), based on a 100-year return period, which is compatible with that adopted in the design of the main shell structures. Hence comparatively high

(a) (b)

FIGURE 41. Laboratory test arrangement for dynamic fatigue loading of horizontal laminated glass beams (560 mm long, 52 mm wide) in four-point bending at room temperature (photographs by the writer): (a) hydraulic loading machine with attached oscilloscope; (b) details of bespoke loading rig

pressure and suction forces must be accommodated by the glass panels and their supporting structure, largely as a consequence of the exposed site location and the complex shape of the building; see also the results of wind tunnel tests on a scale model of the original shell structure [9], partly summarised in section 3.3. Data obtained from the earlier bending tests on small laminated beams were used to determine the glass layer thicknesses of the laminated panels. Although these beam tests demonstrated that the structural performance of laminates with a 'hard' interlayer generally was superior to that for laminates with a 'soft' interlayer, there were manufacturing restrictions imposed by the laminating process which meant that, in practice, the 'soft' interlayer was much to be preferred. Accordingly, this type of interlayer was utilised in the subsequent manufacture of all glass wall laminates.

To provide a detailed physical check on the likely full-scale behaviour of the glazed assembly, several tests were carried out (during February 1970) in which large rectangular glass panels were loaded to failure under sensibly uniform static pressure. The laboratory test equipment was designed and built at the Commonwealth Experimental Building Station in Sydney; see Figure 42(a), where four 1.5 × 2.3 m glass panels essentially formed the main vertical face of the large suction box, located under cover but exposed to external ambient temperatures. The fixing of these panels was designed to simulate, as far as practicable, the proposed *in-situ* glazing system. Thus the (short) vertical glass edges were supported by glazing bars, together with shaped steel pins (with bearing pads) at the bottom corners. The glazing bars were attached to the tubular steel mullions by means of prototype adjustable brackets. The horizontal butt-joints between glass panels were sealed with silicone rubber, as

(a)

(b)

FIGURE 42. Full-scale static loading tests at ambient temperature on prototype glazing system, February 1970 (by courtesy of *Commonwealth Experimental Building Station, Sydney*): (a) suction box with glazed vertical side (four 1.5 × 2.4 m glass panels); (b) crack patterns in adjacent pair of laminated glass panels (annealed clear 'float' glass layers)

were the joints between the glass and the glazing bars. The periphery of the suction box was sealed using a flexible rubber membrane.

The basic test procedure was simply to partially evacuate air from within the box, and to record central and edge deflections for each glass panel until failure occurred in one or more panels. These (horizontal) deflections were measured using dial gauges attached to an auxiliary metal frame. Suction was applied incrementally to facilitate such measurements, and the duration of each test was around 30 minutes. In all such tests, the measured load deflection curves were essentially linear. Typical results are given in Table 1 for monolithic and laminated panels of annealed clear 'float' glass subjected to a uniform normal suction of 958 Pa (20 psf) at rather high ambient temperatures.

Regarding the monolithic glass panels (serving mainly as control specimens), exact analytical results (in terms of infinite series) based on elastic 'small-deflection' theory have been given by Holl [117] for a thin rectangular plate simply supported or 'pinned' on two opposite edges, the other two edges being entirely 'free' or unrestrained. Some of these results are incorrect, as discussed in a detailed re-analysis by the writer summarised in appendix A3, but the errors turn out to be small in the present application. Based on deflection and strain gauge measurements taken during the aforementioned beam loading tests, values of Young's modulus and Poisson's ratio for the glass were taken as 72.4 GPa and 0.22, respectively, and the 'free' span was 2.27 m. Average measured deflections from two tests are given in Table 1, and the agreement with calculated values was better than anticipated considering the embryonic nature of the experimentation.

Despite maximum deflections of around double the plate thickness, membrane effects were reckoned to be negligible, as was the restraint offered by the silicone rubber edge seals. The calculated difference in mid-span deflections at the centre and edge of each plate was slightly higher than the measured value. However, rigid supports are assumed in the theoretical solution, whereas the glass edges rested on neoprene strips fixed to glazing bars, leading to small non-uniform normal displacements along the supported glass edges. This slightly modified response was confirmed in later finite element analyses in which the two opposite rigid line supports were replaced by deformable members to model the finite

TABLE 1. Example comparisons between measured and calculated mid-span deflections (w) of monolithic and laminated rectangular panels (annealed clear 'float' glass, nominal 2.3 × 1.5 m, glass layer thicknesses $t_1 = 6.15$ mm, $t_2 = 9.98$ mm, 'soft' plastic interlayer thickness $c = 0.76$ mm, total thickness T) simply supported on two short edges (other two edges 'free') and subjected to uniform normal suction (958 Pa) at ambient temperature (30–32 °C)

Test specimen	T: mm	Measured: mm		Calculated: mm		Theory
		w(centre)	w(edge)	w(centre)	w(edge)	
Monolithic	12.45	27.0	28.1	27.1	29.1	Plate
				28.3	–	Beam
Laminated	16.89	16.9	17.9	17.0	–	Beam

stiffness of the supporting structure. As indicated in Table 1, the central deflection calculated from elementary beam theory fell between the computed theoretical values at the plate centre and at mid-span on a 'free' edge.

For this laminated panel (glass layer thicknesses $t_1 = 6.15$ mm, $t_2 = 9.98$ mm, thickness of 'soft' plastic interlayer $c = 0.76$ mm), the calculated central deflection was based on closed-form equations given in appendix A1 for a simply supported laminated beam under uniform normal loading over the entire span. Upon referring to earlier test results for laminated beams obtained by the writer [116], and allowing for differences in loading duration and temperature, the interlayer shear modulus was taken as 0.5 MPa (giving the stiffness parameter $\Omega = 3.46$). The acceptable level of agreement between measured and calculated deflections (Table 1) confirmed that, in this application, elementary laminated beam theory would be sufficiently accurate for design purposes.

Several different types of architectural laminate were tested in this manner, and in all cases there was close agreement between measured and predicted panel deflections. When failure did occur, the fractured panels remained intact with only minor spalling. Reassuringly, they also continued to carry an appreciable post-fracture load, probably as a result of membrane action of the intact interlayer combined with mechanical interlocking between adjacent pieces of broken glass. Fracture nearly always initiated at a horizontal ('free') edge of one or more laminated panels, although not necessarily at mid-span as the location of the critical surface flaw in the glass may not have coincided with the point of maximum tensile stress. Figure 42(b) illustrates the crack patterns on completion of one of the tests.

A more detailed description of these full-scale tests is given in appendix A2, where fracture patterns observed in each test are presented, together with comparisons of measured and calculated panel deflections. Here it was noted that the laminated panels always failed in adjacent vertical pairs, highlighting the alarming possibility of *cascade failure* occurring in the main glass walls, as a direct consequence of dispensing with horizontal glazing bars in the architectural design; fortuitously, an outcome considered by the writer to be sufficiently remote in practice. The results appeared to fully justify the choice of laminated glass as a safety-glazing product, together with the novel system of glass panel support. Furthermore, the process of erecting and sealing the test panels served a useful purpose in drawing attention to various handling and construction problems that were likely to be encountered on site.

Additional outdoor tests were performed to check general integrity and weather resistance by exposing a full-scale prototype glazed assembly to a combination of dynamic air pressure and swirling water spray at ambient temperature. In a somewhat outlandish yet effective arrangement (Figure 43), use was made of a modified naval combat aircraft (Hawker Sea Fury FB Mk 11) in a more peaceful role, with its outer folding wings detached and ordnance long since removed. Wind gusting was simulated by the action of the large 5-bladed propeller on this former high-performance piston-engined aircraft (powered by a Bristol Centaurus 18 radial engine rated at 2470 hp, or 1.84 MW), which was securely fixed to a foundation plinth. In the event of engine fire or explosion, the technician operating the throttle was instructed to leap promptly from the cockpit on to a nearby vertical pole and descend rapidly to *terra firma*; whereupon any ground marshals still standing could intervene with the aid of fire extinguishers. Air pressures were measured by means of

FIGURE 43. Test arrangement (1970) based on aircraft propeller thrust (modified Hawker Sea Fury) for full-scale prototype glazed assembly subjected simultaneously to dynamic wind loading and water spray at ambient temperature (photograph by the writer)

Pitot-static tubes mounted on a circular frame attached to the rear of the aircraft fuselage. Although these tests were essentially of a qualitative nature, the entirely satisfactory performance of the glazing under such turbulent environmental conditions extended confidence in the basic design.

5.7. Calculated design bending stresses: primary loading

The initial structural design calculations for the laminated glass panels proceeded on the assumption of idealised boundary conditions (*i.e.* two opposite simply supported edges, the remaining two edges 'free' or unrestrained), the degree of rotational and in-plane edge-fixity imparted to the panels being reckoned negligible. Estimates of maximum bending stress were based on small-deflection theory for a simply supported laminated beam under complete uniform loading; see appendix A1, wherein the required closed-form expressions are readily amenable to hand calculations.

In view of the experimental findings summarised in section 5.4, bending stresses in the glazed laminated panels resulting from self-weight loading were calculated on the assumption that the glass layers would deflect independently, separated at a constant distance but with *zero* shear coupling generated by the plastic interlayer. Total bending stresses resulting from the primary combination of self-weight and wind loading were then determined by superposition, with a *finite* degree of interlayer shear coupling deemed to be applicable under a 3-second gust pressure. For under such transient wind loads, the laminated panel will respond as a composite member having a 'dynamic' interlayer shear modulus appropriate to its temperature.

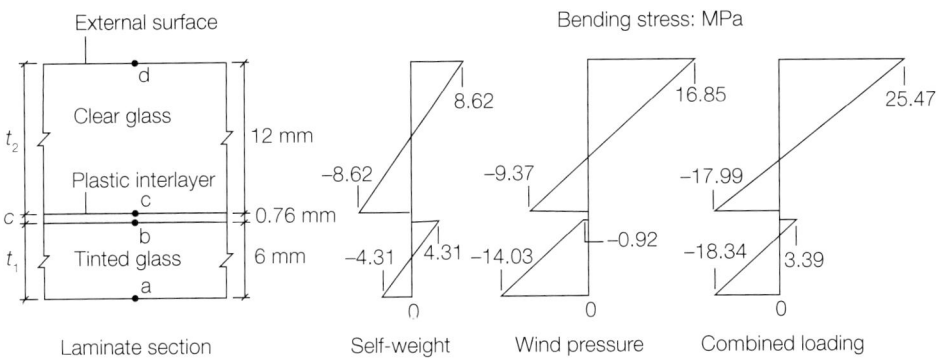

FIGURE 44. Calculated distributions of elastic bending stress (MPa, tension negative) at mid-span from simplified beam analysis of simply supported laminated glass panel in lower cone (A4 glass wall, span 2.1 m) subjected to self-weight loading and wind pressure (1.44 kPa)

Attention herein is focussed firstly on the vulnerable lower cone region (A4 glass wall), where the glass panels are inclined at around 17° to the horizontal and have a maximum 'free' span of approximately 2.1 m. The nominal glass layer thicknesses are $t_1 = 6$ mm and $t_2 = 12$ mm (thicker layer uppermost), although these dimensions vary slightly in practice due to specified manufacturing tolerances. The assumed density of the glass is 2500 kg/m³, and the self-weight of the 0.76 mm thick plastic interlayer is neglected. The elastic moduli are taken as $E_g = 72.4$ GPa for the glass, and $G_p = 0.5$ MPa for the interlayer, while the design wind loading of ±1.44 kPa (±30 psf) is assumed to act normal to the glass surface.

Calculated linear variations of bending stress (tension negative) over the laminate cross-section at mid-span are given in Figure 44, while noting that the rate of change of stress is the same for both glass layers. The maximum permanent tensile stress generated by self-weight loading is −8.62 MPa (−1250 psi) at the *inner* surface of the thicker glass layer, which fortunately is shielded from atmospheric moisture; a distinct strength advantage of such an asymmetric laminate. This substantial tensile stress is more than double that for a monolithic glass beam of the same overall thickness ($T = 18.76$ mm), namely −4.13 MPa (−600 psi). Under a (positive) wind pressure (1.44 kPa), the momentary maximum tensile stress −14.03 MPa (−2030 psi) occurs at the outer surface of the thinner glass layer, while the maximum *combined* tensile stress of −18.34 MPa (−2660 psi) in the thinner layer is only marginally higher than that in the thicker layer. Under a (negative) wind suction of the same intensity, the *combined* stresses within the thinner layer are entirely compressive, while a maximum *combined* tensile stress of −8.23 MPa (−1190 psi) occurs at the outer surface of the thicker layer. The maximum bending stresses in an equivalent monolithic beam ($T = 18.76$ mm) under wind loading are ±13.50 MPa (±1960 psi), giving a maximum *combined* tensile stress of −17.63 MPa (−2560 psi) under (positive) wind pressure. The corresponding maximum calculated bending stresses in a uniformly loaded laminated panel of *finite* length would be slightly higher, depending on the geometric aspect ratio b/a ('free'

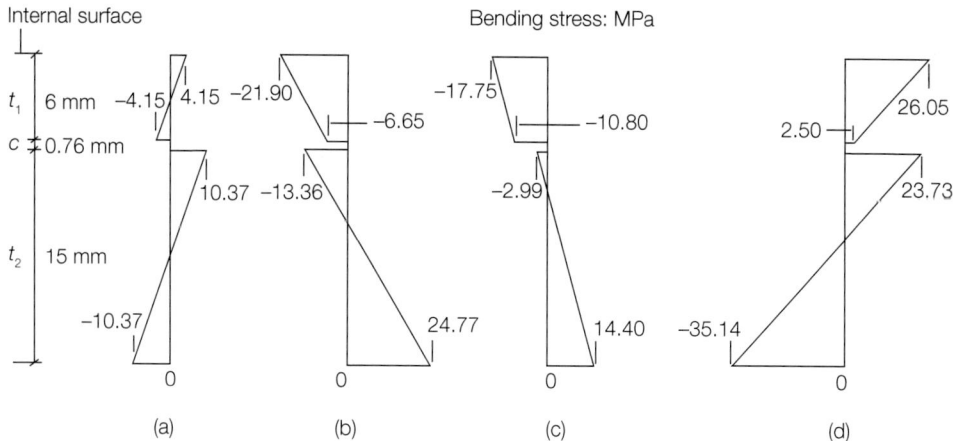

FIGURE 45. Calculated distributions of elastic bending stress (MPa, tension negative) at mid-span from simplified beam analysis of simply supported laminated glass panel in 'view windows' (A4 glass wall, span 3.1 m) subjected to self-weight and wind loading (±1.44 kPa): (a) self-weight; (b) wind pressure; (c) combined self-weight and wind pressure; (d) combined self-weight and wind suction

span a, length b). However, reference to the detailed flexural analysis of a rectangular monolithic plate (appendix A3, Table A3.2) indicates that this increase in stress would be less than 5%.

The aforementioned laminate composition applies to the entire A4 glass wall except for the 'view windows' at podium level, inclined *outwards* from the base at 42° to the vertical, where the thickness of the outer layer of clear glass was increased to 15 mm as a consequence of the much longer 'free' span (typically $a \approx 3.1$ m, length $b \approx 2.1$ m). Several distributions of calculated bending stress at mid-span are given in Figure 45. Under self-weight loading, the maximum tensile bending stress −10.37 MPa (−1500 psi) is now located at the outer surface of the thicker glass layer, while the maximum tension under wind pressure (1.44 kPa) is −21.90 MPa (−3180 psi) at the outer surface of the thinner glass layer. For a monolithic glass beam of the same overall thickness ($T = 21.76$ mm), the maximum tensile stresses are −5.43 MPa (−790 psi) under self-weight loading, and −21.86 MPa (−3170 psi) under wind loading. The *combined* effect of self-weight and wind pressure is to reduce substantially the maximum tension in both glass layers. The most onerous condition occurs under wind suction (−1.44 kPa), where the *combined* maximum tensile stress of −35.14 MPa (−5100 psi) is close to the acceptable limit. For a monolithic beam ($T = 21.76$ mm), the maximum *combined* tensile stress is −27.14 MPa (−3940 psi) under (negative) wind suction. The increase in maximum bending stresses due to plate action ($b/a \approx 0.68$) is likely to be around 2%.

Here it is noted that a lower overall maximum tensile stress is calculated if the thinner glass layer is located externally. Then the bending stress distributions in Figures 45(b)–(d) are reversed, giving a maximum tensile stress of −14.40 MPa (−2090 psi) in the thicker glass layer under *combined* self-weight and wind pressure, while the maximum tensile stress under *combined* self-weight and wind suction is –26.05 MPa (–3780 psi) in the thinner glass layer.

The ultimate strength of glass in most practical applications is governed by the resistance of this brittle material to tensile forces, and reduces significantly with increasing duration of loading. Allowable tensile stresses for annealed flat glass subjected to short- and long-term loading were obtained from several glass manufacturers, based on a given statistical probability of failure under different types of loading, deduced from extensive laboratory testing of numerous specimens of various shape and size. However, the design calculations were based on the generic 'edge strength' of glass rather than the higher 'face strength' to take account of the panel support conditions, whereby the maximum bending stresses occur at the relatively weak 'free' edge of a sawn glass panel. This approach enabled the two primary safety design criteria relating to self-weight and wind loading to be satisfied simultaneously. An additional allowance was made for temporary 'unauthorised' loading during and after construction, especially in the lower cone regions where direct human access (*e.g.* standing on a glass panel), although officially discouraged, was considered inevitable; see Figure 46.

5.8. Calculated design bending stresses: secondary loading

The probability of static loading on the glass walls generated by snow or accumulated hail was estimated to be extremely low, with meteorological records indicating that the last fall of snow ('nearly an inch in depth') to settle upon Sydney was in June 1836. Moreover, simplified dynamic structural analysis strongly suggested that the glass panels were unlikely to be damaged by hail impact or by sonic booms. No experimental testing was carried out to verify this assertion, largely on account of project constraints. Likewise, no such calculations or experiments were made in respect of blast loading or ballistic impact, or to any other

FIGURE 46. Unspecified transverse loading of lower cone glass panel (by courtesy of *Harry Sowden Photography*)

physical form of malignant or accidental occurrence (*e.g.* hand tools dropped from high elevations), as these were regarded as secondary aspects within the broad scope of structural integrity; whilst also assuming that adequate protection would be provided during both construction operations and subsequent routine maintenance.

The prospective resistance of the laminated glass panels to hailstone impact was based on a simplified design analysis of a horizontal laminated beam-strip simply supported at both ends and subjected to a vertical concentrated load at mid-span; see appendix A5. This quasi-static load, deduced from the basic principles of energy conservation, was taken as approximately equivalent to the dynamic impact force imparted by a 10 cm diameter hailstone striking the beam vertically at a terminal velocity of 45 m/s. Even under these conservative assumptions, the resulting maximum bending stress was much lower than that calculated for the design wind loading, as were estimated values of localised tensile stress in the glass contact region generated during the assumed rapid compression and disintegration of the hailstone. The comparatively short duration of such impact loading also permitted the use of a higher design tensile strength for the glass. More onerous environmental conditions such as multiple hailstone impact, possibly accompanied by high (positive) wind-gust pressures, were excluded as their occurrence was deemed to be sufficiently remote.

Consideration was given to possible structural damage to the laminated glass panels resulting from sonic booms generated by high-flying supersonic aircraft. However, the shock-wave pressures generated at ground level depend on so many variables that, at least for present design purposes, detailed calculations could hardly be justified. Measured far-field pressure signatures characteristically resemble an idealised *N*-wave (duration T_s), as sketched in Figure 47, where the maximum pressures $\pm p_s$ (in excess of ambient) vary widely, depending mainly on atmospheric conditions and Mach number, together with the altitude,

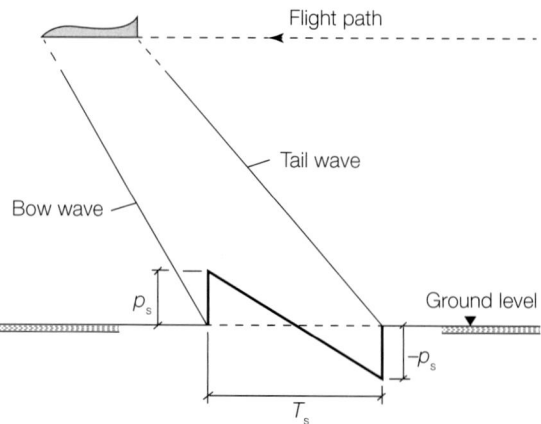

FIGURE 47. Schematic representation of sonic-boom shock waves generated by aircraft in steady supersonic flight, with idealised pressure signature at ground level (*N*-wave, maximum pressures $\pm p_s$, duration T_s)

configuration and flight path of the aircraft. Copious field measurements reported by Parrott [118], Hubbard *et al.* [119] and others indicate that, for steady level fight paths at high altitude, values of p_s are unlikely to exceed ±0.24 kPa (±5 psf), typically with $T_s \approx 0.1$ s; while also noting that in exceptional circumstances (*e.g.* low-altitude flight), the maximum transient air pressures at ground level can be very much higher.

The resulting flexural response of a glass panel will depend to some extent on its natural frequency – see, for example, Crocker and Hudson [120] and Kao [121] – but even allowing for dynamic amplification, this anticipated maximum sonic-boom pressure is sufficiently far below the design wind-gust pressure of ±1.44 kPa (±30 psf) to be regarded as minor secondary loading. Moreover, the airborne shock-wave pressures act over a much shorter time period (*i.e.* by an order of magnitude), thereby justifying an enhanced glass design strength. Accordingly, it was judged highly unlikely that sonic booms would initiate fracture in any of the laminated glass panels, even in the presence of substantial wind loading.

6. Glass colour measurement and control

6.1. Standard glass specimens

The body-tinted *demi-topaze* glass is pale bronze in colour, and was formulated specially for the glass walls. It was manufactured over a two-year period using a pot-casting process whose roots date back to the 17th century; see also section 7.2. The basic constituent materials (*e.g.* sand, soda, lime), together with the colouring oxides, were contained in a refractory pot and heated in a furnace. Upon obtaining the necessary mixing and fusion, the molten glass was poured on to a roller mechanism from which it emerged as a red-hot continuous ribbon. After annealing, the glass was ground and polished. Stringent controls were exercised during the prolonged and intermittent manufacturing period, primarily to ensure that the colour of the glass was uniform throughout the building; see Hooper and Wassall [122]. The resulting chromaticity data might also be highly beneficial in any future manufacture of matching glass, upon exhausting the initial stock of spare laminated panels intended to replace damaged glass.

At the start of production, the writer formally established a glass colour test procedure, arranged in conjunction with the Department of Applied Optics at Imperial College, London. Whence selected 75 mm square specimens of tinted, clear and laminated glass were adopted as colour standards, enabling the initial colorimetric data to be obtained. Subsequently, similar quality-control specimens of laminated glass were supplied by the manufacturers at an approximate rate of one for every 200 m² produced. These specimens were compared both visually and colorimetrically with the laminated glass colour standard. In this connection, the specified thickness tolerances were ±1 mm for the tinted plate glass, ±0.3 mm for the clear 'float' glass, and ±1.4 mm for the laminated glass.

6.2. Chromaticity definitions

Of the various methods of colour measurement then readily available, the conventional trichromatic system was used for the glass walls project. In 1931, the Commission

Internationale de l'Éclairage (C.I.E.) established a framework for colour specification, largely based on pioneering investigations [123], [124] by the British scientists William David Wright (1906–1997) – later a distinguished academic at Imperial College and, by a remarkable coincidence, the writer's first scoutmaster in early youth (155th North London scout group) – and John Guild (1889–1979), working at the National Physical Laboratory, Teddington. By this means, a given colour (C) can be expressed algebraically in terms of three tristimulus values (X, Y, Z) that define the proportions of the red, green and blue reference stimuli; see, for instance, the landmark monograph by Wright [125]. These values are weighted such that Y also represents the percentage light transmission through the glass.

Upon normalising the tristimulus values (x, y, z), their sum becomes unity, enabling any colour to be represented graphically by x and y co-ordinates on a plane diagram. This system is illustrated by the standard C.I.E. chromaticity chart shown in Figure 48, which encompasses all possible colours. The outer curved boundary is the spectrum locus, with wavelengths shown in nanometres, whilst recognising that the precise rendering of the coloured version in Figure 48(b) is partially dependent on parameters embedded in the print reproduction process. Because the locus of spectral colours is everywhere either straight or convex, and the resultant mixture of any two colours lies somewhere along a straight line between those two colours in chromaticity space, all other colours must lie within the area bounded by the spectrum locus and the straight line joining its red and violet extremities

Having plotted the chromaticity co-ordinates (x, y) for a given colour C, as indicated in Figure 48(a), a straight line can be drawn from the white point W (representing the illuminant) through C to intersect the spectrum locus at λ_d. Hence the colour C can also be defined by λ_d, known as the dominant wavelength, together with the ratio of the distance CW to $W\lambda_d$, known as the purity, p. From a subjective viewpoint, this latter definition of colour may be preferable to that of chromaticity co-ordinates, especially when assessments of small colour differences are required. The colours of a Planckian (black-body) radiator at several absolute temperatures are included for reference purposes. The chromaticity values for the C.I.E. standard illuminants S_A, S_B, S_C and D_{6500} (corresponding to correlated colour temperatures of 2856, 4874, 6774 and 6504 K, respectively) are seen to lie close to the Planckian locus. The same applies to the equal-energy white, E, on which the three reference stimuli are based.

One of the advantages of colorimetric analyses performed according to the C.I.E. system is that measurements are obtained in an unambiguous form, which besides having an internationally agreed meaning, can be related to subjective descriptions of colour. Such analyses also avoid certain problems associated with the visual assessment of colour quality (*e.g.* different colour rendering characteristics of various illuminants; colour-defective vision of the observer; differences in visual discrimination between observers with normal colour vision), especially when such assessments are required over a long period of time. However, it is emphasised that the colorimetric analyses relating to the present project were designed to augment rather than to replace the visual judgement of colour quality.

An additional benefit derived from the spectrophotometric analyses of laminated glass was that, by extending measurements to the near-infrared region of the spectrum, the necessary

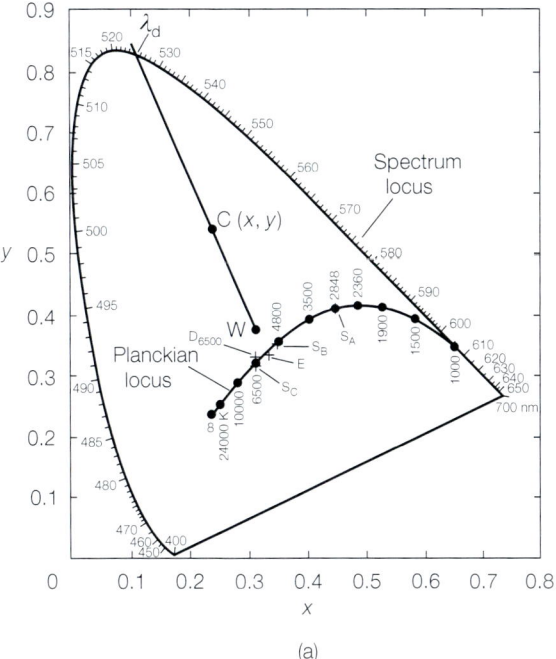

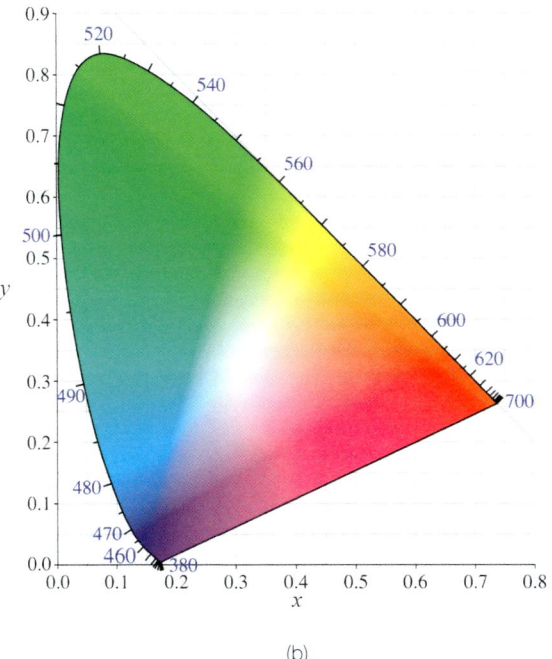

FIGURE 48. Standard 1931 C.I.E. chromaticity chart: (a) monochrome version; (b) coloured version

data were obtained from which to estimate the moisture content of the plastic interlayer; a quantity that is directly related to the degree of adhesion at the glass–plastic interface, and thereby to the structural integrity of the laminate.

6.3. Measured results

The glass colour measurements were carried out at room temperature using a modified version of the Wright spectrophotometer [126], illustrated in Figure 49. The cross-sectional area of the beam incident upon the glass specimens was approximately 10×8 mm, and readings were taken at discrete wavelengths throughout the spectrum. The instrument was carefully calibrated prior to testing, with numerous steps taken to first isolate and then minimise all likely sources of measurement error.

Although it is normal practice to quote chromaticity co-ordinates to four places of decimals, the fourth figure cannot be regarded as exact. Even using a high-precision instrument such as the one referred to above, there are certain random fluctuations that are unavoidable, leading to small errors. Following preliminary tests in which repeated measurements were made on the same specimen, it was estimated that around 95% of all normalised tristimulus values for the glass control samples should be accurate to within ±0.0002.

The spectral transmission curves for the standard quality-control specimens of tinted and laminated glass are given in Figure 50(a), while the corresponding transmission and chromaticity values are listed in Table 2. These values relate to the standard illuminant D_{6500}, which generally is considered to be the one most representative of average daylight conditions. The corresponding C.I.E. coefficients (x, y) are plotted on the chromaticity chart depicted in Figure 50(b), although the values of dominant wavelength (λ_d) give a clearer indication of the colour difference between the two specimens.

FIGURE 49. Modified Wright spectrophotometer, with covers removed (photograph by the writer)

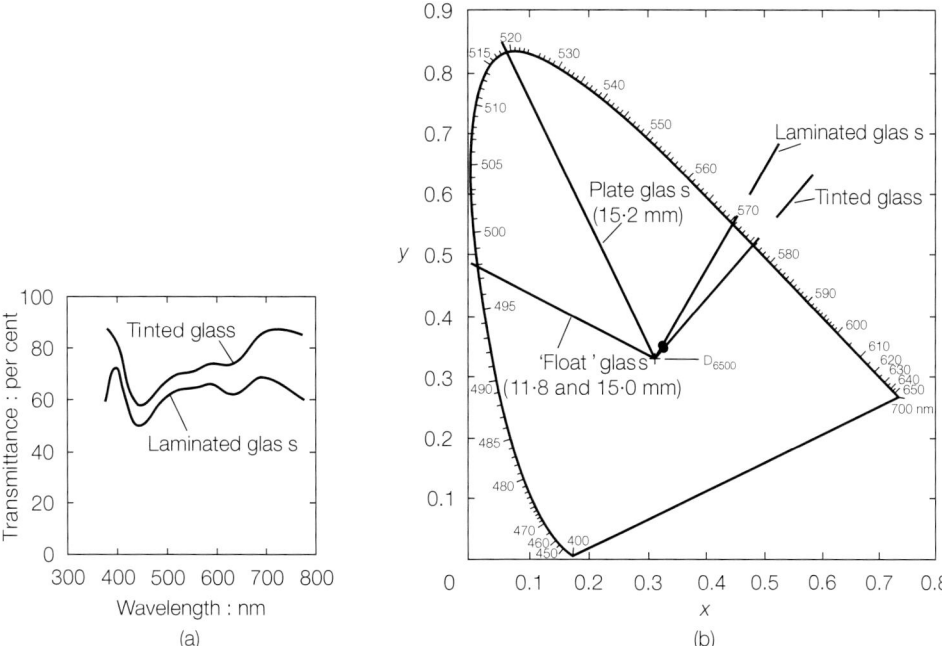

FIGURE 50. Measured data for glass colour quality-control standard specimens, after Hooper and Wassall [122]: (a) spectral transmission; (b) chromaticity values

Data for clear plate and 'float' monolithic glass specimens also are included in Figure 50(b) and Table 2. The chromaticity coefficients are similar to those for the illuminant, but there is a marked difference between the dominant wavelengths of the two types of glass. Whence the 'green' of the plate glass and the 'blue-green' of the 'float' glass are plainly visible to the naked eye when viewed 'end-on' (*i.e.* through the 75 mm 'thickness'). However, the degree of colour saturation in both glasses is very low.

During the production period, measurements were made on 26 quality-control samples of laminated glass. The percentage transmission (Y) varied from 63.5 to 68.9, with an average

TABLE 2. Measured transmission and chromaticity values for glass colour quality-control standard specimens (illuminant D_{6500}), after Hooper and Wassall [122]

Specimen	Thickness: mm	Y: %	x	y	λ_d: nm	p
Tinted plate glass	5.7	71.3	0.3284	0.3467	575	0.09
Clear 'float' glass	11.8	87.4	0.3090	0.3310	498	0.02
Clear 'float' glass	15.0	86.0	0.3079	0.3316	498	0.02
Clear plate glass	15.2	86.8	0.3107	0.3332	519	0.01
Laminated glass	18.8	64.1	0.3264	0.3521	570	0.11

of 65.9. Values of x ranged from 0.3244 to 0.3279, with an average of 0.3264; for y the range was 0.3493 to 0.3539, with an average of 0.3517. Based on these average values of x and y, the dominant wavelength is 571 nm and the colour purity is 0.11. Evidently these average chromaticity values are almost identical to those for the standard laminate (Table 2), although such close agreement might be fortuitous.

6.4. Interpretation of measured data

If a variable colour is matched visually with a colour of fixed chromaticity, there will be a point at which it is just possible to distinguish one colour from the other. This particular mismatch in colour is termed the 'just noticeable difference', usually abbreviated to 'jnd', and can be depicted as a short line or bar on the C.I.E. chromaticity chart. By changing the colour mixture of the variable colour, and again matching up with the fixed colour, a two-dimensional representation of 'jnd' values is obtained. The results of a large number of colour matching trials reported by MacAdam [127] indicated that the locus of 'jnd' points around a given reference colour is approximately elliptical.

MacAdam's perceptibility ellipses are plotted in Figure 51(a), and represent the standard deviation of colour mismatch multiplied by a factor of ten. Strictly these ellipses are cross-sections of the 'jnd' ellipsoids within the colour solid corresponding to constant spectral transmittance, but the method of plotting 'jnd' values on a plane chromaticity chart remains a simple and effective way of assessing small colour differences. In particular, any large variations in the magnitude of 'jnd' values in any region of the chart become immediately self-evident.

Numerical differences (multiplied by a factor of ten) between measured chromaticity values for the standard glass laminate (x_s, y_s) and those for the subsequent quality-control specimens (x, y) are plotted in Figure 51(b). An approximate bounding ellipse, symmetrical about the co-ordinate origin, is also sketched. By superposing this ellipse on to Figure 51(a), and comparing its size with that of others in nearby regions of the chromaticity chart, it is reasonable to postulate that there should be no noticeable difference in colour between the standard glass laminate and any of the quality-control specimens. In support of this assertion, visual comparisons of the specimens carried out under room lighting conditions (fluorescent tubes) did not reveal any perceptible colour differences.

The foregoing test results based on small laboratory specimens are highly satisfactory, but the efficacy of colour control ultimately depends on whether or not colour variations can be seen in the glass walls themselves. Here the comments of Wright [125] are pertinent:

> 'Our powers of discrimination are at their best when we are looking at two adjacent areas of colour with a sharp boundary line between them, when the areas are large, when the illumination is good, and when we are using unrestricted binocular vision'.

These conditions are very similar to those prevailing in many of the glass wall regions, particularly for an observer inside the building looking out. In the case of the inclined 'view windows', for example, only a thin bead of silicone rubber separates the largest glass panels. Despite these exacting field conditions, there appeared to be no visible colour variations in any part of the completed glass walls.

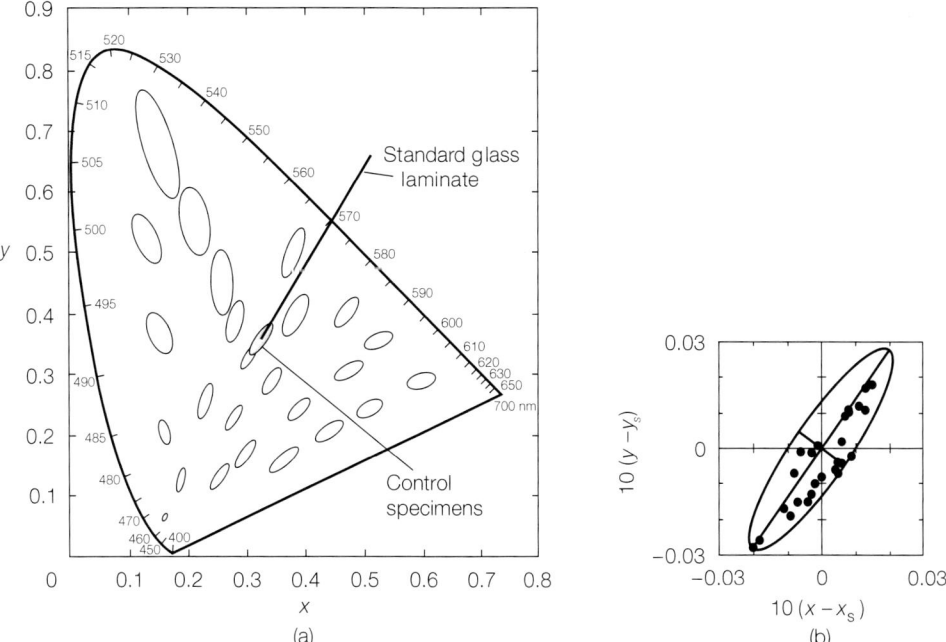

FIGURE 51. Observed deviations in chromaticity from indicated standards, after Hooper and Wassall [122]: (a) MacAdam's perceptibility ellipses, represented ten times actual scale on C.I.E. chromaticity chart, with glass wall data superposed; (b) measured differences between standard glass laminate and colour quality-control specimens (subscript 's' denotes standard colour specimen)

7. Construction of main glass walls

7.1. Erection of support structure

Field surveys of the spatial positions of the critical arched ribs revealed excellent agreement with the theoretical geometry used in the original shell design, the maximum recorded deviation from the specified position being approximately 25 mm. This remarkable accuracy in construction enabled site work on the glass walls to proceed on the basis of their theoretical geometry, with the connections between mullions and ribs being detailed to take up any discrepancies. Figure 52(a) illustrates work in progress in erecting the two main glass walls A4 and B4, while noting the temporary glass processing factory in front of the opera hall building.

During construction of the concrete shells, selected ribs had been strengthened and extra holes cast into them to allow for subsequent fixing of the glass walls; but to avoid damaging the highly stressed reinforcing cables, any form of drilling into such ribs was prohibited. To facilitate the mullion–rib connection, a 'continuous strip' of reinforced concrete was designed to be cast *in-situ*, thickened out into corbels at specified intervals and bolted to the shell rib *via* the existing holes. On site, the shuttering and steel reinforcement to the 'continuous strip' was prefabricated in short lengths to the calculated dimensions, then lifted and attached to the

FIGURE 52. Details of glazing support structure for glass wall A4 (by courtesy of *Harry Sowden Photography*): (a) overview during construction; (b) connection of steel mullions to concrete corbels; (c) adjustable bronze fixing bracket connecting bronze glazing bar to steelwork; (d) junction of vertical and inclined glass surfaces

shell rib. Survey stations on each mullion plane had been established, enabling the formwork to be moved up or down the rib until the vertical face of the corbel lay in the mullion plane. This portion of concrete was then cast, and the entire procedure repeated at other rib locations.

Upon completion of the 'continuous strip', each corbel bolt was surveyed, allowing the corresponding hole in the steel mullion to be drilled in its correct position; see Figure 52(b). The prefabricated steelwork could then be erected and would automatically be located in its theoretical position, independent of deviations in the shell rib. The steel ties between the mullions were adjustable in length to allow for tolerances, and to assist in aligning the mullions into the surveyed planes, as they were quite flexible in the lateral direction.

After the steelwork had been painted, the adjustable fixing brackets and glazing bars were attached; see Figure 52(c). The mullions and glazing bars were shop-drilled to accommodate the fixings. The junction of the vertical and inclined glass surfaces on wall A4 is illustrated in Figure 52(d). Further construction details of the glazing support structure for glass wall A4 (March 1971) are illustrated in Figure 53. An external bronze monorail was bolted on to the concrete 'continuous strip' to support a carriage and suspended cradle, thereby providing a moveable working platform from which to clean and otherwise maintain the upper region of the glass wall. However, this mechanical contrivance turned out to be of limited practical use, and was replaced by abseiler access systems.

7.2. Manufacture of laminated glass panels

The layer of annealed clear glass (12 or 15 mm nominal thickness) used in the laminated panels was manufactured by the so-called 'float' process, originally developed by the Pilkington company in England in the 1950s. Here the molten glass floats on a bath of molten tin before being drawn horizontally into a continuous ribbon and passed through a long annealing lehr, after which it is automatically cut into large rectangular sheets (*i.e.* using a cutter head that rapidly traverses the glass in an inclined plan direction largely dependent on ribbon speed and width). This revolutionary process soon became adopted worldwide to produce 'float' glass in various standard thicknesses (usually 'clear', but occasionally coloured if required in sufficiently large quantities), while also having a naturally flat fire-polished surface with a high-quality smooth finish (thereby eliminating the cumbersome grinding and polishing processes).

A notable departure from modern technology took place in the manufacture of the body-tinted *demi-topaze* plate glass at Boussois-sur-Sambre in northern France. In this traditional 'Bicheroux process', dating from the early 1900s and reportedly named after the Belgian engineer Emil Bicheroux, the raw materials were first mixed in an oval refractory clay pot that was made by hand; see Figure 54(a). The open pot was heated in one of several small furnaces (maximum capacity 4 pots) for around two days, and then withdrawn on a bogie running on track rails; see Figures 54(b) and (c). The pot was lifted and the molten glass poured on to the upper pair of horizontal rollers of the casting machine, initially emerging from the base as a red-hot tongue of glass of sensibly uniform thickness, before being carried further on a moving floor platform; see Figure 54(d). Each pot contained about one tonne of molten glass, enough to produce approximately 28 m^2 of 6 mm thick plate glass.

(a)

(b)

FIGURE 53. Construction of glazing support structure for glass wall A4, March 1971 (photographs by the writer): (a) eastern end, from above; (b) western end, from above

(c) (d)

FIGURE 53. (continued) Construction of glazing support structure for glass wall A4, March 1971 (photographs by the writer): (c) vertical steel mullions bolted to concrete corbels, with bronze glazing bars attached by adjustable bronze brackets; (d) eastern end, from above

After initial cooling and subsequent annealing, the rough-cast glass sheets were cut to form rectangular blanks of various sizes that were mounted on the surface of a large circular steel table (approximate diameter 12 m); see Figure 55(a). The upper glass surface was ground and polished by means of a pair of rotating discs pressed against the lower revolving table, using progressively finer abrasives with water and finishing with wet 'rouge'; see Figure 55(b). Upon reversing the glass sheets, the entire grinding and polishing process was repeated for the opposite surface. In total, around 6500 m^2 of this special polished plate glass was produced. It is recorded with considerable regret that these glass manufacturing facilities no longer exist, so that any future production of replacement glass may not be entirely straightforward. However, the known approximate chemical composition of the glass, together with the aforementioned results of colorimetric measurements, would be most useful in obtaining a matching product.

The laminating procedure was carried out according to well-established industry standards. Initially a thin flexible sheet of clear plastic interlayer (plasticised polyvinyl butyral, 0.76 mm thick) is placed between the two glass layers in an air-conditioned 'clean room', as illustrated in Figure 56(a), and the loose composite panel is passed between heated rollers to give partial adhesion. Full lamination is achieved under conditions of modest temperature and pressure

(a)

(b)

FIGURE 54. Production of pot-cast body-tinted flat glass at Boussois-sur-Sambre (photographs by the writer): (a) initial hand-crafting of refractory clay pot; (b) firing of pots in furnace (door raised)

(c)

(d)

FIGURE 54. (continued) Production of pot-cast body-tinted flat glass at Boussois-sur-Sambre (photographs by the writer): (c) withdrawal of pot from furnace; (d) molten glass poured on to roller casting machine

(a) (b)

FIGURE 55. Grinding and polishing of body-tinted flat plate glass at Boussois-sur-Sambre (photographs by the writer): (a) rough cast glass sheets mounted on circular steel table, prior to grinding; (b) final polishing phase using wet 'rouge'

(a) (b)

FIGURE 56. Glass lamination process (photographs by the writer): (a) initial placement of plastic interlayer in 'clean room'; (b) loading of autoclave

within a large cylindrical autoclave, whose size incidentally governs the maximum available panel dimensions; see Figure 56(b).

7.3. Shaping of laminated glass panels

Given the basic glass wall geometry, the theoretical size and shape of each panel could be computed according to the chosen architectural layout. Although not an essential requirement, the laminated glass panels could then be ordered from the manufacturer in only slightly oversize rectangular blanks, resulting in minimum wastage. Except in a few special cases, computer generated schedules were produced (*i.e.* one for each panel, giving principal dimensions), forming an extension of the geometrical software already developed by the

Engineer for the supporting structure (*e.g.* steel mullions and bronze glazing bars). However, because the dimensional tolerances required in general glazing practice are much tighter than the likely differences between theoretical and as-built locations in the complex glazing sub-structure, each panel opening was measured accurately prior to glass processing on site, enabling the necessary final adjustments to the idealised computer output.

One of the major site production problems was how best to shape the laminated glass just prior to glazing, particularly as hundreds of different-sized panels were required, often in the form of irregular quadrilaterals. Reference to Figure 18, for instance, indicates that almost every glass panel to either side of the vertical plane of symmetry on wall A4 is a different shape; see also Figure 57 for the corresponding developed surfaces on one side of the vertical plane of symmetry, which depict the true shape of each panel located above the 'view windows'. Moreover, pairs of panels located symmetrically about the central plane are not identical because of the asymmetric layer configuration of the glass laminate.

Two methods of shaping the laminated glass were appraised; namely by sawing, using a diamond-impregnated circular blade, and by the traditional process of cutting and breaking. In the latter, both glass layers are scribed and broken in turn, followed by the application of heat along the line of fracture. With the plastic interlayer thus softened, the two laminated

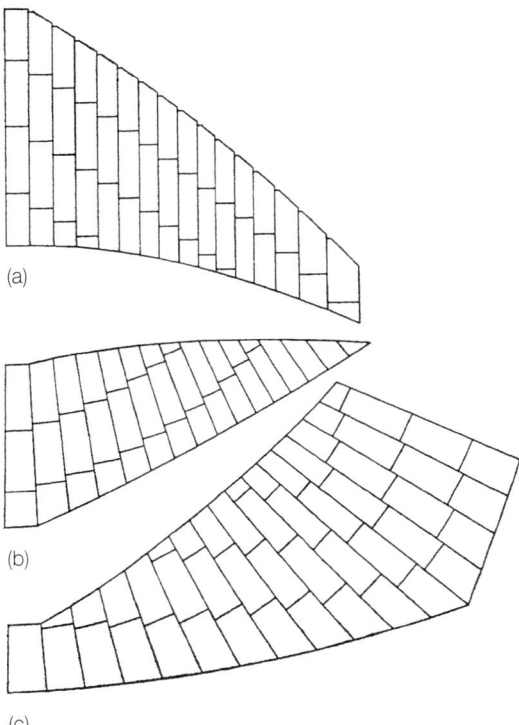

FIGURE 57. Developed surfaces of laminated glass panels for glass wall A4: (a) elliptic cylinder; (b) upper cone; (c) lower cone

sections are pulled slightly apart and finally separated using a thin sharp blade. However, this method becomes cumbersome when dealing with large panels and thick laminates, and the resulting stepped glass edges generally require substantial finishing work.

In contrast, rotary sawing can give a glass edge that is both straight and flat, and which only needs arrising to render it suitable for glazing. This method was therefore adopted, and with a dearth of commercially available equipment, two purpose-built sawing machines were designed and developed by local specialist sub-contractors. A substantial temporary structure resembling a production-line factory was established on site to deal solely with the storage of crated blank glass panels supplied by the manufacturer, together with the subsequent processing and handling of the glass prior to glazing. This building structure appears in Figure 52(a), to the left foreground, and housed equipment that included a gantry crane with a vacuum cup rig, a tilting setting-out table, and two L-shaped cutting tables capable of accommodating a 4.3 m long saw cut; see Figure 58(a).

Part of one of the sawing machines is illustrated in Figure 58(b). The two end-frames were connected by a heavy cross-member of box-type construction that accommodated the movable head. Pneumatically operated castors were built into the working table to facilitate easy handling of the glass. During sawing, with the castors retracted, the glass could be firmly held down by suction pads. Each edge of every laminated glass panel was sawn in a single pass along the prescribed straight line, using a copious supply of cooling water mixed with cutting fluid (1% Castrol B); Figure 58(c). Upon pneumatically reversing the heavy machine chuck, the sawn edge was arrised on the return pass by abrasive wheels (*i.e.* 1.6 mm arris at 45° on upper and lower faces); Figure 58(d). An additional benefit of this shaping process was that any localised edge damage or delamination in the glass blanks delivered to site was automatically removed. Finally, the edges were cleaned with water followed by solvent (methyl ethyl ketone) to remove cutting fluid and particulates, and then air dried before sealing the interlayer temporarily with a thin protective coating of silicone rubber.

Early problems were encountered as a result of the laminate overheating, leading to partial combustion of the plastic interlayer and to excessive glass spalling. Moreover, machine vibrations occasionally were of sufficient magnitude to crack the glass. An initial stop-gap measure (implemented by the writer) was to bed the glass panel on a layer of plaster of Paris, but fortunately this makeshift approach was rapidly superseded; much to the relief of the sawing machine operators and also the local chemist, who was unfamiliar with supplying such material in one-hundredweight sacks. After further experimentation on site, these difficulties were overcome, and the sawing operation proved to be entirely successful in shaping the laminated glass panels to the required high standard. Linear feed rates of up to 15 mm/s were attained using a 350 mm diameter diamond-tipped saw, but the rate subsequently adopted for general production purposes was around 8 mm/s. To circumvent what appeared to be residual stress problems sometimes encountered when making large diagonal cuts, a hot-wire technique was used initially to reduce the glass blank roughly to size. The panel was then accurately sawn to the final shape in the established manner. More generally, a piece of waste glass was butted against the corner of the laminated panel to avoid splitting near the end of a cut.

FIGURE 58. Shaping and arrising of laminated glass panels: (a) site factory equipment layout; (b) part L-shaped cutting table; (c) sawing phase (by courtesy of *Lawson Publications*); (d) arrising phase (by courtesy of *Harry Sowden Photography*)

7.4. Glazing operations

Glazing was carried out from the top downward, based on wide working platforms supported on scaffolding; see Figures 59(a) and (b). The laminated glass panels, having been sawn to shape on site and mounted on an A-frame, were hoisted on to the platform by a tower crane, usually in groups of three or four; as depicted in Figure 59(c), accompanied in a casual manner by a team member whose sterling attributes clearly included a head for heights and a firm grip whilst in slow but intermittent aerial motion; see also Figure 60(a). Various representatives were present on the elevated working platform to witness the initial glazing operations, and Figure 60(b) shows the writer in conversation with his older colleague Robert Kelman, whose main technical concern was the bronze supporting metalwork.

On the platform, individual panels were moved using a purpose-built mobile crane with suction lifting equipment. Figure 61(a) illustrates the glazing of the first laminated glass panel of the entire project (March 1971), at the apex of glass wall A4, while at podium level, the first inclined glass panel in the 'view window' was installed in October 1971; see

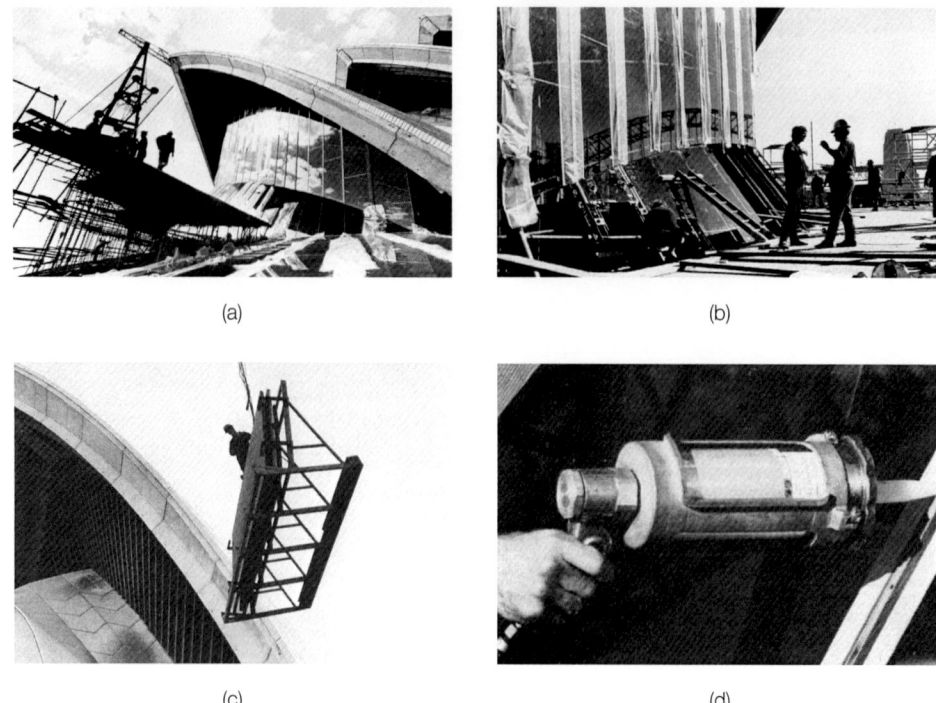

FIGURE 59. Glazing operations on glass wall A4 (by courtesy of *Harry Sowden Photography*): (a) general view from podium level; (b) working platform level near junction of vertical and upper cone glass surfaces; (c) hoisting of glass panels on to platform; (d) application of silicone rubber sealant to inner glazing bar joint

Figure 61(b). The imposing photographs in Figure 62 taken on the A4 glass wall working platform in March 1971 portray part of the glazing sequence for an irregular shaped panel adjacent to the central rectangular apex panel.

As a most useful aid to the glazing installation process, particularly regarding the necessity of meeting comparatively fine construction tolerances, special mechanical jigs were designed and developed in conjunction with local external consultant Leslie Dixon from Hawker de Havilland, whose wide practical experience prior to emigrating from England to Australia included extensive design work on the legendary Comet jet airliner. Here the main objective was to accurately position the pair of stainless-steel pins (including a bonded fibre load-bearing pad) permanently supporting the in-plane weight of each laminated glass panel; see the prototypes in Figure 63. Firstly, two jigs were clamped on to adjacent glazing bars to provide temporary support to the horizontal lower glass edge in the corner regions. The correct in-plane alignment of the glass panel was accomplished using a screw mechanism forming part of the support cradles; see Figure 64(a) for an example of vertical glazing. Each jig incorporated a pivoted drill attachment that could be swung around to form a hole in the glazing bar flange to accept a push-fit pin, thereby automatically located in the exact position by means of a standard template forming part of the jig. Upon slightly lowering the glass

FIGURE 60. Initial glazing of glass wall A4, March 1971: (a) supply of glass panels on swinging A-frame (photograph by the writer); (b) on-site discussions (by courtesy of *Harry Sowden Photography*)

panel on to the two horizontal pins by releasing the vertical screws, the jigs were removed and made ready for re-use. With the panel now supported in its correct position and held against the glazing bars using temporary clamps, as in Figure 64(b), the joints were sealed in line with the details sketched in Figure 23.

Being a one-part sealant, silicone rubber is generally easy to apply. Compressed-air guns were used on site, as illustrated in Figure 59(d). Initial trials were conducted (by others) to optimise the sequence of sealing operations, as the rate of curing of silicone rubber markedly depends upon ambient temperature and relative humidity, and upon the cross-sectional geometry of the joint. Its deployment as a construction sealant also can pose substantial difficulties because the substrates need to be cleaned and prepared to stringent standards, frequently under harsh site conditions. Thus the sawn edges of the laminated glass panels were provided with protective covers during handling and storage, and cleaned just prior to glazing. Additionally, the bronze glazing bar surfaces were abraded, solvent cleaned, primed and air dried for around 18 hours before applying the sealant. Some *ad hoc* experimentation was required in selecting the most appropriate primer; it had to be insensitive to surface

(a)

(b)

FIGURE 61. Glazing landmarks for glass wall A4: (a) installation of first vertical glass panel at apex of shell, March 1971 (photograph by the writer); (b) installation of first inclined glass panel in 'view window', October 1971 (by courtesy of *Harry Sowden Photography*)

(a)

(b)

FIGURE 62. Part glazing sequence for irregular shaped panel near apex of A4 glass wall, March 1971 (photographs by the writer)

(c)   (d)

FIGURE 62. (continued) Part glazing sequence for irregular shaped panel near apex of A4 glass wall, March 1971 (photographs by the writer)

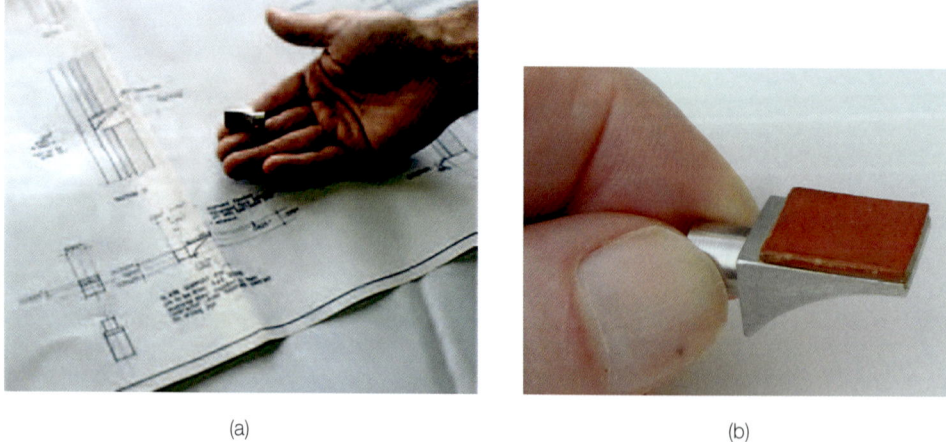

(a)   (b)

FIGURE 63. Permanent stainless-steel push-fit pins (including a bonded fibre load-bearing pad) supporting self-weight of laminated glass panels (photographs by the writer)

(a) (b)

FIGURE 64. Vertical support of laminated glass panel (photographs by the writer): (a) temporary clamps with drill attachment; (b) permanent support by horizontal corner pins in glazing bar flanges

moisture, and also resistant to attack from acetic acid given off by the silicone rubber during curing. Moreover, the various joint materials needed to be chemically compatible. On one occasion, following joint failures on site, the writer was dispatched urgently from London to central France for technical discussions with the sealant manufacturer, who soon concluded that because another material had been substituted for the specified inert foam spacer rod, the resulting chemical reaction was almost certain to have degraded the bond between the silicone rubber and the substrates.

On the main glass wall surfaces, the primary seal is provided by the horizontal butt-joints between panels, and by the vertical joints between the glass and the web of the glazing bar T-section. As a second line of defence along the vertical joints, where initial teething problems were encountered, the outer glazing bar flange or cover piece was assembled in such a way as to generate a form of mechanical seal. In this arrangement, the flange was left slightly loose during application of the neoprene D-ring strips and the outer silicone rubber seal. When the silicone had cured sufficiently (typically after 10 days), the flange was fully screwed down on to the web of the T-section, thereby causing compression of the silicone and creating a more effective external gasket. However, the longevity of the imposed compression remained unknown. These vertical joints were covered loosely with clear plastic sheeting during the initial three-week curing period, to give a measure of weather protection. At joint locations with no cover piece, such as those near the intersection of the principal glass surfaces, the web of the glazing bar was cut back to enable the silicone rubber to bridge between adjacent glass edges; see Figure 52(d), which also includes a similar vertical butt-joint over a short length with no glazing bar.

Slight edge delamination in a few panels was observed shortly after glazing, partially resulting from chemical reactions between the plastic interlayer and the coolant fluid used in machining the glass panels, and also from direct contact with the acidic silicone rubber sealant during the curing process. Here the long-term behaviour was unknown, and difficult to predict. The same applied to the anticipated level of joint adhesion, where degradation could occur over time, partly due to the influence of natural ultraviolet radiation in the case of glass-to-glass butt joints. Without any available evidence to the contrary, it was optimistically assumed that edge delamination was a short-term effect, whereas eventual breakdown of interfacial adhesion between silicone rubber and the principal substrates was considered inevitable.

7.5. Construction of remaining glass walls

Most of the preceding construction details for glass wall A4 applied equally to the other main walls, while the layout of the side walls essentially reduced to a modified 'view window' with comparatively little elevated glazing. Figures 65–70 are photographs of the glass walls taken by the writer in September 1973, shortly before the building was first opened to the general public.

8. Post-construction assessment

Upon completion of the glass walls, the potential problem of thermal fracture in the tinted glass was re-visited on a more detailed basis. This investigation was prompted by a Client proposal that adjustable shading blinds might subsequently be installed inside the main north-facing glass walls, primarily to reduce solar heat gain and air-conditioning loads within a large space having little natural ventilation. These concerns had been raised much earlier when specifying the glass laminate details, but assurance was given that the proposed air-conditioning plant could readily cope with the anticipated thermal loading; an assertion perhaps encouraged by the abundant supply of cooling water surrounding the building. Here it is noted that edge cracks would tend to initiate in the central region of the 'vertical' edges of a glass panel, due mainly to thermal gradients between the exposed glass and that covered by the outer glazing bar flange or cover piece.

Accordingly, in September 1973, *in-situ* glass temperatures were continuously recorded by the writer over several days on a small region of the lower cone on wall A4, fitted with a temporary inner blind; see Figure 71(a), showing the recording apparatus connected to thermocouples attached to the interior tinted glass surface. Despite the incidence of high levels of solar radiation during the measurement period, the increased glass temperatures caused by the presence of the blind were judged to be within acceptable limits and, based largely on the results of a numerical thermoelastic stress analysis, unlikely to cause thermal fracture. To date, no such solar blinds have been installed, not least to avoid what might be described as a disturbing visual intrusion.

Likewise, the opportunity was taken to measure differential temperatures across the internal face of several glass panels that were subjected to partial external shading by the roof

FIGURE 65. Internal views in northern concert hall main foyer, glass wall A4 (photographs by the writer, September 1973): (a) looking eastward; (b) eastern end, looking towards glass wall B4; (c) looking upward through lower cone; (d) looking westward

structure. Figure 71(b) portrays such measurements in progress on side wall A8 (west), close to the concrete pedestal. Fortunately, elastic finite element analyses based on the observed temperature gradients confirmed the initial design calculations, which indicated that thermal stresses resulting from this form of partial shading would be acceptably low.

On this occasion it was also possible to carry out a cursory visual inspection of the silicone rubber butt-joints at podium level, which had then been in place for around two years. Some

(a)

(b)

FIGURE 66. South facing glass walls (photographs by the writer, September 1973): (a) A1 and C11; (b) A1 and B1

FIGURE 67. Juxtaposition of glass wall B1 with shell roof and concrete canopy structure (photograph by the writer, September 1973)

(a)

FIGURE 68. Restaurant structure (photographs by the writer, September 1973): (a) glass walls C11 and east side C13

(b)   (c)

FIGURE 68. (continued) Restaurant structure (photographs by the writer, September 1973): (b) glass wall C12, looking eastward; (c) C12 'view windows', looking southward through entire structure

(a)

FIGURE 69. External views of side shell glass walls (photographs by the writer, September 1973): (a) A7 west

GLASS WALLS, SYDNEY OPERA HOUSE

(b)

FIGURE 69. (continued) External views of side shell glass walls (photographs by the writer, September 1973): (b) A5 west

(a)

FIGURE 70. Internal views of side shell glass walls (photographs by the writer, September 1973): (a) A5 and A6 west

(b)

(c)   (d)

FIGURE 70. (continued) Internal views of side shell glass walls (photographs by the writer, September 1973): (b) A8 west; (c) A5 east; (d) A7 and A8 west

GLASS WALLS, SYDNEY OPERA HOUSE

(e)

FIGURE 70. (continued) Internal views of side shell glass walls (photographs by the writer, September 1973): (e) A5 west

(a)   (b)

FIGURE 71. *In-situ* measurement of glass temperature (photographs by the writer, September 1973): (a) lower cone on glass wall A4, with temporary blind installed; (b) partially shaded region of west side glass wall A8

minor edge-delamination was evident in the glass panels, but appeared not to have increased beyond that observed during the first few weeks following the original glazing; a conclusion that fortunately remains valid to the present day. More generally, the ageing silicone rubber sealant in all joints has needed replacement periodically over the past decades, occasionally in more pressing response to minor localised water ingress usually caused by loss of substrate adhesion. But following this prolonged exposure of the plastic interlayer to solar heating and irradiation, there appears to have been no significant degradation in its optical quality and other physico-chemical properties, although laboratory testing of irradiated samples of laminated glass would be needed to detect and quantify any such changes.

Further visual inspection was directed toward assessing the clarity of distant objects experienced by an internal observer at podium level. The respectable optical quality of the 'view windows' is partly due to the high degree of surface flatness of the annealed laminated glass, despite the inner layer being polished plate glass rather than naturally flat 'float' glass. Moreover, the effect of out-of-plane curvature of the inclined glass panels due to self-weight loading appears to be negligible. The visual perception of distant images when looking outward perpendicular to the plane of a glass panel is excellent. There is little loss in definition when the horizontal viewing angle is moderately oblique, as indicated in Figure 72(a) for the A4 glass wall. Only at much greater viewing angles does optical distortion becomes noticeable, as in the rather extreme example illustrated in Figure 72(b). However, the latter effect is due largely to the curved plan outline of the 'view windows'. Where the plan curvature is less pronounced, especially for the lower side shell glass walls, the corresponding image quality is good; see, for example, Figure 70(e). Here and elsewhere, the laminated glass has withstood the vagaries of extreme weather up to the present time without inherent structural failure or discoloration, with replacements generally confined to cracked panels resulting from accidental impact damage. Moreover, the absence of any visible signs of distress in

(a)   (b)

FIGURE 72. Images through A4 'view window' to illustrate levels of optical clarity (photographs by the writer, September 1973): (a) good definition at low angle of incidence; (b) optical distortion at high angle of incidence

the concentrated pin-loaded regions has vindicated this unique method of providing in-plane support to the laminated glass panels.

The projecting corner regions of the two northern glass walls – sometimes described as 'horns' – are considered ungainly by some observers (including the writer), both externally (especially when viewed from a broadly southerly location) and internally, burdened with an awkward support structure; as illustrated in Figure 73, and in marked contrast to the elegance of the restaurant glass walls (C11 and C12). This outcome partly stems from what might be regarded as a questionable architectural brief (seemingly of obscure origin): namely that the entire base line of the 'view windows' should follow the outer irregular plan shape of the podium structure, as indicated in Figure 18(b). In hindsight, it appears from a personal commentary [55] in 1990 that Peter Hall also regarded this particular design aspect with misgivings. Referring to the two northern glass walls, he stated: *'Looked at from the inside they are fine, but from the outside the corners stick out too far'*. And in a document [69] referring to an earlier statement dated June 2000, Utzon declared: *'The glass wall, which has been built, is in family with the glass wall we arrived at with its feeling of hanging from the shell, but the old solution was not splaying it out, as has been done.'* Any major refurbishment might wish to correct this anomaly, while also forming perforations in the steel mullion web plates to lighten the visual aspect of the main supporting structure.

Following its opening to the public, the building undoubtedly has been greatly admired, as exemplified by the panoramic aerial view in Figure 74(a), and the newly occupied northern foyer beneath the A4 glass wall illustrated in Figure 74(b). Having been separated from the building since its completion, it is not possible for the writer to comment further on long-term performance levels attained in practice. But as far as can reasonably be ascertained, and notwithstanding the continuing controversy surrounding their architectural merit, the glass walls have successfully fulfilled their demanding technical role during the past fifty years. It would seem churlish not to regard this achievement as a resounding endorsement of the strenuous efforts made by individuals and organisations from many countries closely associated with their design and construction.

In a pair of contemporary internal views, Figure 75(a) emphasises the stark angularity of the corner regions of the northern glass walls, noting also the modest edge delamination in the A8(east) 'view window', while the foyer ambience in Figure 75(b) is very similar to that depicted in Figure 74(b) when first opened to the public. Some companion external views are shown in Figure 76, for comparison with previous photographs taken some fifty years ago. Finally, as perhaps a visual precursor toward continuing good fortune during the next half-century, Figure 77 shows recent dramatic images of the Sydney Opera House against the setting sun.

Acknowledgements

The research and development work on glass and glazing benefited from the broad support and genial encouragement of various engineering personnel, especially Ove Arup and Jack Zunz in London, together with Leslie Dixon, Robert Kelman, Michael Lewis and John Nutt in Sydney. Numerous on-site discussions with Peter Hall ensured that proposed technical solutions were aesthetically acceptable,

(a)

(b)

FIGURE 73. Perceived maladroit visual aspects of corner regions of northern glass walls: (a) external corner projection (A4) viewed from south-west; (b) internal view (B4), looking east

(a)

(b)

FIGURE 74. Completed building: (a) aerial view, *circa* 2005 (by courtesy of *Sydney Opera House Trust*); (b) northern foyer to concert hall soon after public opening, September 1973 (by courtesy of *State Government of New South Wales*)

(a)

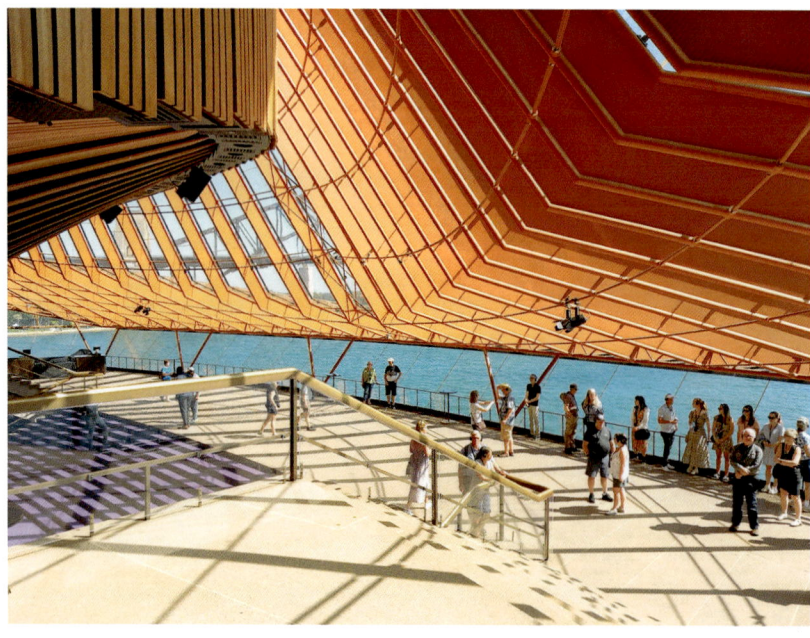

(b)

FIGURE 75. Contemporary internal views, March 2023 (photographs by the writer): (a) corner region of B4 glass wall, through A8(east) 'view window'; (b) northern foyer to concert hall

(a)

(b)

FIGURE 76. Contemporary external views, Sydney Opera House, March 2023 (photographs by the writer)

(c)

(d)

FIGURE 76. (continued) Contemporary external views, Sydney Opera House, March 2023 (photographs by the writer)

(e)

(f)

FIGURE 76. (continued) Contemporary external views, Sydney Opera House, March 2023 (photographs by the writer)

(a)

(b)

FIGURE 77. Sydney Opera House at sunset, March 2023 (photographs by the writer)

while more general contributions were made by Patrick Bourrier, Wally Doig, David Fischer, Bruce Perdriau and Harry Savage. Generous advice on monitoring the colour quality of the pot-cast tinted glass during manufacture was given by Michael Wassall and David Wright in London.

Client: Government of New South Wales. *Architect (Stage 3)*: Hall, Todd & Littlemore, Sydney. *Consulting structural engineer*: Ove Arup & Partners, London & Sydney. *Main contractor (Stage 3)*: M. R. Hornibrook Pty Ltd. *Main subcontractors for glass walls*: J. W. Broomhead Pty Ltd (steelwork); Permasteel Pty Ltd (metalwork); Hawker de Havilland (Austr.) Pty Ltd (fixing brackets, glazing jigs, loading test apparatus); Quick-Steel Engineering Pty Ltd (sawing machinery, glass handling and maintenance equipment); VASOB Glass Pty Ltd (glazing). *Main suppliers for glass walls*: Austral Bronze Crane Copper Pty Ltd (bronze extrusions); Boussois Souchon Neuvesel and Société Industrielle Triplex (glass); Rhône Poulenc (silicone rubber sealant). *Test facilities*: National Materials Handling Bureau, Sydney (patch loading tests); Materials Testing Laboratory, New South Wales Public Works Department, Sydney (static beam loading tests); Civil Engineering Department, University of Sydney (dynamic beam loading tests); Commonwealth Scientific and Industrial Research Organisation (C.S.I.R.O.), Division of Food Preservation, Sydney (creep-loading tests); Commonwealth Experimental Building Station, Sydney (full-scale static and dynamic loading tests); Department of Applied Optics, Imperial College, London (glass chromaticity measurements).

9. References

1. HEMSLEY, J. A. *Glass in engineering science.* Society of Glass Technology, Sheffield, 2016, Vol. 2 (*Glass under load*), Ch. 1 (*Glass walls, Sydney Opera House*), 1–59.

2. SYDNEY OPERA HOUSE. *An international competition for a National Opera House at Bennelong Point, Sydney, New South Wales, Australia: conditions and programme (The Brown Book).* Government of State of New South Wales, Sydney, Dec. 1955, 28 pp.

3. SYDNEY OPERA HOUSE. *Competition drawings submitted by Jørn Utzon to the Opera House Committee.* New South Wales Public Works Department, Sydney, Dec. 1956.

4. SYDNEY OPERA HOUSE. *Opera House competition: memorandum to competitors.* New South Wales Public Works Department, Sydney, Feb. 1957, 6 pp.

5. UTZON, J. O. *Sydney National Opera House (The Red Book).* Atelier Elektra, Copenhagen, Mar. 1958, 55 pp.

6. SYDNEY OPERA HOUSE. *Ceremony to commemorate the commencement of the building of the Sydney Opera House (The Gold Book).* New South Wales Public Works Department, Sydney, Mar. 1959, 30 pp.

7. UTZON, J. O. *Sydney Opera House (The Yellow Book).* New South Wales Public Works Department, Sydney, Jan. 1962.

8. UTZON, J. O. *Sydney Opera House (The Blue Book).* New South Wales Public Works Department, Sydney, Nov. 1962. (*Recommendations for car park, bus terminus and pedestrian approach*).

9. WHITBREAD, R. E. and PACKER, M. A. *Wind pressure measurements on a model of the proposed new Sydney Opera House.* National Physical Laboratory, Teddington, Aerodynamics Div., Dec. 1962, N.P.L. Aero. Rep. 1049, 30 pp.

10. UTZON, J. O. *Descriptive narrative, Sydney Opera House.* New South Wales Public Works Department, Sydney, Jan. 1965, State Archives NRS 12709 (2/8644A-C), 99 pp.

11. WESTCOTT, R. and WESTCOTT, P. *The Sydney Opera House.* Ure Smith Pty Ltd, Sydney, Oct. 1965, 52 pp.

12. UTZON, J. O. The Sydney Opera House. In: *Zodiac 14: a review for contemporary architecture.* Edizioni di Comunità, Milano, 1965, 36–93.

13. UTZON, J. O. Sydney Opera House: the roof tiles. *Architecture in Australia*, 1965, Dec., 85–92.

14. NUTT, J. G. Influence of corrosion on some aspects of design of the Sydney Opera House. *Proc. 7th Annual Conf. Australasian Corrosion Assoc. (Practical aspects of corrosion control), Manly, New South Wales, Australia, Nov. 1966.* Paper presented at Conference but not published in Proceedings. In: *The Arup Journal*, 1973, 8, 3 (Oct.), 52–54.

15. NUTT, J. G. Grouting prestressing ducts at the Sydney Opera House. *Australian Prestressed Concrete Group*, Aug. 1967. In: *The Arup Journal*, 1973, 8, 3 (Oct.), 42–44.

16. ARUP, O. N. and JENKINS, R. S. The evolution and design of the concourse at the Sydney Opera House. *Proc. Instn Civ. Engrs*, 1968, 39, Apr., 541–565. Discussion, 1968, 41, Nov., 613–619. In: *The Arup Journal*, 1973, 8, 3 (Oct.), 22–27.

17. BLANCHARD, J. C. Model tests for the Sydney Opera House. *The Arup Journal*, 1968, 3, 3 (May), 60.

18. SONDHEIMER, H. *Sydney Opera House opera theatre: a survey study from a user's point of view.* Hans Sondheimer, Sydney, 1968.

19. NUTSCH, J. *Report concerning the results of the consultations on acoustical matters related to Jørn Utzon's interior design of the Sydney Opera House.* New South Wales Public Works Department, Sydney, Dec. 1968.

20. ARUP, O. N. and ZUNZ, G. J. Sydney Opera House. *Struct. Engr*, 1969, 47, 3 (Mar.), 99–132. Discussion, 1969, 47, 10 (Oct.), 419–425. In: *The Arup Journal*, 1973, 8, 3 (Oct.), 4–19.

21. ANONYMOUS. Consortium to undertake glazing of the Opera House. *Architectural Glass & Aluminium*, 1970, 7, 8 (Mar.), 6–7.

22. DOWRICK, D. J. The prestressing of Sydney Opera House roof. *Proc 6th Int. Congr. F.I.P. (Fédération Internationale de la Précontrainte), Prague, Czechoslovakia, Jun. 1970.* Paper presented at Congress but not published in Proceedings. Available as separate document, 19 pp.

23. ANONYMOUS. Glazing at the Sydney Opera House: a progress report. *Architectural Glass & Aluminium*, 1971, 8, 6 (Nov.), 10–11.

24. COMPAGNONI, P. The Sydney Opera House glass walls. *Architecture in Australia*, 1972, 61, 3 (Jun.), 333.

19. BREWER, C. Sydney Opera House glass walls. *Builder N.S.W.*, 1972, Oct., 4–11.

26. O'BRIEN, T. and NUTT, J. G. Adhesives for structural jointing. *The Arup Journal*, 1973, 8, 3 (Oct.), 48–49.

27. PERKIN, G. and ZUNZ, G. J. A matter of silhouette: an editorial comment on Sydney Opera House, with a personal viewpoint by Jack Zunz. *Concrete Quart. (Cement & Concrete Assoc.)*, 1973, 99, Oct.–Dec., 40–48.

28. HOARE, H. R. The Sydney Opera House. The contractor and some aspects of Stage III. *J. & Proc. Royal Soc. New South Wales*, 1973, 106, Nov., 3–17.

29. LEWIS, M. Roof cladding of the Sydney Opera House. *J. & Proc. Royal Soc. New South Wales*, 1973, 106, Nov., 18–32.

30. JORDAN, V. L. Acoustical design considerations of the Sydney Opera House. *J. & Proc. Royal Soc. New South Wales*, 1973, 106, Nov., 33–53.

31. HALL, P. B. The design of the Concert Hall of the Sydney Opera House. *J. & Proc. Royal Soc. New South Wales*, 1973, 106, Nov., 54–69.

32. SHARP, R. The Grand Organ in the Sydney Opera House. *J. & Proc. Royal Soc. New South Wales*, 1973, 106, Nov., 70–80.
33. LITTLEMORE, D. S. *Sydney Opera House. Anatomy of stage three construction and completion: a general index (The Green Book).* New South Wales Public Works Department, Sydney, circa 1973, 324 pp.
34. WALDRAM, J. M. Little visible equipment at Sydney Opera House. *Light & Lighting*, 1974, 67, 1 (Jan.–Feb.), 26–32.
35. ANONYMOUS. The Sydney Opera House: a description of the engineering services. *Building Services Engr*, 1974, 42, Apr., A22–A23.
36. WESTCOTT, R. *The Sydney Opera House*. Ure Smith Pty Ltd, Sydney, 1965 (with P. Westcott), 1st edn; 1968, 2nd edn.
37. BAUME, M. *The Sydney Opera House affair*. Thomas Nelson, Melbourne, 1967 (epilogue by P. Hall).
38. CURTIS, R. E. *A vision takes form: a graphic record of the building of the Sydney Opera House during Stages One and Two.* A. H. & A. W. Reed, Sydney, 1967 (with H. I. Ashworth and R. Covell).
39. DUEK-COHEN, E. (Ed.). *Utzon and the Sydney Opera House: statement in the public interest.* Morgan Publ., Sydney, 1967.
40. YEOMANS, J. S. *The other Taj Mahal: what happened to the Sydney Opera House*. Longmans, Green & Co., London, 1968.
41. YEOMANS, J. S. (Ed.). *Building the Sydney Opera House*. Hornibrook Group, Sydney, 1973, 28 pp.
42. YEOMANS, J. S. *A guide to the Sydney Opera House*. Sydney Opera House Trust, Sydney, 1973.
43. MILLER, P. (Ed.). *Sydney Opera House*. Land Printers Pty Ltd, Lidcombe, 1973 (with R. Hill).
44. ZIEGLER, O. L. (Ed.). *Sydney builds an opera house*. Oswald Ziegler Publ., Sydney, 1973 (with M. Dupain and A. Boothroyd).
45. ZIEGLER, O. L. (Ed.). *Sydney has an opera house*. Oswald Ziegler Publ., Sydney, 1974 (with A. D. Ziegler).
46. SMITH, V. *The Sydney Opera House*. Paul Hamlyn Pty Ltd, Sydney, 1974. (Series: *Summit books*).
47. FUTAGAWA, Y. (Ed.). *Sydney Opera House, Sydney, Australia, 1957–73: Jørn Utzon*. A.D.A. Edita, Tokyo, 1980 (with C. Norberg-Schulz; text in Japanese and English). (Series: *Global architecture*, Vol. 54).
48. JORDAN, V. L. *Acoustical design of concert halls and theatres: a personal account*. Appl. Sci. Publ., London, 1980.
49. SIM, J. *The Sydney Opera House: the first decade*. View Productions, Sydney, 1983.
50. HUBBLE, A. *The Sydney Opera House: more than meets the eye*. Lansdowne Press, Sydney, 1983.
51. HUBBLE, A. *More than an Opera House*. Lansdowne Press, Sydney, 1983.
52. SMITH, M. P. *Sydney Opera House: how it was built and why it is so*. William Collins Pty Ltd, Sydney, 1984.
53. ROWE, J. and HUBBLE, A. *Sydney Opera House grand organ*. Sydney Opera House Trust, Sydney, 1987.
54. HUBBLE, A. *The strange case of Eugene Goossens and other tales from the Opera House*. Collins, Sydney, 1988.

55. HALL, P. B. *Sydney Opera House: the design approach to the building with recommendations on its conservation.* Sydney Opera House Trust, Sydney, 1990, 212 pp.
56. SYKES, J. *Sydney Opera House: from the outside in.* Playbill, Pymble, 1993.
57. KERR, J. S. *Sydney Opera House: an interim plan for the conservation of the Sydney Opera House and its site.* Sydney Opera House Trust, Sydney, Dec. 1993, 1st edn, 68 pp.
58. NOBIS, P. *Utzon's interiors for the Sydney Opera House: the design development of the major and minor halls 1958–1966.* B.Arch. Dissertation, University of Technology, Sydney, 1994, 183 pp.
59. DREW, P. *Sydney Opera House: Jørn Utzon.* Phaidon Press, London, 1995 (with A. Browell and L. Atkin), 1st edn; 2002, new edn. (Series: *Architecture in detail*).
60. MESSENT, D. *Opera house act one.* David Messent Photography, Sydney, 1997.
61. FROMONOT, F. *Jørn Utzon: the Sydney Opera House.* Gingko Press Inc., Corte Madera (Electa, Milan), 1998 (transl. C. Thompson).
62. KERR, J. S. *Sydney Opera House: a revised plan for the conservation of the Sydney Opera House and its site.* Sydney Opera House Trust, Sydney, May 1999, 2nd edn; Jun. 2003, 3rd edn, 113 pp.
63. DREW, P. *The masterpiece: Jørn Utzon; a secret life.* Hardie Grant Books, South Yarra, Victoria, 1999, 1st edn; 2001, new edn.
64. FLOYD, C. and COLLINGWOOD, J. *The Sydney Opera House.* Frenchs Forest, N.S.W; New Holland, London, 2000.
65. DREW, P. *Utzon and the Sydney Opera House.* Inspire Press, Annandale, 2000.
66. PRIP-BUUS, M. *Letters from Sydney: an unedited version. The Sydney Opera House saga seen through the eyes of Utzon's chief assistant, Mogens Prip-Buus.* Edition Bløndal, Hellerup, 2000 (transl. W. Glyn Jones).
67. MIKAMI, Y. *Utzon's sphere, Sydney Opera House: how it was designed and built.* Shokokusha, Tokyo, 2001 (with O. Murai).
68. WESTON, R. *Utzon: inspiration, vision, architecture.* Edition Bløndal, Hellerup, 2002.
69. UTZON, J. O. *Sydney Opera House: Utzon design principles.* Sydney Opera House Trust, Sydney, May 2002, 90 pp.
70. JENKINS, S. *The Sydney Opera House.* Pearson Education Australia Pty Ltd, Melbourne, 2002, 16 pp.
71. PEMBER, B. *The Sydney Opera House.* Cambridge Univ. Press, Cambridge, 2003. (Series: *Livewire investigates*).
72. MURRAY, P. *The saga of Sydney Opera House: the dramatic story of the design and construction of the icon of modern Australia.* Taylor & Francis, London, 2003.
73. HOLM, M. J., KJELDSEN, K. and MARCUS, M. (Eds). *Jørn Utzon: the architect's universe.* Exhibition catalogue, Louisiana Museum of Modern Art, Humlebaek, 2004. (Series: *Louisiana Revy*, Vol. 44, No. 2, Apr. 2004).
74. PARKS, P. J. *The Sydney Opera House: building world landmarks.* Blackbirch Press, Farmington Hills, 2004.
75. SHOFNER, S. *Sydney Opera House: modern wonders of the world.* Creative Education, Mankato, 2006.
76. COMPAGNONI, P. and BUHRICH, C. *Rethinking the glass walls of Sydney Opera House.* Peter Compagnoni, Epping, 2006, 62 pp
77. WATSON, A. J. (Ed.). *Building a masterpiece: the Sydney Opera House.* Powerhouse Publ., Sydney, 2006, 1st edn; 2013, 40th anniversary edn.

78. TAFFS, D. Computers and the Opera House: pioneering a new technology. In: *Building a masterpiece: the Sydney Opera House.* Powerhouse Publ., Sydney, 2006, 1st edn (ed. A. J. Watson), 84–101.

79. NUTT, J. G. Constructing a legacy: technological innovation and achievements. In: *Building a masterpiece: the Sydney Opera House.* Powerhouse Publ., Sydney, 2006, 1st edn (ed. A. J. Watson), 104–121.

80. NOBIS, P. 'Great strength with extreme lightness': Utzon's use of plywood. In: *Building a masterpiece: the Sydney Opera House.* Powerhouse Publ., Sydney, 2006, 1st edn (ed. A. J. Watson), 136–151.

81. COMMONWEALTH OF AUSTRALIA. *Sydney Opera House: nomination by the Government of Australia for inscription on the World Heritage List. Prepared by the Australian Government and the New South Wales Government.* Australian Government Department of the Environment and Heritage, Canberra, 2006, 122 pp.

82. MOY, M. *Sydney Opera House: idea to icon.* Alpha Orion Press, Ashgrove, 2007.

83. KEIDING, M. (Ed.). *A tribute to Jørn Utzon.* Danish Architectural Press, Copenhagen, 2008.

84. MATTHEWS, S. *Sydney Opera House: structural wonders.* Weigl Publ., New York, 2008.

85. STÜBE, K. and UTZON, J. O. *A tribute to Jørn Utzon: Sydney Opera House.* Revealbooks, Potts Point, 2009.

86. WOOLLEY, K. *Reviewing the performance: the design of the Sydney Opera House.* Watermark Press, Sydney, 2010.

87. PELLS, P. *The Sydney Opera House car park and the double helix.* Writelight Pty Ltd, Blackheath, N.S.W., 2011.

88. GUBLER, F. *Great, grand and famous opera houses: where art and drama meet.* Arbon Publishing, Crows Nest, N.S.W., 2012.

89. DELLORA, D. *Utzon and the Sydney Opera House: how Australia's most famous building was designed, how it was actually built, and why the difference is important.* Penguin Books, Melbourne, 2013.

90. WEBBER, P. *Peter Hall architect: the phantom of the opera house.* Watermark Press, Sydney, 2013.

91. ANDERSEN, M. A. (Ed.). *Jørn Utzon, drawings and buildings.* Princeton Architectural Press, New York, 2013.

92. DELOITTE AUSTRALIA. *How do you value an icon? The Sydney Opera House: economic, cultural and digital value.* Deloitte Touche Tohmatsu, Sydney, 2013.

93. CROKER, A. (Ed.). *Respecting the vision: Sydney Opera House – a conservation management plan.* Sydney Opera House Trust, Sydney, Jul. 2017, 4th edn, 307 pp.

94. WATSON, A. J. *The poisoned chalice: Peter Hall and the Sydney Opera House.* OpusSOH Inc., Ballina, N.S.W., 2017.

95. PITT, H. *The House: the dramatic story of the Sydney Opera House and the people who made it.* Allen & Unwin, Sydney, 2018.

96. FREEMAN, C. G. *Participatory culture and the social value of an architectural icon: Sydney Opera House.* Routledge, New York, 2018.

97. COMPAGNONI, P. T. *Sydney Opera House halls and glass walls.* Peter Compagnoni, Epping, N.S.W., 2019.

98. SYDNEY OPERA HOUSE. *Annual Report: Financial Year 2018–19.* New South Wales Government, Sydney, 2019, 214 pp.

99. KUNER, J. *Raising the sails: the story behind the building of the Sydney Opera House roof.* Logueville Media, Haberfield, N.S.W., 2020.

100. FITZSIMONS, P. *The Opera House: the extraordinary story of the building that symbolises Australia — the people, the secrets, the scandals and the sheer genius.* Hachette Australia, Sydney, 2022.

101. DOUST, S. and GIBSON, P. *Sydney Opera House.* Scala Arts & Heritage Publ., Woodbridge, 2022.

102. JENKINS, R. S. *Theory and design of cylindrical shell structures.* Ove Arup & Partners, London, 1947, 80 pp.

103. JENKINS, R. S. *Matrix methods in structural mechanics.* University of Nottingham, 1961, 99 pp. (Series: *Taylor Woodrow Foundation Lectures*).

104. BLANCHARD, J. C. (Tech. ed.). Ronald Jenkins Memorial Issue. *The Arup Journal*, 1976, 11, 1 (Apr.), 44 pp.

105. WHITTINGHAM, H. E. *Extreme wind gusts in Australia.* Bureau of Meteorology, Melbourne, Feb. 1964, Bull. 46, 133 pp.

106. SOWDEN, H. (Ed.). *Sydney Opera House glass walls.* John Sands, Sydney, 1972. Contributing writers: Michael Lewis, David Croft, John Hooper, Robert Kelman, Peter Hall.

107. CROFT, D. D. The Sydney Opera House glass walls. *Proc. 4th Austr. Conf. Mechanics of Structures & Materials, Queensland Univ., Brisbane, Australia, Aug. 1973*, 60–67.

108. CROFT, D. D. and HOOPER, J. A. The Sydney Opera House glass walls. *Struct. Engr*, 1973, 51, 9 (Sept.), 311–322. Discussion, 1974, 52, 6 (Jun.), 216–220.

109. CROFT, D. D. and HOOPER, J. A. The Sydney Opera House glass walls. *The Arup Journal*, 1973, 8, 3 (Oct.), 30–41.

110. PILGRIM, D. H., CORDERY, I. and FRENCH, R. Temporal patterns of design rainfall for Sydney. *Civil Engng Trans, Instn of Engrs, Austr.*, 1969, CE 11, 1 (Apr.), 9–14.

111. HOOPER, J. A. The failure of glass cylinders in diametral compression. *J. Mech. Phys Solids*, 1971, 19, 4 (Aug.), 179–200.

112. HUNT, R. On the coloured glass employed in glazing the new Palm House in the Royal Botanic Garden at Kew. *Rep. 17th Mtg Brit. Assoc. Adv. Sci., Oxford, Jun. 1847.* John Murray, London, 1848, Pt 2 (notices and abstracts), 51–52.

113. GAFFIELD, T. *Action of sunlight on glass.* Tuttle, Morehouse & Taylor, New Haven, 1867, 19 pp (from *American Journal of Science and Arts*, Sept. & Nov. 1867).

114. HOOPER, J. A. *Sydney Opera House: results of preliminary flexural tests on annealed plate and laminated glass.* Ove Arup & Partners, Sydney, May 1969, 29 pp.

115. HEMSLEY, J. A. *Glass in engineering science.* Society of Glass Technology, Sheffield, 2015, Vol. 1 (*Optical birefringence in glass*), Ch. 4 (*Birefringent glass transducers*), 389–483.

116. HOOPER, J. A. On the bending of architectural laminated glass. *Int. J. Mech. Sci.*, 1973, 15, 4 (Apr.), 309–323.

117. HOLL, D. L. *Analysis of thin rectangular plates supported on opposite edges.* Iowa State College of Agriculture & Mechanic Arts, Ames, Iowa, U.S.A., Vol 35, No. 30, Dec. 1936; Iowa Engineering Experiment Station, Bull. 129, 100 pp.

118. PARROTT, T. L. Experimental studies of glass breakage due to sonic booms. *Sound*, 1962, 1, 3 (May), 18–21.

119. HUBBARD, H. H., MAGLIERI, D. J., HUCKEL, V. and HILTON, D. A. *Ground measurements of sonic-boom pressures for the altitude range of 10000 to 75000 feet*. National Aeronautics and Space Administration, Washington, D.C., U.S.A., Jul. 1964, NASA TR R-198, 45 pp.

120. CROCKER, M. J. and HUDSON, R. R. Structural response to sonic booms. *J. Sound & Vibr.*, 1969, 9, 3 (May), 454–468.

121. KAO, G. C. *An experimental study to determine the effects of repetitive sonic booms on glass breakage*. Federal Aviation Administration, Washington, D.C., U.S.A., Jun. 1970, FAA-NO-70-13, 85 pp.

122. HOOPER, J. A. and WASSALL, M. P. Sydney Opera House glass walls: colour measurement and control. *Building Sci.*, 1975, 10, 1 (Mar.), 65–72.

123. WRIGHT, W. D. A re-determination of the trichromatic coefficients of the spectral colours. *Trans Optical Soc.*, 1929, 30, 4 (Mar.), 141–164.

124. GUILD, J. The colorimetric properties of the spectrum. *Phil. Trans Royal Soc. (Lond.)*, 1931, A230, Jun., 149–187.

125. WRIGHT, W. D. *The measurement of colour*. Adam Hilger, London, 1969, 4th edn.

126. WRIGHT, W. D. A photoelectric spectrophotometer and tristimulus colorimeter designed for teaching and research. *Optica Acta*, 1954, 1, 2 (Sept.), 102–107.

127. MACADAM, D. L. Visual sensitivities to colour differences in daylight. *J. Opt. Soc. Amer.*, 1942, 32, 5 (May), 247–274.

A1. ELASTIC FLEXURE OF SIMPLY SUPPORTED LAMINATED GLASS BEAM

A1.1. Introduction

The structural requirements for beams and plates of sandwich-type construction often are satisfied by a pair of thin outer faces or layers, generally of equal thickness, and a relatively thick core of lightweight honeycomb material. The flexural resistance of such sandwich elements stems mainly from the membrane action of the thin outer faces, and much theoretical and experimental work has been directed towards this type of laminate.

In the case of 3-ply architectural laminated glass, however, the outer glass layers are much thicker and stiffer than the plastic interlayer, and are sometimes of unequal thickness. Accordingly, the flexural behaviour of such laminates is strongly dependent upon the bending resistance of the outer glass layers about their own middle planes, and the following analysis takes this into account.

The assumptions made in the present theory are as follows.
(a) The horizontal beam is simply supported equidistant from its centre, and is subjected to symmetric distributions of vertical applied loading.
(b) The beam supports are rigid and frictionless.
(c) The laminated materials are homogeneous, isotropic and linearly elastic.
(d) The laminate comprises two outer glass layers having the same material properties, but not necessarily the same thickness, in complete adhesive contact with a thin plastic interlayer.
(e) The thickness of all three layers is constant prior to loading.
(f) No slip occurs at either interface between glass and plastic.
(g) The plastic interlayer offers no resistance to bending, other than to prevent relative normal displacements of the adjoining glass faces.
(h) Shear stresses are constant throughout the depth of the plastic interlayer, enabling the latter to be classed as an antiplane core.
(i) Shear strains, other than those within the plastic interlayer, are small and their contribution to the distortion of the glass layers is negligible.
(j) Normal strains are small, so that the laminate thickness remains constant and each constituent layer follows the same deflection profile.

(k) Beam deflections are relatively small, thereby avoiding the need to invoke large-deflection theory.

(l) The transverse loads acting on the laminated glass are applied statically.

In practice, assumption (d) is consistent with most architectural glazing, although more elaborate multi-layered laminates are used for security glazing and other special applications. Assumption (l) strictly applies to self-weight or similar sustained loading, and generally is encompassed within the design process to include transient wind loading. Conversely, assumption (c) essentially holds for both glass and plastic under transient loading, although the plastic interlayer usually undergoes substantial creep deformation under sustained loading. In most applications, this inelastic behaviour of the interlayer can readily be accommodated by retaining elastic theory and assigning a lower value of shear modulus to the plastic material.

The method used here for the flexural analysis of laminated glass beams partly stems from an elastic solution by Chitty [A1.1] to a problem concerning the bending of parallel beams connected by cross-bars. In this analysis, the discrete assemblage of cross-members was replaced by a continuous elastic medium of equivalent stiffness, the medium itself being securely attached to the beams at each interface. This latter arrangement is precisely that pertaining to architectural laminated glass, in which a relatively soft continuous layer of thermoplastic material is confined between two glass layers and remains in adhesive contact with them during flexure. Thus the distribution of interfacial shear force derived in such an analysis can be used to determine both the longitudinal stresses within the glass layers and the shear stresses transmitted by the plastic interlayer. It can also be incorporated within the standard moment–curvature relationship for homogeneous beams, enabling slopes and deflections to be calculated.

Similar methods of analysis have been developed independently over many decades in several disparate fields of engineering science. These include the early solutions of van der Neut [A1.2] and March [A1.3] for the flexure of wooden box beams and rectangular plywood strips, respectively. Other studies were made by Hoff and Mautner [A1.4], March and Smith [A1.5] and Norris et al. [A1.6] on the flexure of structural sandwich components used in aircraft construction and elsewhere. The method described by Chitty [A1.1] was itself based on a solution by Pippard and Francis [A1.7] to a related structural problem associated with radially spoked wire wheels, and was applied by Chitty and Wan [A1.8] to the analysis of multistorey building frames subjected to lateral wind loading. Subsequently, the method was used by Schulz [A1.9], Eriksson [A1.10], Beck [A1.11], Rosman [A1.12] and many others in the structural analysis of plane coupled shear walls of a type often encountered in tall buildings of reinforced concrete construction. In this latter application, it is the discrete system of lintel beams or floor slabs that is replaced by an equivalent continuous medium.

Although most of the present theoretical analysis was carried out by the writer some fifty years ago, and revisited herein to produce a more comprehensive set of results, the original four-fold main objectives remain apposite. Firstly, to simulate the loading arrangements used in the flexural testing of laminated glass beams, thereby facilitating the interpretation of experimental results. Secondly, to provide exact closed-form solutions against which the results of approximate numerical analysis can readily be compared, prior to undertaking more complex problems. Thirdly, to be of practical assistance during preliminary design, especially in feasibility studies to establish the basic structural characteristics of various glass laminates, and rapidly to assess their sensitivity to the main parameters. Fourthly, for rectangular panels supported only along two opposite edges, the beam solutions often can be used directly in final design calculations. Under certain conditions, the solutions also give a useful first approximation for such panels simply supported along all four edges, rather than appealing to complex laminated plate theory.

A1.2. Governing equations

Referring to Figure A1.1(a), consider a laminated glass beam of length $2L$ with simple end-supports subjected to symmetric vertical loading. The beam cross-section, width B, has outer glass layers of thicknesses t_1 and t_2, where $t_1 \leq t_2$, and a plastic interlayer of thickness c, as indicated in Figure A1.1(b). The longitudinal axis of the beam is taken as horizontal and coincident with the x-axis, the co-ordinate origin being located at the left end of the beam. Downward loads acting on the outer face of the upper glass layer are reckoned positive.

If the laminated beam is subjected to some specified distribution of applied vertical load, corresponding to a bending moment distribution $M(x)$, then the governing ordinary differential equation for the interfacial shear force $S(x)$ between inner and outer layers is

$$\frac{d^2 S}{dx^2} + \frac{\alpha^2}{\mu h}(M - \mu h S) = 0 \tag{A1.1}$$

where

$$\alpha = h\left(\frac{\mu B G_p}{c E_g I_h}\right)^{1/2}, \quad \mu = 1 + \frac{(t_1 + t_2)I_h}{B t_1 t_2 h^2} \tag{A1.2}$$

Also, E_g denotes the Young's modulus of the outer glass layers, G_p denotes the shear modulus of the plastic interlayer, $h = c + (t_1 + t_2)/2$ and $I_h = B(t_1^3 + t_2^3)/12$. Hence μ is a dimensionless parameter, while α may be interpreted as the reciprocal of the characteristic length of the system.

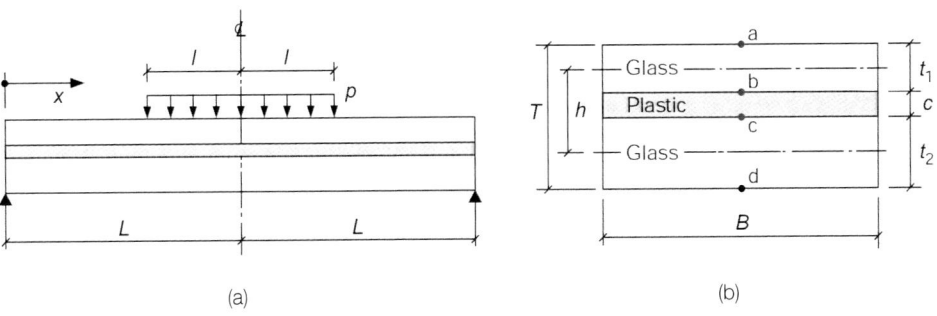

FIGURE A1.1. Notation used in flexural analysis of horizontal laminated glass beam subjected to symmetric central partial uniform vertical pressure ($\xi = x/L$, $\lambda = l/L$): (a) elevation; (b) cross-section

The general solution of equation (A1.1) takes the form

$$S(x) = A_j \sinh(\alpha x) + B_j \cosh(\alpha x) + \frac{1}{\mu h}\left[M(x) + \sum_{m=2}^{\infty} \frac{1}{\alpha^m} \frac{d^m M}{dx^m}\right] \quad (A1.3)$$

where A_j and B_j are arbitrary constants (with j a positive integer), and where the summation is taken over the m even integers required to give a zero m-th derivative of $M(x)$. For simple types of applied loading, the arbitrary constants are determined from the boundary conditions in the usual way, although the integral transform approach can be used to advantage in cases of more complex loading patterns.

The distribution of interfacial stress $\tau(x)$ can be derived from the standard relation

$$\tau(x) = \frac{1}{B}\frac{dS}{dx} \quad (A1.4)$$

although it also forms part of the solution for $S(x)$. The longitudinal flexural stresses at the surface of each glass layer are given by

$$\left.\begin{array}{l}\sigma_a(x)\\ \sigma_b(x)\end{array}\right\} = \pm\frac{t_1}{2I_h}[M(x) - hS(x)] + \frac{S(x)}{Bt_1} \quad (A1.5)$$

$$\left.\begin{array}{l}\sigma_c(x)\\ \sigma_d(x)\end{array}\right\} = \pm\frac{t_2}{2I_h}[M(x) - hS(x)] - \frac{S(x)}{Bt_2} \quad (A1.6)$$

where the subscripts 'a'–'d' relate to the positions on the laminate cross-section shown in Figure A1.1(b). Here sagging bending moments and compressive stresses are reckoned positive. It is clear from these last two expressions that $S(x)$ may also be interpreted as the membrane force in the outer glass layers.

The beam deflection profile $w(x)$ and the corresponding slope $\phi(x)$ can be obtained from

$$E_g I_h \frac{d^2 w}{dx^2} = E_g I_h \frac{d\phi}{dx} = -(M - hS) \quad (A1.7)$$

where downward deflections are reckoned positive. These equations either can be integrated directly or, in cases of complex applied loading, solved more concisely using the integral transform method.

A1.3. Beam under symmetric partial uniform loading

A1.3.1. General case

Consider the case of a simply supported laminated glass beam of span $2L$ and width B carrying a uniformly distributed transverse load p per unit area along a length l on either side of the beam centre; see Figure A1.1(a). Because of symmetry, only one-half of the beam need be considered in detail, with results herein given for $0 \leq x \leq L$. Using the principle of superposition, this solution can usefully be deployed to determine the approximate beam flexure under the action of other symmetric distributions of transverse applied loading; see section A1.3.4.

ELASTIC FLEXURE OF LAMINATED GLASS BEAM

For the present case, the distribution of bending moment within the two loading zones is given by

$$M(x) = \frac{pB}{2} \begin{cases} 2lx & 0 \le x \le L-l \\ x(2L-x) - (L-l)^2 & L-l < x \le L \end{cases} \quad \text{(A1.8)}$$

The corresponding distribution $S(x)$ is determined from equation (A1.3), the four arbitrary constants (*i.e.* two in each zone, $j = 1, 2$) being determined from the boundary conditions. These are $S(0) = S'(L) = 0$, where a prime signifies the first derivative with respect to the independent variable, together with the continuity requirements for $S(L-l)$ and $S'(L-l)$ as both shear force and shear stress must be single-valued at this junction. Upon writing $\xi = x/L$ and $\lambda = l/L$, and introducing the dimensionless stiffness parameter $\Omega = \alpha L$, the distributions of interfacial shear force and the corresponding shear stress can be expressed in the form

$$S(\xi) = \frac{pBL^2}{\mu h} F_1(\xi, \lambda), \quad \tau(\xi) = \frac{pL}{\mu h} F_2(\xi, \lambda) \quad \text{(A1.9)}$$

where the influence factors are

$$F_1(\xi, \lambda) = \frac{1}{\Omega^2} \begin{cases} \Omega^2 \lambda \xi - \text{sech}\,\Omega \sinh(\Omega\lambda) \sinh(\Omega\xi) & 0 \le \xi \le 1-\lambda \\ (\Omega^2/2)[\xi(2-\xi) - (1-\lambda)^2] - 1 \\ + \text{sech}\,\Omega \cosh[\Omega(1-\lambda)] \cosh[\Omega(1-\xi)] & 1-\lambda < \xi \le 1 \end{cases} \quad \text{(A1.10)}$$

$$F_2(\xi, \lambda) = \frac{1}{\Omega} \begin{cases} \Omega\lambda - \text{sech}\,\Omega \sinh(\Omega\lambda) \cosh(\Omega\xi) & 0 \le \xi \le 1-\lambda \\ \Omega(1-\xi) - \text{sech}\,\Omega \cosh[\Omega(1-\lambda)] \sinh[\Omega(1-\xi)] & 1-\lambda < \xi \le 1 \end{cases} \quad \text{(A1.11)}$$

For any value of λ, the maximum values of $F_1(\xi, \lambda)$ and $F_2(\xi, \lambda)$ occur at $\xi = 1$ and $\xi = 0$, respectively. Thus, with such maxima signified by an asterisk,

$$F_1^*(1, \lambda) = \frac{1}{\Omega^2}\left\{\frac{\lambda}{2}(2-\lambda)\Omega^2 + \text{sech}\,\Omega \cosh[\Omega(1-\lambda)] - 1\right\} \quad \text{(A1.12)}$$

$$F_2^*(0, \lambda) = \frac{1}{\Omega}[\Omega\lambda - \text{sech}\,\Omega \sinh(\Omega\lambda)] \quad \text{(A1.13)}$$

The beam deflection profile $w(\xi)$ is obtained using equation (A1.7) in conjunction with the boundary conditions $w(0) = w'(1) = 0$, together with the required continuity in slope and deflection at $\xi = 1 - \lambda$. The result is

$$w(\xi) = \frac{5pBL^4}{24E_g I_h} F_3(\xi, \lambda) \quad \text{(A1.14)}$$

where

$$F_3(\xi, \lambda) = \frac{1}{5\mu} \begin{cases} 4\lambda\xi(\mu-1)(3-\lambda^2-\xi^2) + (24/\Omega^2)F_1(\xi, \lambda) & 0 \le \xi \le 1-\lambda \\ (\mu-1)\{(1-\lambda)^4 + \xi(2-\xi)[6\lambda(2-\lambda) \\ +\xi(2-\xi) - 2]\} + (24/\Omega^2)F_1(\xi, \lambda) & 1-\lambda < \xi \le 1 \end{cases} \quad \text{(A1.15)}$$

which clearly is a maximum at $\xi = 1$. Hence

$$F_3^*(1,\lambda) = \frac{\mu-1}{5\mu}[(1-\lambda)^4 + 6\lambda(2-\lambda)-1] + \frac{24}{5\mu\Omega^2}F_1^*(1,\lambda) \tag{A1.16}$$

Profiles of the above general influence factors within the range $0 \le \Omega \le 10$ are shown in Figure A1.2 for the case $\lambda = 0.1$, taking $\mu = 1.5$ for $F_3(\xi, \lambda)$. Likewise, the maximum values of these factors for other loading widths are given in Figure A1.3.

The corresponding result for beam slope $\phi(\xi)$ is

$$\phi(\xi) = \frac{pBL^3}{6E_g I_h} F_4(\xi,\lambda) \tag{A1.17}$$

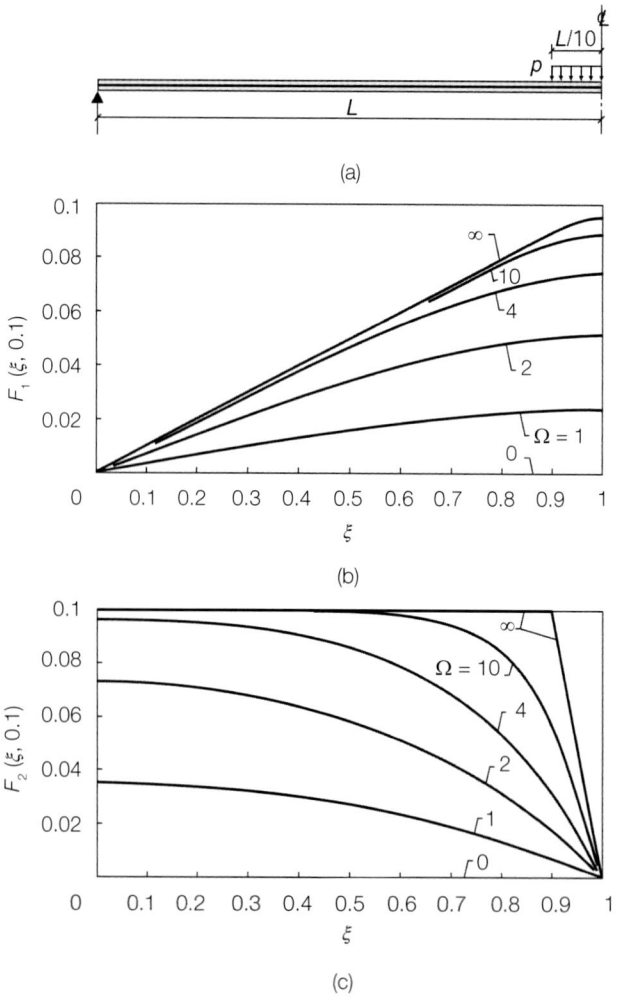

FIGURE A1.2. Influence factors $F_i(\xi,\lambda)$ for elastic flexure of horizontal laminated glass beam subjected to symmetric central partial uniform vertical pressure ($\lambda = 0.1$, $\mu = 1.5$): (a) half-beam elevation; (b) $F_1(\xi,0.1)$; (c) $F_2(\xi,0.1)$

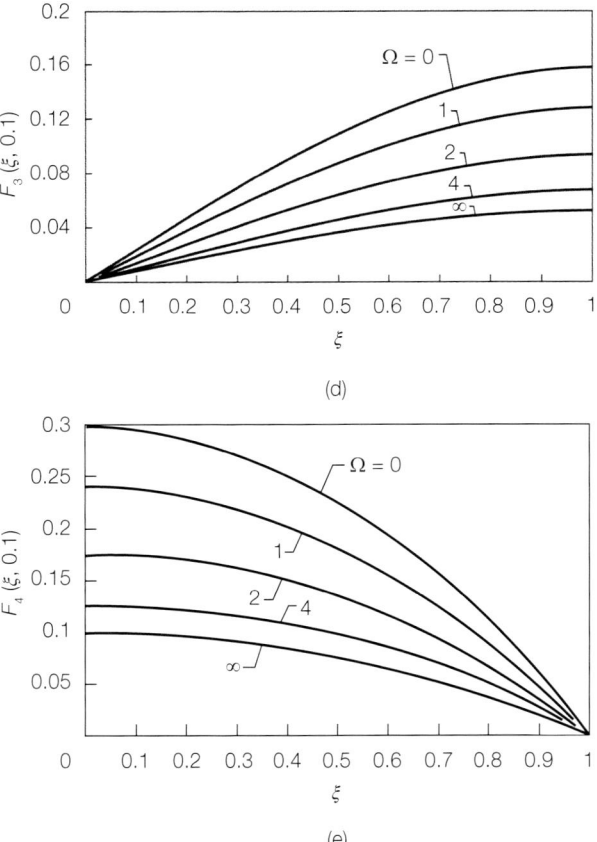

FIGURE A1.2. (continued) Influence factors $F_i(\xi,\lambda)$ for elastic flexure of horizontal laminated glass beam subjected to symmetric central partial uniform vertical pressure ($\lambda = 0.1$, $\mu = 1.5$): (d) $F_3(\xi,0.1)$; (e) $F_4(\xi,0.1)$

where

$$F_4(\xi,\lambda) = \frac{1}{\mu}\begin{cases} \lambda(\mu-1)(3-\lambda^2-3\xi^2)+(6/\Omega^2)F_2(\xi,\lambda) & 0 \leq \xi \leq 1-\lambda \\ (\mu-1)(1-\xi)[3\lambda(2-\lambda)-(1-\xi)^2] \\ +(6/\Omega^2)F_2(\xi,\lambda) & 1-\lambda < \xi \leq 1 \end{cases} \quad (A1.18)$$

while the maximum values at the beam support are given by

$$F_4^*(0,\lambda) = \frac{\mu-1}{\mu}\lambda(3-\lambda^2)+\frac{6}{\mu\Omega^2}F_2^*(0,\lambda) \quad (A1.19)$$

Calculated influence factors for beam slope are included in Figures A1.2 and A1.3, while observing that $F_4^*(0,\lambda) \approx 2F_3^*(1,\lambda)$ for all Ω.

The longitudinal stresses at the surfaces of the glass layers are obtained using equations (A1.5) and (A1.6). With $\gamma_1 = t_1/(t_1+t_2)$ and $\gamma_2 = t_2/(t_1+t_2)$, the resulting expressions are

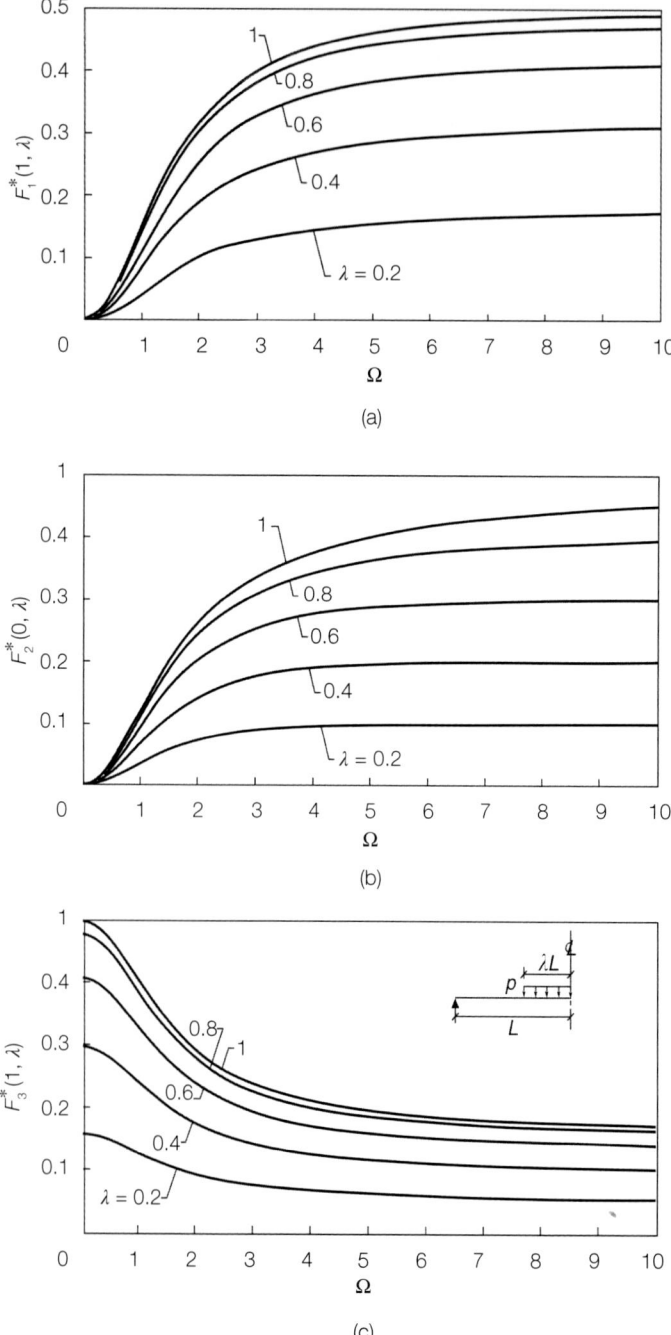

FIGURE A1.3. Maximum influence factors $F_i^*(\xi,\lambda)$ for elastic flexure of horizontal laminated glass beam subjected to symmetric central partial uniform vertical pressure over various loading widths ($\mu = 1.5$): (a) $F_1^*(1,\lambda)$; (b) $F_2^*(0,\lambda)$; (c) $F_3^*(1,\lambda)$

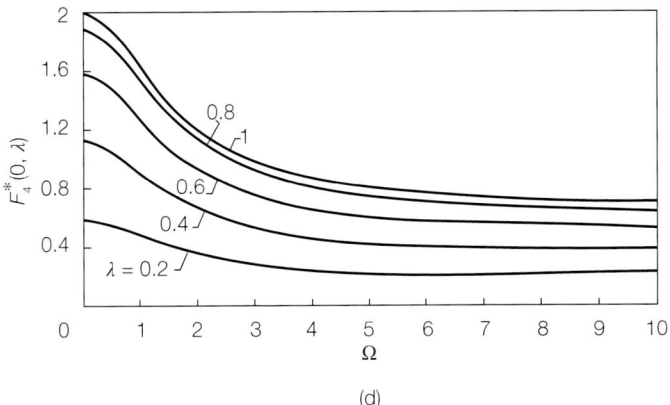

FIGURE A1.3. (continued) Maximum influence factors $F_i^*(\xi,\lambda)$ for elastic flexure of horizontal laminated glass beam subjected to symmetric central partial uniform vertical pressure over various loading widths ($\mu = 1.5$): (d) $F_4^*(0,\lambda)$

$$\left.\begin{array}{c}\sigma_a(\xi)\\ \sigma_b(\xi)\end{array}\right\} = \frac{pBL^2}{4\mu I_h}\left\{\begin{array}{c}\mu t_1 F_5(\xi,\lambda) - 2[t_1 - 2\gamma_2 h(\mu-1)]F_1(\xi,\lambda)\\ -\mu t_1 F_5(\xi,\lambda) + 2[t_1 + 2\gamma_2 h(\mu-1)]F_1(\xi,\lambda)\end{array}\right. \quad (A1.20)$$

$$\left.\begin{array}{c}\sigma_c(\xi)\\ \sigma_d(\xi)\end{array}\right\} = \frac{pBL^2}{4\mu I_h}\left\{\begin{array}{c}\mu t_2 F_5(\xi,\lambda) - 2[t_2 + 2\gamma_1 h(\mu-1)]F_1(\xi,\lambda)\\ -\mu t_2 F_5(\xi,\lambda) + 2[t_2 - 2\gamma_1 h(\mu-1)]F_1(\xi,\lambda)\end{array}\right. \quad (A1.21)$$

where

$$F_5(\xi,\lambda) = \begin{cases}2\lambda\xi & 0 \le \xi \le 1-\lambda\\ \xi(2-\xi)-(1-\lambda)^2 & 1-\lambda < \xi \le 1\end{cases} \quad (A1.22)$$

Because the solutions for beam flexure strongly depend on the non-dimensional parameter Ω, it is worthwhile to consider the effect of interlayer thickness on the stiffness parameter α under a given set of material properties and loading conditions. For simplicity, assume that $t_1 = t_2 = t$, noting that most architectural laminated glass is symmetric and confined to the range $0.03 < c/t < 0.3$. Then it is easily shown, for example, that

$$\alpha t\left(\frac{E_g}{G_p}\right)^{1/2} = \left\{\left(\frac{2t}{c}\right)\left[1+\frac{3}{t^2}(c+t)^2\right]\right\}^{1/2} \quad (A1.23)$$

This relation is plotted in Figure A1.4, and has a minimum value of $2(3+2\sqrt{3})^{1/2} \approx 5.085$ which occurs at $c/t = 2/\sqrt{3} \approx 1.155$.

When laminated glass is subjected to sustained transverse loading (e.g. self-weight loading in non-vertical glazing), creep deformation within the plastic interlayer can reduce substantially the long-term shear coupling between the glass layers. In these circumstances, it is useful to recall solutions for the limiting case $\Omega = 0$, whereby the outer glass layers become uncoupled. Whence the influence factors for beam deflection and slope reduce to

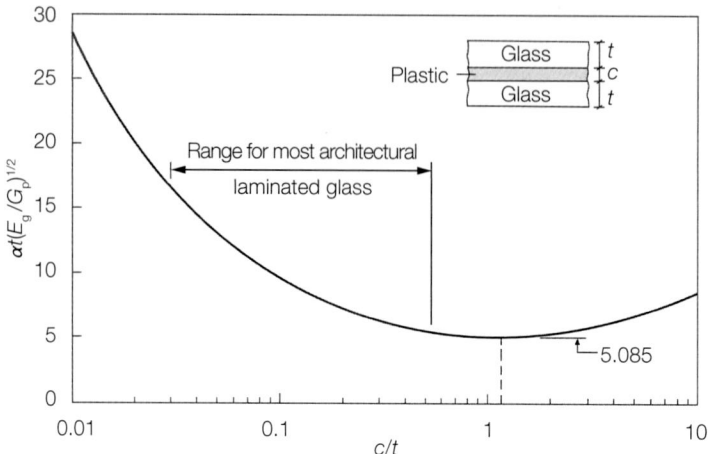

FIGURE A1.4. Effect of interlayer thickness on dimensionless stiffness parameter $\alpha t(E_g/G_p)^{1/2}$ for elastic flexure of laminated glass beam of symmetric cross-section

$$F_3(\xi,\lambda)_0 = \frac{1}{5}\begin{cases} 4\lambda\xi(3-\lambda^2-\xi^2) & 0\leq\xi\leq 1-\lambda \\ (1-\lambda)^4+\xi(2-\xi)[6\lambda(2-\lambda)+\xi(2-\xi)-2] & 1-\lambda<\xi\leq 1 \end{cases} \quad (A1.24)$$

$$F_4(\xi,\lambda)_0 = \begin{cases} \lambda(3-\lambda^2-3\xi^2) & 0\leq\xi\leq 1-\lambda \\ (1-\xi)[3\lambda(2-\lambda)-(1-\xi)^2] & 1-\lambda<\xi\leq 1 \end{cases} \quad (A1.25)$$

while the longitudinal bending stresses are given by

$$\left.\begin{array}{c}\sigma_a(\xi)_0 \\ \sigma_b(\xi)_0\end{array}\right\} = \pm\frac{pBL^2 t_1}{4I_h}F_5(\xi,\lambda), \quad \left.\begin{array}{c}\sigma_c(\xi)_0 \\ \sigma_d(\xi)_0\end{array}\right\} = \pm\frac{pBL^2 t_2}{4I_h}F_5(\xi,\lambda) \quad (A1.26)$$

To briefly examine the influence of glass layer configuration when $\Omega = 0$, consider a general laminated section subjected to a given applied moment. As $t_1 \leq t_2$, the maximum tensile stress at location 'd' in Figure A1.1(b) normally will govern the design for flexural strength. Let σ_d (lam) denote this bending stress, and let σ_d (mono) denote the corresponding stress for a monolithic beam of thickness $t_1 + t_2$, as the interlayer thickness c no longer enters the calculations. Then elementary beam theory shows that

$$\frac{\sigma_d(\text{lam})}{\sigma_d(\text{mono})} = \frac{1-\gamma_1}{1-3\gamma_1(1-\gamma_1)} \quad (A1.27)$$

This latter ratio is plotted in Figure A1.5, and has a maximum value of $1/(2\sqrt{3}-3) \approx 2.155$ which occurs at $\gamma_1 = 1-1/\sqrt{3} \approx 0.423$. Likewise, let w (lam) and w (mono) denote the beam deflections based on the laminated and monolithic sections, respectively. Then

$$\frac{w(\text{lam})}{w(\text{mono})} = \frac{1}{1-3\gamma_1(1-\gamma_1)} \quad (A1.28)$$

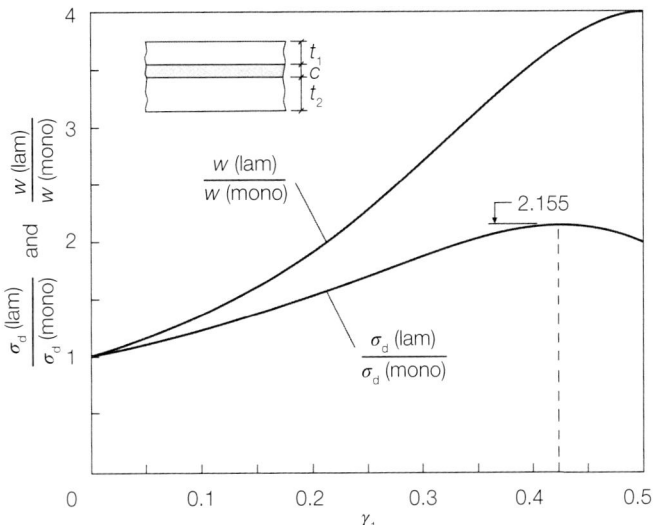

FIGURE A1.5. Amplification of bending stress and deflection for elastic flexure of laminated glass beam of general cross-section in limiting case of zero interlayer shear stiffness ($\Omega = 0$), with $\gamma_1 = t_1/(t_1 + t_2)$

In contrast to the previous ratio for stresses, Figure A1.5 shows that the deflection ratio increases monotonically from unity, attaining a value of four at $\gamma_1 = 1/2$.

General solutions for variable λ corresponding to the opposite limit $\Omega \to \infty$ are also useful in design, especially as they provide an upper bound on the interfacial shear stress induced within the laminate. Whence $S(\xi)_\infty$ is obtained by setting $F_1(\xi,\lambda)_\infty = F_5(\xi,\lambda)/2$, from which it also follows that

$$F_2(\xi,\lambda)_\infty = \begin{cases} \lambda & 0 \le \xi \le 1-\lambda \\ 1-\xi & 1-\lambda < \xi \le 1 \end{cases} \tag{A1.29}$$

to give the shear stress distribution $\tau(\xi)_\infty$; see Figure A1.6. The longitudinal bending stresses then become

$$\left.\begin{matrix}\sigma_a(\xi)_\infty \\ \sigma_b(\xi)_\infty\end{matrix}\right\} = \pm \frac{pBL^2(\mu-1)}{4\mu I_h} F_5(\xi,\lambda) \begin{cases} t_1 + 2\gamma_2 h \\ t_1 - 2\gamma_2 h \end{cases} \tag{A1.30}$$

$$\left.\begin{matrix}\sigma_c(\xi)_\infty \\ \sigma_d(\xi)_\infty\end{matrix}\right\} = \pm \frac{pBL^2(\mu-1)}{4\mu I_h} F_5(\xi,\lambda) \begin{cases} t_2 - 2\gamma_1 h \\ t_2 + 2\gamma_1 h \end{cases} \tag{A1.31}$$

The beam deflections and slopes are obtained simply by replacing I_h by $\mu I_h/(\mu-1)$ in equations (A1.14) and (A1.17), respectively, and using the factors $F_3(\xi,\lambda)_0$ and $F_4(\xi,\lambda)_0$ given by equations (A1.24) and (A1.25).

The stresses calculated from equations (A1.30) and (A1.31) are identical to those obtained using this modified second moment of area in conjunction with the elementary theory for monolithic beams; for if such a beam has the same overall thickness T as the laminate, the bending stresses are

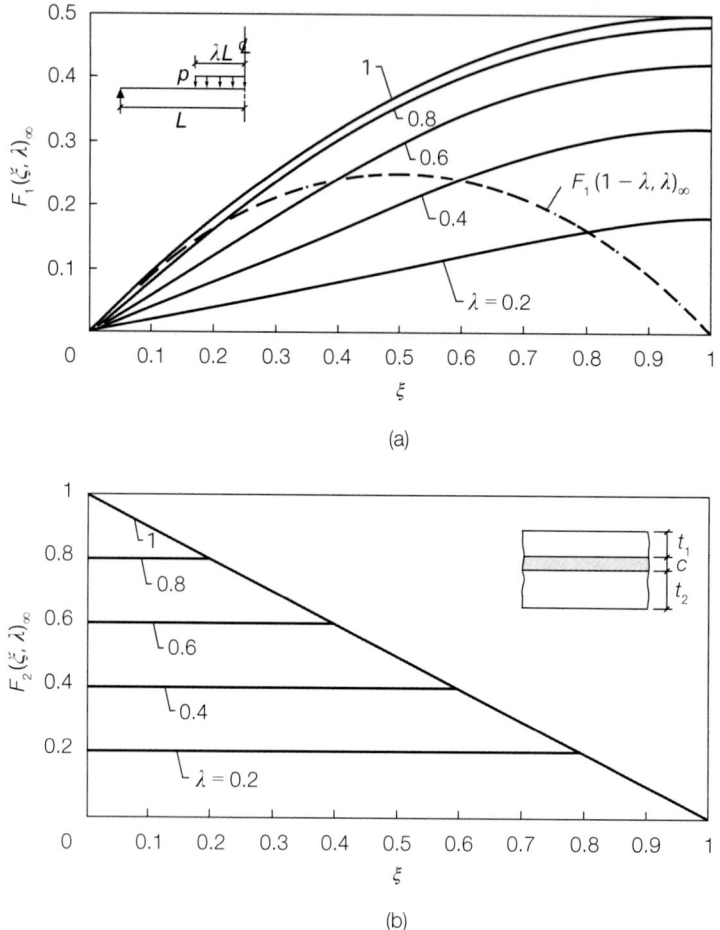

FIGURE A1.6. Influence factors for elastic flexure of laminated glass beam subjected to symmetric central partial uniform vertical pressure in limiting case of infinite interlayer shear stiffness ($\Omega \to \infty$): (a) $F_1(\xi,\lambda)_\infty$; (b) $F_2(\xi,\lambda)_\infty$

$$\left.\begin{array}{l}\sigma_a(\xi)_T \\ \sigma_d(\xi)_T\end{array}\right\} = \pm\frac{3pL^2}{T^2}F_5(\xi,\lambda) \tag{A1.32}$$

The corresponding deflections and slopes are calculated using the expressions for the case $\Omega = 0$ with I_h replaced by $I_T = BT^3/12$, giving

$$w(\xi)_T = \frac{5pBL^4}{24E_g I_T}F_3(\xi,\lambda)_0, \quad \phi(\xi)_T = \frac{pBL^3}{6E_g I_T}F_4(\xi,\lambda)_0 \tag{A1.33}$$

A1.3.2. Limiting case of complete uniform loading

The first limiting case of particular interest is that where the uniform loading is distributed over the entire span. Thus, with $\lambda = 1$,

$$F_1(\xi,1) = \frac{1}{\Omega^2}\left\{\frac{\xi}{2}(2-\xi)\Omega^2 + \operatorname{sech}\Omega\cosh[\Omega(1-\xi)] - 1\right\} \quad \text{(A1.34)}$$

$$F_2(\xi,1) = \frac{1}{\Omega}\{\Omega(1-\xi) - \operatorname{sech}\Omega\sinh[\Omega(1-\xi)]\} \quad \text{(A1.35)}$$

$$F_3(\xi,1) = \frac{1}{5\mu}\left[\xi(\mu-1)(\xi^3 - 4\xi^2 + 8) + \frac{24}{\Omega^2}F_1(\xi,1)\right] \quad \text{(A1.36)}$$

$$F_4(\xi,1) = \frac{1}{\mu}\left[(\mu-1)(\xi^3 - 3\xi^2 + 2) + \frac{6}{\Omega^2}F_2(\xi,1)\right] \quad \text{(A1.37)}$$

and examples of the corresponding influence factor profiles are shown in Figure A1.7, taking $\mu = 1.5$ for $F_3(\xi, 1)$ and $F_4(\xi, 1)$. The maximum values are

$$F_1^*(1,1) = \frac{1}{\Omega^2}\left(\frac{\Omega^2}{2} + \operatorname{sech}\Omega - 1\right) \quad \text{(A1.38)}$$

$$F_2^*(0,1) = 1 - \frac{\tanh\Omega}{\Omega} \quad \text{(A1.39)}$$

$$F_3^*(1,1) = \frac{1}{\mu}\left[\mu - 1 + \frac{24}{5\Omega^2}F_1^*(1,1)\right] \quad \text{(A1.40)}$$

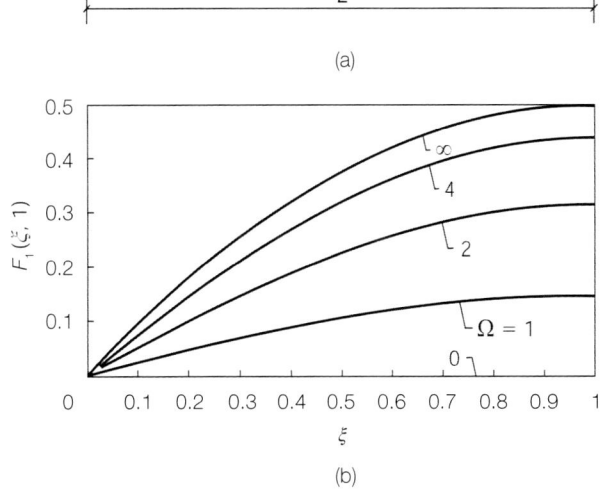

FIGURE A1.7. Influence factors $F_i(\xi,\lambda)$ for elastic flexure of horizontal laminated glass beam subjected to complete uniform vertical pressure ($\lambda = 1$, $\mu = 1.5$): (a) half-beam elevation; (b) $F_1(\xi,1)$

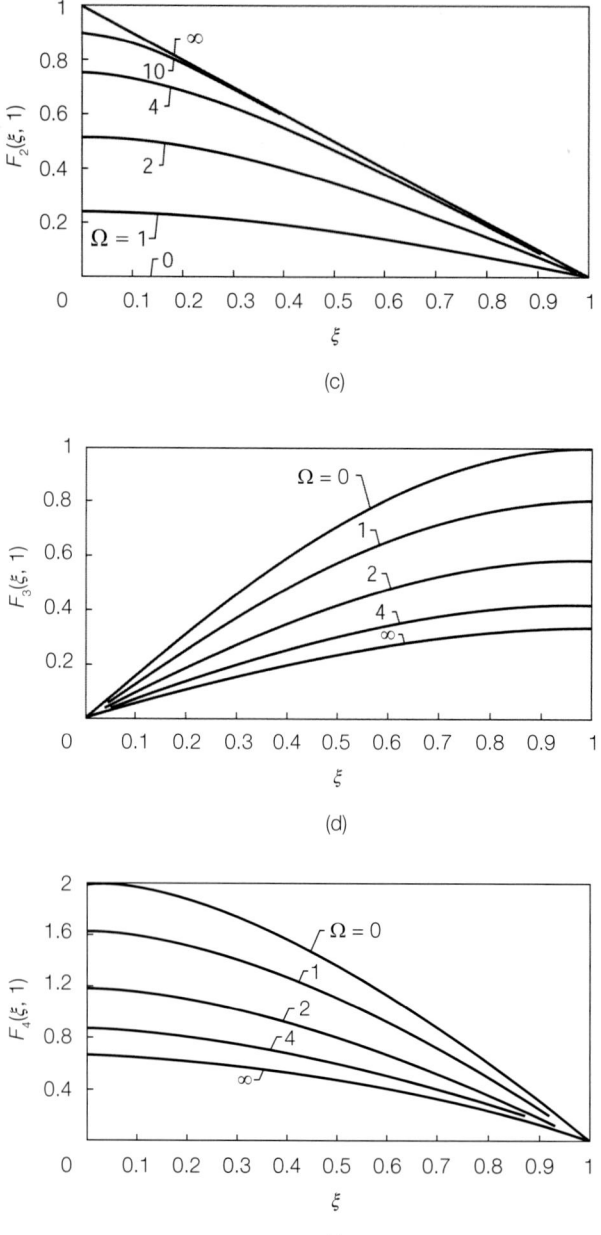

FIGURE A1.7. (continued) Influence factors $F_i(\xi,\lambda)$ for elastic flexure of horizontal laminated glass beam subjected to complete uniform vertical pressure ($\lambda = 1$, $\mu = 1.5$): (c) $F_2(\xi,1)$; (d) $F_3(\xi,1)$; (e) $F_4(\xi,1)$

$$F_4^*(0,1) = \frac{1}{\mu}\left[\mu - 1 + \frac{6}{\Omega^2}F_2^*(0,1)\right] \qquad (A1.41)$$

Numerical values of $F_1^*(1, 1)$ and $F_2^*(0, 1)$ are included in Figure A1.3, while values of $F_3^*(1, 1)$ and $F_4^*(0, 1)$ within the range $1 \leq \mu \leq 2$ are combined in Figure A1.8 as they differ in magnitude by close to a factor of 2. For a solid glass beam of thickness T, the deflection and slope are given by

$$w(\xi)_T = \frac{pBL^4}{24E_g I_T}[\xi(2-\xi)(4+2\xi-\xi^2)] \qquad (A1.42)$$

$$\phi(\xi)_T = \frac{pBL^3}{6E_g I_T}(2 - 3\xi^2 + \xi^3) \qquad (A1.43)$$

with maximum values

$$w^*(1)_T = \frac{5pBL^4}{24E_g I_T}, \quad \phi^*(0)_T = \frac{pBL^3}{3E_g I_T} \qquad (A1.44)$$

A1.3.3. Limiting case of central line load

A second limiting case of special interest is that where the width of the uniformly loaded region approaches zero and a vertical line load of magnitude P acts at the beam centre. Thus, on setting $P = 2pBl$ and determining the limiting values as $\lambda \to 0$ in the previous equations for a partially loaded beam, the required influence factors can be expressed in the form

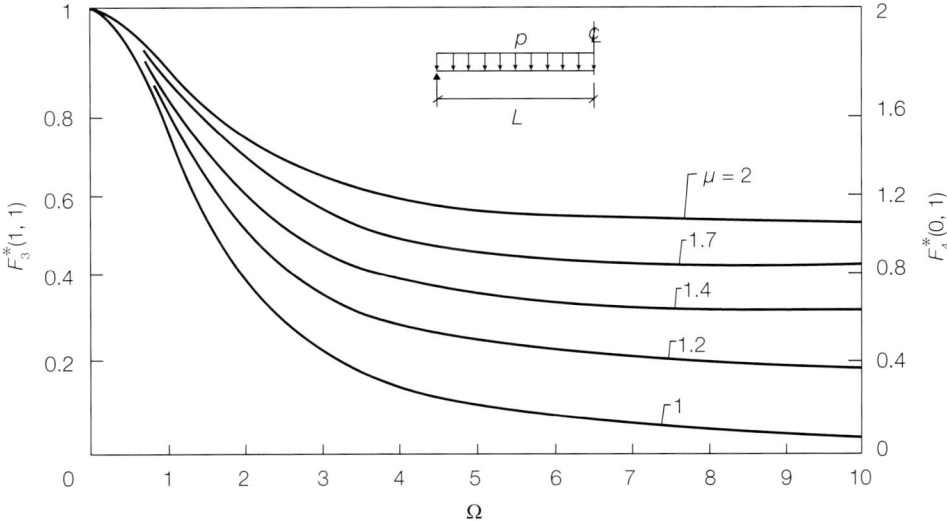

FIGURE A1.8. Maximum influence factors $F_3^*(1, 1)$ and $F_4^*(0, 1)$ for elastic flexure of horizontal laminated glass beam subjected to complete uniform vertical pressure ($\lambda = 1$)

$$S(\xi) = \frac{PL}{\mu h} F_1(\xi,0), \quad \tau(\xi) = \frac{P}{B\mu h} F_2(\xi,0) \tag{A1.45}$$

$$w(\xi) = \frac{PL^3}{6E_g I_h} F_3(\xi,0), \quad \phi(\xi) = \frac{PL^2}{6E_g I_h} F_4(\xi,0) \tag{A1.46}$$

where

$$F_1(\xi,0) = \frac{1}{2\Omega}[\Omega\xi - \operatorname{sech}\Omega \sinh(\Omega\xi)] \tag{A1.47}$$

$$F_2(\xi,0) = \frac{1}{2}[1 - \operatorname{sech}\Omega \cosh(\Omega\xi)] \tag{A1.48}$$

$$F_3(\xi,0) = \frac{1}{2\mu}\left[\xi(\mu-1)(3-\xi^2) + \frac{12}{\Omega^2} F_1(\xi,0)\right] \tag{A1.49}$$

$$F_4(\xi,0) = \frac{3}{2\mu}\left[(\mu-1)(1-\xi^2) + \frac{4}{\Omega^2} F_2(\xi,0)\right] \tag{A1.50}$$

These influence factors are shown graphically in Figure A1.9, taking $\mu = 1.5$ for $F_3(\xi, 0)$ and $F_4(\xi, 0)$. The corresponding maximum values are given by

$$F_1^*(1,0) = \frac{1}{2\Omega}(\Omega - \tanh\Omega) \tag{A1.51}$$

$$F_2^*(0,0) = \frac{1}{2}(1 - \operatorname{sech}\Omega) \tag{A1.52}$$

$$F_3^*(1,0) = \frac{1}{\mu}\left[\mu - 1 + \frac{6}{\Omega^2} F_1^*(1,0)\right] \tag{A1.53}$$

$$F_4^*(0,0) = \frac{3}{2\mu}\left[\mu - 1 + \frac{4}{\Omega^2} F_2^*(0,0)\right] \tag{A1.54}$$

and are plotted in Figure A1.10 for the dual ranges $0 \le \Omega \le 10$ and $1 \le \mu \le 2$. The numerical values of $F_3^*(1, 0)$ and $F_4^*(0, 0)$ turn out to be very similar to those corresponding to the partial uniform loading case $\lambda = 0.1$ considered earlier, when normalised to the same total applied load.

The longitudinal surface stresses in the glass layers are given by

$$\left.\begin{array}{c}\sigma_a(\xi)\\ \sigma_b(\xi)\end{array}\right\} = \frac{PL}{4\mu I_h} \left\{\begin{array}{c}\mu\xi t_1 - 2[t_1 - 2\gamma_2 h(\mu-1)]F_1(\xi,0)\\ -\mu\xi t_1 + 2[t_1 + 2\gamma_2 h(\mu-1)]F_1(\xi,0)\end{array}\right. \tag{A1.55}$$

$$\left.\begin{array}{c}\sigma_c(\xi)\\ \sigma_d(\xi)\end{array}\right\} = \frac{PL}{4\mu I_h} \left\{\begin{array}{c}\mu\xi t_2 - 2[t_2 + 2\gamma_1 h(\mu-1)]F_1(\xi,0)\\ -\mu\xi t_2 + 2[t_2 - 2\gamma_1 h(\mu-1)]F_1(\xi,0)\end{array}\right. \tag{A1.56}$$

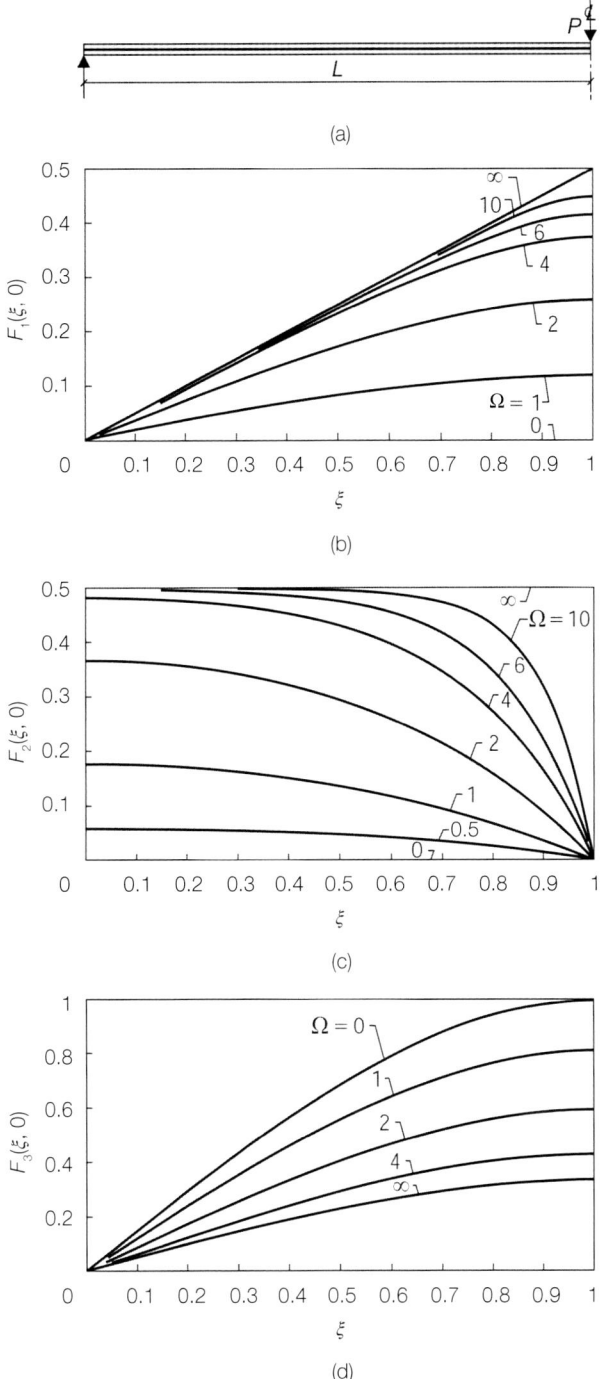

FIGURE A1.9. Influence factors $F_i(\xi,\lambda)$ for elastic flexure of horizontal laminated glass beam subjected to central vertical line load ($\lambda \to 0$, $\mu = 1.5$): (a) half-beam elevation; (b) $F_1(\xi,0)$; (c) $F_2(\xi,0)$; (d) $F_3(\xi,0)$

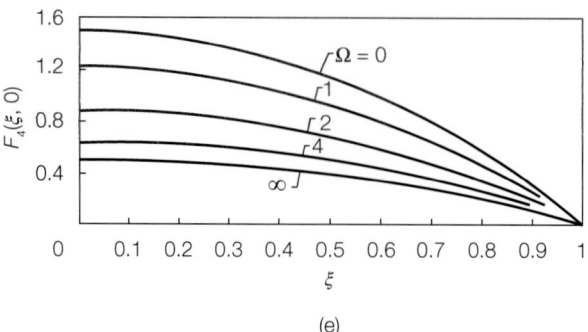

(e)

FIGURE A1.9. (continued) Influence factors $F_i(\xi,\lambda)$ for elastic flexure of horizontal laminated glass beam subjected to central vertical line load ($\lambda \to 0$, $\mu = 1.5$): (e) $F_4(\xi,0)$

When $\Omega = 0$ (zero interlayer stiffness),

$$F_1(\xi,0)_0 = F_2(\xi,0)_0 = 0, \quad F_3(\xi,0)_0 = \frac{\xi}{2}(3-\xi^2), \quad F_4(\xi,0)_0 = \frac{3}{2}(1-\xi^2) \qquad (A1.57)$$

$$\left.\begin{array}{l}\sigma_a(\xi)_0 \\ \sigma_b(\xi)_0\end{array}\right\} = \pm\frac{PL\xi t_1}{4I_h}, \quad \left.\begin{array}{l}\sigma_c(\xi)_0 \\ \sigma_d(\xi)_0\end{array}\right\} = \pm\frac{PL\xi t_2}{4I_h} \qquad (A1.58)$$

When $\Omega \to \infty$ (infinite interlayer stiffness), $F_1(\xi,0)_\infty = \xi/2$ and $F_2(\xi,0)_\infty = 1/2$ to give

$$\left.\begin{array}{l}\sigma_a(\xi)_\infty \\ \sigma_b(\xi)_\infty\end{array}\right\} = \pm\frac{PL\xi}{4\mu I_h}(\mu-1)\left\{\begin{array}{l}t_1 + 2\gamma_2 h \\ t_1 - 2\gamma_2 h\end{array}\right. \qquad (A1.59)$$

$$\left.\begin{array}{l}\sigma_c(\xi)_\infty \\ \sigma_d(\xi)_\infty\end{array}\right\} = \pm\frac{PL\xi}{4\mu I_h}(\mu-1)\left\{\begin{array}{l}t_2 - 2\gamma_1 h \\ t_2 + 2\gamma_1 h\end{array}\right. \qquad (A1.60)$$

while the deflections and slopes are given by equations (A1.46) with the influence factors based on $\Omega = 0$, and with I_h replaced by $\mu I_h/(\mu-1)$.

The surface bending stresses in a monolithic beam of thickness $T = t_1 + t_2 + c$ are

$$\left.\begin{array}{l}\sigma_a(\xi)_T \\ \sigma_d(\xi)_T\end{array}\right\} = \pm\frac{3PL\xi}{BT^2} \qquad (A1.61)$$

and the corresponding deflections and slopes are

$$w(\xi)_T = \frac{PL^3}{12E_g I_T}\xi(3-\xi^2), \quad \phi(\xi)_T = \frac{PL^2}{4E_g I_T}(1-\xi^2) \qquad (A1.62)$$

with maximum values

$$w^*(1)_T = \frac{PL^3}{6E_g I_T}, \quad \phi^*(0)_T = \frac{PL^2}{4E_g I_T} \qquad (A1.63)$$

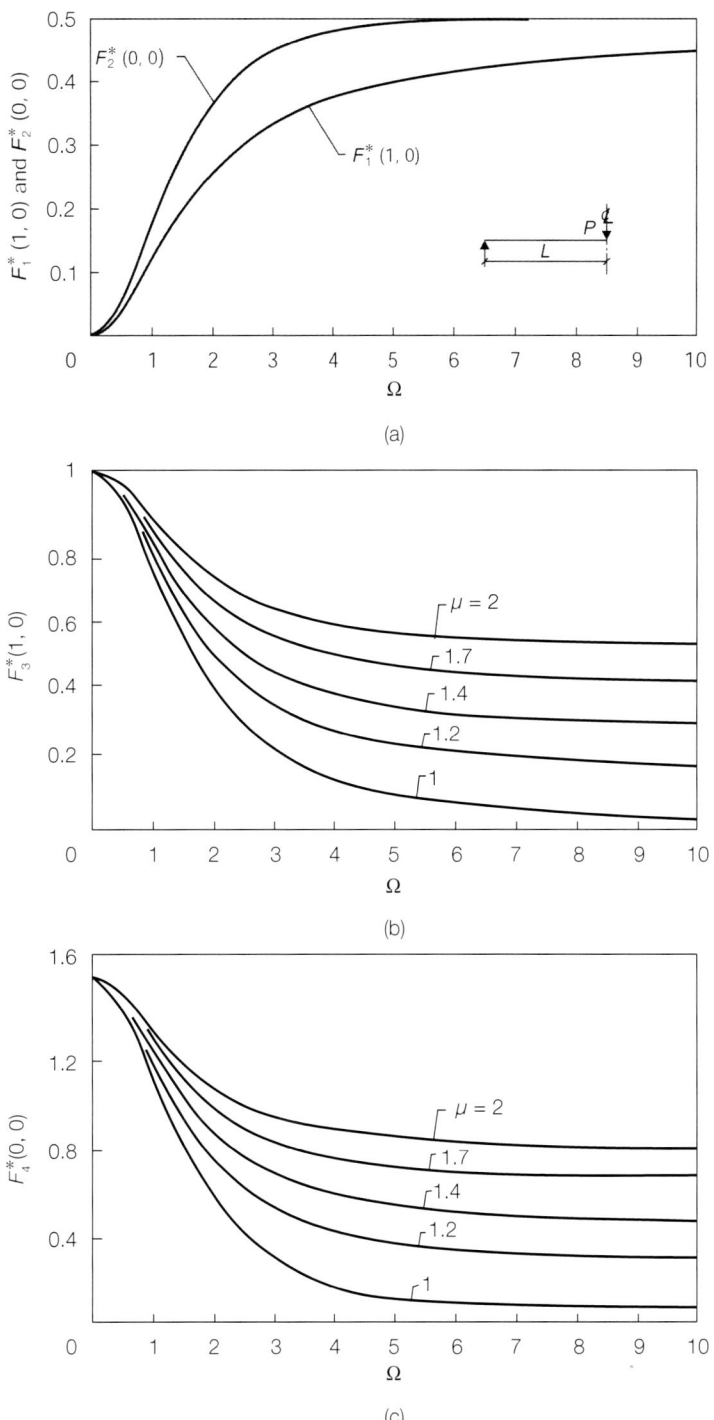

FIGURE A1.10. Maximum influence factors $F_i^*(\xi,\lambda)$ for elastic flexure of horizontal laminated glass beam subjected to central vertical line load ($\lambda \to 0$, $\mu = 1.5$): (a) $F_1^*(1,0)$ and $F_2^*(0,0)$; (b) $F_3^*(1,0)$; (c) $F_4^*(0,0)$

A1.3.4. Other load cases

Where appropriate, results for other symmetric load cases can readily be obtained by deploying the previous solutions for partial uniform loading in conjunction with the principle of superposition. Supposing, for example, the beam is subjected to a pair of *non-central* uniform loads (vertical pressure p) extending horizontal distances l_1 and l_2 from the beam centre ($l_1 > l_2$). Then with $\lambda_1 = l_1/L$ and $\lambda_2 = l_2/L$, the combined general solution for beam flexure becomes

$$S(\xi) = \frac{pBL^2}{\mu h}[F_1(\xi,\lambda_1) - F_1(\xi,\lambda_2)], \quad \tau(\xi) = \frac{pL}{\mu h}[F_2(\xi,\lambda_1) - F_2(\xi,\lambda_2)] \tag{A1.64}$$

$$w(\xi) = \frac{5pBL^4}{24E_g I_h}[F_3(\xi,\lambda_1) - F_3(\xi,\lambda_2)], \quad \phi(\xi) = \frac{pBL^3}{6E_g I_h}[F_4(\xi,\lambda_1) - F_4(\xi,\lambda_2)] \tag{A1.65}$$

For illustrative purposes, Figure A1.11 shows results for the particular case where uniform applied loads of width $L/10$ are centred at the two quarter points of the beam, taking $\mu = 1.5$ and $\Omega = 4$. Calculated values of interfacial shear, slope and deflection are very close to those given in the following section for symmetric four-point line loading applied to a beam of length $2L$ (i.e. no overhang), normalised to give the same total applied load ($P \equiv pBL/5$). In a similar manner, non-uniform distributed loading can often be modelled by superposing solutions for uniform loading over different span widths, giving a useful first approximation to the exact theoretical solution.

A1.4. Beam without overhang under symmetric four-point line loading

With particular reference to laboratory testing, consider the case where the laminated beam (span $2L$, no overhang) is subjected to a symmetric pair of vertical line loads ($P/2$) applied at a distance l from the beam centre. Then with the previous general notation, and with $\lambda = l/L$ as before, the solution for interfacial shear gives

$$S(\xi) = \frac{PL}{\mu h} H_1(\xi,\lambda), \quad \tau(\xi) = \frac{P}{B\mu h} H_2(\xi,\lambda) \tag{A1.66}$$

where

$$H_1(\xi,\lambda) = \frac{1}{2\Omega}\begin{cases} \Omega\xi - \text{sech}\,\Omega\cosh(\Omega\lambda)\sinh(\Omega\xi) & 0 \le \xi \le 1-\lambda \\ \Omega(1-\lambda) + \sinh[\Omega(\xi+\lambda-1)] & \\ -\text{sech}\,\Omega\cosh(\Omega\lambda)\sinh(\Omega\xi) & 1-\lambda < \xi \le 1 \end{cases} \tag{A1.67}$$

$$H_2(\xi,\lambda) = \frac{1}{2}\begin{cases} 1 - \text{sech}\,\Omega\cosh(\Omega\lambda)\cosh(\Omega\xi) & 0 \le \xi \le 1-\lambda \\ \cosh[\Omega(\xi+\lambda-1)] & \\ -\text{sech}\,\Omega\cosh(\Omega\lambda)\cosh(\Omega\xi) & 1-\lambda < \xi \le 1 \end{cases} \tag{A1.68}$$

while noting that for a central line load (P), $H_1(\xi, 0) = F_1(\xi, 0)$ and $H_2(\xi, 0) = F_2(\xi, 0)$. The corresponding maximum values are

ELASTIC FLEXURE OF LAMINATED GLASS BEAM 135

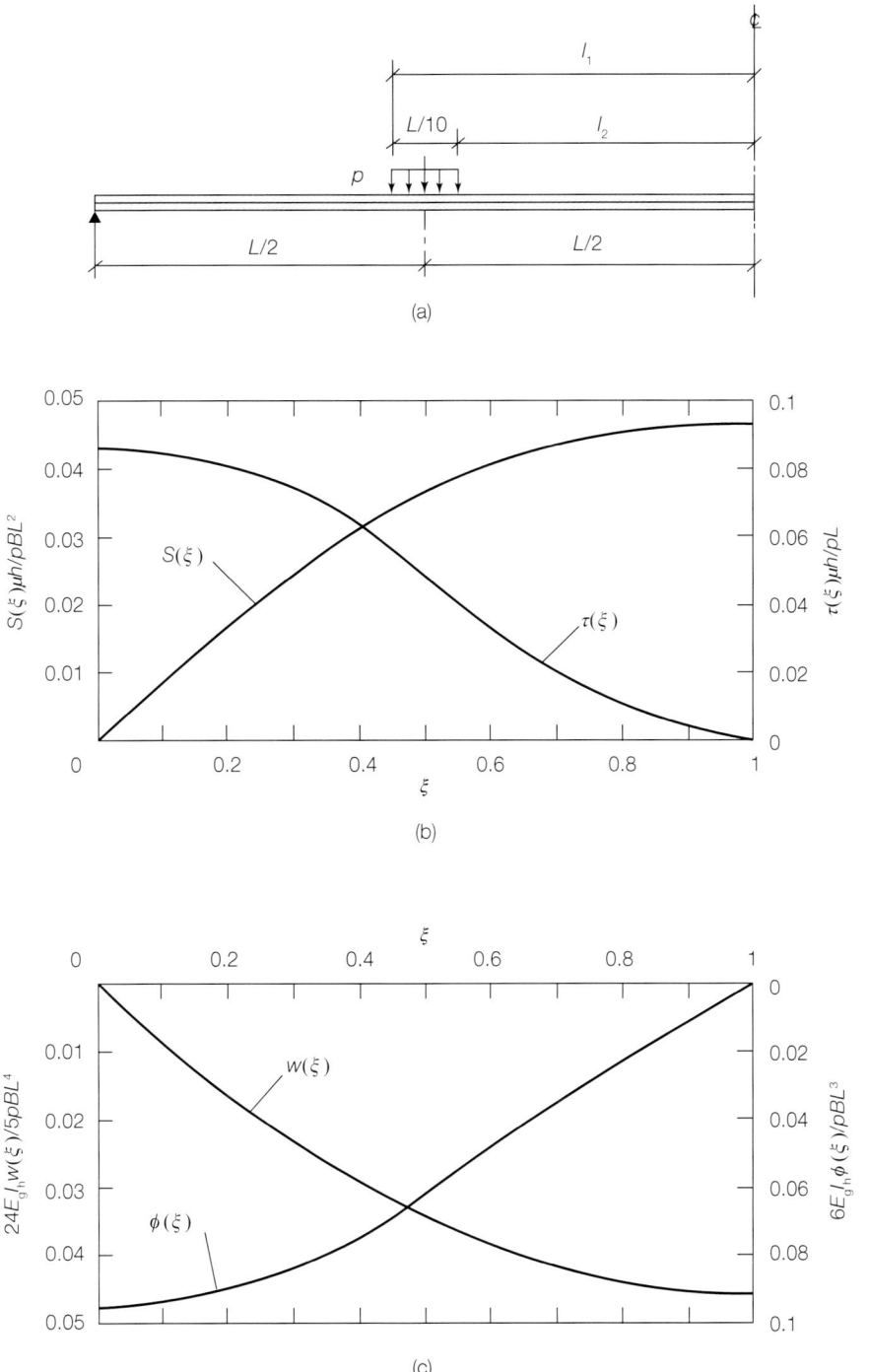

FIGURE A1.11. Elastic flexure of horizontal laminated glass beam subjected to symmetric pair of non-central uniform vertical loads (pressure p) distributed over widths $L/10$ centred at quarter points ($\mu = 1.5$, $\Omega = 4$): (a) half-beam elevation; (b) interfacial shear; (c) slope and deflection

$$H_1^*(1,\lambda) = \frac{1}{2\Omega}[\Omega(1-\lambda) + \sinh(\Omega\lambda) - \tanh\Omega\cosh(\Omega\lambda)] \tag{A1.69}$$

$$H_2^*(0,\lambda) = \frac{1}{2}[1 - \mathrm{sech}\,\Omega\cosh(\Omega\lambda)] \tag{A1.70}$$

In the limiting case of an infinite interlayer shear modulus ($\Omega \to \infty$), $H_1(\xi,\lambda)_\infty = H_5(\xi,\lambda)/2$, while $H_2(\xi,\lambda)_\infty = 1/2$ on $0 \le \xi \le 1 - \lambda$ and zero elsewhere.

The expressions for beam deflection and slope are

$$w(\xi) = \frac{PL^3}{12E_g I_h} H_3(\xi,\lambda), \quad \phi(\xi) = \frac{PL^2}{4E_g I_h} H_4(\xi,\lambda) \tag{A1.71}$$

where

$$H_3(\xi,\lambda) = \frac{1}{\mu}\begin{cases} \xi[3(1-\lambda^2) - \xi^2](\mu-1) + (12/\Omega^2)H_1(\xi,\lambda) & 0 \le \xi \le 1-\lambda \\ (\mu-1)(1-\lambda)[3\xi(2-\xi) - (1-\lambda)^2] \\ +(12/\Omega^2)H_1(\xi,\lambda) & 1-\lambda < \xi \le 1 \end{cases} \tag{A1.72}$$

$$H_4(\xi,\lambda) = \frac{1}{\mu}\begin{cases} (\mu-1)(1-\lambda^2 - \xi^2) + (4/\Omega^2)H_2(\xi,\lambda) & 0 \le \xi \le 1-\lambda \\ 2(\mu-1)(1-\lambda)(1-\xi) + (4/\Omega^2)H_2(\xi,\lambda) & 1-\lambda < \xi \le 1 \end{cases} \tag{A1.73}$$

which results again reduce to those give earlier for the limiting case of central line loading. The maximum values are

$$H_3^*(1,\lambda) = \frac{1}{\mu}\left\{(\mu-1)(1-\lambda)[3-(1-\lambda)^2] + \frac{12}{\Omega^2}H_1^*(1,\lambda)\right\} \tag{A1.74}$$

$$H_4^*(0,\lambda) = \frac{1}{\mu}\left[(\mu-1)(1-\lambda^2) + \frac{4}{\Omega^2}H_2^*(0,\lambda)\right] \tag{A1.75}$$

For the two limiting cases of interlayer shear modulus,

$$H_3(\xi,\lambda)_0 = \begin{cases} \xi[3(1-\lambda^2) - \xi^2] & 0 \le \xi \le 1-\lambda \\ (1-\lambda)[3\xi(2-\xi) - (1-\lambda)^2] & 1-\lambda < \xi \le 1 \end{cases} \tag{A1.76}$$

$$H_4(\xi,\lambda)_0 = \begin{cases} 1-\lambda^2 - \xi^2 & 0 \le \xi \le 1-\lambda \\ 2(1-\lambda)(1-\xi) & 1-\lambda < \xi \le 1 \end{cases} \tag{A1.77}$$

$$\begin{Bmatrix} H_3(\xi,\lambda)_\infty \\ H_4(\xi,\lambda)_\infty \end{Bmatrix} = \frac{\mu-1}{\mu}\begin{Bmatrix} H_3(\xi,\lambda)_0 \\ H_4(\xi,\lambda)_0 \end{Bmatrix} \tag{A1.78}$$

With $\mu = 1.5$, Figure A1.12 depicts the spanwise variation in flexural characteristics over the entire range of beam stiffness for a particular load case ($\lambda = 0.5$), while Figure A1.13 indicates how the maximum influence factors vary with loading position.

The surface bending stresses across the laminate section are given by

$$\left.\begin{array}{c}\sigma_a(\xi)\\ \sigma_b(\xi)\end{array}\right\} = \frac{PL}{4\mu I_h}\left\{\begin{array}{c}\mu t_1 H_5(\xi,\lambda) - 2[t_1 - 2\gamma_2 h(\mu-1)]H_1(\xi,\lambda)\\ -\mu t_1 H_5(\xi,\lambda) + 2[t_1 + 2\gamma_2 h(\mu-1)]H_1(\xi,\lambda)\end{array}\right. \quad (A1.79)$$

$$\left.\begin{array}{c}\sigma_c(\xi)\\ \sigma_d(\xi)\end{array}\right\} = \frac{PL}{4\mu I_h}\left\{\begin{array}{c}\mu t_2 H_5(\xi,\lambda) - 2[t_2 + 2\gamma_1 h(\mu-1)]H_1(\xi,\lambda)\\ -\mu t_2 H_5(\xi,\lambda) + 2[t_2 - 2\gamma_1 h(\mu-1)]H_1(\xi,\lambda)\end{array}\right. \quad (A1.80)$$

where

$$H_5(\xi,\lambda) = \begin{cases} \xi & 0 \leq \xi \leq 1-\lambda \\ 1-\lambda & 1-\lambda < \xi \leq 1 \end{cases} \quad (A1.81)$$

For the case of zero shear modulus,

$$\left.\begin{array}{c}\sigma_a(\xi)_0\\ \sigma_b(\xi)_0\end{array}\right\} = \pm\frac{PLt_1}{4I_h}H_5(\xi,\lambda), \quad \left.\begin{array}{c}\sigma_c(\xi)_0\\ \sigma_d(\xi)_0\end{array}\right\} = \pm\frac{PLt_2}{4I_h}H_5(\xi,\lambda) \quad (A1.82)$$

For the case of infinite shear modulus,

$$\left.\begin{array}{c}\sigma_a(\xi)_\infty\\ \sigma_b(\xi)_\infty\end{array}\right\} = \pm\frac{PL(\mu-1)}{4\mu I_h}H_5(\xi,\lambda)\begin{cases}t_1 + 2\gamma_2 h\\ t_1 - 2\gamma_2 h\end{cases} \quad (A1.83)$$

$$\left.\begin{array}{c}\sigma_c(\xi)_\infty\\ \sigma_d(\xi)_\infty\end{array}\right\} = \pm\frac{PL(\mu-1)}{4\mu I_h}H_5(\xi,\lambda)\begin{cases}t_2 - 2\gamma_1 h\\ t_2 + 2\gamma_1 h\end{cases} \quad (A1.84)$$

For a monolithic beam (thickness T),

$$\left.\begin{array}{c}\sigma_a(\xi)_T\\ \sigma_d(\xi)_T\end{array}\right\} = \pm\frac{3PL}{BT^2}H_5(\xi,\lambda) \quad (A1.85)$$

while the deflection and slope are given by

$$w(\xi)_T = \frac{PL^3}{12E_g I_T}H_3(\xi,\lambda)_0, \quad \phi(\xi)_T = \frac{PL^2}{4E_g I_T}H_4(\xi,\lambda)_0 \quad (A1.86)$$

with maximum values

$$w^*(1)_T = \frac{PL^3}{12E_g I_T}(1-\lambda)[3-(1-\lambda)^2], \quad \phi^*(0)_T = \frac{PL^2}{4E_g I_T}(1-\lambda^2) \quad (A1.87)$$

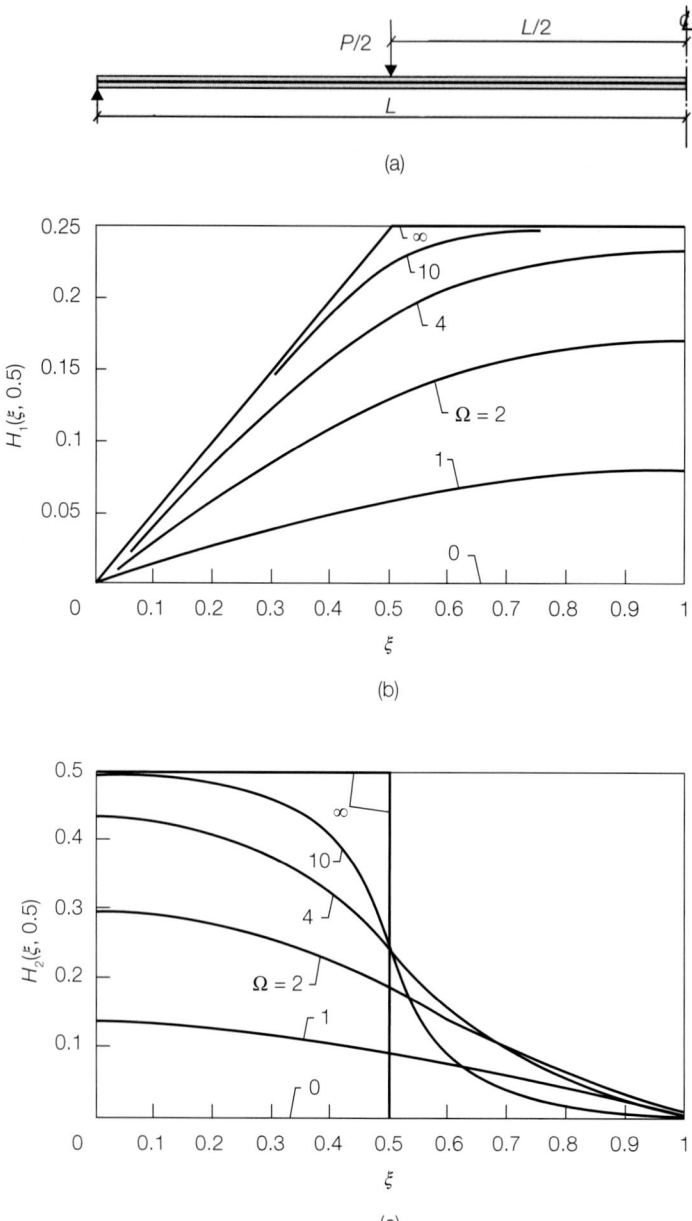

FIGURE A1.12. Influence factors $H_i(\xi, \lambda)$ for elastic flexure of horizontal laminated glass beam subjected to symmetric vertical four-point line loading ($\lambda = 0.5$, $\mu = 1.5$): (a) half-beam elevation; (b) $H_1(\xi, 0.5)$; (c) $H_2(\xi, 0.5)$

ELASTIC FLEXURE OF LAMINATED GLASS BEAM

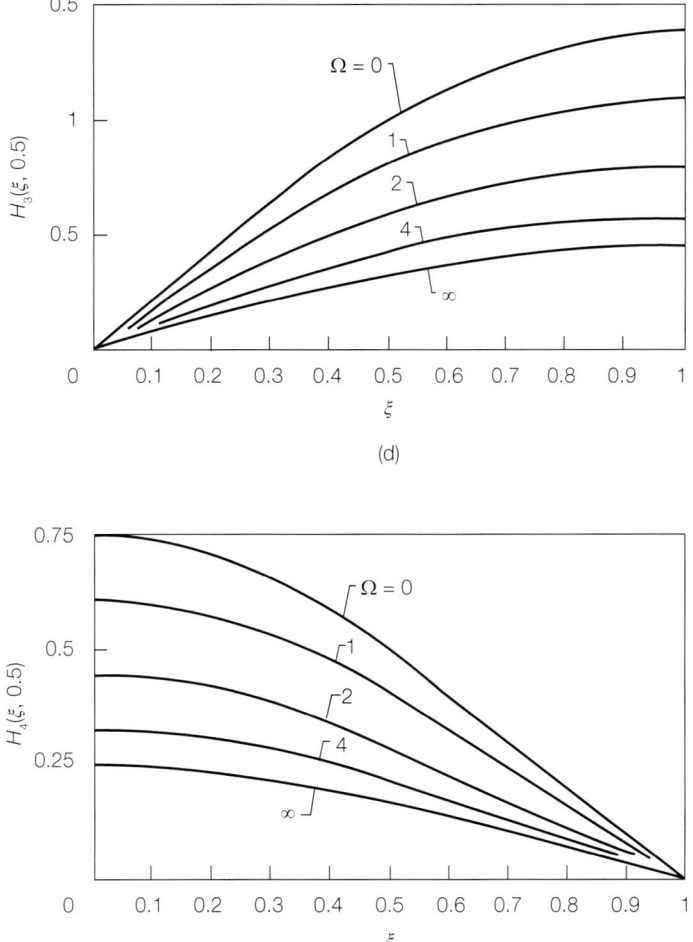

FIGURE A1.12. (continued) Influence factors $H_i(\xi, \lambda)$ for elastic flexure of horizontal laminated glass beam subjected to symmetric vertical four-point line loading ($\lambda = 0.5$, $\mu = 1.5$): (d) $H_3(\xi, 0.5)$; (e) $H_4(\xi, 0.5)$

A1.5. Beam with overhang under symmetric four-point line loading

Consider the case shown in Figure A1.14 in which two vertical line loads of magnitude $P/2$ are applied symmetrically to a simply supported horizontal beam having a clear span $2L$ and a cantilever of length a beyond each support. This arrangement conveniently gives a constant bending moment in the central region between the two loads, and represents a most useful laboratory test for determining the basic flexural properties of laminates; see, for example, Hooper [A1.13]. The beam cross-section remains as shown in Figure A1.1(b).

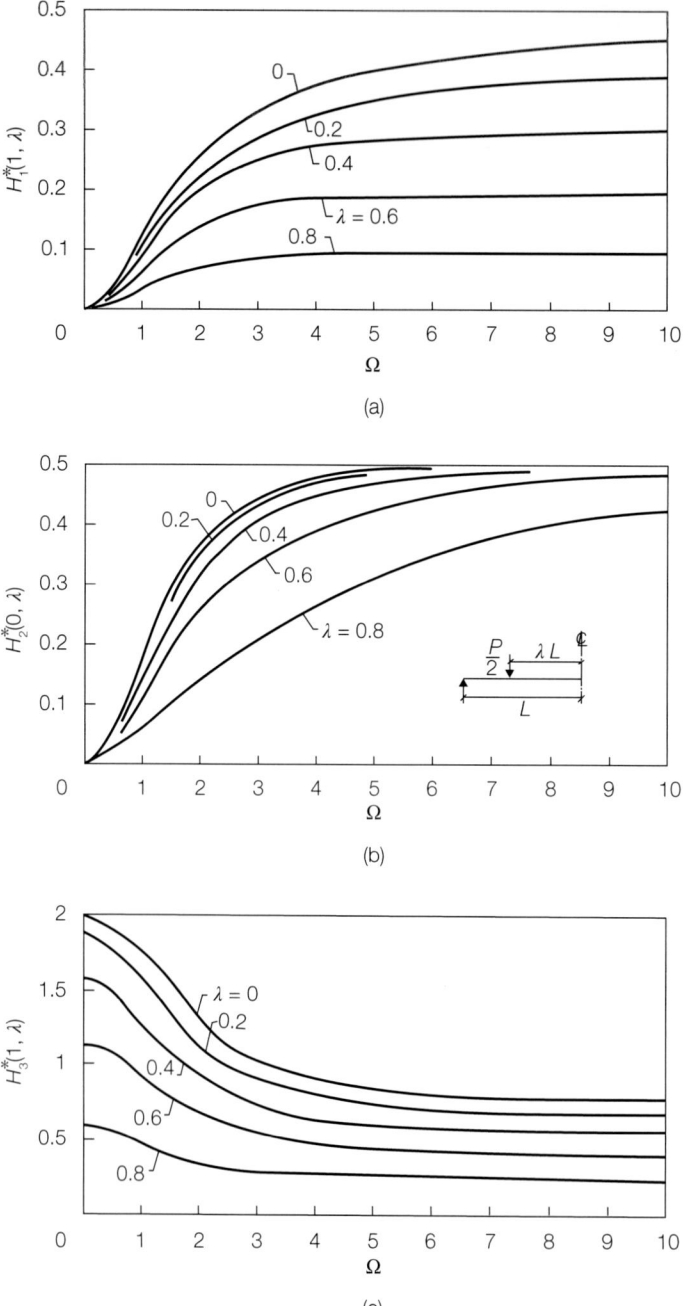

FIGURE A1.13. Maximum influence factors $H_i^*(\xi, \lambda)$ for elastic flexure of horizontal laminated glass beam subjected to symmetric vertical four-point line loads ($P/2$) at various distances (l) from beam centre ($\lambda = l/L$, $\mu = 1.5$): (a) $H_1^*(1, \lambda)$; (b) $H_2^*(0, \lambda)$; (c) $H_3^*(1, \lambda)$

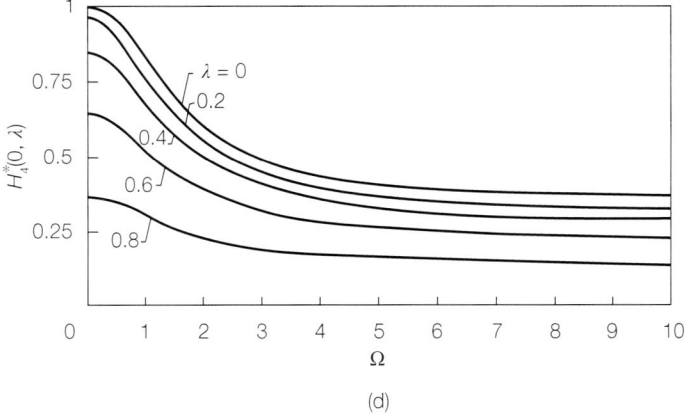

(d)

FIGURE A1.13. (continued) Maximum influence factors $H_i^*(\xi,\lambda)$ for elastic flexure of horizontal laminated glass beam subjected to symmetric vertical four-point line loads ($P/2$) at various distances (l) from beam centre ($\lambda = l/L$, $\mu = 1.5$): (d) $H_4^*(0, \lambda)$

In this example of beam flexure under discontinuous applied loading, it is advantageous to express the distribution of bending moment within the left side of the beam in the operational form

$$M(x) = \frac{P}{2}[(x-a)\mathrm{H}(x-a) - (x-b)\mathrm{H}(x-b)] \tag{A1.88}$$

where the Heaviside unit step function is defined as

$$\mathrm{H}(x-\tilde{x}) = \begin{cases} 0 & x < \tilde{x} \\ 1 & x > \tilde{x} \end{cases} \tag{A1.89}$$

with $\tilde{x}$ denoting a specific value of x. Substituting equation (A1.88) into equation (A1.1) and taking the Laplace transform of the latter gives

$$k^2\hat{S}(k) - kS(0) - S'(0) - \alpha^2\hat{S}(k) + \frac{P\alpha^2}{2\mu h k^2}[\exp(-ak) - \exp(-bk)] = 0 \tag{A1.90}$$

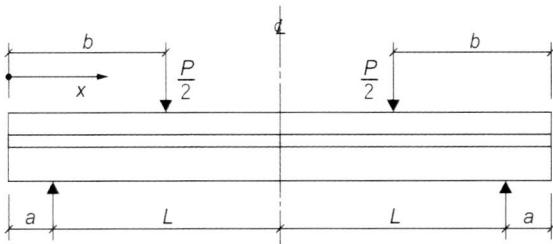

FIGURE A1.14. Notation used in flexural analysis of horizontal laminated glass beam with overhang subjected to symmetric vertical four-point line loading [$\xi = x/L$, $\eta = a/L$, $\delta = (b-a)/L$]

where k denotes the transform parameter, and where the Laplace transform $\hat{f}(k)$ of any suitably continuous function $f(x)$ is defined as

$$\hat{f}(k) = \int_0^\infty f(x)\exp(-kx)dx \tag{A1.91}$$

With the boundary condition $S(0) = 0$, equation (A1.90) can be inverted to give

$$S(x) = \frac{S'(0)}{\alpha}\sinh(\alpha x) - \frac{P}{\alpha\mu h}\{\sinh[\alpha(x-a)] - \sinh[\alpha(x-b)] - \alpha(b-a)\} \tag{A1.92}$$

The remaining unknown parameter is readily determined from the second boundary condition $S'(a + L) = 0$.

Hence, with $\xi = x/L$, $\eta = a/L$ and $\delta = (b - a)/L$, the distribution of interfacial shear force can be expressed in the form

$$S(\xi) = \frac{PL}{\mu h} G_1(\xi, \eta, \delta) \tag{A1.93}$$

where

$$G_1(\xi,\eta,\delta) = \frac{1}{2\Omega}\begin{cases} \Phi(\eta,\delta)\sinh(\Omega\xi) & 0 \le \xi \le \eta \\ \Phi(\eta,\delta)\sinh(\Omega\xi) - \sinh[\Omega(\xi-\eta)] + \Omega(\xi-\eta) & \eta < \xi \le \eta+\delta \\ \Phi(\eta,\delta)\sinh(\Omega\xi) - \sinh[\Omega(\xi-\eta)] \\ \quad + \sinh[\Omega(\xi-\eta-\delta)] + \Omega\delta & \eta+\delta < \xi \le \eta+1 \end{cases} \tag{A1.94}$$

and where the auxiliary factor is defined as

$$\Phi(\eta,\delta) = \frac{\cosh\Omega - \cosh[\Omega(1-\delta)]}{\cosh[\Omega(1+\eta)]} \tag{A1.95}$$

Likewise, the distribution of interfacial shear stress is given by

$$\tau(\xi) = \frac{P}{B\mu h} G_2(\xi, \eta, \delta) \tag{A1.96}$$

where

$$G_2(\xi,\eta,\delta) = \frac{1}{2}\begin{cases} \Phi(\eta,\delta)\cosh(\Omega\xi) & 0 \le \xi \le \eta \\ \Phi(\eta,\delta)\cosh(\Omega\xi) - \cosh[\Omega(\xi-\eta)] + 1 & \eta < \xi \le \eta+\delta \\ \Phi(\eta,\delta)\cosh(\Omega\xi) - \cosh[\Omega(\xi-\eta)] \\ \quad + \cosh[\Omega(\xi-\eta-\delta)] & \eta+\delta < \xi \le \eta+1 \end{cases} \tag{A1.97}$$

Of particular interest is the case where $\delta = 0.5$, which is often referred to as 'standard' four-point loading (*i.e.* line loads applied at the quarter points between the beam supports), and values of the above pair of influence factors over the entire stiffness range are plotted in Figures A1.15(b) and (c), taking $\eta = 0.1$. The maximum interfacial shear force always occurs at the beam centre, so that

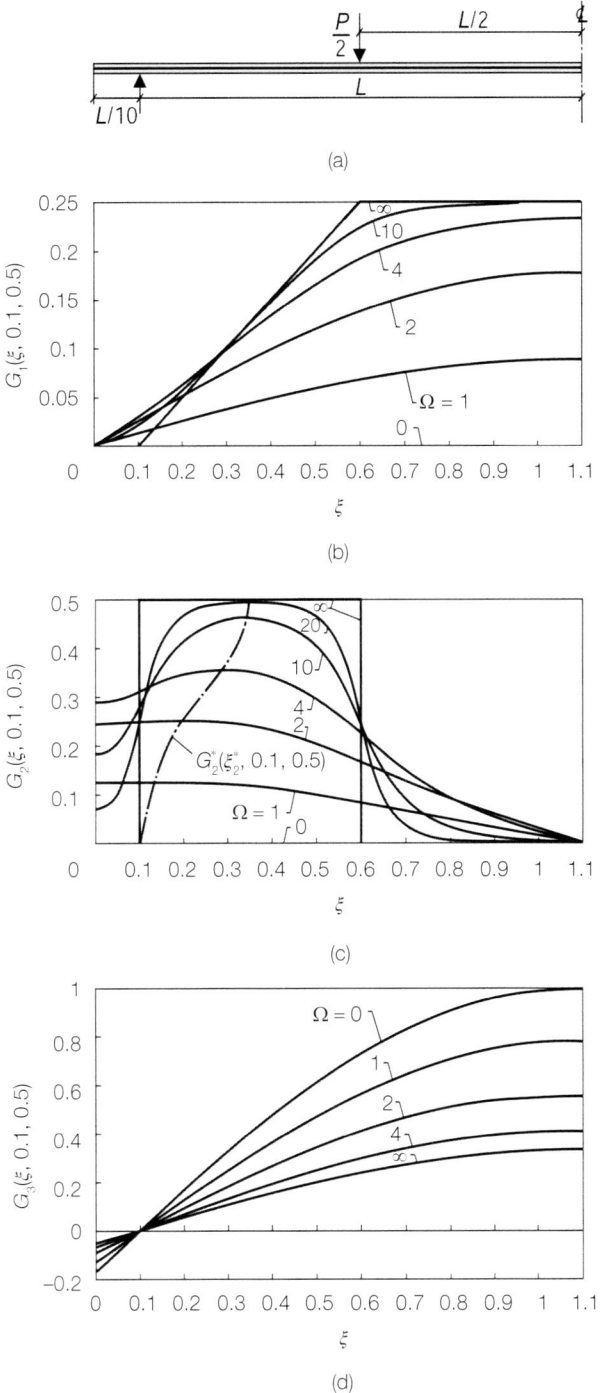

FIGURE A1.15. Influence factors $G_i(\xi, \eta, \delta)$ for elastic flexure of horizontal laminated glass beam with overhang subjected to symmetric vertical four-point line loading ($\eta = 0.1$, $\delta = 0.5$, $\mu = 1.5$): (a) half-beam elevation; (b) $G_1(\xi, 0.1, 0.5)$; (c) $G_2(\xi, 0.1, 0.5)$; (d) $G_3(\xi, 0.1, 0.5)$

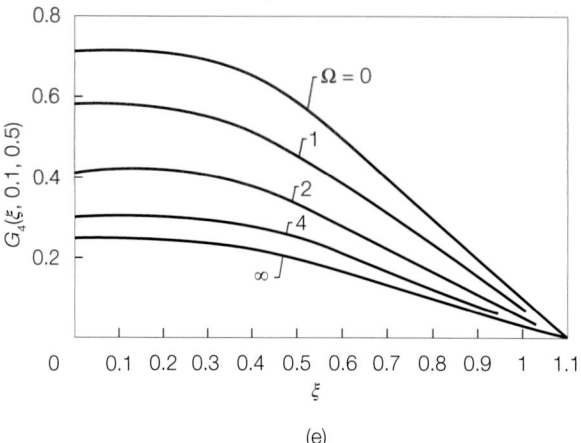

FIGURE A1.15. (continued) Influence factors $G_i(\xi, \eta, \delta)$ for elastic flexure of horizontal laminated glass beam with overhang subjected to symmetric vertical four-point line loading ($\eta = 0.1$, $\delta = 0.5$, $\mu = 1.5$): (e) $G_4(\xi, 0.1, 0.5)$

$$G_1^*(1+\eta,\eta,\delta) = \frac{1}{2\Omega}\{\Phi(\eta,\delta)\sinh[\Omega(1+\eta)] - \sinh\Omega + \sinh[\Omega(1-\delta)] + \Omega\delta\} \quad \text{(A1.98)}$$

In contrast, the maximum interfacial shear stress always occurs within the zone $\eta \leq \xi \leq \eta + \delta$, but its precise location ξ_2^* depends upon Ω, η and δ. Hence the condition for $\tau(\xi)$ to be a maximum is satisfied when

$$\Phi(\eta,\delta)\sinh(\Omega\xi_2^*) - \sinh[\Omega(\xi_2^* - \eta)] = 0 \quad \text{(A1.99)}$$

from which it follows that the required location is given by

$$\xi_2^* = \frac{1}{2\Omega}\ln\left[\frac{\Phi(\eta,\delta) - \exp(\Omega\eta)}{\Phi(\eta,\delta) - \exp(-\Omega\eta)}\right] \quad \text{(A1.100)}$$

The locus of $G_2^*(\xi_2^*, 0.1, 0.5)$, for example, is shown by the chain-dotted line in Figure A1.15(c). More generally, values of $G_1^*(1.1, 0.1, 0.5)$ and $G_2^*(\xi_2^*, 0.1, 0.5)$ are plotted in Figure A1.16(a) for the stiffness range $0 \leq \Omega \leq 10$.

To obtain beam deflections, the integral transform method once again can usefully be deployed to give a more concise solution than that derived by conventional integration of the moment–curvature relation. Whence the operational form of the expression for longitudinal shear force becomes

$$S(x) = \frac{P}{2\alpha\mu h}\langle \Phi(\eta,\delta)\sinh(\alpha x) + \{\alpha(x-a) - \sinh[\alpha(x-a)]\}H(x-a)$$
$$- \{\alpha(x-b) - \sinh[\alpha(x-b)]\}H(x-b)\rangle \quad \text{(A1.101)}$$

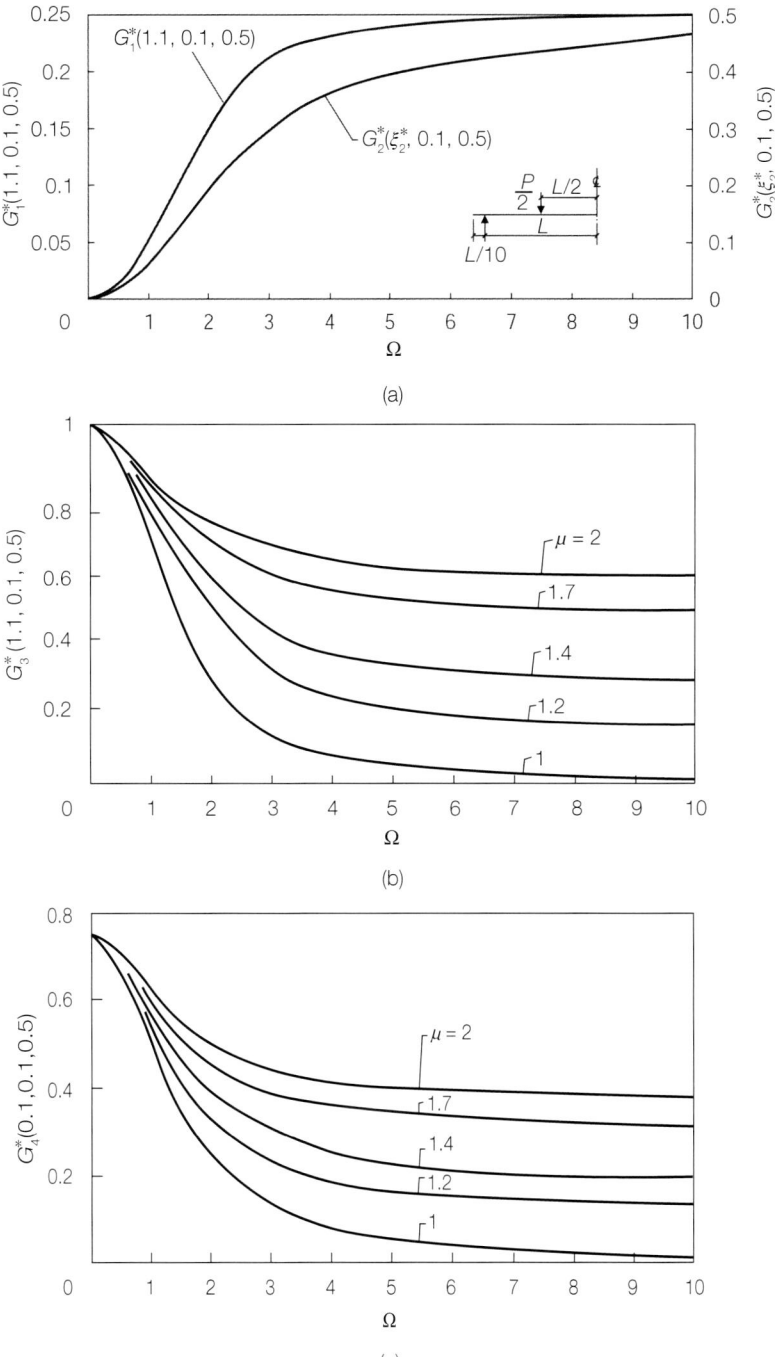

FIGURE A1.16. Maximum influence factors $G_i^*(\xi,\eta,\delta)$ for elastic flexure of horizontal laminated glass beam with overhang subjected to symmetric vertical four-point line loading ($\eta = 0.1$, $\delta = 0.5$): (a) $G_1^*(1.1, 0.1, 0.5)$ and $G_2^*(\xi_2^*, 0.1, 0.5)$; (b) $G_3^*(1.1, 0.1, 0.5)$; (c) $G_4^*(0.1, 0.1, 0.5)$

Substituting equations (A1.88) and (A1.101) into equation (A1.7) and taking the Laplace transform of the latter then gives

$$k^2 \hat{w}(k) - kw(0) - w'(0) = \frac{P}{2\mu E_g I_h} \left\{ \frac{\Phi(\eta,\delta)}{k^2 - \alpha^2} \right. \tag{A1.102}$$
$$\left. - \left(\frac{\mu-1}{k^2} + \frac{1}{k^2 - \alpha^2} \right) [\exp(-ak) - \exp(-bk)] \right\}$$

Inversion of this last equation gives

$$w(x) = \frac{P}{12\mu E_g I_h} \left\langle (\mu-1)[(x-b)^3 - (x-a)^3] - \frac{6}{\alpha^3} \{\Phi(\eta,\delta)[\alpha x - \sinh(\alpha x)] \right. \tag{A1.103}$$
$$\left. + \sinh[\alpha(x-a)] - \sinh[\alpha(x-b)] - \alpha(b-a)\} \right\rangle + w(0) + xw'(0)$$

while the two constants are determined from the boundary conditions $w(a) = w'(a+L) = 0$. The final expression for beam deflection then becomes

$$w(\xi) = \frac{11PL^3}{96 E_g I_h} G_3(\xi,\eta,\delta) \tag{A1.104}$$

where

$$G_3(\xi,\eta,\delta) = \frac{8}{11\mu} \begin{cases} \Theta(\xi,\eta,\delta) + 3\delta(\mu-1)(2-\delta)(\xi-\eta) & 0 \le \xi \le \eta \\ \Theta(\xi,\eta,\delta) \\ +(\mu-1)(\xi-\eta)[3\delta(2-\delta) - (\xi-\eta)^2] & \eta < \xi \le \eta+\delta \\ \Theta(\xi,\eta,\delta) + \delta(\mu-1)[3-\delta^2 - 3(1+\eta-\xi)^2] & \eta+\delta < \xi \le \eta+1 \end{cases} \tag{A1.105}$$

and where the auxiliary function is defined as

$$\Theta(\xi,\eta,\delta) = \frac{6}{\Omega^3}[2\Omega G_1(\xi,\eta,\delta) - \Phi(\eta,\delta)\sinh(\Omega\eta)] \tag{A1.106}$$

Various profiles of normalised beam deflection are shown in Figure A1.15(d), taking $\eta = 0.1$, $\delta = 0.5$ and $\mu = 1.5$. The maximum deflection can be obtained using the relation

$$G_3^*(1+\eta,\eta,\delta) = \frac{8}{11\mu} [\Theta(1+\eta,\eta,\delta) + \delta(\mu-1)(3-\delta^2)] \tag{A1.107}$$

and values of this factor within the dual ranges $0 \le \Omega \le 10$ and $1 \le \mu \le 2$ are plotted in Figure A1.16(b) for the case $\eta = 0.1$, $\delta = 0.5$.

The corresponding profiles of beam slope are given by

$$\phi(\xi) = \frac{PL^2}{4 E_g I_h} G_4(\xi,\eta,\delta) \tag{A1.108}$$

where the influence factor is

$$G_4(\xi,\eta,\delta) = \frac{1}{\mu} \begin{cases} \delta(\mu-1)(2-\delta) + (4/\Omega^2) G_2(\xi,\eta,\delta) & 0 \le \xi \le \eta \\ (\mu-1)[\delta(2-\delta) - (\xi-\eta)^2] \\ +(4/\Omega^2) G_2(\xi,\eta,\delta) & \eta < \xi \le \eta+\delta \\ 2\delta(\mu-1)(1+\eta-\xi) + (4/\Omega^2) G_2(\xi,\eta,\delta) & \eta+\delta < \xi \le \eta+1 \end{cases} \tag{A1.109}$$

ELASTIC FLEXURE OF LAMINATED GLASS BEAM 147

The slope at the left beam support then becomes

$$G_4^*(\eta,\eta,\delta) = \frac{1}{\mu}[\delta(\mu-1)(2-\delta) + (2/\Omega^2)\Phi(\eta,\delta)\cosh(\Omega\eta)] \quad (A1.110)$$

Values of the latter two factors are included in Figures A1.15 and A1.16, while the general effect of beam overhang is indicated in Table A1.1 in terms of maximum influence factors. Upon setting $\eta = 0$, the above equations of flexure reduce to those given earlier for a simply supported beam with zero overhang subjected either to a central line load (P), or to a symmetric pair of line loads ($P/2$).

The expressions for longitudinal stress at the surfaces of each glass layer are

$$\left.\begin{array}{c}\sigma_a(\xi)\\ \sigma_b(\xi)\end{array}\right\} = \frac{PL}{4\mu I_h}\left\{\begin{array}{c}\mu t_1 G_5(\xi,\eta,\delta) - 2[t_1 - 2\gamma_2 h(\mu-1)]G_1(\xi,\eta,\delta)\\ -\mu t_1 G_5(\xi,\eta,\delta) + 2[t_1 + 2\gamma_2 h(\mu-1)]G_1(\xi,\eta,\delta)\end{array}\right. \quad (A1.111)$$

$$\left.\begin{array}{c}\sigma_c(\xi)\\ \sigma_d(\xi)\end{array}\right\} = \frac{PL}{4\mu I_h}\left\{\begin{array}{c}\mu t_2 G_5(\xi,\eta,\delta) - 2[t_2 + 2\gamma_1 h(\mu-1)]G_1(\xi,\eta,\delta)\\ -\mu t_2 G_5(\xi,\eta,\delta) + 2[t_2 - 2\gamma_1 h(\mu-1)]G_1(\xi,\eta,\delta)\end{array}\right. \quad (A1.112)$$

where

$$G_5(\xi,\eta,\delta) = \begin{cases} 0 & 0 \leq \xi \leq \eta \\ \xi - \eta & \eta < \xi \leq \eta + \delta \\ \delta/2 & \eta + \delta < \xi \leq \eta + 1 \end{cases} \quad (A1.113)$$

TABLE A1.1. Effect of overhang (η) on elastic flexure of horizontal laminated glass beam subjected to vertical line loads at quarter points ($\delta = 0.5$, $\mu = 1.5$): maximum influence factors $G_i^*(\xi, \eta, \delta)$

η	Ω	$G_1^*(1+\eta, \eta, 0.5)$	$G_2^*(\xi_2^*, \eta, 0.5)$	$G_3^*(1+\eta, \eta, 0.5)$	$G_4^*(\eta, \eta, 0.5)$
0	1	0.0812	0.1346	0.8055	0.6090
0	2	0.1719	0.2949	0.5834	0.4466
0	4	0.2334	0.4311	0.4182	0.3219
0.1	1	0.0892	0.1253	0.7799	0.5837
0.1	2	0.1784	0.2526	0.5573	0.4152
0.1	4	0.2345	0.3548	0.4078	0.3021
0.2	1	0.0961	0.1177	0.7582	0.5621
0.2	2	0.1828	0.2280	0.5396	0.3939
0.2	4	0.2350	0.3315	0.4031	0.2932
0.3	1	0.1020	0.1115	0.7398	0.5438
0.3	2	0.1858	0.2127	0.5277	0.3795
0.3	4	0.2352	0.3220	0.4010	0.2892

When $\Omega = 0$, corresponding to zero interlayer stiffness, these latter equations simplify to

$$\left.\begin{array}{l}\sigma_a(\xi)_0 \\ \sigma_b(\xi)_0\end{array}\right\} = \pm\frac{PLt_1}{4I_h}G_5(\xi,\eta,\delta), \quad \left.\begin{array}{l}\sigma_c(\xi)_0 \\ \sigma_d(\xi)_0\end{array}\right\} = \pm\frac{PLt_2}{4I_h}G_5(\xi,\eta,\delta) \qquad (A1.114)$$

and the expressions for beam deflection and slope reduce to

$$w(\xi)_0 = \frac{PL^3}{12E_gI_h}\begin{cases} 3\delta(2-\delta)(\xi-\eta) & 0 \leq \xi \leq \eta \\ (\xi-\eta)[3\delta(2-\delta)-(\xi-\eta)^2] & \eta < \xi \leq \eta+\delta \\ \delta[3-\delta^2-3(1+\eta-\xi)^2] & \eta+\delta < \xi \leq \eta+1 \end{cases} \qquad (A1.115)$$

$$\phi(\xi)_0 = \frac{PL^2}{4E_gI_h}\begin{cases} \delta(2-\delta) & 0 \leq \xi \leq \eta \\ \delta(2-\delta)-(\xi-\eta)^2 & \eta < \xi \leq \eta+\delta \\ 2\delta(1+\eta-\xi) & \eta+\delta < \xi \leq \eta+1 \end{cases} \qquad (A1.116)$$

When $\Omega \to \infty$, corresponding to infinite interlayer stiffness, the distribution of $S(\xi)_\infty$ is obtained simply by setting $G_1(\xi,\eta,\delta)_\infty = G_5(\xi,\eta,\delta)/2$ in equation (A1.93), which in turn leads to discontinuities in $\tau(\xi)_\infty$ at $\xi = \eta$ and $\xi = \eta+\delta$; see Figure A1.15. The corresponding longitudinal stresses in the two glass layers then become

$$\left.\begin{array}{l}\sigma_a(\xi)_\infty \\ \sigma_b(\xi)_\infty\end{array}\right\} = \pm\frac{PL(\mu-1)}{4\mu I_h}G_5(\xi,\eta,\delta)\begin{cases}t_1+2\gamma_2 h \\ t_1-2\gamma_2 h\end{cases} \qquad (A1.117)$$

$$\left.\begin{array}{l}\sigma_c(\xi)_\infty \\ \sigma_d(\xi)_\infty\end{array}\right\} = \pm\frac{PL(\mu-1)}{4\mu I_h}G_5(\xi,\eta,\delta)\begin{cases}t_2-2\gamma_1 h \\ t_2+2\gamma_1 h\end{cases} \qquad (A1.118)$$

and the beam deflections and slopes are obtained by replacing I_h by $\mu I_h/(\mu-1)$ in equations (A1.115) and (A1.116). In the case of a monolithic glass beam of thickness T, the bending stresses are

$$\left.\begin{array}{l}\sigma_a(\xi)_T \\ \sigma_d(\xi)_T\end{array}\right\} = \pm\frac{3PL}{BT^2}G_5(\xi,\eta,\delta) \qquad (A1.119)$$

and the deflections and slopes are obtained upon replacing I_h by I_T in equations (A1.115) and (A1.116).

A1.6. References

A1.1. CHITTY, L. On the cantilever composed of a number of parallel beams interconnected by cross-bars. *Phil. Mag.*, 1947, Ser. 7, 38, No. 285 (Oct.), 685–699.

A1.2. VAN DER NEUT, A. *De 3-punts-buigproef by doosliggers*. Nationaal Lucht- en Ruimtevaartlaboratorium, Amsterdam, Netherlands, May 1933, N.L.R. Tech. Rep. S72, 7 pp.

A1.3. MARCH, H. W. Bending of a centrally loaded rectangular strip of plywood. *Physics*, 1936, 7, 1 (Jan.), 32–41.

A1.4. HOFF, N. J. and MAUTNER, S. E. Bending and buckling of sandwich beams. *J. Aero. Sci.*, 1948, 15, 12 (Dec.), 707–720.

A1.5. MARCH, H. W. and SMITH, C. B. *Flexural rigidity of a rectangular strip of sandwich construction*. U.S. Dept Agriculture, Madison, Wisconsin, U.S.A., Feb. 1944, Forest Products Lab. Rep. 1505. Revised Feb. 1955, 21 pp.

A1.6. NORRIS, C. B., ERICKSEN, W. S. and KOMMERS, W. J. *Flexural rigidity of a rectangular strip of sandwich construction: comparison between mathematical analysis and results of tests*. U.S. Dept Agriculture, Madison, Wisconsin, U.S.A., May 1952, Forest Products Lab. Rep. 1505A (Supplement to Rep. 1505), 55 pp.

A1.7. PIPPARD, A. J. S. and FRANCIS, W. E. On a theoretical and experimental investigation of the stresses in a radially spoked wire wheel under loads applied to the rim. *Phil. Mag.*, 1931, Ser. 7, 11, No. 69 (Feb.), 233–285.

A1.8. CHITTY, L. and WAN, W-Y. Tall building structures under wind load. *Proc. 7th Int. Congr. Appl. Mech., London, England, Sept. 1948*, 1, 254–268.

A1.9. SCHULZ, M. Analysis of reinforced concrete walls with openings. *Indian Concr. J.*, 1961, 35, 11 (Nov.), 432–433.

A1.10. ERIKSSON, O. Analysis of wind bracing walls in multistorey housing. *Ingeniøren (Int. edn)*, 1961, 5, 4 (Dec.), 115–124.

A1.11. BECK, H. Contribution to the analysis of coupled shear walls. *J. Amer. Concr. Inst.*, 1962, 59, 8 (Aug.), 1055–1070.

A1.12. ROSMAN, R. Approximate analysis of shear walls subject to lateral loads. *J. Amer. Concr. Inst.*, 1964, 61, 6 (Jun.), 717–733.

A1.13. HOOPER, J. A. On the bending of architectural laminated glass. *Int. J. Mech. Sci.*, 1973, 15, 4 (Apr.), 309–323.

A2. FULL-SCALE LOADING TESTS ON GLASS PANELS

A2.1. Introduction

Following the aforementioned experimental work on small-scale laboratory specimens, mostly comprising short glass beams under static and dynamic four-point bending, several full-scale loading tests principally devised and specified by the writer were carried out during February 1970 at the Commonwealth Experimental Building Station located in North Ryde, Sydney. Flat rectangular panels of monolithic and laminated glass, approximately 2.3 × 1.5 m in size and of various cross-sections, were assembled to form the front vertical face of a large suction box, as illustrated in Figure 42 (main text). The basic glazing bar details and glass support system used in these experiments were similar to those finally adopted for most segments of the glass walls of the Opera House, as depicted in Figure 23 (main text). The structural support along the vertical edge of each test panel comprised an I-section steel glazing bar attached to an internal circular steel tube by four prototype adjustable fixings, while the horizontal glass-to-glass butt joints were formed in silicone rubber. Thin adhesive rubber strips provided an air-tight seal between the sides of the suction box and the outer edges of the glazed assembly.

It is emphasised that these tests were undertaken in some haste, as part of a wider investigation directed towards producing a final structural and architectural design for the glass walls, whose construction lay on the critical path to weatherproof the open shell structures and allow internal fit-out to proceed. Similar tests carried out under present-day conditions doubtless would encompass more elaborate instrumentation, with test specimens festooned with the latest electronic gadgetry. In the prevailing circumstances, however, the adopted test procedures were reckoned to be an adequate compromise in providing sufficient vindication of the proposed final design.

The principal objectives of the test series were:
(a) to assess possible construction problems in what was a new and innovative glazing system;
(b) to measure glass panel deflections and compare with theoretical values for quasi-static applied loading;
(c) to observe any possible adverse dynamic effects in the structural response of glass panels;
(d) to examine the post-failure condition and structural integrity of cracked laminated panels.

Upon a satisfactory conclusion of this test programme, the full manufacturing and supply process for the various glazing components could then commence.

A2.2. Equations of flexure

For monolithic glass panels, the calculated deflections and bending stresses are based on exact analytical results derived from the classical small-deflection theory of thin elastic isotropic plates given in appendix A3, which includes corrections to previous solutions published in the technical literature. Referring to Figure A2.1(a), a rectangular panel of width a, length b and thickness T is simply supported on two opposite edges ($x = 0, a$), the other edges being 'free' or unrestrained. The supports are assumed to be rigid, and to remain in complete contact with the plate during loading. A static uniform normal pressure (p) is applied over an entire face. The elastic parameters for the glass are Young's modulus E_g and Poisson's ratio v_g, with the flexural stiffness of the plate defined as $D = E_g T^3/12(1 - v_g^2)$. At the plate centre, the deflection (reckoned positive in the direction of loading) is given by

$$w(a/2,0) = \frac{pa^4}{D}\left[\frac{5}{384} + \frac{4v_g}{\pi^5(1-v_g)}\sum_{n=1}^{\infty}\frac{(-1)^{(n-1)/2} A_n^+}{n^5 \Delta_n^+}\right] \quad (A2.1)$$

while at the centre of a 'free' edge,

$$w(a/2,b/2) = \frac{pa^4}{D(1-v_g)}\left[\frac{5}{384} - \frac{8v_g}{\pi^5}\sum_{n=1}^{\infty}\frac{(-1)^{(n-1)/2}}{n^5 \Delta_n^+}\right] \quad (A2.2)$$

The non-dimensional coefficients are

$$A_n^+ = (1+v_g)\operatorname{sech}(\alpha_n b/2) - (1-v_g)(\alpha_n b/2)\operatorname{cosech}(\alpha_n b/2) \quad (A2.3)$$

$$\Delta_n^+ = (3+v_g) - (1-v_g)(\alpha_n b)\operatorname{cosech}(\alpha_n b) \quad (A2.4)$$

where $\alpha_n = n\pi/a$, and n is a positive *odd* integer. The corresponding equations for transverse bending moments per unit width ('sagging' positive) are

$$M_x(a/2,0) = pa^2\left\{\frac{1}{8} + \frac{4v_g}{\pi^3}\sum_{n=1}^{\infty}\frac{(-1)^{(n-1)/2}}{n^3 \Delta_n^+}[A_n^+ - 2v_g \operatorname{sech}(\alpha_n b/2)]\right\} \quad (A2.5)$$

$$M_x(a/2,b/2) = pa^2(1+v_g)\left[\frac{1}{8} - \frac{8v_g}{\pi^3}\sum_{n=1}^{\infty}\frac{(-1)^{(n-1)/2}}{n^3 \Delta_n^+}\right] \quad (A2.6)$$

while the maximum bending stresses (tension negative) are $\sigma = \pm 6M_x/T^2$. Comparative equations for a simply supported beam (span a) are

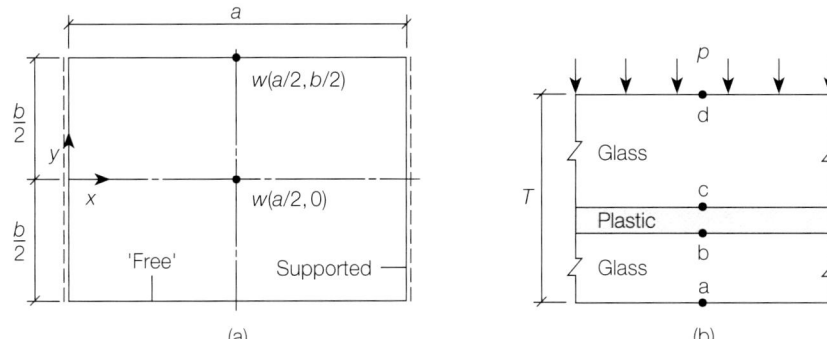

FIGURE A2.1. Notation used in analysis of full-scale loading tests on rectangular glass panels: (a) vertical elevation (single panel); (b) laminate cross-section ($t_1 \leq t_2$)

$$w(a/2) = \frac{5pa^4}{384E_g I}, \qquad M(a/2) = \frac{pa^2}{8} \tag{A2.7}$$

where the second moment of area $I = T^3/12$ for a beam of unit width.

For laminated glass panels under the same support conditions, where there appears to be no exact theoretical solution, the equations given in appendix A1 for the flexure of a laminated glass beam can give results of sufficient accuracy for present purposes. The ends of a beam of length a and unit width are simply supported, and uniform normal loading (p) is applied over the entire span. With the cross-section notation given in Figure A2.1(b), and with $t_1 \leq t_2$, the basic dimensionless parameters are defined as

$$\mu = 1 + \frac{(t_1 + t_2)(t_1^3 + t_2^3)}{12 t_1 t_2 h^2}, \qquad \Omega = ah\left[\frac{3\mu G_p}{cE_g(t_1^3 + t_2^3)}\right]^{1/2} \tag{A2.8}$$

where $h = c + (t_1 + t_2)/2$, and where G_p denotes the shear modulus of the plastic interlayer. Whence the mid-span deflection becomes

$$w(a/2) = \frac{5pa^4}{32\mu E_g(t_1^3 + t_2^3)}\left\{\mu - 1 + \frac{12}{5\Omega^2}\left[1 + \frac{2}{\Omega^2}(\operatorname{sech}\Omega - 1)\right]\right\} \tag{A2.9}$$

and the corresponding bending stresses at the surface of each glass layer are

$$\left.\begin{array}{c}\sigma_a(a/2)\\ \sigma_b(a/2)\end{array}\right\} = \frac{3pa^2}{4\mu(t_1^3 + t_2^3)}\left\{\mp \mu t_1 \pm \left[t_1 \mp \frac{2t_2 h(\mu-1)}{t_1 + t_2}\right]\left[1 - \frac{2}{\Omega^2}(1 - \operatorname{sech}\Omega)\right]\right\} \tag{A2.10}$$

$$\left.\begin{array}{c}\sigma_c(a/2)\\ \sigma_d(a/2)\end{array}\right\} = \frac{3pa^2}{4\mu(t_1^3 + t_2^3)}\left\{\mp \mu t_2 \pm \left[t_2 \pm \frac{2t_1 h(\mu-1)}{t_1 + t_2}\right]\left[1 - \frac{2}{\Omega^2}(1 - \operatorname{sech}\Omega)\right]\right\} \tag{A2.11}$$

while noting herein that the thinner glass layer is positioned on the 'tension face' of the laminate, and that the positive pressure indicated in Figure A2.1(b) equally relates to suction applied to the opposite outer surface of the laminate. In the limiting case where the interlayer shear modulus effectively reduces to zero ($\Omega = 0$), as applicable to architectural laminated glass under long-term sustained loading,

$$\left.\begin{array}{l}\sigma_a(a/2)\\ \sigma_b(a/2)\end{array}\right\} = \mp \frac{3pa^2t_1}{4(t_1^3 + t_2^3)}, \quad \left.\begin{array}{l}\sigma_c(a/2)\\ \sigma_d(a/2)\end{array}\right\} = \mp \frac{3pa^2t_2}{4(t_1^3 + t_2^3)} \tag{A2.12}$$

A2.3. Test arrangement

Nine loading tests were carried out, each comprising four glass panels of the same rectangular shape but not necessarily the same thickness. The first two tests deployed identical monolithic glass panels, partly acting as control specimens to calibrate the test procedure. Seven further tests were carried out using laminated glass of various layer configurations. To minimise the number of laminated test specimens, both left-side panels were identical although sometimes differing from the two identical right-side panels. For asymmetric laminates, the thinner glass layer was positioned internally, as in the Opera House glass walls. In each test, the applied (internal) suction (reckoned positive, as for an external pressure) was increased until failure occurred in one of the glass panels. When a monolithic panel failed, large glass shards became detached, and the suction immediately reduced to zero. When one or more laminated panels failed, the suction was maintained for a short period to observe the performance of the cracked panels, and then reduced to zero.

Standard dial gauges were affixed to external frame members attached to the sides of the suction box, mostly positioned at mid-span (centre and edge) of each glass panel and close to the vertical edge supports; see Figure A2.2. Deflection readings were taken manually by two individuals at selected load increments (typically eight) up to 958 Pa (20 psf), after which the loading was increased progressively up to failure. As a consequence, and despite some brisk human endeavour during repeated ascents of a pair of step ladders, the duration of each test was approximately half-an-hour, thereby resulting in appreciable sustained loading. The test programme was conducted over a three-week period, during which ambient temperatures in the range 25–33 °C were recorded. This working schedule allowed for the partial or full re-glazing of each glass assembly, including sufficient time for the silicone rubber sealant to achieve an adequate cure.

Clear 'float' glass of basic soda–lime–silica composition was used throughout. The monolithic glass test specimens were manufactured in France by Boussois Souchon Neuvesel (BSN Groupe S.A.), and glass from the same source was used by the French company Société Industrielle Triplex to produce laminated glass test panels with a 'soft' interlayer. In the glass walls, however, a special 6 mm layer of body-tinted polished plate glass was incorporated within the laminated panels produced by the same manufacturers; see section 7.2 (main text). The test panels laminated with a 'hard' interlayer were manufactured by Shatterprufe (Pty) Ltd in Port Elisabeth, South Africa, using glass imported from Pilkington Bros Ltd in England.

FIGURE A2.2. Manual recording of glass panel deflections during full-scale suction-box loading test

The synthetic polymer interlayer comprised a 'pvb' resin (polyvinyl butyral, $(C_8H_{14}O_2)_k$) containing an ester plasticiser (triethylene glycol di-2-ethylbutyrate, $C_{18}H_{34}O_6$), generally manufactured as a flexible sheet of standard thickness (either 0.38 mm or 0.76 mm, although multiple layers can be combined to form thicker glass interlayers), initially opaque but becoming transparent during the lamination process. Two types of interlayer were utilised in the test specimens, containing either 21 or 41 parts by weight of plasticiser per 100 parts of 'pvb' resin, and designated herein as 'hard' and 'soft' interlayers, respectively. Using near-infrared spectroscopy, measurements on corresponding beam specimens revealed an interlayer moisture content of around 0.4%, indicating a high level of adhesion between the glass and the plastic interlayer. At that time, the 'hard' interlayer was deployed widely in the transportation industry, especially in the manufacture of laminated windshields, while the 'soft' interlayer was used for architectural laminated glass. By the late 1960s, this polymer interlayer had been in widespread use for over thirty years, with a proven record of general excellence in various practical applications of glass laminates.

A2.4. Test results

An essential prerequisite in the interpretation of these test results is the selection of material properties, to be incorporated within a retrospective elastic analysis. Based upon results obtained by the writer from preliminary four-point bending tests on small strain-gauged monolithic beams, as illustrated in Figure 35 (main text), the elastic parameters (Young's modulus and Poisson's ratio) for glass were taken as $E_g = 72.4$ GPa, $v_g = 0.22$. In the analysis of glass panels laminated using a 'soft' interlayer, the experimental results summarised in Figure 39 (main text) for beams under short-term loading suggested a shear modulus $G_p = 0.5$ GPa to be apposite, while also recognising that minimal creep deformation during test loading would be anticipated.

The choice of an appropriate interlayer shear modulus is more difficult for the panels with a 'hard' interlayer, because an appreciable reduction in shear stiffness is to be expected during each complete loading cycle. Figure 39 (main text) gives values of shear modulus an order of magnitude higher than those for a 'soft' interlayer, within the temperature range of the tests (e.g. varying from approximately 8 GPa at 25 °C to 3 GPa at 35 °C). Moreover, beam tests under short-term loading conducted at 21 °C indicated that the 'effective' modulus increased substantially when the interlayer thickness was decreased from 0.76 mm to 0.38 mm, possibly due to some form of boundary-layer effect. But any such increase was likely to be more than offset in the present tests by interlayer creep deformation during each full-scale loading cycle. For beams tested at 25 °C, for example, Figure 40 (main text) indicates an increase in central deflection of about 85% during the initial 30-minute loading period for the case of a 0.76 mm thick 'hard' interlayer, in contrast to a corresponding increase of around 5% for a beam with a 'soft' interlayer. Even greater differences would likely be observed at the somewhat higher ambient temperatures recorded during the full-scale tests, although allowance for the effect of incremental loading rather than constant sustained loading then becomes necessary. Accordingly, a nominal shear modulus $G_p = 1$ GPa was used in the retrospective elastic analysis of these laminated glass panels.

In possible applications of laminated glass with a 'hard' interlayer to building structures, the appropriate choice of shear modulus is not entirely clear. For although relatively high values would generally be relevant for a standard 3-second wind gust, the cumulative effect of a prolonged period of stormy weather might reduce such values appreciably, especially in hot climates. Even higher sustained loading would be experienced in transportation windshield applications, for which this type of laminate was originally developed. In practice, therefore, some of the notional benefits of a 'hard' interlayer can become greatly diminished, at least from a structural perspective. With the notation of Figure A2.1(b), the section properties of the full-scale test specimens are given in Table A2.1, where the 'laminate type' refers to the nominal glass and interlayer thicknesses (mm), with labels 'S' and 'H' denoting 'soft' and 'hard' interlayers, respectively.

For calculation purposes, with $v_g = 0.22$ and $b/a = 0.7$, the deflections and bending moments for the monolithic glass test panels are based on equations A2.1–A2.6, together with the computed non-dimensional influence factors listed in Tables A3.1 and A3.2 in

TABLE A2.1. Section properties of specimens used in full-scale testing of rectangular panels of monolithic glass ($E_g = 72.4$ GPa, $v_g = 0.22$) and laminated glass (shear modulus G_p), where S and H refer to 'soft' and 'hard' interlayers, respectively

Test	t_1: mm	c: mm	t_2: mm	T: mm	G_p: MPa	Nominal type
1	–	–	–	12.45	–	Monolithic
2	–	–	–	12.45	–	Monolithic
3	6.15	0.76	9.98	16.89	0.5	6/0.76S/10
4	6.50	0.76	9.98	17.24	0.5	6/0.76S/10
5	6.07	0.38	9.75	16.20	1.0	6/0.38H/10
6	6.07	0.38	8.08	14.53	1.0	6/0.38H/8
7	6.07	0.38	8.08	14.53	1.0	6/0.38H/8
8	6.10	0.38	6.10	12.58	1.0	6/0.38H/6
9	5.87	0.38	5.87	12.12	1.0	6/0.38H/6

appendix A3. Whence the mid-span deflections at the centre and edge of a panel are given by

$$w(a/2,0) = 0.0131 \frac{pa^4}{D}, \quad w(a/2,b/2) = 0.0141 \frac{pa^4}{D} \quad \text{(A2.13)}$$

while the corresponding bending moments are

$$M_x(a/2,0) = 0.1233 pa^2, \quad M_x(a/2,b/2) = 0.1287 pa^2 \quad \text{(A2.14)}$$

The maximum tensile stress at the two locations is given by

$$\sigma_x(a/2,0) = -\frac{6}{T^2} M_x(a/2,0), \quad \sigma_x(a/2,b/2) = -\frac{6}{T^2} M_x(a/2,b/2) \quad \text{(A2.15)}$$

For the laminated panels, in the absence of an exact theoretical solution, calculations are based on equations A2.8–A2.12 for a simply supported beam under uniform loading. In view of the close relationship between the calculated elastic flexure of a monolithic plate and a 'beam-strip' under quasi-static transverse loading, it is not unreasonable to assume a broadly similar parity for the corresponding laminated structures. The additional complication implicit in an exact theoretical solution would be appreciable – see, for example, an analysis of the less demanding case [A2.1] of a rectangular laminated glass panel simply supported on all four edges – and is perhaps difficult to justify in this application, especially in view of the errors and uncertainties associated with some of the material input parameters.

Table A2.2 summarises the measured deflections at the centre of a panel $w(a/2,0)$ and at the centre of an unsupported ('free') edge $w(a/2,b/2)$, as indicated in Figure A2.1(a) for an individual glass panel within the test assembly. Comparative calculated deflections are also listed, together with measured ambient test temperatures. These results correspond to

TABLE A2.2. Measured and calculated mid-span deflections (w) at centre and edge of rectangular panels of monolithic and laminated glass (total thickness T) simply supported on two opposite edges (span $a = 2.266$ m, length $b \approx 1.5$ m) and subjected to uniform (internal) suction (958 Pa) at given ambient temperature

Test	Measured: mm		Calculated: mm		$w(a/2, 0)/T$	Temp: °C	Theory
	$w(a/2, 0)$	$w(a/2, b/2)$	$w(a/2, 0)$	$w(a/2, b/2)$			
1	26.8	28.1	27.1	29.1	2.2	30	Plate
			28.3	–	2.3		Beam
2	27.2	28.1	27.1	29.1	2.2	30	Plate
			28.3	–	2.3		Beam
3	16.9	17.9	17.0	–	1.0	32	Beam
4	16.5	17.8	16.2	–	0.9	30	Beam
5	14.3	–	14.6	–	0.9	33	Beam
6	20.1	–	20.1	–	1.4	30	Beam
7	19.1	–	20.1	–	1.4	25	Beam
8	29.8	–	30.6	–	2.4	31	Beam
9	32.1	–	34.1	–	2.8	27	Beam

an external applied pressure of 958 Pa (20 psf), while noting that a design wind loading of ±1436 Pa (±30 psf) was adopted for the glass walls, in accordance with prevailing engineering practice. All measured load–deflection curves were sensibly linear. However, by the time the tests were conducted, it appeared unlikely from design considerations that laminated glass with a 'hard' interlayer would be specified for the glass walls, and fewer deflections were recorded for these test panels. As a guide to experimental repeatability, the measured central deflection of an intact Test 4 glass panel was 16.5 ± 0.5 mm during Tests 4–9, conducted in the ambient temperature range 25–33 °C.

Here there is some merit in comparing measured deflections with those calculated using simple elastic theory, thereby enabling the essential features of the structural response to remain unfettered by complicating factors of questionable validity. For the monolithic glass panels, where the material properties are linearly elastic, transverse shear displacements could be taken into account, and recourse also made to some form of large-deflection theory. Regarding the former, the calculated increase in deflection would be negligible, since $T/a < 0.01$. Regarding the latter, a collateral nonlinear numerical analysis would depend markedly on the assumed degree of in-plane restraint, which is likely to be of low magnitude in the present application. Indeed, the provision of such minimal edge restraint is one of the guiding principles in most forms of architectural glazing, not least to allow in-plane thermal displacements to take place unhindered. Accordingly, although Table A2.1 gives maximum observed deflections that are approximately double the panel thickness, the corresponding transverse slopes along the edge supports are small ($\approx 2.3°$), and calculations based on small-deflection theory provide a good first approximation.

Likewise, it is possible nowadays to carry out sophisticated numerical analysis of laminated panels under quasi-static transverse loading (*e.g.* a nonlinear viscoelastic analysis of a laminated plate on deformable edge supports, incorporating large-deflection theory and allowing for direct and shear strains normal to the panel surface), even including the post-failure response of cracked panels. Altogether more formidable would be a corresponding dynamic analysis, especially to model the complex interaction between adjacent panels following initial glass fracture. An impressive array of advanced software is currently available, and the latent opportunities for computational gymnastics are endless. But any such numerical results would be of limited value without recourse to a corresponding set of high-quality experimental data for the full-scale tests, including near-instantaneous measurements of deflection and copious strain-gauge records. A further impediment to such advanced analysis being of tangible benefit would rest upon adequately representing physical properties and composite material behaviour. In particular, extensive laboratory testing would initially be required to establish a realistic constitutive model for the thermoplastic interlayer in both shear and tensile modes, leading to empirical relationships dependent upon load intensity, elapsed time and ambient temperature.

Figure A2.3 depicts the calculated distribution of cross-sectional bending stress (tension negative) at the centre of a panel for each test under a uniform (internal) suction of 958 Pa,

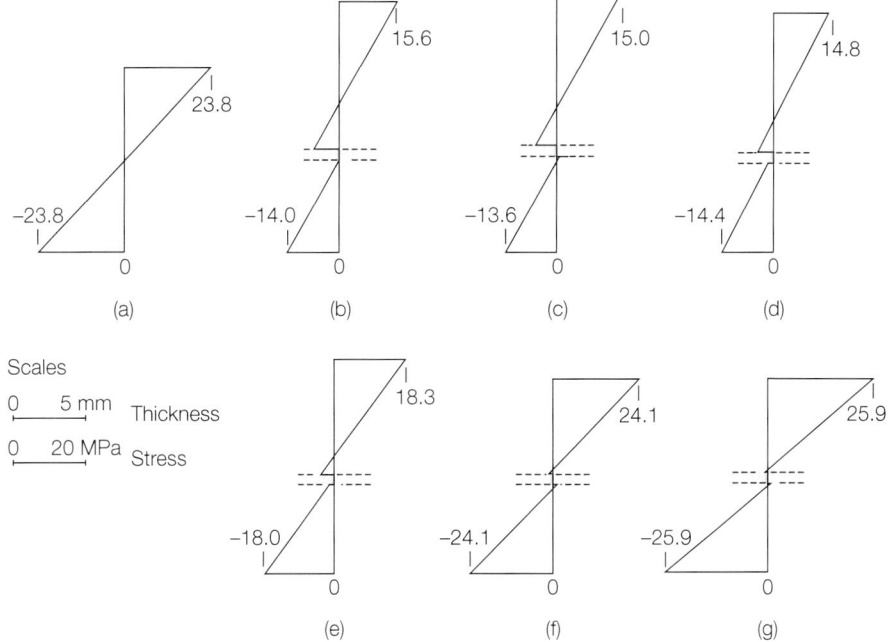

FIGURE A2.3. Calculated distribution of bending stress (tension negative) through specimen thickness at centre of rectangular panels (nominal 2.3 × 1.5 m) of monolithic and laminated glass (total thickness T) simply supported on two opposite edges (span $a = 2.266$ m) and subjected to uniform (internal) suction (958 Pa), based on beam theory: (a) Tests 1,2; (b) Test 3; (c) Test 4; (d) Test 5; (e) Tests 6,7; (f) Test 8; (g) Test 9

based on simple elastic beam theory. Here it is noted that for asymmetric laminates, the maximum tensile stress is somewhat reduced if the thinner glass layer is positioned on the 'tension face'. For the monolithic panels deployed in Tests 1 and 2, the maximum bending stresses calculated from classical elastic plate theory are ±23.5 MPa at the plate centre, and ±24.5 MPa at the centre of a 'free' edge.

Upon recalling the presence of almost horizontal laminated panels in the lower cone areas of the north-facing glass walls on the opera and concert hall buildings, it is of interest to demonstrate an additional benefit of positioning the thicker glass layer uppermost. Consider, for instance, a horizontal laminated glass beam simply supported at both ends, subjected to uniform vertical loading (downward positive) over the entire span ($a = 2.266$ m). The long-term response to permanent self-weight loading (395 Pa) is calculated using equation (A2.12), as the interlayer shear modulus can be taken as zero. Then the combined stress resultants generated by an additional loading representing short-term positive wind pressure or negative suction (±958 Pa) can be determined by superposition using equations (A2.10) and (A2.11). The results are summarised in Figure A2.4, based on the Test 3 data listed in Table A2.1. Of particular note is that the maximum combined tensile stress of −28.0 MPa in Figure A2.4(a), with the thinner layer uppermost, well exceeds the corresponding value of −21.6 MPa in Figure A2.4(c), where the thicker layer is uppermost.

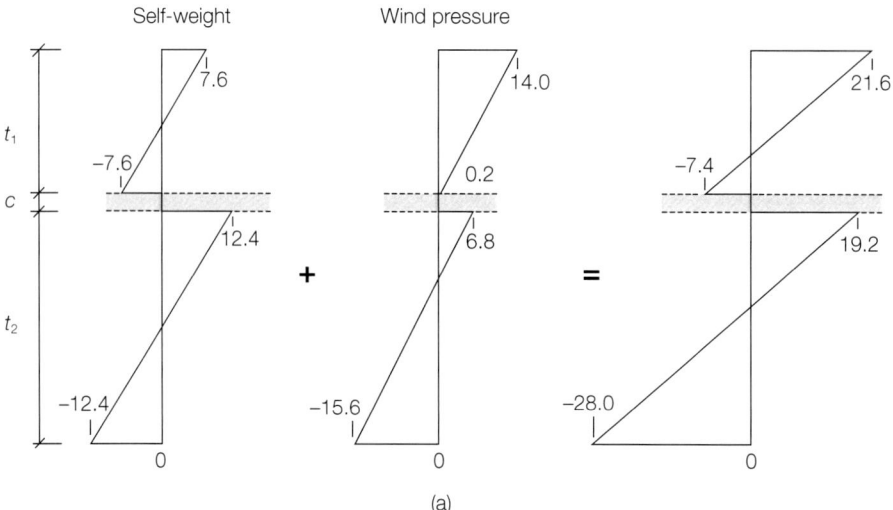

(a)

FIGURE A2.4. Effect of glass layer orientation on calculated distribution of bending stress (MPa, tension negative) at mid-span of asymmetric laminated beam simply supported horizontally at both ends, subjected to combined self-weight loading (395 Pa) and uniform wind pressure or suction (±958 Pa), based on beam theory and Test 3 data: (a) thinner layer uppermost, pressure

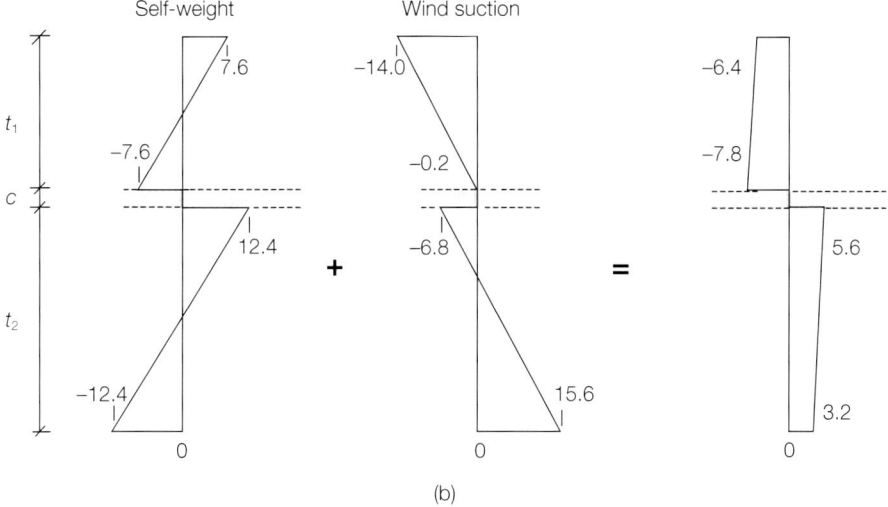

(b)

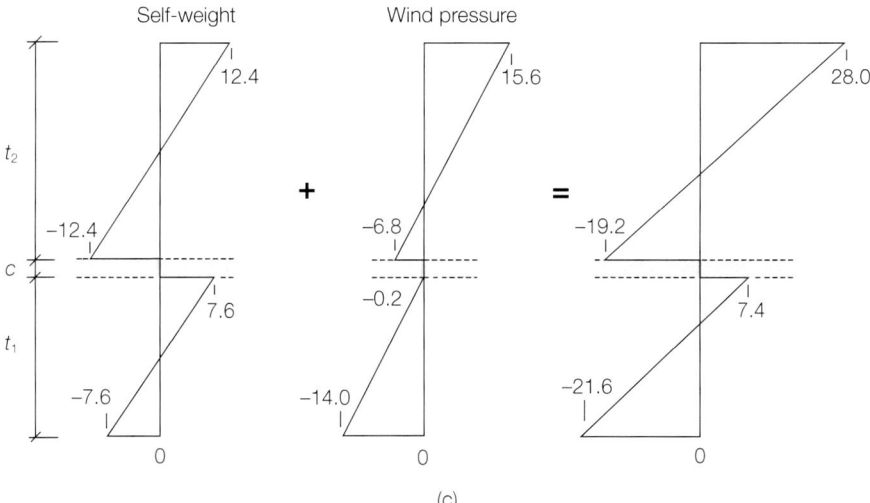

(c)

FIGURE A2.4. (continued) Effect of glass layer orientation on calculated distribution of bending stress (MPa, tension negative) at mid-span of asymmetric laminated beam simply supported horizontally at both ends, subjected to combined self-weight loading (395 Pa) and uniform wind pressure or suction (±958 Pa), based on beam theory and Test 3 data: (b) thinner layer uppermost, suction; (c) thicker layer uppermost, pressure

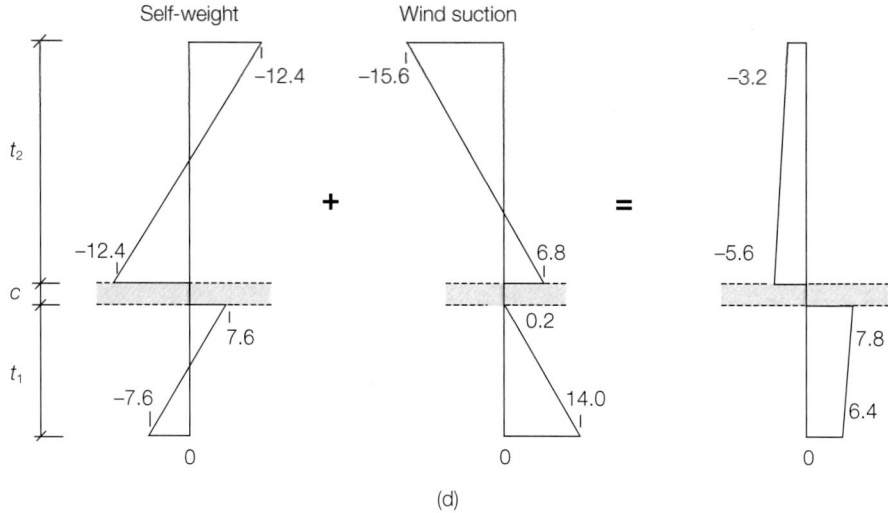

(d)

FIGURE A2.4. (continued) Effect of glass layer orientation on calculated distribution of bending stress (MPa, tension negative) at mid-span of asymmetric laminated beam simply supported horizontally at both ends, subjected to combined self-weight loading (395 Pa) and uniform wind pressure or suction (±958 Pa), based on beam theory and Test 3 data: (d) thicker layer uppermost, suction

Post-failure photographs taken during the first two tests on monolithic glass panels of the same thickness are shown in Figure A2.5. In both cases, fracture initiated at some distance from mid-span, and only a *single* panel failed. In Test 1, fracture initiated at or near the top edge of the upper right panel (when viewed externally, facing the glazed area), and the crack patterns are typical of those encountered in high-energy glass fracture. It is reasonable to postulate that the initial crack propagated normal to the glass edge before the onset of bifurcation, rapidly accelerating to a terminal speed of approximately 1500 m/s and thereby traversing the length of the panel in around 0.001s. Uniquely in this test series, the fracture pattern in the lower right panel in Test 2 has the characteristics of a low-energy fracture, probably initiating at a surface flaw located well away from a 'free' edge.

By contrast, in the subsequent tests on laminated glass, failure *always* occurred almost simultaneously in *two* vertically adjacent panels, in each case with crack patterns exhibiting the multiple bifurcation associated with rapid energy dissipation. In view of the correspondingly high crack speeds within these laminated panels (*i.e.* comparable to those in Test 1), the panel failure sequence was not discernible to the naked eye. The photographs in Figure A2.6 illustrate the failure mode of similar specimens of laminated glass with a 'soft' interlayer (6/0.76S/10), while noting both here and subsequently that, as a consequence of directional lighting, projected shadows of crack patterns appear on the rear face of the suction box. In Test 3, failure occurred in the two right-side panels, and for Test 4 in the opposite pair. In both tests, fracture initiated at *non-adjacent* 'free' edges.

FIGURE A2.5. Fracture patterns in 12 mm thick monolithic glass panels: (a) Test 1; (b) Test 2

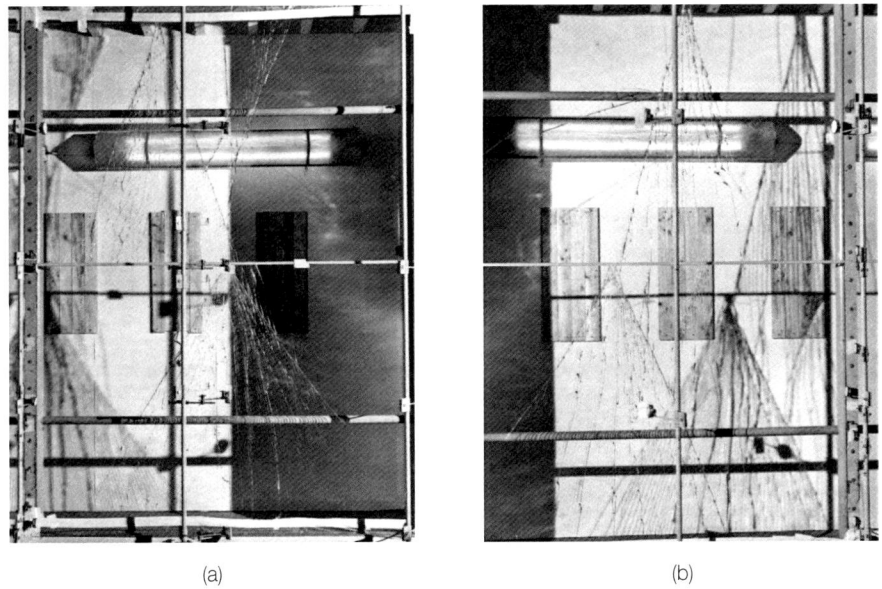

FIGURE A2.6. Fracture patterns in laminated glass panels with 'soft' interlayer (6/0.76S/10): (a) Test 3; (b) Test 4

Figures A2.7–A2.12 relate to glass laminates with a 'hard' interlayer. Moreover, only two such identical panels were deployed in each test, located on the left-side of the suction box; the right-side pair being those from Test 4, which fortuitously remained intact during the entire test series. Figure A2.7 highlights the extensive crack patterns for the thickest

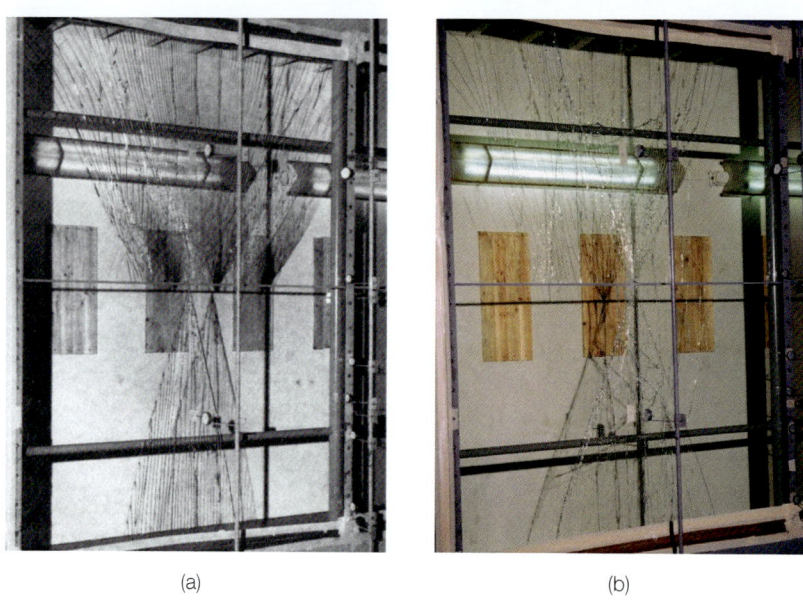

FIGURE A2.7. Fracture patterns in laminated glass panels with 'hard' interlayer (6/0.38H/10): Test 5

FIGURE A2.8. Fracture patterns in laminated glass panels with 'hard' interlayer (6/0.38H/8): Test 6

FULL-SCALE LOADING TESTS ON GLASS PANELS

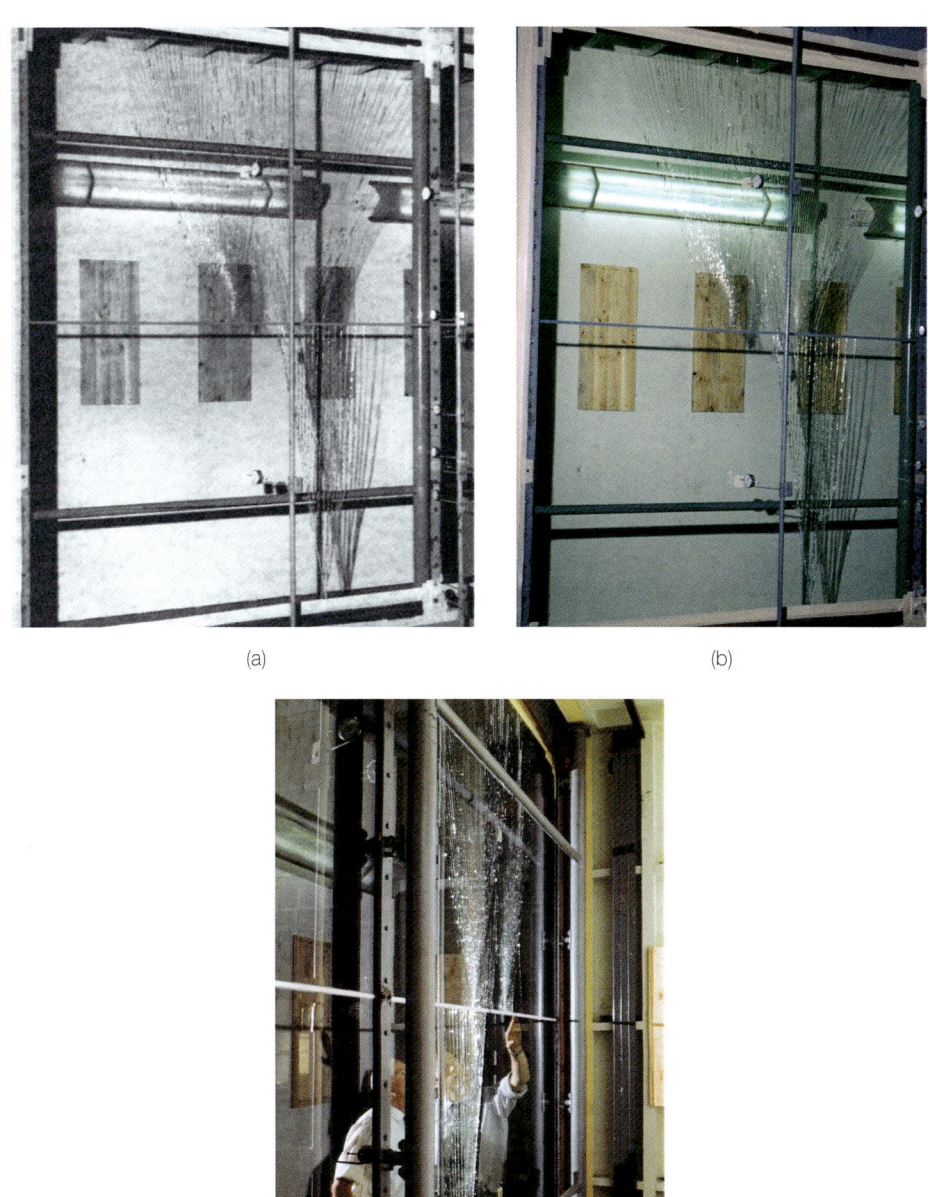

FIGURE A2.9. Fracture patterns in laminated glass panels with 'hard' interlayer (6/0.38H/8): Test 7

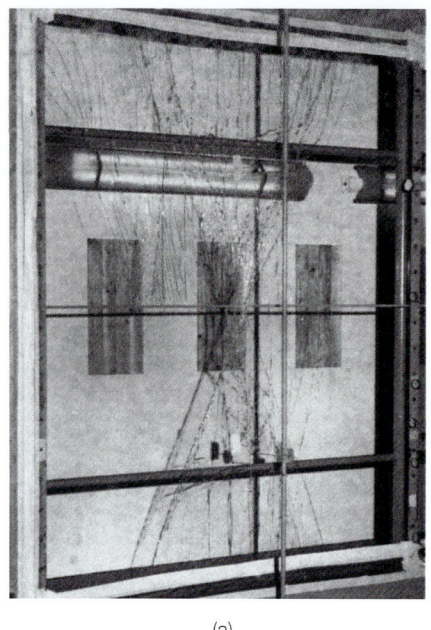 

(a)   (b)

FIGURE A2.10. Fracture patterns in laminated glass panels with 'hard' interlayer (6/0.38H/6): Test 8

such laminate (6/0.38H/10) in Test 5, where failure in the two left-side panels initiated at *adjacent* 'free' edges. Identical laminated panels (6/0.38H/8) were deployed in Tests 6 and 7; see Figures A2.8 and A2.9. However, fracture initiated at *adjacent* 'free' edges in Test 6, and at *non-adjacent* 'free' edges in Test 7. The same observation applies to Tests 8 and 9, respectively, on similar panels of the thinnest laminate (6/0.38H/6); see Figures A2.10 and A2.11. The deformed shape of the pair of failed panels in Test 8, shortly after the removal of load, is illustrated in Figure A2.12. However, in this and all other tests on laminated panels, there was very little glass debris within the suction box upon completion, and the fractured panels remained airtight during the entire loading cycle.

To supplement the still photography, segments of Tests 4 and 8 on laminated glass panels were recorded on standard 16 mm colour cine film at a nominal 24 frames per second. The ambient lighting was far from ideal, as was the quality of the film image, but Figure A2.13 depicts sequential frames taken during Test 4 (6/0.76S/10). In Figure A2.13(a), failure had just occurred in the upper left panel, initiating at or close to the top edge. In the following frame, taken some 0.04 s later and shown in Figure A2.13(b), the adjacent lower left panel also had failed. Here the cracking also initiated at the top edge, probably reflecting the localised transfer of load from upper to lower panel, with the applied suction unchanged. Two similar cine frames were obtained for Test 8 (6/0.38H/6), with initial fracture occurring at the top edge of the lower left panel, followed by cracks propagating from the bottom edge of the upper left panel. However, these frames are omitted herein as the quality of the printed images is inadequate.

FULL-SCALE LOADING TESTS ON GLASS PANELS 167

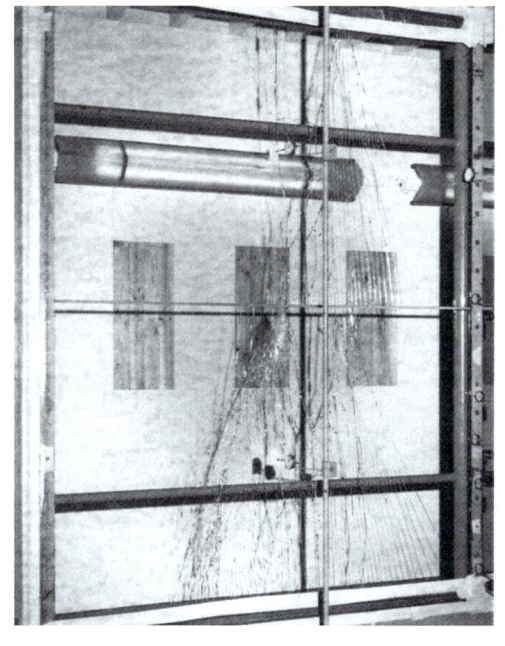

(a)

(b)

(c)

FIGURE A2.11. Fracture patterns in laminated glass panels with 'hard' interlayer (6/0.38H/6): Test 9

FIGURE A2.12. Deformed shape of failed pair of laminated glass panels with 'hard' interlayer (6/0.38H/6): Test 8

(a)   (b)

FIGURE A2.13. Sequential cine frames from Test 4 demonstrating cascade failure of laminated glass panels with 'soft' interlayer (6/0.76S/10): (a) initial failure of upper left panel; (b) subsequent instantaneous failure of lower left panel

These latter observations indicated the daunting possibility of *cascade failure* occurring on the main glass walls, where several vertical or inclined laminated glass panels are confined between a given pair of continuous edge supports. For in the absence of horizontal glazing bars – an essential feature of the innovative glazing system – adjacent glass panels are directly linked structurally. The detailed failure mechanism in such a series of coupled glass panels is doubtless complex, but would encompass a rapid partial load transfer from the first cracked panel to a much stiffer adjoining intact panel, together with additional dynamic components in the form of stress wave resultants generated during fast fracture. Fortunately, any occurrence of this onerous form of progressive failure was deemed unlikely under service conditions in view of the considerable intrinsic redundancy inherent in the structural design of the glass panels. For practical purposes, the laminate thickness was specified to be constant throughout (except for the inclined 'view windows'), and because the mullion spacing changes with elevation on the north-facing glass walls, the design was based on the panel having the longest span.

Acknowledgement

The writer is pleased to recall the cheerful and proficient assistance afforded by the resident staff at the Commonwealth Experimental Building Station (C.E.B.S.), Sydney, in conducting this short but exigent experimental programme utilising their newly established test facilities. All half-tone photographs in this appendix are by courtesy of C.E.B.S., as are one colour photograph, Figure A2.9(c), and two cine frames in Figure A2.13; the remaining colour images are by the writer.

A2.5. Reference

A2.1.   HEMSLEY, J. A. *Glass in engineering science.* Society of Glass Technology, Sheffield, 2016, Vol. 2 (*Glass under load*), App. 2 (*Elastic flexure of rectangular laminated glass panel*), 175–202.

A3. ELASTIC FLEXURE OF UNIFORMLY LOADED THIN MONOLITHIC RECTANGULAR PLATE SIMPLY SUPPORTED ON OPPOSITE EDGES

A3.1. Introduction

With particular regard to the large-scale loading tests on glass panels described briefly in the main text and in further detail in appendix 2, and also more generally in cognisance of historical developments in applied mechanics, it is instructive to re-examine the theoretical flexure of a thin monolithic elastic plate under loading and boundary conditions approximating to those prevailing in the experimental work. The resulting exact solution for a uniformly loaded rectangular plate supported only on two opposite edges may be considered as an essential prerequisite in correlating theoretical and observed behaviour, as well as providing benchmark numerical values for use in wider studies of plate flexure. Even under the simplifying assumptions associated with classical plate theory, a comprehensive solution to this relatively straightforward elastostatic problem has many practical applications, not least in the design of the Sydney Opera House glass walls.

In the present context, these applications include the theoretical limiting cases of an architectural laminate in which the interlayer properties are characterised either by zero shear stiffness, or by effectively approaching those of the adjacent glass. Other limiting cases occur where the opposite edges are simply supported by beams or frame members of infinite stiffness, or where the 'free' edges are supported by beams of vanishingly small bending and torsional stiffness. Copious graphical and tabulated results in a non-dimensional format are presented herein for general reference purposes. These results are rather more detailed than those hitherto available in the technical literature, and several errors in published solutions are rectified.

A3.2. General solution

Consider an initially flat rectangular plate (width a) simply supported along two opposite edges (length b), with the other two edges entirely free of restraint; see Figure A3.1 where, for convenience in the following summary, the plate is taken to be located in the horizontal

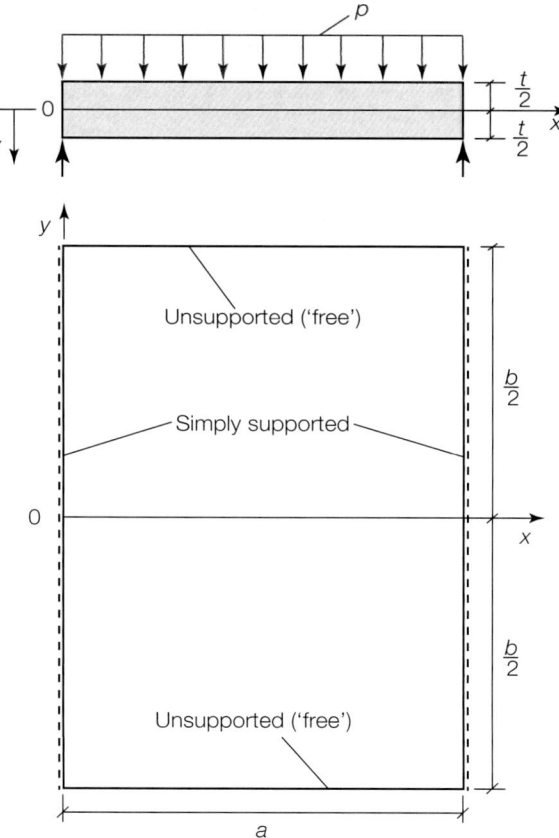

FIGURE A3.1. General notation in flexural analysis of thin rectangular plate (width a, length b, thickness t, flexural stiffness D) of elastic material (Young's modulus E, Poisson's ratio v) simply supported on two opposite edges ($x = 0, a$) with two edges 'free', subjected to uniform normal applied loading (p)

plane. The continuous edge supports are taken to be rigid, and to remain in complete contact with the plate throughout the loading period (*i.e.* no partial separation at the contact interface). Normal displacements relative to the middle surface are neglected, as are in-plane displacements within the middle surface, thereby giving zero membrane forces. The plate is assumed to be homogeneous, isotropic and linearly elastic (Young's modulus E, Poisson's ratio v) with a uniform thickness (t) that is small compared to its other dimensions. In the general case, a distributed static vertical loading $p(x, y)$ per unit area is applied downward to the upper plate surface ($z = t/2$), while body forces are neglected.

Then according to classical small-deflection theory for thin monolithic plates, the downward (positive) vertical deflection (w) at any point (x, y) on the middle plane ($z = 0$) is governed by the nonhomogeneous biharmonic equation

ELASTIC FLEXURE OF MONOLITHIC RECTANGULAR PLATE

$$\nabla^4 w = \nabla^2(\nabla^2 w) = \left(\frac{\partial^2}{\partial x^2} + \frac{\partial^2}{\partial y^2}\right)\left(\frac{\partial^2 w}{\partial x^2} + \frac{\partial^2 w}{\partial y^2}\right)$$
$$= \left(\frac{\partial^4 w}{\partial x^4} + 2\frac{\partial^4 w}{\partial x^2 \partial y^2} + \frac{\partial^4 w}{\partial y^4}\right) = \frac{p(x,y)}{D} \tag{A3.1}$$

where ∇^2 denotes the two-dimensional Laplacian operator in Cartesian co-ordinates, and with the flexural stiffness of the plate D defined as

$$D = \frac{Et^3}{12(1-v^2)} \tag{A3.2}$$

According to Todhunter and Pearson [A3.1] in their majestic historical treatise, this equation first appeared – without a derivation – as long ago as 1811, having been discovered posthumously among the papers of the Italian-born mathematician and astronomer Joseph Lagrange (1736–1813). It was contained in a personal note [A3.2] written when examining and correcting a prize competition memoir submitted to the French Academy of Science in 1811 by Sophie Germain (1776–1831) on the broader subject of plate vibration, and in particular to explain or predict the experimental nodal patterns formed on resonating plates lightly covered with fine sand, originally observed by the German physicist Ernst Chladni (1756–1827). She was the sole competitor yet initially unsuccessful. Following two further attempts in 1813 and 1815 during what became an extended scientific contest, she finally was awarded the prize in 1816, but declined to attend the award ceremony; a decision that doubtless stemmed from the prevailing ambience of endemic prejudice and controversy during her protracted quest for formal recognition. The resulting memoir was published independently [A3.3] in 1821, followed by lengthy corollaries [A3.4], [A3.5] some years later.

Around this time, various theoretical aspects of the work were discussed in a series of published correspondence (*Annales de Chimie et de Physique, Paris*) by the celebrated French elasticians Siméon Poisson (1781–1840) and Claude Navier (1785–1836), the latter quoting part of Lagrange's original note in one of the letters [A3.6]. Several more general commentaries and reviews have appeared in recent years, including an entire volume by Bucciarelli and Dworsky [A3.7] devoted to this prolonged and intriguing melodrama; one with distinct parallels encountered much later during the unfolding gestation period of the Sydney Opera House.

Solutions to equation (A3.1) are simplified considerably if, as in the present case, two opposite edges are simply supported and the vertical applied loading is invariant along any line parallel to the *y*-axis. The four constants of integration are determined from the boundary conditions in the usual manner, but two of them are identically zero because the deflection and transverse bending moment vanish at a simply supported edge. The two remaining constants are determined from the conditions on an unsupported ('free') edge, where the longitudinal moment and edge reaction are zero.

Acting promptly upon a suggested general solution to equation (A3.1) proposed by Lévy [A3.8] in 1899, in which the much earlier double Fourier series method developed by Navier was replaced by a single series approach, several plate bending problems were examined by Estanave [A3.9], including the current example. These analyses were based on the somewhat restrictive condition that $\lambda = \mu = 1$, where the familiar Lamé elastic constants are defined as

$$\lambda = \frac{\nu E}{(1+\nu)(1-2\nu)}, \qquad \mu = G = \frac{E}{2(1+\nu)} \tag{A3.3}$$

with G also denoting the shear modulus. Upon recalling the standard relation

$$\nu = \frac{\lambda}{2(\lambda + \mu)} \tag{A3.4}$$

it follows that all results obtained by Estanave [A3.9] relate specifically to $\nu = 1/4$. However, this value is positioned mid-way within the entire positive range of Poisson's ratio, and is adequately representative of many engineering materials.

The applied loading is expressed in terms of the single Fourier series

$$p(x) = \sum_{n=1}^{\infty} F_n \sin(\alpha_n x) \tag{A3.5}$$

where n denotes a positive *odd* integer, and $\alpha_n = n\pi/a$. The Fourier coefficients are given by

$$F_n = \frac{2}{a} \int_0^a p(x) \sin(\alpha_n x) dx \tag{A3.6}$$

which leads to a simple expression for most practical distributions of applied loading. Indeed, it is usually possible to express the summation term in closed form, with verification obtained merely by recalling the corresponding standard expression for a simply supported beam, or more strictly for an infinitely long plate.

In the analysis described by Estanave [A3.9], which is confined to the determination of deflections, the co-ordinate origin is taken at the corner of a plate occupying the first quadrant, in contrast to that shown in Figure A3.1, which configuration is often advantageous in general analysis. However, with changes in notation and some simplification, Estanave's solution can be written as

$$w(x,y) = \frac{5}{16EI} \sum_{n=1}^{\infty} \frac{F_n}{\alpha_n^4} \left[3 + \frac{E_n}{13\sinh(\alpha_n b) - 3\alpha_n b} \right] \sin(\alpha_n x) \tag{A3.7}$$

in the present co-ordinate system, where $I = t^3/12$ denotes the second moment of area per unit length, and where

$$\begin{aligned} E_n = &\; 2\sinh(\alpha_n b/2)\{5\cosh(\alpha_n y) + 3\alpha_n[(b/2) + y]\sinh(\alpha_n y)\} \\ & - 3\alpha_n b \cosh\{\alpha_n[(b/2) + y]\} \end{aligned} \tag{A3.8}$$

Following further studies summarised by Nádai [A3.10], [A3.11], Marcus [A3.12], Galerkin [A3.13] and others, a more detailed analysis of the present class of problems was described by Holl [A3.14] in 1936 for plates with an arbitrary Poisson's ratio. Again, with changes in notation, the shortened general solution to equation (A3.1) takes the form

$$w(x,y) = \frac{1}{D}\sum_{n=1}^{\infty}\frac{1}{\alpha_n^4}[F_n + A_n^*\cosh(\alpha_n y) + B_n^*\alpha_n y\sinh(\alpha_n y)]\sin(\alpha_n x) \quad (A3.9)$$

Here it is convenient to write

$$A_n^* = \frac{\nu F_n A_n}{(1-\nu)\Delta_n}, \quad B_n^* = \frac{\nu F_n}{\Delta_n} \quad (A3.10)$$

where the coefficients

$$A_n = 1 + \nu - (1-\nu)(\alpha_n b/2)\coth(\alpha_n b/2) \quad (A3.11)$$

$$\Delta_n = (3+\nu)\cosh(\alpha_n b/2) - (1-\nu)(\alpha_n b/2)\operatorname{cosech}(\alpha_n b/2) \quad (A3.12)$$

depend solely on the plate aspect ratio (b/a) and Poisson's ratio. Upon re-deriving this solution, it is evident that the equivalent expressions for A_n^* and B_n^* given by Holl [A3.14] are both in error by a factor of four. However, with two notable exceptions (see below), it appears that the related numerical calculations were based on the correct formulae, albeit using only three terms ($n = 1, 3, 5$) in evaluating the various infinite series.

A3.3. Uniformly loaded plate

In the case of a uniform vertical pressure distributed over the entire upper surface of the plate,

$$p(x) = p, \quad F_n = \frac{4p}{n\pi} \quad (A3.13)$$

while elementary considerations of the theory of Fourier series give the required summation

$$\sum_{n=1}^{\infty}\frac{1}{n^5}\sin(n\pi x/a) = \frac{\pi^5 x}{96a^4}(x^3 - 2ax^2 + a^3) \quad 0 \le x/a \le 1 \quad (A3.14)$$

Similar series summations used for the remaining quantities of flexure can readily be obtained by differentiation.

With the loading terms specified by equations (A3.13) substituted in Estanave's restricted solution ($\nu = 1/4$) given by equation (A3.7), the deflection at the plate centre becomes

$$w(a/2,0) = \frac{5pa^4}{4EI\pi^5}\sum_{n=1}^{\infty}\frac{(-1)^{(n-1)/2}}{n^5}\left[3 + \frac{10\sinh(\alpha_n b/2) - 3\alpha_n b\cosh(\alpha_n b/2)}{13\sinh(\alpha_n b) - 3\alpha_n b}\right] \quad (A3.15)$$

and at the centre of a 'free' edge,

$$w(a/2,b/2) = \frac{5pa^4}{EI\pi^5}\sum_{n=1}^{\infty}\frac{(-1)^{(n-1)/2}}{n^5}\left[\frac{11\sinh(\alpha_n b) - 3\alpha_n b}{13\sinh(\alpha_n b) - 3\alpha_n b}\right] \quad (A3.16)$$

In Holl's unrestricted solution ($0 \leq v \leq 1/2$) represented by equation (A3.9), the full-field plate deflections are given by

$$w(x,y) = \frac{px}{24D}(x^3 - 2ax^2 + a^3) + \frac{4v\,pa^4}{\pi^5 D(1-v)} \sum_{n=1}^{\infty} \frac{1}{n^5 \Delta_n}[A_n \cosh(\alpha_n y) \\ + (1-v)\alpha_n y \sinh(\alpha_n y)]\sin(\alpha_n x) \qquad (A3.17)$$

which represents the anticlastic middle surface of the deformed plate. In this last expression, the first full term corresponds to the cylindrical deflection profile of an infinitely long plate, upon recognising that because all y-derivatives then become zero, equation (A3.1) at once reduces to the ordinary fourth-order differential equation encountered in elementary beam theory, with D replaced by EI. At the opposite limiting case of a notional 'one-dimensional beam strip', $v = 0$ and the deflections are those for a transverse beam with flexural stiffness $D = EI$. The same general 'first term' principle also applies to the other quantities of plate flexure given below.

These theoretical solutions are exact within the bounds of classical elasticity, yet despite the inherent simplicity of the resulting equations of flexure, a modicum of care is required to obtain accurate numerical results when evaluating the Fourier series. This is because very large numbers occur for hyperbolic sine and cosine functions of large argument, possibly resulting in computational overflow. Some form of normalisation is required, and a simple remedy in the present application is to divide the numerator and denominator of the series expression in equation (A3.17) by $\cosh(\alpha_n b/2)$, thereby giving

$$w(x,y) = \frac{px}{24D}(x^3 - 2ax^2 + a^3) + \frac{4v\,pa^4}{\pi^5 D(1-v)} \sum_{n=1}^{\infty} \frac{1}{n^5 \Delta_n^+}[A_n \mathbb{C}(\alpha_n y) \\ + (1-v)\alpha_n y \mathbb{S}(\alpha_n y)]\sin(\alpha_n x) \qquad (A3.18)$$

where

$$\left.\begin{array}{c}\mathbb{S}(\alpha_n y) \\ \mathbb{C}(\alpha_n y)\end{array}\right\} = \frac{\exp[-\alpha_n(b-2y)/2] \mp \exp[-\alpha_n(b+2y)/2]}{1 + \exp(-\alpha_n b)} \qquad (A3.19)$$

$$\Delta_n^+ = (3+v) - (1-v)\alpha_n b \operatorname{cosech}(\alpha_n b) \qquad (A3.20)$$

In other words, the functions $\sinh(\alpha_n y)$ and $\cosh(\alpha_n y)$ in equation (A3.17) are replaced by the bespoke exponential functions $\mathbb{S}(\alpha_n y)$ and $\mathbb{C}(\alpha_n y)$, respectively, while the denominator coefficient Δ_n is replaced by Δ_n^+. This procedure is followed in deriving all subsequent full-field equations of plate flexure.

At the plate centre,

$$w(a/2, 0) = \frac{5pa^4}{384D} + \frac{4v\,pa^4}{\pi^5 D(1-v)} \sum_{n=1}^{\infty} \frac{(-1)^{(n-1)/2} A_n^+}{n^5 \Delta_n^+} \qquad (A3.21)$$

while at the centre of a 'free' edge,

$$w(a/2,b/2) = \frac{5pa^4}{384D(1-v)} - \frac{8v\, pa^4}{\pi^5 D(1-v)} \sum_{n=1}^{\infty} \frac{(-1)^{(n-1)/2}}{n^5 \Delta_n^+} \quad (A3.22)$$

where the modified coefficient in equation (A3.21) is

$$A_n^+ = (1+v)\operatorname{sech}(\alpha_n b/2) - (1-v)(\alpha_n b/2)\operatorname{cosech}(\alpha_n b/2) \quad (A3.23)$$

It is encouraging to note that if $v = 1/4$, equations (A3.21) and (A3.22) become identical to equations (A3.15) and (A3.16), respectively. However, the expression equivalent to equation (A3.22) given by Holl [A3.14] appears to be incorrect.

The plate slopes are given by

$$\phi_x = \frac{\partial w}{\partial x}, \quad \phi_y = \frac{\partial w}{\partial y} \quad (A3.24)$$

and are reckoned positive in cases where the downward deflection increases with increasing positive values of x or y. Whence

$$\phi_x(x,y) = \frac{p}{24D}(4x^3 - 6ax^2 + a^3) + \frac{4v\, pa^3}{\pi^4 D(1-v)} \sum_{n=1}^{\infty} \frac{1}{n^4 \Delta_n^+} [A_n \mathbb{C}(\alpha_n y) \\ + (1-v)\alpha_n y \mathbb{S}(\alpha_n y)]\cos(\alpha_n x) \quad (A3.25)$$

$$\phi_y(x,y) = \frac{4v\, pa^3}{\pi^4 D(1-v)} \sum_{n=1}^{\infty} \frac{1}{n^4 \Delta_n^+} [(A_n + 1-v)\mathbb{S}(\alpha_n y) \\ + (1-v)\alpha_n y \mathbb{C}(\alpha_n y)]\sin(\alpha_n x) \quad (A3.26)$$

Along a supported edge,

$$\phi_x(0,0) = \frac{pa^3}{24D} + \frac{4v\, pa^3}{\pi^4 D(1-v)} \sum_{n=1}^{\infty} \frac{A_n^+}{n^4 \Delta_n^+} \quad (A3.27)$$

$$\phi_x(0,b/2) = \frac{pa^3}{24D(1-v)} - \frac{8v\, pa^3}{\pi^4 D(1-v)} \sum_{n=1}^{\infty} \frac{1}{n^4 \Delta_n^+} \quad (A3.28)$$

while at mid-span of a 'free' edge,

$$\phi_y(a/2,b/2) = \frac{8v\, pa^3}{\pi^4 D(1-v)} \sum_{n=1}^{\infty} \frac{(-1)^{(n-1)/2}}{n^4 \Delta_n^+} \tanh(\alpha_n b/2) \quad (A3.29)$$

Plate bending moments per unit length are obtained using the standard relations

$$M_x = -D\left(\frac{\partial^2 w}{\partial x^2} + v\frac{\partial^2 w}{\partial y^2}\right), \quad M_y = -D\left(\frac{\partial^2 w}{\partial y^2} + v\frac{\partial^2 w}{\partial x^2}\right) \quad (A3.30)$$

where positive values are assigned to 'sagging moments' (*i.e.* compression at the upper surface). Hence it follows from equations (A3.17) and (A3.18) that

$$M_x(x,y) = \frac{px}{2}(a-x) + \frac{4\nu pa^2}{\pi^3}\sum_{n=1}^{\infty}\frac{1}{n^3\Delta_n^+}[(A_n - 2\nu)\mathbb{C}(\alpha_n y) \\ + (1-\nu)\alpha_n y \mathbb{S}(\alpha_n y)]\sin(\alpha_n x)$$ (A3.31)

$$M_y(x,y) = \frac{\nu px}{2}(a-x) - \frac{4\nu pa^2}{\pi^3}\sum_{n=1}^{\infty}\frac{1}{n^3\Delta_n^+}[(A_n + 2)\mathbb{C}(\alpha_n y) \\ + (1-\nu)\alpha_n y \mathbb{S}(\alpha_n y)]\sin(\alpha_n x)$$ (A3.32)

At the plate centre,

$$M_x(a/2,0) = \frac{pa^2}{8} + \frac{4\nu pa^2}{\pi^3}\sum_{n=1}^{\infty}\frac{(-1)^{(n-1)/2}}{n^3\Delta_n^+}[A_n^+ - 2\nu \operatorname{sech}(\alpha_n b/2)]$$ (A3.33)

$$M_y(a/2,0) = \frac{\nu pa^2}{8} - \frac{4\nu pa^2}{\pi^3}\sum_{n=1}^{\infty}\frac{(-1)^{(n-1)/2}}{n^3\Delta_n^+}[A_n^+ + 2\operatorname{sech}(\alpha_n b/2)]$$ (A3.34)

At the mid-span of a 'free' edge, $M_y = 0$ and

$$M_x(a/2,b/2) = \frac{pa^2}{8}(1+\nu) - \frac{8\nu(1+\nu)pa^2}{\pi^3}\sum_{n=1}^{\infty}\frac{(-1)^{(n-1)/2}}{n^3\Delta_n^+}$$ (A3.35)

while noting that the corresponding equation given by Holl [A3.14] appears to be incorrect. The twisting moments per unit length are given by

$$M_{xy} = -M_{yx} = D(1-\nu)\frac{\partial^2 w}{\partial x \partial y}$$ (A3.36)

and therefore become

$$M_{xy}(x,y) = \frac{4\nu pa^2}{\pi^3}\sum_{n=1}^{\infty}\frac{1}{n^3\Delta_n^+}[(A_n + 1 - \nu)\mathbb{S}(\alpha_n y) \\ + (1-\nu)\alpha_n y \mathbb{C}(\alpha_n y)]\cos(\alpha_n x)$$ (A3.37)

At a plate corner,

$$M_{xy}(0,b/2) = \frac{8\nu pa^2}{\pi^3}\sum_{n=1}^{\infty}\frac{\tanh(\alpha_n b/2)}{n^3\Delta_n^+}$$ (A3.38)

The major and minor principal moments M_1 and M_2, respectively, can be expressed as

$$\left.\begin{array}{r}M_1\\M_2\end{array}\right\} = \frac{M_x + M_y}{2} \pm \frac{1}{2}[(M_x - M_y)^2 + 4M_{xy}^2]^{1/2} \qquad (A3.39)$$

and act in the orthogonal directions defined by

$$\theta_1 = \frac{1}{2}\tan^{-1}\left(\frac{2M_{xy}}{M_x - M_y}\right), \qquad \theta_2 = \theta_1 \pm 90° \qquad (A3.40)$$

where positive angles are measured clockwise from the x-axis. Along the supported edges, $M_1 = -M_2 = M_{xy}$ and reach maximum values at the plate corners; while also noting that as both M_x and M_y are zero, $\theta_1 = \pm 45°$ ($y > 0$) or zero ($y = 0$).

The maximum twisting moments are proportional to the difference in principal moments, so that

$$M_{xy}^* = \frac{M_1 - M_2}{2} = \frac{1}{2}[(M_x - M_y)^2 + 4M_{xy}^2]^{1/2} \qquad (A3.41)$$

with the maximum value occurring at the centre of a 'free' edge. The corresponding average moment sum, namely

$$\bar{M} = \frac{M_1 + M_2}{2} = \frac{M_x + M_y}{2} \qquad (A3.42)$$

is invariant upon rotation of the co-ordinate axes and reaches a maximum value at the plate centre. Equations (A3.39)–(A3.42) are broadly analogous to those encountered in conventional two-dimensional photoelasticity, encompassing principal and maximum shear stresses, together with isochromatic, isoclinic and isopachic lines or fringes. For although the optical retardation of a normal ray of incident polarised light would be nullified when passing through a such plate under flexure, the specialised technique of magnetophotoelasticity can be deployed to obtain the required experimental data; see, for instance, the introductory notes by Hemsley [A3.15]. It is also recognised from equations (A3.30) that

$$M_x + M_y = -D(1+v)\left(\frac{\partial^2 w}{\partial x^2} + \frac{\partial^2 w}{\partial y^2}\right) \qquad (A3.43)$$

Hence the governing biharmonic equation (A3.1) can be represented by the pair of harmonic equations

$$\nabla^2 \bar{M} = -\frac{1+v}{2}p(x,y), \qquad \nabla^2 w = -\frac{2\bar{M}}{D(1+v)} \qquad (A3.44)$$

which separated form can be deployed to advantage under favourable boundary conditions.

The vertical shear forces per unit length are given by the standard relations

$$V_x = -D\frac{\partial}{\partial x}(\nabla^2 w) = -D\frac{\partial}{\partial x}\left(\frac{\partial^2 w}{\partial x^2} + \frac{\partial^2 w}{\partial y^2}\right),$$

$$V_y = -D\frac{\partial}{\partial y}(\nabla^2 w) = -D\frac{\partial}{\partial y}\left(\frac{\partial^2 w}{\partial x^2} + \frac{\partial^2 w}{\partial y^2}\right)$$

(A3.45)

so that

$$V_x(x,y) = \frac{p}{2}(a-2x) - \frac{8v\,pa}{\pi^2}\sum_{n=1}^{\infty}\frac{1}{n^2\Delta_n^+}\mathbb{C}(\alpha_n y)\cos(\alpha_n x) \qquad (A3.46)$$

$$V_y(x,y) = -\frac{8v\,pa}{\pi^2}\sum_{n=1}^{\infty}\frac{1}{n^2\Delta_n^+}\mathbb{S}(\alpha_n y)\sin(\alpha_n x) \qquad (A3.47)$$

Having regard to the different axes of antisymmetry of these last two quantities, the total vertical shear force acting on the given quarter-plate is given by

$$V^T = \int_0^{b/2} V_x(0,y)\,dy - \int_0^{a/2} V_y(x,b/2)\,dx \qquad (A3.48)$$

Hence

$$V^T = \frac{pab}{4} - \frac{8v\,pa}{\pi^2}\sum_{n=1}^{\infty}\frac{1}{n^2\Delta_n^+}\left[\int_0^{b/2}\cosh(\alpha_n y)\,dy - \sinh(\alpha_n b/2)\int_0^{a/2}\sin(\alpha_n x)\,dx\right]$$

$$= \frac{pab}{4}$$

(A3.49)

so that vertical force equilibrium is verified. At an edge support,

$$V_x(0,0) = \frac{pa}{2} - \frac{8v\,pa}{\pi^2}\sum_{n=1}^{\infty}\frac{\mathrm{sech}(\alpha_n b/2)}{n^2\Delta_n^+} \qquad (A3.50)$$

$$V_x(0,b/2) = \frac{pa}{2} - \frac{8v\,pa}{\pi^2}\sum_{n=1}^{\infty}\frac{1}{n^2\Delta_n^+} \qquad (A3.51)$$

while the maximum orthogonal shear force occurs at the centre of a 'free' edge, given by

$$V_y(a/2,b/2) = -\frac{8v\,pa}{\pi^2}\sum_{n=1}^{\infty}\frac{(-1)^{(n-1)/2}}{n^2\Delta_n^+}\tanh(\alpha_n b/2) \qquad (A3.52)$$

Likewise, the vertical edge reactions per unit length are given by

$$R_x = \left(V_x - \frac{\partial M_{xy}}{\partial y}\right)_{x=0} = -D\left[\frac{\partial^3 w}{\partial x^3} + (2-v)\frac{\partial^3 w}{\partial x \partial y^2}\right]_{x=0} \qquad (A3.53)$$

which leads to

$$R_x(0, y) = \frac{pa}{2} - \frac{4\nu pa}{\pi^2} \sum_{n=1}^{\infty} \frac{1}{n^2 \Delta_n^+} \{[A_n + 2(2-\nu)]\mathbb{C}(\alpha_n y) \qquad \text{(A3.54)}$$
$$+ (1-\nu)\alpha_n y \mathbb{S}(\alpha_n y)\}$$

At the centre of a plate support,

$$R_x(0,0) = \frac{pa}{2} - \frac{4\nu pa}{\pi^2} \sum_{n=1}^{\infty} \frac{1}{n^2 \Delta_n^+} [A_n^+ + 2(2-\nu)\operatorname{sech}(\alpha_n b/2)] \qquad \text{(A3.55)}$$

while at an end location,

$$R_x(0, b/2) = \frac{pa}{2}(1-\nu) - \frac{8\nu pa}{\pi^2}(1-\nu)\sum_{n=1}^{\infty} \frac{1}{n^2 \Delta_n^+} = (1-\nu)V_x(0, b/2) \qquad \text{(A3.56)}$$

Finally, from considerations of the assumed boundary forces and couples, the concentrated corner forces are

$$P_c = 2M_{xy}(0, b/2) = \frac{16\nu pa^2}{\pi^3} \sum_{n=1}^{\infty} \frac{\tanh(\alpha_n b/2)}{n^3 \Delta_n^+} \qquad \text{(A3.57)}$$

It follows that the supported plate edges need to be restrained vertically to prevent partial contact separation or lift-off, which in turn can generate substantial *negative* ('hogging') bending moments in the corner regions. The last equation can be verified by noting that the total vertical reaction along the supported edge of a quarter-plate, acting downward on the support and upward on the plate, is given by

$$R_x^T = \int_0^{b/2} R_x(0, y) \, dy = \frac{pab}{4} - \frac{4\nu pa}{\pi^2} \sum_{n=1}^{\infty} \frac{1}{n^2 \Delta_n^+}$$
$$\left\{ [A_n + 2(2-\nu)] \int_0^{b/2} \cosh(\alpha_n y) \, dy + (1-\nu) \int_0^{b/2} \alpha_n y \sinh(\alpha_n y) \, dy \right\} \qquad \text{(A3.58)}$$

Whence

$$R_x^T = \frac{pab}{4} - \frac{16\nu pa^2}{\pi^3} \sum_{n=1}^{\infty} \frac{\tanh(\alpha_n b/2)}{n^3 \Delta_n^+} = \frac{pab}{4} - P_c \qquad \text{(A3.59)}$$

thereby demonstrating that, because the total uniform applied load on the quarter-plate is $pab/4$, an additional edge force of magnitude P_c and of opposite sign to R_x^T is needed to maintain vertical equilibrium. From symmetry, the same downward force P_c acts at all four corners of the plate.

This discontinuous form of vertical edge reaction stems from simplifications implicit in the so-called Kirchhoff boundary conditions deployed in classical thin-plate theory, whereby two of the three boundary conditions that strictly need to be satisfied at a 'free' edge are combined to give a single equivalent condition; see, for example, Timoshenko and Woinowsky-Krieger [A3.16]. Improved estimates of plate flexure can be obtained in more refined theories that take account of the out-of-plane shear deformation through the plate thickness, perhaps augmented by direct normal strains, so that the corner forces are re-distributed along the edge supports, mostly in the end regions. However, with $t/a < 0.01$ in the present applications, the corresponding effect on mid-span plate deflections and bending moments is negligible; as confirmed, for example, in the analyses described by Salerno and Goldberg [A3.17] and Voyiadjis *et al.* [A3.18], notwithstanding possible inaccuracies in tabulated values given in the latter study due to insufficient terms taken in the summation of several infinite series. Further quantitative guidance might have been expected from the exact eigenvalue solution obtained by Zhong *et al.* [A3.19], which includes computed values of the increased central deflection for a moderately thick square plate with various ratios of shear to flexural stiffness. However, the derived final deflection equation is dimensionally incorrect, and computed values given in the paper appear not to converge to the limiting classical value for a thin plate.

Whereas exact analytical solutions can be derived for moderately thick elastic plates undergoing small deflections, usually in the form of infinite series, solutions based on large-deflection theories can only be obtained using some type of numerical analysis, even for thin elastic plates. Corresponding problems for thin beams are simpler, but closed-form analytical solutions are available only for restricted loading and edge conditions, as explained by Frisch-Fay [A3.20] and others. In either category, solutions will depend crucially upon the assumed in-plane boundary conditions at the plate supports. Here it is noted that in most architectural glazing, including the present applications, the in-plane restraint is likely to be negligible. Hence it would seem reasonable in any corresponding large-deflection analysis to assume such restraint to be zero at the two supported edges, at least under design loading (*e.g.* as if the plate edges were held between a pair of rigid frictionless rollers mounted on the two opposite rigid supports). Under these circumstances, the analyses and numerical results pertaining to a simply supported beam under uniform loading reported by Iyengar and Rao [A3.21], Wang *et al.* [A3.22] and Wang [A3.23] indicate that large-deflection effects can be neglected in retrospective analyses of the aforementioned loading tests on monolithic glass plates (appendix A2). This conclusion is also compatible with the potential-energy solution described by Stippes [A3.24] for rectangular plates simply supported on two parallel edges, although numerical results are given only for a uniformly loaded square plate simply supported on all four edges; in which case, incidentally, the maximum deflection according to classical theory is less than one-third of that for the same plate simply supported only on two parallel edges and subjected to the same uniform pressure.

A3.4. Numerical results

In their basic form, some of the above infinite series are well-conditioned and rapidly convergent, while others are much slower to converge. Moreover, the variation in magnitude of any given quantity of flexure with aspect ratio does not always converge monotonically to the limiting value for an infinitely long plate. However, any computational difficulties are greatly reduced by adopting the aforementioned remedial measures based on normalisation. The numerical values tabulated herein are believed to be correct to the given accuracy, although vigilance is ever necessary as manual post-production errors can occur with alarming facility. Generally small differences with corresponding results in various text books and other technical publications may be noted. Although such differences are of little direct consequence in design applications, they can obscure the validation of approximate numerical methods that are widely used in structural analysis, such as those based on boundary or finite elements, finite differences, or on potential- or strain-energy principles.

Because the plate boundary conditions depend upon Poisson's ratio, so do all quantities of flexure in a non-trivial manner. Accordingly, the graphical and numerical values reported herein encompass the full positive range of Poisson's ratio, as it is this parameter which governs the extent of transverse flexure; in contrast to the equivalent plane analysis of a free-spanning beam in which Poisson's ratio effectively is taken as zero. Such coverage encompasses almost all homogeneous isotropic elastic materials, for although negative values of Poisson's ratio are admissible theoretically from strain-energy considerations, their occurrence in practice is assumed to be either non-existent or exceeding rare: notwithstanding the more recent development of auxetic materials, which invariably appear to be heterogeneous; see, for instance, Greaves *et al.* [A3.25].

In order to examine the limiting behaviour of the various quantities of flexure, it is useful initially to recall some classical results for the sum of reciprocal powers. Thus, with $k, s, = 1, 2, 3, 4, \ldots$,

$$\Upsilon(s) = \sum_{k=1}^{\infty} \frac{1}{(2k-1)^s} = 1 + \frac{1}{3^s} + \frac{1}{5^s} + \frac{1}{7^s} + \ldots = \left(1 - \frac{1}{2^s}\right)\zeta(s) \tag{A3.60}$$

Here $\zeta(s)$ denotes the Riemann Zeta function, which for a real variable is defined as

$$\zeta(s) = \sum_{k=1}^{\infty} k^{-s} = 1 + \frac{1}{2^s} + \frac{1}{3^s} + \frac{1}{4^s} + \ldots \tag{A3.61}$$

while noting that $\zeta(1) \to \infty$. For some *even* powers, these summations can be expressed in simple closed form; whence, for example,

$$\zeta(2) = \frac{\pi^2}{6}, \quad \zeta(4) = \frac{\pi^4}{90}, \quad \Upsilon(2) = \frac{\pi^2}{8}, \quad \Upsilon(4) = \frac{\pi^4}{96} \tag{A3.62}$$

The corresponding alternating series is

$$\Lambda(s) = \sum_{k=1}^{\infty} \frac{(-1)^{k-1}}{(2k-1)^s} = 1 - \frac{1}{3^s} + \frac{1}{5^s} - \frac{1}{7^s} + \ldots \tag{A3.63}$$

and closed results for some *odd* powers are

$$\Lambda(1) = \frac{\pi}{4}, \quad \Lambda(3) = \frac{\pi^3}{32}, \quad \Lambda(5) = \frac{5\pi^5}{1536} \tag{A3.64}$$

Moreover, the special value $\Lambda(2) = G$, where G denotes Catalan's constant ($G \approx 0.915965594$). Numerical values of all such reciprocal powers are tabulated by Abramowitz and Stegun [A3.26], while noting for present purposes that $\Upsilon(3) \approx 1.051799790$ and $\Lambda(4) \approx 0.988944552$.

Among the few numerical results computed by Estanave [A3.9] are truncated two-term values ($n = 1, 3$) of mid-span deflection for a square plate, tabulated for several boundary conditions. For the present case, taking $v = 1/4$ as assumed in the original analysis, these are $w(a/2,0)EI/pa^4 = 0.01218$ and $w(a/2,b/2)EI/pa^4 = 0.01360$, which are equivalent to $w(a/2,0)D/pa^4 = 0.01299$ and $w(a/2,b/2)D/pa^4 = 0.01451$, respectively. Because equations (A3.15) and (A3.16) are rapidly convergent, these latter values are very close to the 'exact' results computed herein, namely $w(a/2,0)D/pa^4 = 0.0130014$ and $w(a/2,b/2)D/pa^4 = 0.0145106$.

Non-dimensional values of deflection at the plate centre and at mid-span on a 'free' edge are shown in Figure A3.2 for a wide range of parameters, while noting that the corresponding graphical results presented by Holl [A3.14] are incorrect for small values of plate aspect ratio. In the computation of benchmark numerical values, the alternating infinite series in the deflection equations converge rapidly in view of the n^5-term in the denominator. Selected results are listed in Tables A3.1 and A3.2, taking $v = 0.22$ both here and subsequently (in this section) as being representative of architectural soda-lime glass. For an infinitely short plate ($b/a \to 0$),

$$w(a/2,0)_0 = w(a/2,b/2)_0 = \frac{5pa^4}{384D(1-v^2)} = \frac{5pa^4}{384EI} \tag{A3.65}$$

and for an infinitely long plate ($b/a \to \infty$),

$$w(a/2,0)_\infty = \frac{5pa^4}{384D}, \quad w(a/2,b/2)_\infty = \frac{5pa^4}{384D(1-v)}\left(\frac{3-v}{3+v}\right) \tag{A3.66}$$

Calculations indicate that when the aspect ratio exceeds two, the mid-span deflections are close to those for an infinitely long plate. This observation also holds for the remaining quantities of plate flexure. First approximations to these mid-span deflections for a plate of general aspect ratio are

$$w(a/2,0) \approx w(a/2,0)_\infty + \frac{4v\,pa^4 A_1^+}{\pi^5 D(1-v)\Delta_1^+} \tag{A3.67}$$

$$w(a/2,b/2) \approx w(a/2,b/2)_\infty - \frac{8v\,pa^4}{\pi^5 D(1-v)}\left(\frac{1}{\Delta_1^+} - \frac{1}{3+v}\right) \tag{A3.68}$$

Thus, here and subsequently, the approximations can be expressed as perturbations on the limiting case of an infinitely long plate, the latter signified by the appropriate subscript.

ELASTIC FLEXURE OF MONOLITHIC RECTANGULAR PLATE

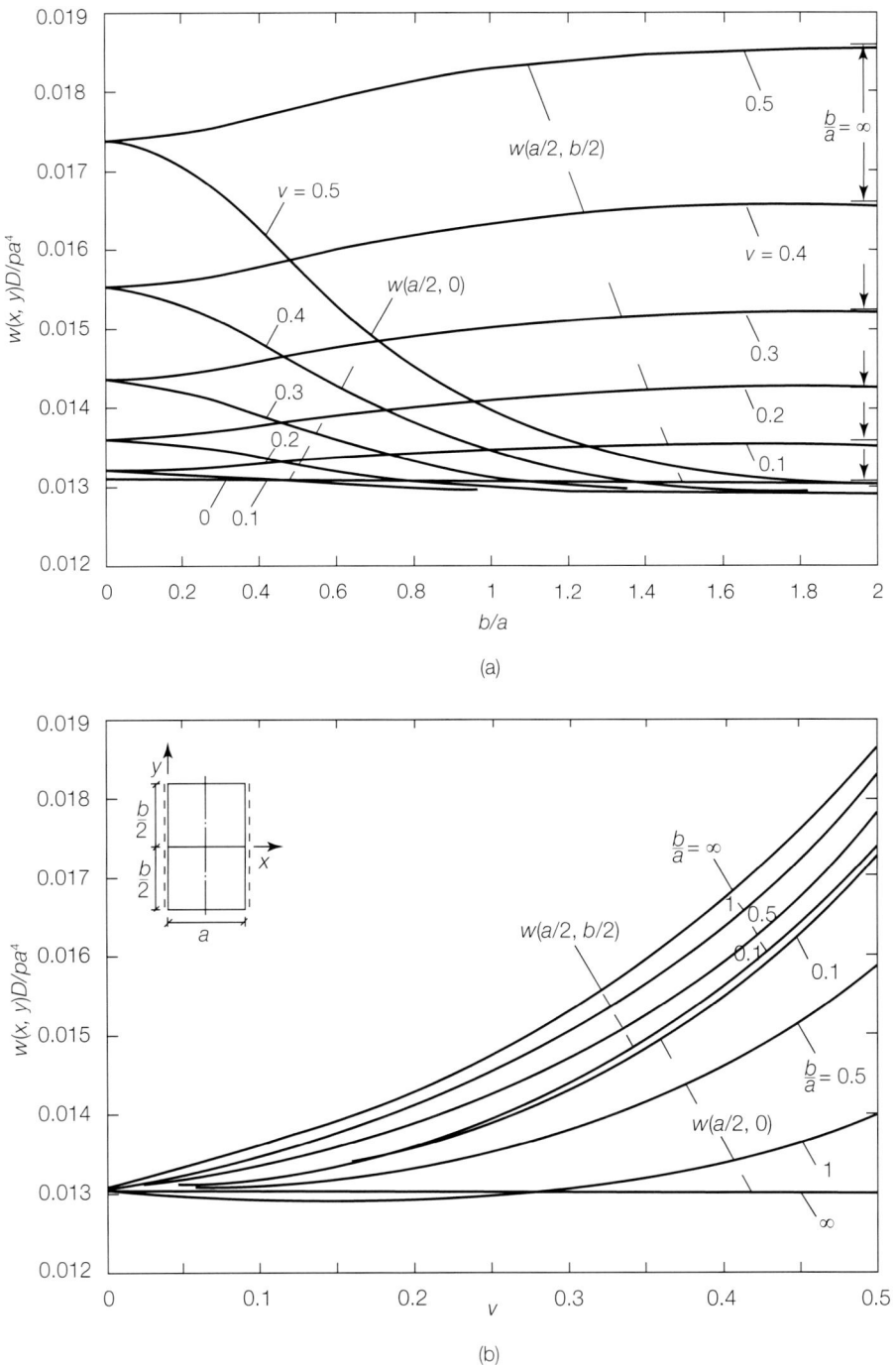

FIGURE A3.2. Deflections of thin rectangular plate (width a, length b, Poisson's ratio v, flexural stiffness D) simply supported on two opposite edges ($x = 0, a$) with two edges 'free', subjected to uniform loading (p): (a) dependence on aspect ratio; (b) dependence on Poisson's ratio

TABLE A3.1. Deflection and bending moments at centre of thin rectangular plate (width a, length b, flexural stiffness D, Poisson's ratio $v = 0.22$) simply supported on two opposite edges ($x = 0, a$) with two edges 'free', subjected to uniform loading (p)

b/a	$\dfrac{w(a/2,0)D}{pa^4}$	$\dfrac{M_x(a/2,0)}{pa^2}$	$\dfrac{M_y(a/2,0)}{pa^2}$
0	0.0136831	0.125	0
0.1	0.0136627	0.124941	0.000451
0.2	0.0136049	0.124767	0.001796
0.3	0.0135195	0.124493	0.003926
0.4	0.0134197	0.124165	0.006554
0.5	0.0133174	0.123834	0.009360
0.7	0.0131370	0.123291	0.014620
1	0.0129644	0.122926	0.020412
1.5	0.0128779	0.123158	0.025218
2	0.0128977	0.123716	0.026918
3	0.0129703	0.124541	0.027572
5	0.0130165	0.124963	0.027520
∞	0.0130208	0.125	0.0275

TABLE A3.2. Deflection, slope, bending moment and shear force at centre of 'free' edge of thin rectangular plate (width a, length b, flexural stiffness D, Poisson's ratio $v = 0.22$) simply supported on two opposite edges ($x = 0, a$) with two edges 'free', subjected to uniform loading (p)

b/a	$\dfrac{w(a/2,b/2)D}{pa^4}$	$\dfrac{\phi_y(a/2,b/2)D}{pa^3}$	$\dfrac{M_x(a/2,b/2)}{pa^2}$	$\dfrac{V_y(a/2,b/2)}{pa}$
0	0.0136831	0	0.125	0
0.1	0.0136983	0.0014291	0.125117	−0.009016
0.2	0.0137413	0.0027640	0.125467	−0.017973
0.3	0.0138052	0.0039217	0.126020	−0.026334
0.4	0.0138806	0.0048578	0.126699	−0.033376
0.5	0.0139589	0.0055725	0.127418	−0.038843
0.7	0.0141012	0.0064602	0.128743	−0.045682
1	0.0142508	0.0069838	0.130145	−0.049720
1.5	0.0143643	0.0071342	0.131211	−0.050880
2	0.0143992	0.0071285	0.131539	−0.050836
3	0.0144114	0.0071158	0.131654	−0.050738
5	0.0144123	0.0071143	0.131661	−0.050727
∞	0.0144123	0.0071143	0.131661	−0.050727

Likewise, and with a similar subscript notation, the limiting values of plate slope are

$$\phi_x(0,0)_0 = \phi_x(0,b/2)_0 = \frac{pa^3}{24D(1-v^2)} = \frac{pa^3}{24EI} \qquad (A3.69)$$

$$\phi_x(0,0)_\infty = \frac{pa^3}{24D}, \quad \phi_x(0,b/2)_\infty = \frac{pa^3}{24D(1-v)}\left(\frac{3-v}{3+v}\right) \qquad (A3.70)$$

$$\phi_y(a/2,b/2)_\infty = \frac{8v\,pa^3\Lambda(4)}{\pi^4 D(1-v)(3+v)} \qquad (A3.71)$$

Convergence of those series in the general expressions for slope is fairly rapid, largely on account of the n^4-term in the denominator, and first approximations are given by

$$\phi_x(0,0) \approx \phi_x(0,0)_\infty + \frac{4v\,pa^3 A_1^+}{\pi^4 D(1-v)\Delta_1^+} \qquad (A3.72)$$

$$\phi_x(0,b/2) \approx \phi_x(0,b/2)_\infty - \frac{8v\,pa^3}{\pi^4 D(1-v)}\left(\frac{1}{\Delta_1^+} - \frac{1}{3+v}\right) \qquad (A3.73)$$

$$\phi_y(a/2,b/2) \approx \phi_y(a/2,b/2)_\infty + \frac{8v\,pa^3}{\pi^4 D(1-v)}\left[\frac{\tanh(\pi b/2a)}{\Delta_1^+} - \frac{1}{3+v}\right] \qquad (A3.74)$$

Variations of plate slope at the centre and end of a supported edge, and at mid-span of a 'free' edge, are depicted in Figure A3.3, while numerical values are included in Tables A3.2 and A3.3.

Figure A3.4 shows the bending moments at the same plate locations as for deflections, and selected numerical values are included in Tables A3.1 and A3.2. Here the equivalent term in the denominator becomes n^3, and far more terms in the infinite series may be needed, especially at a 'free' edge. The limiting values at the plate centre are

$$M_x(a/2,0)_0 = \frac{pa^2}{8}, \quad M_y(a/2,0)_0 = 0 \qquad (A3.75)$$

$$M_x(a/2,0)_\infty = \frac{pa^2}{8}, \quad M_y(a/2,0)_\infty = \frac{v\,pa^2}{8} \qquad (A3.76)$$

and the first approximations become

$$M_x(a/2,0) \approx M_x(a/2,0)_\infty - \frac{4v\,pa^2}{\pi^3 \Delta_1^+}[A_1^+ - 2v\,\text{sech}(\pi b/2a)] \qquad (A3.77)$$

$$M_y(a/2,0) \approx M_y(a/2,0)_\infty - \frac{4v\,pa^2}{\pi^3 \Delta_1^+}[A_1^+ + 2\,\text{sech}(\pi b/2a)] \qquad (A3.78)$$

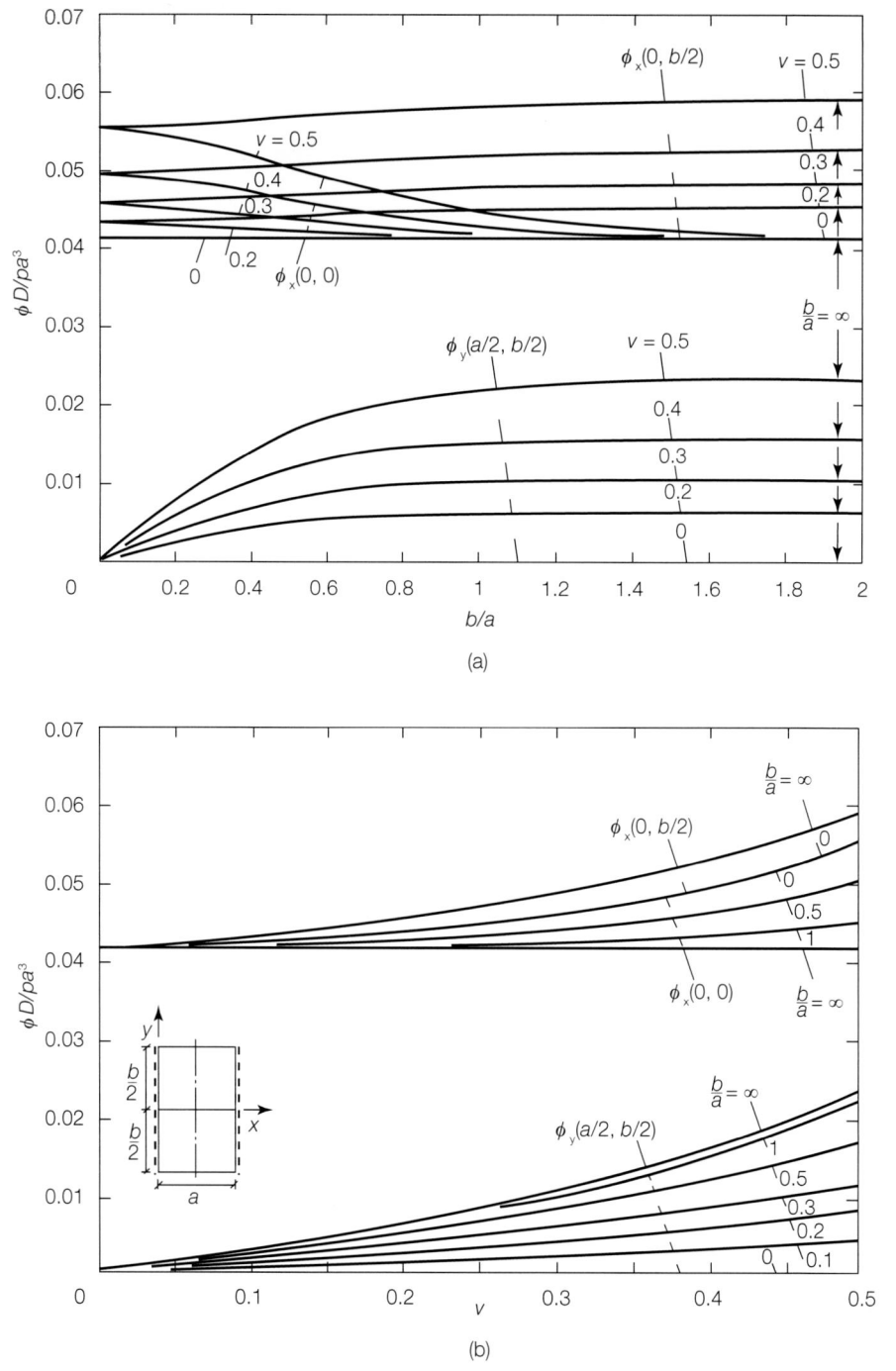

FIGURE A3.3. Slopes of thin rectangular plate (width a, length b, Poisson's ratio v, flexural stiffness D) simply supported on two opposite edges ($x = 0, a$) with two edges 'free', subjected to uniform loading (p): (a) dependence on aspect ratio; (b) dependence on Poisson's ratio

ELASTIC FLEXURE OF MONOLITHIC RECTANGULAR PLATE

TABLE A3.3. Slopes, shear force, reactions and corner force at edge support for thin rectangular plate (width a, length b, flexural stiffness D, Poisson's ratio $v = 0.22$) simply supported on two opposite edges ($x = 0, a$) with two edges 'free', subjected to uniform loading (p)

b/a	$\dfrac{\phi_x(0,0)D}{pa^3}$	$\dfrac{\phi_x(0,b/2)D}{pa^3}$	$\dfrac{V_x(0,0)}{pa}$	$\dfrac{R_x(0,0)}{pa}$	$\dfrac{R_x(0,b/2)}{pa}$	$\dfrac{P_c}{pa^2}$
0	0.0437859	0.0437859	0.409836	0.319672	0.319672	0
0.1	0.0437101	0.0438424	0.417704	0.336357	0.321375	0.008386
0.2	0.0435112	0.0439912	0.425571	0.353042	0.323077	0.015512
0.3	0.0432319	0.0442015	0.433427	0.369701	0.324776	0.021378
0.4	0.0429132	0.0444438	0.441181	0.386132	0.326450	0.026019
0.5	0.0425906	0.0446924	0.448631	0.401885	0.328048	0.029531
0.7	0.0420255	0.0451414	0.461810	0.429564	0.330817	0.033880
1	0.0414868	0.0456118	0.476260	0.459251	0.333659	0.036443
1.5	0.0412171	0.0459687	0.489367	0.484554	0.335804	0.037180
2	0.0412798	0.0460782	0.495194	0.494427	0.336461	0.037152
3	0.0415079	0.0461167	0.499005	0.499451	0.336692	0.037090
5	0.0416530	0.0461193	0.499957	0.500029	0.336708	0.037083
∞	0.0416667	0.0461193	0.5	0.5	0.336708	0.037083

Likewise, at the centre of a 'free' edge,

$$M_x(a/2,b/2)_0 = \frac{pa^2}{8}, \quad M_x(a/2,b/2)_\infty = \frac{pa^2}{8}(1+v)\left(\frac{3-v}{3+v}\right) \qquad (A3.79)$$

$$M_x(a/2,b/2) \approx M_x(a/2,b/2)_\infty - \frac{8v(1+v)pa^2}{\pi^3}\left(\frac{1}{\Delta_1^+} - \frac{1}{3+v}\right) \qquad (A3.80)$$

At the end of a supported edge,

$$M_{xy}(0,b/2)_0 = 0, \quad M_{xy}(0,b/2)_\infty = \frac{8v\,pa^2\,\Upsilon(3)}{\pi^3(3+v)} \qquad (A3.81)$$

$$M_{xy}(0,b/2) \approx M_{xy}(0,b/2)_\infty + \frac{8v\,pa^2}{\pi^3}\left[\frac{\tanh(\pi b/2a)}{\Delta_1^+} - \frac{1}{3+v}\right] \qquad (A3.82)$$

In equation (A3.82), for example, obtained simply by neglecting the second term in Δ_n^+ for $n > 1$, the accuracy largely depends on the plate aspect ratio: the error is less than 1% when $b/a = 0.3$, and decreases markedly thereafter for higher aspect ratios. Because the corner twisting moment is exactly one-half of the corresponding corner force, only numerical values of the latter quantity are tabulated herein (see below).

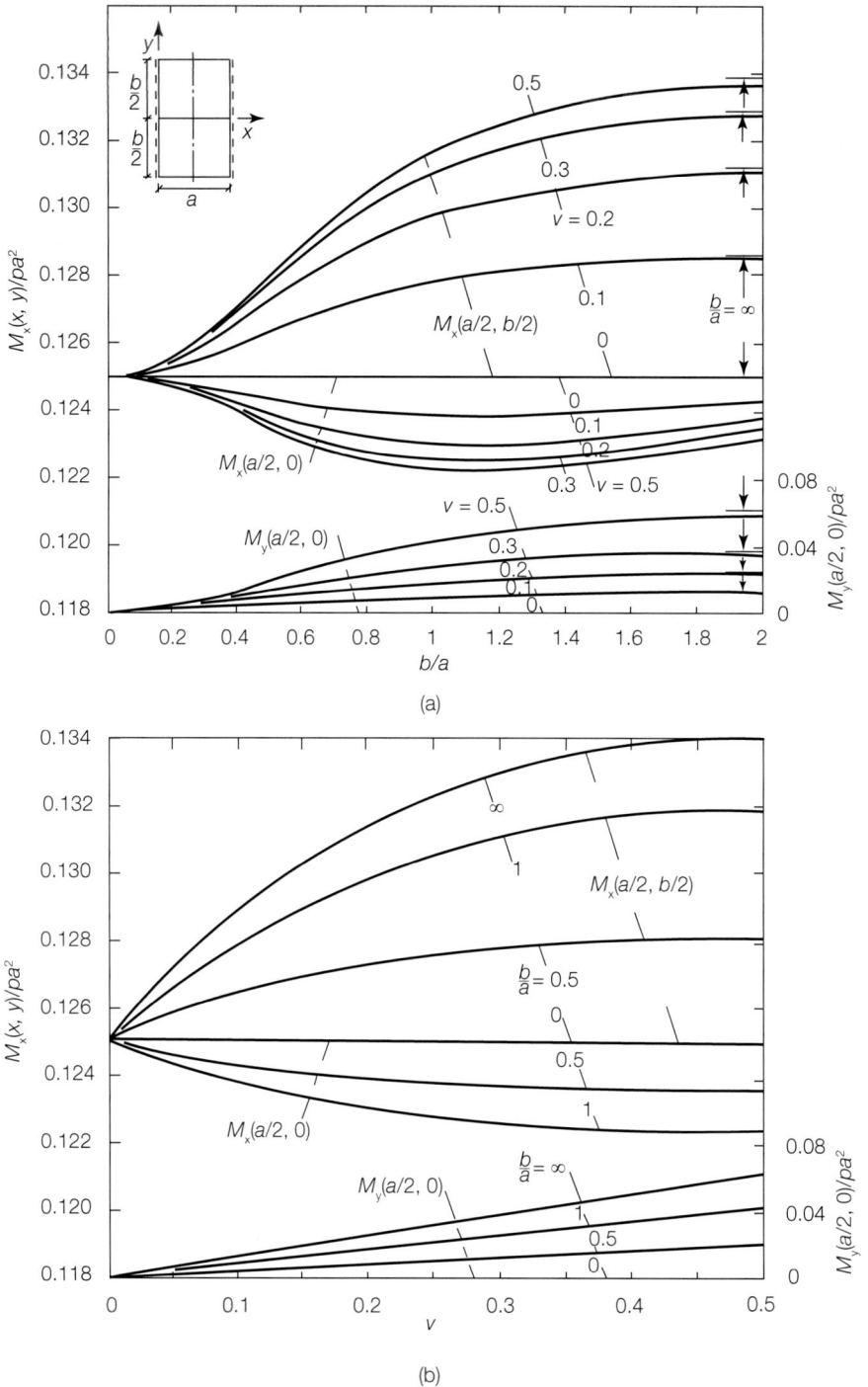

FIGURE A3.4. Bending moments in thin rectangular plate (width a, length b, Poisson's ratio v) simply supported on two opposite edges ($x = 0, a$) with two edges 'free', subjected to uniform loading (p): (a) dependence on aspect ratio; (b) dependence on Poisson's ratio

The vertical shear force at the end of a plate support given by equation (A3.51) is even slower to converge, with n^2 in the denominator. For an infinitely short plate,

$$V_x(0,0)_0 = V_x(0,b/2)_0 = \frac{pa}{2(1+v)} \tag{A3.83}$$

and for an infinitely long plate,

$$V_x(0,0)_\infty = \frac{pa}{2}, \quad V_x(0,b/2)_\infty = \frac{pa}{2}\left(\frac{3-v}{3+v}\right) \tag{A3.84}$$

At the end of a supported edge, the first-term approximation is

$$V_x(0,b/2) \approx V_x(0,b/2)_\infty - \frac{8v\,pa}{\pi^2}\left(\frac{1}{\Delta_1^+} - \frac{1}{3+v}\right) \tag{A3.85}$$

The values given by equation (A3.85) deviate from the exact values by less than 1% when $b/a > 0.1$. Numerical values of shear force at the centre of a simply supported edge are given in Table A3.3, while the corresponding end-values differ from those for the end-reaction only by a linear factor, as indicated by equation (A3.56).

Slow convergence also is encountered for the vertical shear force at the centre of a 'free' edge given by equation (A3.52), based on an alternating infinite series. The limiting values are

$$V_y(a/2,b/2)_0 = 0, \quad V_y(a/2,b/2)_\infty = -\frac{8v\,paG}{\pi^2(3+v)} \tag{A3.86}$$

while the first-term approximation becomes

$$V_y(a/2,b/2) \approx V_y(a/2,b/2)_\infty - \frac{8v\,pa}{\pi^2}\left[\frac{\tanh(\pi b/2a)}{\Delta_1^+} - \frac{1}{3+v}\right] \tag{A3.87}$$

Errors obtained using equation (A3.87) are less than 1% when $b/a = 0.4$, and decrease rapidly with increasing aspect ratio, compared to the benchmark values listed in Table A3.2.

The convergence characteristics of the equations giving the plate support reactions are similar to those for the vertical shear force along an edge support. For an infinitely short plate,

$$R_x(0,0)_0 = R_x(0,b/2)_0 = \frac{pa}{2}\left(\frac{1-v}{1+v}\right) \tag{A3.88}$$

and for an infinitely long plate,

$$R_x(0,0)_\infty = \frac{pa}{2}, \quad R_x(0,b/2)_\infty = \frac{pa(1-v)}{2}\left(\frac{3-v}{3+v}\right) \tag{A3.89}$$

Whence the first-term approximation at an end location becomes

$$R_x(0, b/2) \approx R_x(0, b/2)_\infty - \frac{8 v p a (1-v)}{\pi^2} \left(\frac{1}{\Delta_1^+} - \frac{1}{3+v} \right) \qquad (A3.90)$$

Plate reactions at the centre and end of a simply supported edge (including the corner force) are shown in Figure A3.5, while observing that the corresponding graphical results presented by Holl [A3.14] for the central reaction are in error for small values of plate aspect ratio. Benchmark numerical values are given in Table A3.3. Also listed are corresponding values of the corner force based on equation (A3.57), which are simply double the magnitude of those computed for the corner twisting moment given by equation (A3.38).

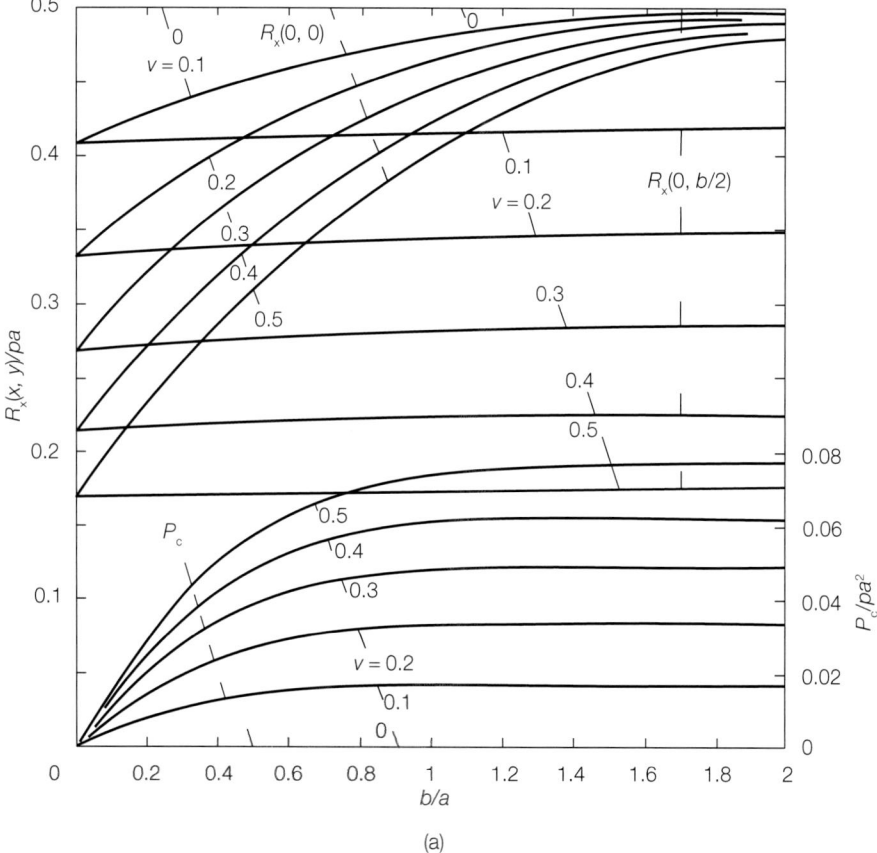

FIGURE A3.5. Reactions and corner force for thin rectangular plate (width a, length b, Poisson's ratio v) simply supported on two opposite edges ($x = 0, a$) with two edges 'free', subjected to uniform loading (p): (a) dependence on aspect ratio

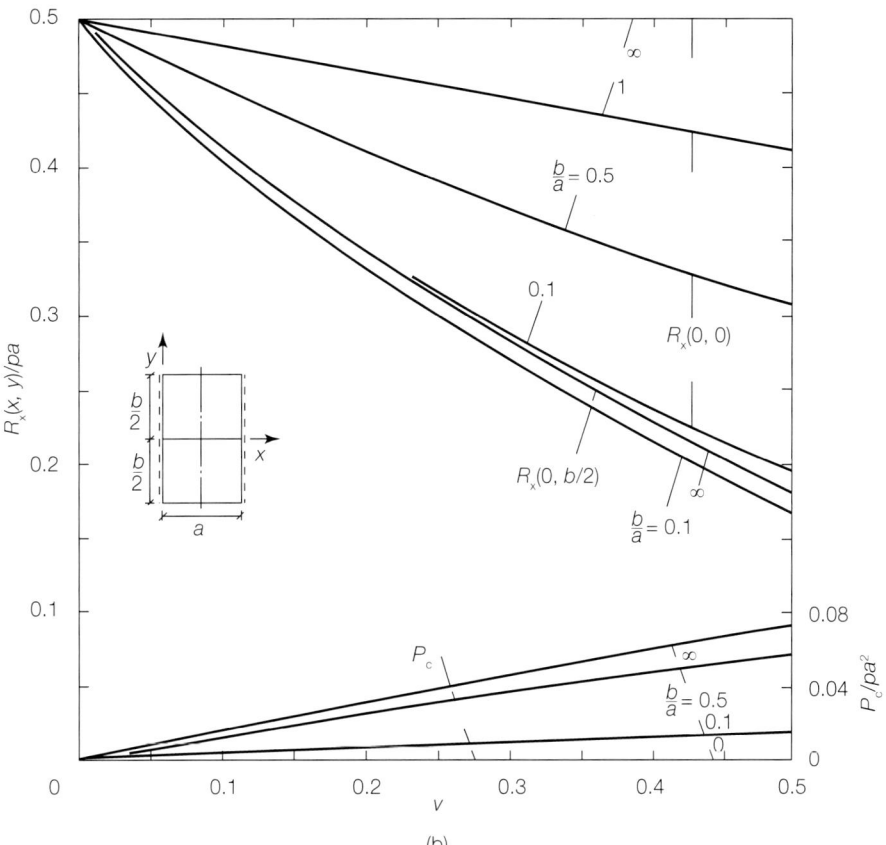

FIGURE A3.5. (continued) Reactions and corner force for thin rectangular plate (width a, length b, Poisson's ratio v) simply supported on two opposite edges ($x = 0, a$) with two edges 'free', subjected to uniform loading (p): (b) dependence on Poisson's ratio

A broad perspective of flexural response is illustrated in Figure A3.6 for a glass plate with an aspect ratio of 0.7, which is close to that for specimens deployed in the full-scale pressure loading tests referred to in both the main text and appendix A2. Deflections and bending moments are plotted along both centre lines and along the 'free' edge, while twisting moments are shown along a support and a 'free' edge. Values of principal moment are plotted along a plate diagonal, which direction generally does not coincide with their orientation, as depicted erroneously by Holl [A3.14]; at the centre of the given quadrant, for example, $\theta_1 = 2.9°$. Also shown are distributions of vertical shear force and edge reaction. Further details of computed flexure within a plate quadrant are represented by the extensive contour diagrams presented in Figures A3.7–A3.12, augmented by the principal moment trajectories shown in Figure A3.11(b). Of particular note are the *negative* bending moments (M_2) in the corner region plotted on a diagonal in Figure A3.6, and illustrated in more detail in Figure A3.10(b), where the shaded portion of the quarter-plate contour diagram represents the entire negative zone.

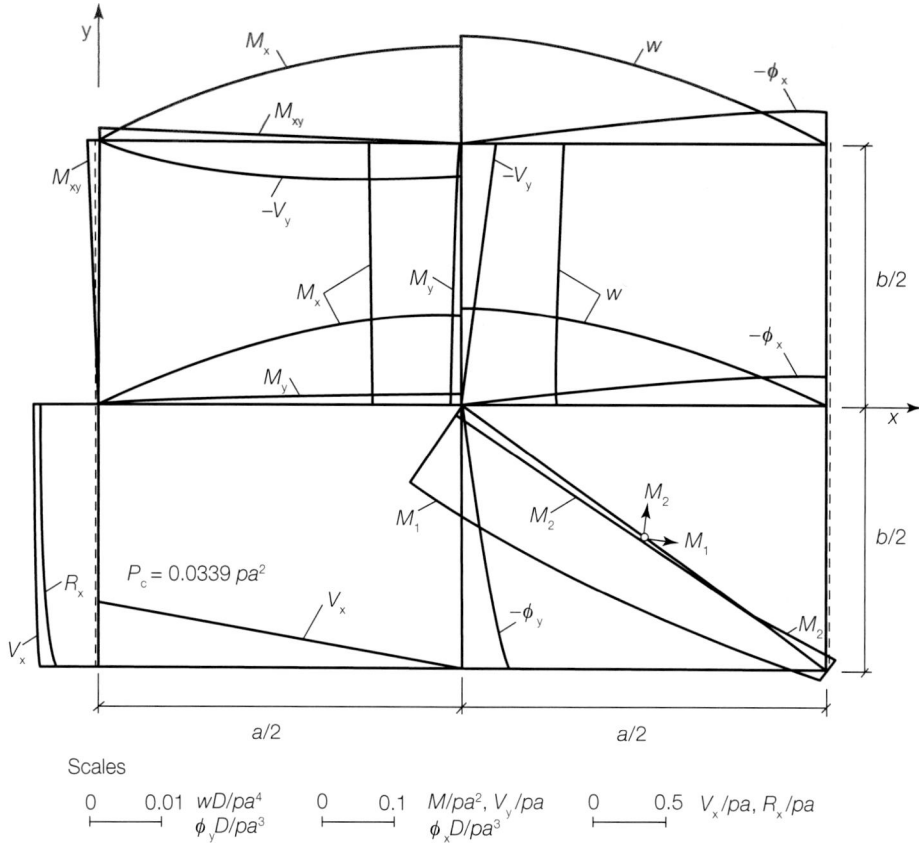

FIGURE A3.6. Elastic flexure of thin rectangular plate (width a, length $b = 0.7a$, Poisson's ratio $v = 0.22$, flexural stiffness D) simply supported on two opposite edges ($x = 0, a$) with two edges 'free', subjected to uniform loading (p)

A3.5. Supplementary computed results

Some further computed results are appended herein, based on the preceding theoretical solution, in order to provide a wider coverage of the topic for reference purposes. Non-dimensional values of particular quantities of flexure are listed for plates with a Poisson's ratio of 0.2, 0.25 and 0.3, as comparative results based on these parameters are often quoted in the literature; see Tables A3.4–A3.6. Values of plate deflection and bending moment given in the standard text by Timoshenko and Woinowsky-Krieger [A3.16] are in agreement with those listed in Table A3.4 for the case $v = 0.3$, except their tabulated result for $M_y(a/2,0)$ at $b = 0.5a$, which is substantially lower than the correct value. Graphical results given in Figures A3.13–A3.26 illustrate the flexure of plates having aspect ratios of 0.5 and 1, taking $v = 0.25$.

ELASTIC FLEXURE OF MONOLITHIC RECTANGULAR PLATE

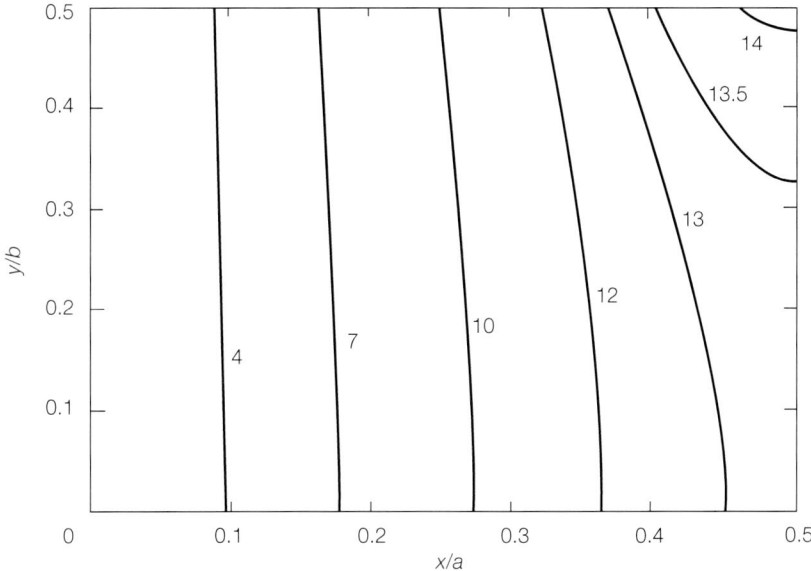

FIGURE A3.7. Deflection contours $w(x,y)D \times 10^3/pa^4$ within quadrant of thin rectangular plate (width a, length $b = 0.7a$, Poisson's ratio $\nu = 0.22$, flexural stiffness D) simply supported on two opposite edges ($x = 0, a$) with two edges 'free', subjected to uniform loading (p)

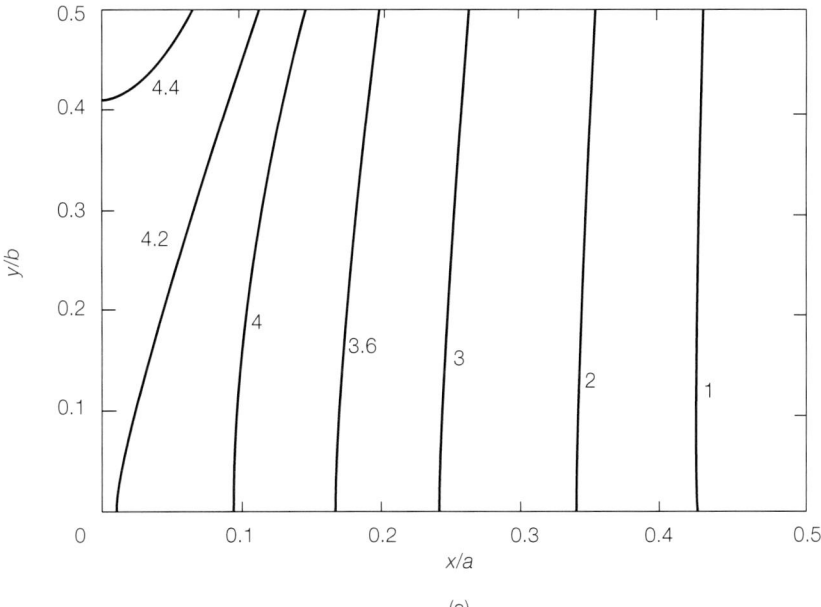

(a)

FIGURE A3.8. Contours of slope within quadrant of thin rectangular plate (width a, length $b = 0.7a$, Poisson's ratio $\nu = 0.22$, flexural stiffness D) simply supported on two opposite edges ($x = 0, a$) with two edges 'free', subjected to uniform loading (p): (a) $\phi_x(x,y)D \times 10^2/pa^3$

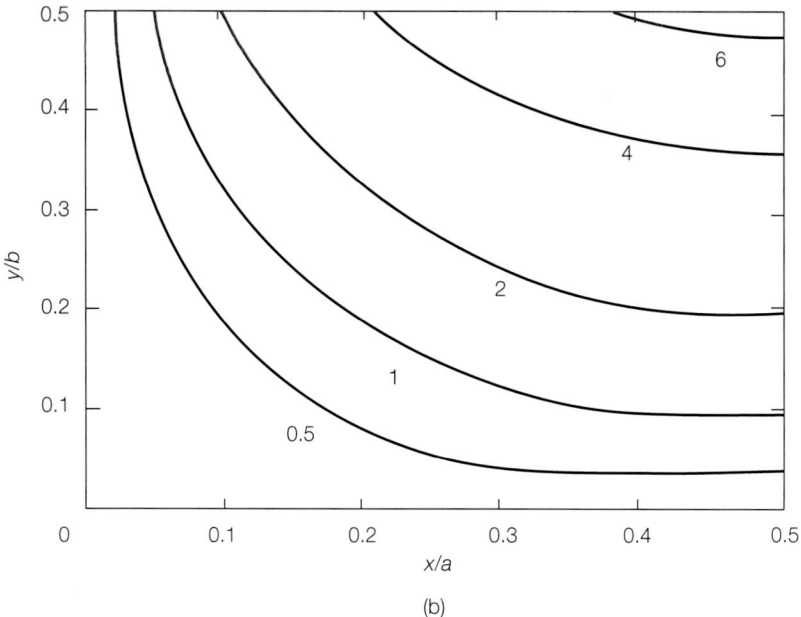

FIGURE A3.8. (continued) Contours of slope within quadrant of thin rectangular plate (width a, length $b = 0.7a$, Poisson's ratio $v = 0.22$, flexural stiffness D) simply supported on two opposite edges ($x = 0, a$) with two edges 'free', subjected to uniform loading (p): (b) $\phi_y(x,y)D \times 10^3/pa^3$

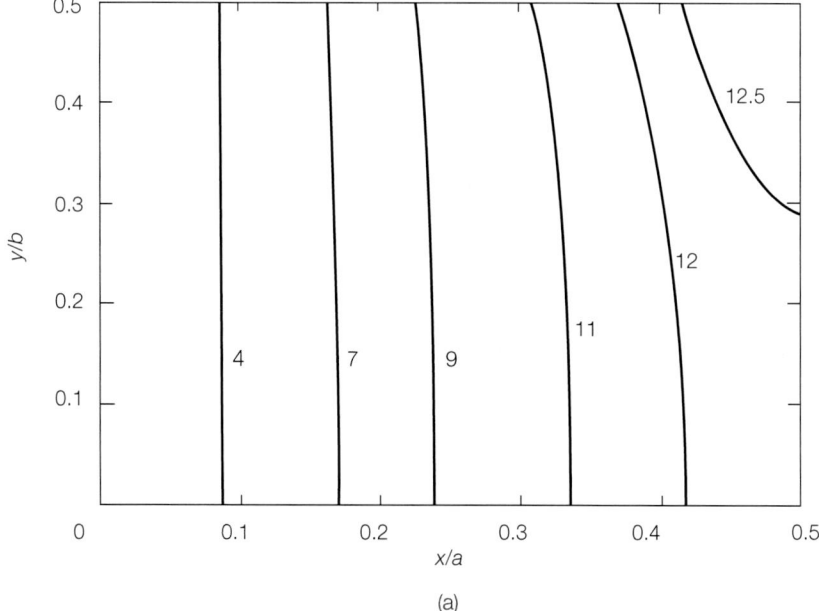

FIGURE A3.9. Contours of bending and twisting moments within quadrant of thin rectangular plate (width a, length $b = 0.7a$, Poisson's ratio $v = 0.22$) simply supported on two opposite edges ($x = 0, a$) with two edges 'free', subjected to uniform loading (p): (a) $M_x(x,y) \times 10^2/pa^2$

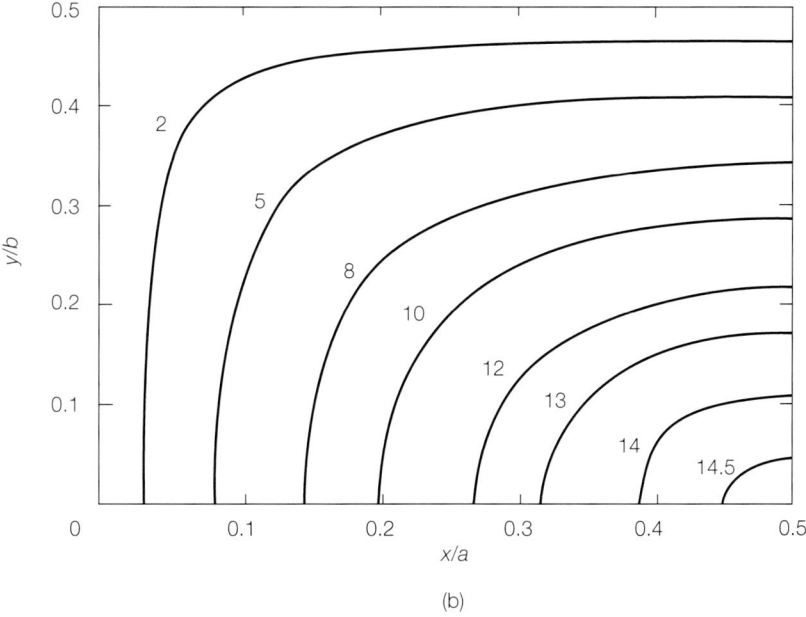

(b)

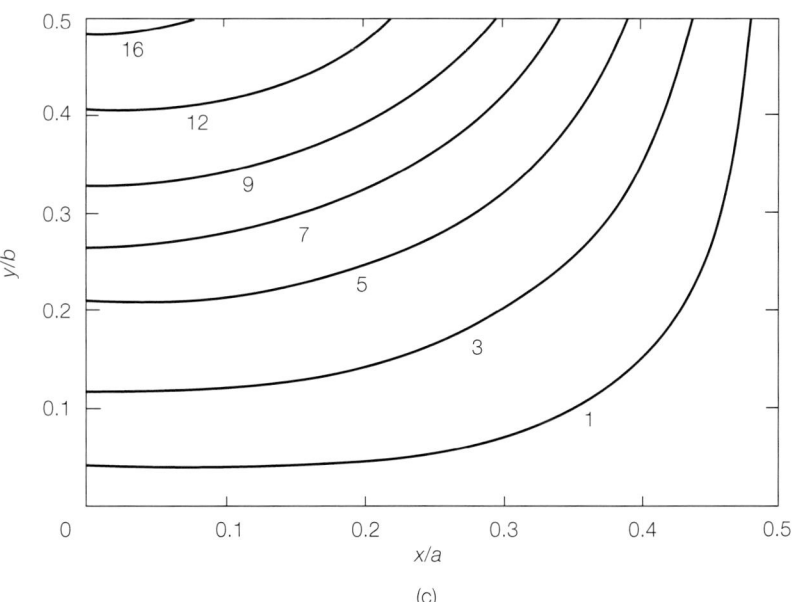

(c)

FIGURE A3.9. (continued) Contours of bending and twisting moments within quadrant of thin rectangular plate (width a, length $b = 0.7a$, Poisson's ratio $v = 0.22$) simply supported on two opposite edges ($x = 0, a$) with two edges 'free', subjected to uniform loading (p): (b) $M_y(x,y) \times 10^3 / pa^2$; (c) $M_{xy}(x,y) \times 10^3 / pa^2$

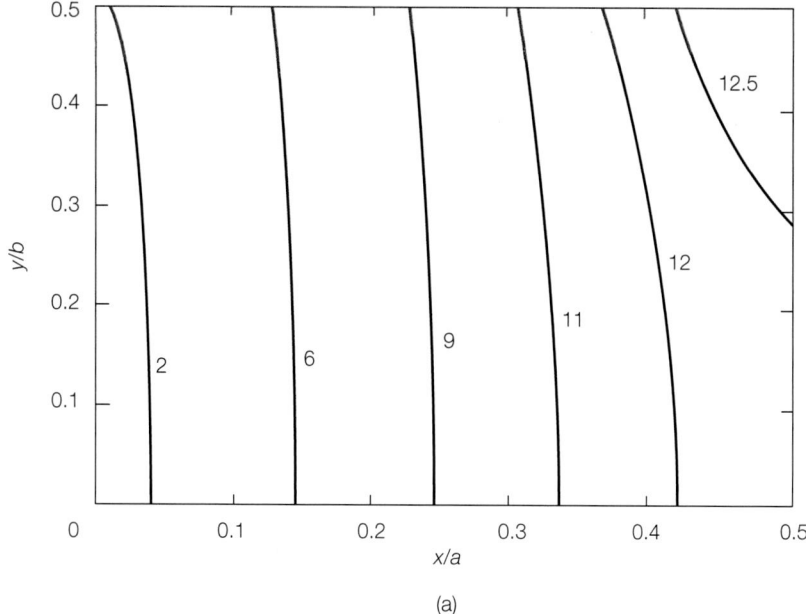

(a)

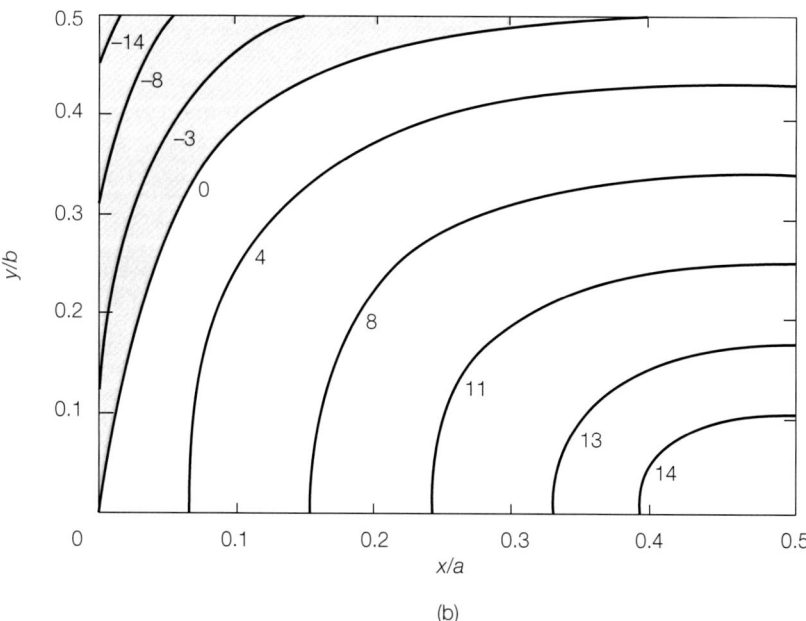

(b)

FIGURE A3.10. Contours of major and minor principal moments within quadrant of thin rectangular plate (width a, length $b = 0.7a$, Poisson's ratio $v = 0.22$) simply supported on two opposite edges ($x = 0, a$) with two edges 'free', subjected to uniform loading (p): (a) $M_1(x, y) \times 10^2 / pa^2$; (b) $M_2(x, y) \times 10^3 / pa^2$

ELASTIC FLEXURE OF MONOLITHIC RECTANGULAR PLATE

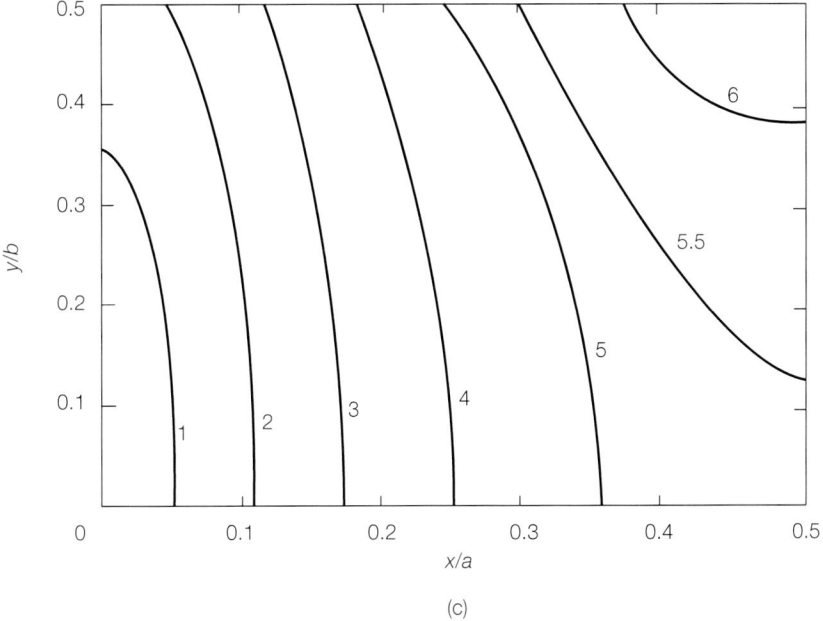

(c)

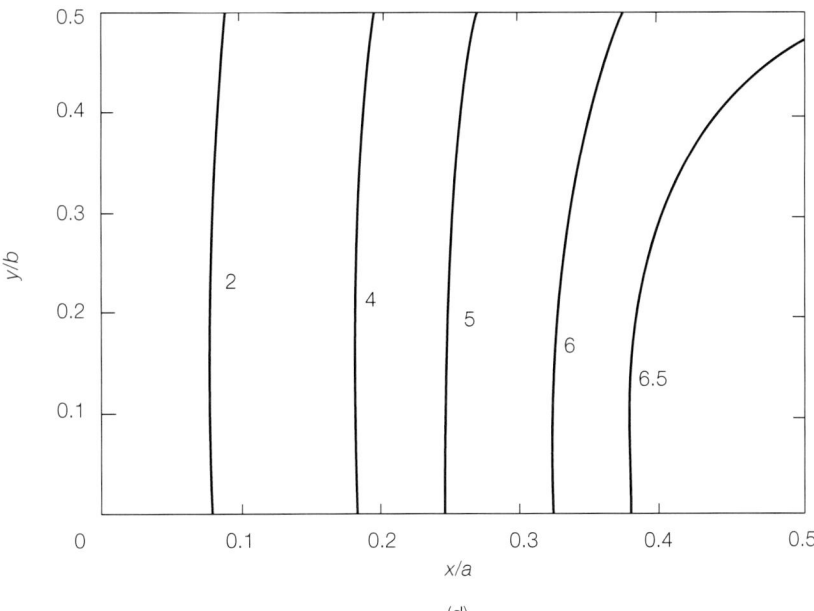

(d)

FIGURE A3.10. (continued) Contours of major and minor principal moments within quadrant of thin rectangular plate (width a, length $b = 0.7a$, Poisson's ratio $v = 0.22$) simply supported on two opposite edges ($x = 0$, a) with two edges 'free', subjected to uniform loading (p): (c) $M^*_{xy}(x,y) \times 10^2 / pa^2$; (d) $\bar{M}(x,y) \times 10^2 / pa^2$

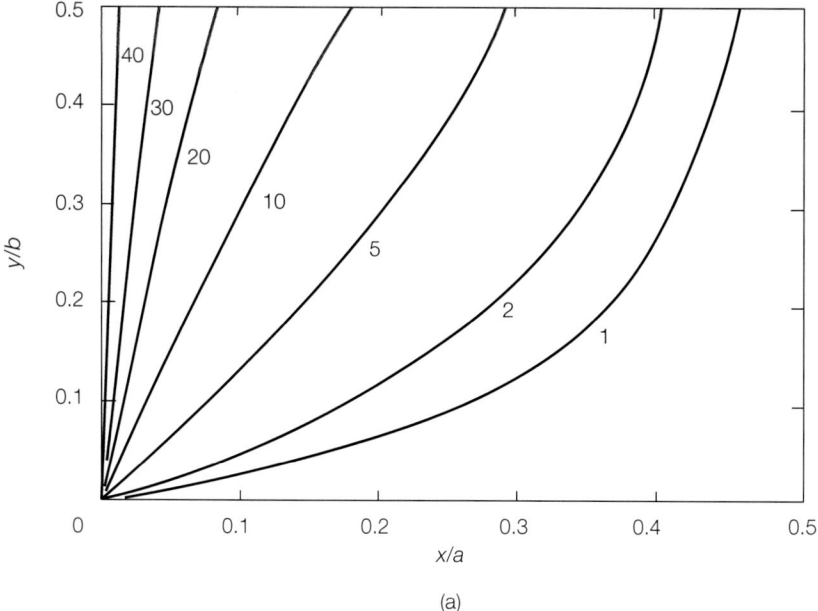

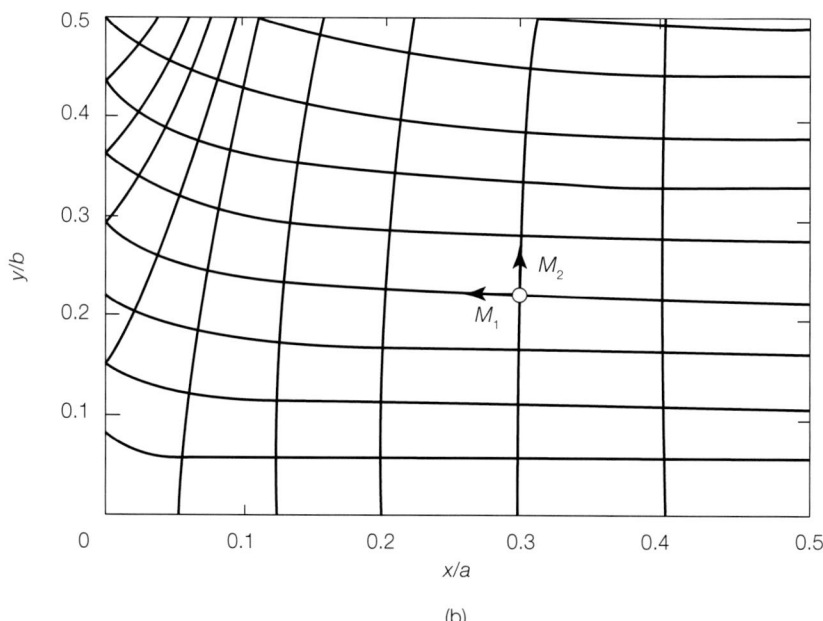

FIGURE A3.11. Orientation of principal bending moments within quadrant of thin rectangular plate (width a, length $b = 0.7a$, Poisson's ratio $v = 0.22$) simply supported on two opposite edges ($x = 0, a$) with two edges 'free', subjected to uniform loading (p): (a) contours of $\theta_1°$; (b) principal moment trajectories

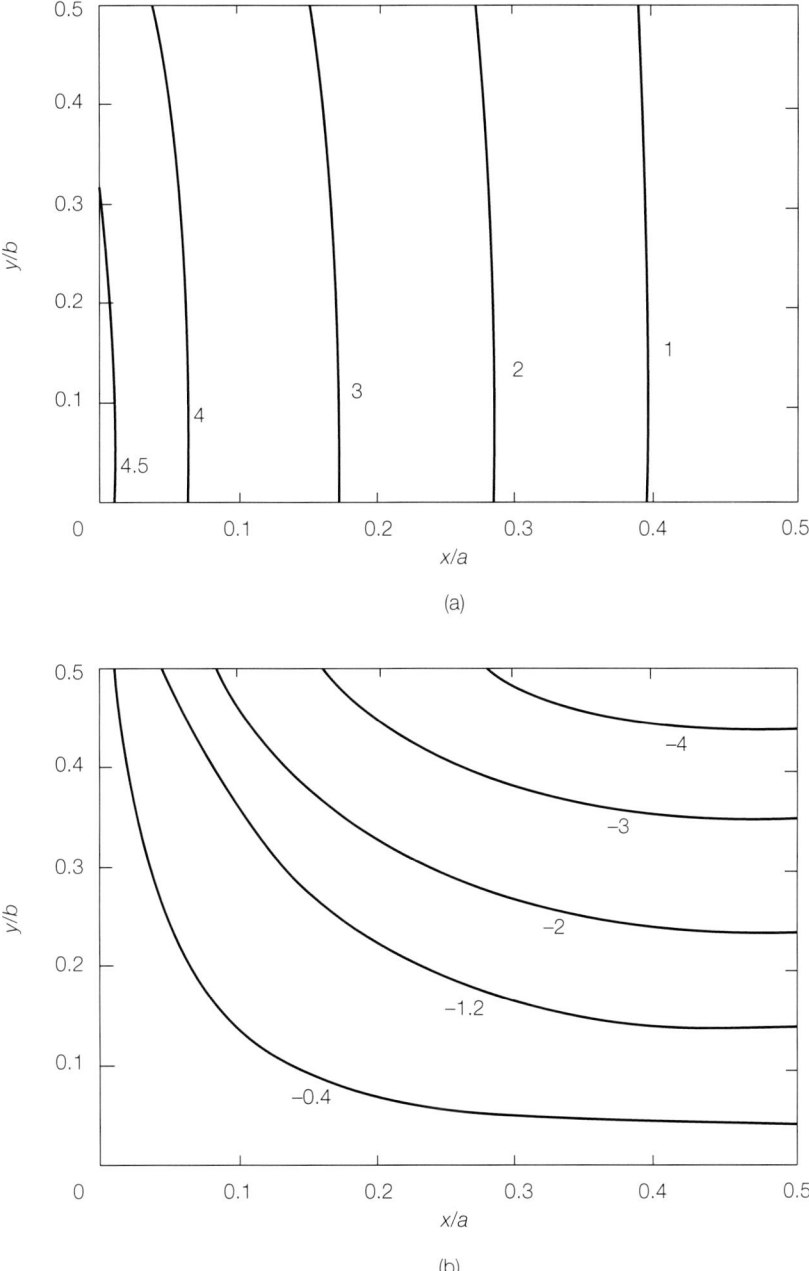

FIGURE A3.12. Contours of vertical shear force within quadrant of thin rectangular plate (width a, length $b = 0.7a$, Poisson's ratio $v = 0.22$) simply supported on two opposite edges ($x = 0, a$) with two edges 'free', subjected to uniform loading (p): (a) $V_x(x,y) \times 10/pa$; (b) $V_y(x,y) \times 10^2/pa$

TABLE A3.4. Deflection and bending moments at centre of thin rectangular plate (width a, length b, flexural stiffness D, Poisson's ratio v) simply supported on two opposite edges ($x = 0, a$) with two edges 'free', subjected to uniform loading (p)

b/a	$\dfrac{w(a/2,0)D}{pa^4}$	$\dfrac{M_x(a/2,0)}{pa^2}$	$\dfrac{M_y(a/2,0)}{pa^2}$	v
0	0.0135634	0.125	0	
0.1	0.0135455	0.124944	0.000417	
0.2	0.0134952	0.124779	0.001660	
0.3	0.0134210	0.124520	0.003625	
0.4	0.0133345	0.124211	0.006044	
0.5	0.0132462	0.123898	0.008619	
0.7	0.0130918	0.123390	0.013426	0.2
1	0.0129467	0.123050	0.018688	
1.5	0.0128799	0.123271	0.023017	
2	0.0129036	0.123795	0.024526	
3	0.0129737	0.124570	0.025083	
5	0.0130168	0.124966	0.025019	
∞	0.0130208	0.125	0.025	
0	0.0138889	0.125	0	
0.1	0.0138642	0.124938	0.000500	
0.2	0.0137944	0.124751	0.001992	
0.3	0.0136909	0.124459	0.004361	
0.4	0.0135693	0.124107	0.007294	
0.5	0.0134440	0.123750	0.010438	
0.7	0.0132205	0.123162	0.016367	0.25
1	0.0130014	0.122763	0.022954	
1.5	0.0128798	0.123008	0.028490	
2	0.0128912	0.123610	0.030488	
3	0.0129658	0.124503	0.031301	
5	0.0130160	0.124960	0.031270	
∞	0.0130208	0.125	0.03125	
0	0.0143086	0.125	0	
0.1	0.0142760	0.124933	0.000577	
0.2	0.0141835	0.124732	0.002300	
0.3	0.0140457	0.124416	0.005043	
0.4	0.0138826	0.124033	0.008461	
0.5	0.0137131	0.123642	0.012148	
0.7	0.0134059	0.122994	0.019167	0.3
1	0.0130937	0.122545	0.027078	
1.5	0.0128977	0.122806	0.033862	
2	0.0128873	0.123468	0.036389	
3	0.0129598	0.124452	0.037500	
5	0.0130153	0.124956	0.037520	
∞	0.0130208	0.125	0.0375	

ELASTIC FLEXURE OF MONOLITHIC RECTANGULAR PLATE

TABLE A3.5. Deflection, slope, bending moment and shear force at centre of 'free' edge of thin rectangular plate (width a, length b, flexural stiffness D, Poisson's ratio v) simply supported on two opposite edges ($x = 0, a$) with two edges 'free', subjected to uniform loading (p)

b/a	$\dfrac{w(a/2, b/2)D}{pa^4}$	$\dfrac{\phi_y(a/2, b/2)D}{pa^3}$	$\dfrac{M_x(a/2, b/2)}{pa^2}$	$\dfrac{V_y(a/2, b/2)}{pa}$	v
0	0.0135634	0	0.125	0	
0.1	0.0135776	0.0012876	0.125111	−0.008333	
0.2	0.0136181	0.0024889	0.125442	−0.016609	
0.3	0.0136780	0.0035282	0.125966	−0.024317	
0.4	0.0137485	0.0043657	0.126607	−0.030785	
0.5	0.0138215	0.0050023	0.127284	−0.035784	
0.7	0.0139539	0.0057874	0.128528	−0.041989	0.2
1	0.0140925	0.0062431	0.129838	−0.045595	
1.5	0.0141974	0.0063673	0.130832	−0.046577	
2	0.0142295	0.0063591	0.131136	−0.046512	
3	0.0142408	0.0063467	0.131243	−0.046414	
5	0.0142415	0.0063453	0.131250	−0.046403	
∞	0.0142415	0.0063453	0.13125	−0.046403	
0	0.0138889	0	0.125	0	
0.1	0.0139053	0.0016489	0.125125	−0.010000	
0.2	0.0139520	0.0031916	0.125498	−0.019938	
0.3	0.0140214	0.0045343	0.126089	−0.029244	
0.4	0.0141036	0.0056254	0.126817	−0.037125	
0.5	0.0141891	0.0064635	0.127591	−0.043284	
0.7	0.0143454	0.0075154	0.129024	−0.051071	0.25
1	0.0145106	0.0081498	0.130550	−0.055775	
1.5	0.0146367	0.0083451	0.131717	−0.057222	
2	0.0146756	0.0083445	0.132076	−0.057218	
3	0.0146892	0.0083318	0.132203	−0.057124	
5	0.0146902	0.0083303	0.132212	−0.057112	
∞	0.0146902	0.0083302	0.132212	−0.057112	
0	0.0143086	0	0.125	0	
0.1	0.0143269	0.0020393	0.125135	−0.011538	
0.2	0.0143788	0.0039524	0.125537	−0.023014	
0.3	0.0144562	0.0056265	0.126175	−0.033810	
0.4	0.0145483	0.0069976	0.126967	−0.043032	
0.5	0.0146446	0.0080609	0.127813	−0.050313	
0.7	0.0148219	0.0094172	0.129391	−0.059675	0.3
1	0.0150113	0.0102634	0.131088	−0.065528	
1.5	0.0151571	0.0105497	0.132397	−0.067507	
2	0.0152022	0.0105614	0.132802	−0.067588	
3	0.0152181	0.0105498	0.132945	−0.067508	
5	0.0152192	0.0105480	0.132955	−0.067496	
∞	0.0152192	0.0105480	0.132955	−0.067496	

TABLE A3.6. Slopes, shear force, reactions and corner force at edge support for thin rectangular plate (width a, length b, flexural stiffness D, Poisson's ratio v) simply supported on two opposite edges ($x = 0, a$) with two edges 'free', subjected to uniform loading (p)

b/a	$\dfrac{\phi_x(0,0)D}{pa^3}$	$\dfrac{\phi_x(0,b/2)D}{pa^3}$	$\dfrac{V_x(0,0)}{pa}$	$\dfrac{R_x(0,0)}{pa}$	$\dfrac{R_x(0,b/2)}{pa}$	$\dfrac{P_c}{pa^2}$	v
0	0.0434028	0.0434028	0.416667	0.333333	0.333333	0	
0.1	0.0433367	0.0434558	0.423981	0.348859	0.335007	0.007746	
0.2	0.0431636	0.0435954	0.431296	0.364384	0.336680	0.014316	
0.3	0.0429211	0.0437925	0.438599	0.379884	0.338350	0.019714	
0.4	0.0426452	0.0440191	0.445803	0.395163	0.339994	0.023969	
0.5	0.0423672	0.0442510	0.452715	0.409792	0.341561	0.027177	
0.7	0.0418837	0.0446688	0.464914	0.435427	0.344271	0.031121	0.2
1	0.0414314	0.0451047	0.478234	0.462798	0.347041	0.033409	
1.5	0.0412236	0.0454342	0.490267	0.485998	0.349124	0.034033	
2	0.0412984	0.0455351	0.495603	0.494996	0.349761	0.033992	
3	0.0415185	0.0455705	0.499090	0.499531	0.349985	0.033929	
5	0.0416540	0.0455729	0.499961	0.500029	0.350000	0.033922	
∞	0.0416667	0.0455729	0.5	0.5	0.35	0.033922	
0	0.0444444	0.0444444	0.4	0.3	0.3	0	
0.1	0.0443529	0.0445056	0.408651	0.318321	0.301718	0.009310	
0.2	0.0441122	0.0446670	0.417303	0.336643	0.303435	0.017239	
0.3	0.0437729	0.0448958	0.425942	0.354938	0.305150	0.023790	
0.4	0.0433841	0.0451599	0.434476	0.372998	0.306841	0.028995	
0.5	0.0429883	0.0454316	0.442690	0.390346	0.308458	0.032957	
0.7	0.0422876	0.0459247	0.457274	0.420945	0.311271	0.037913	0.25
1	0.0416027	0.0464443	0.473360	0.453984	0.314174	0.040901	
1.5	0.0412231	0.0468407	0.488041	0.482376	0.316376	0.041821	
2	0.0412593	0.0469627	0.494591	0.493552	0.317053	0.041818	
3	0.0414937	0.0470056	0.498879	0.499319	0.317292	0.041758	
5	0.0416515	0.0470085	0.499952	0.500028	0.317308	0.041750	
∞	0.0416667	0.0470085	0.5	0.5	0.317308	0.041750	
0	0.0457875	0.0457875	0.384615	0.269231	0.269231	0	
0.1	0.0456664	0.0458555	0.394462	0.290031	0.270913	0.010758	
0.2	0.0453466	0.0460352	0.404308	0.310830	0.272595	0.019955	
0.3	0.0448937	0.0462907	0.414142	0.331605	0.274275	0.027594	
0.4	0.0443712	0.0465869	0.423869	0.352138	0.275933	0.033707	
0.5	0.0438347	0.0468931	0.433255	0.371919	0.277524	0.038403	
0.7	0.0428698	0.0474527	0.450019	0.407027	0.280306	0.044370	0.3
1	0.0418923	0.0480481	0.468685	0.445340	0.283203	0.048090	
1.5	0.0412793	0.0485062	0.485889	0.478710	0.285420	0.049349	
2	0.0412471	0.0486479	0.493610	0.492039	0.286106	0.049400	
3	0.0414751	0.0486979	0.498676	0.499073	0.286347	0.049349	
5	0.0416493	0.0487013	0.499943	0.500023	0.286364	0.049341	
∞	0.0416667	0.0487013	0.5	0.5	0.286364	0.049341	

ELASTIC FLEXURE OF MONOLITHIC RECTANGULAR PLATE 205

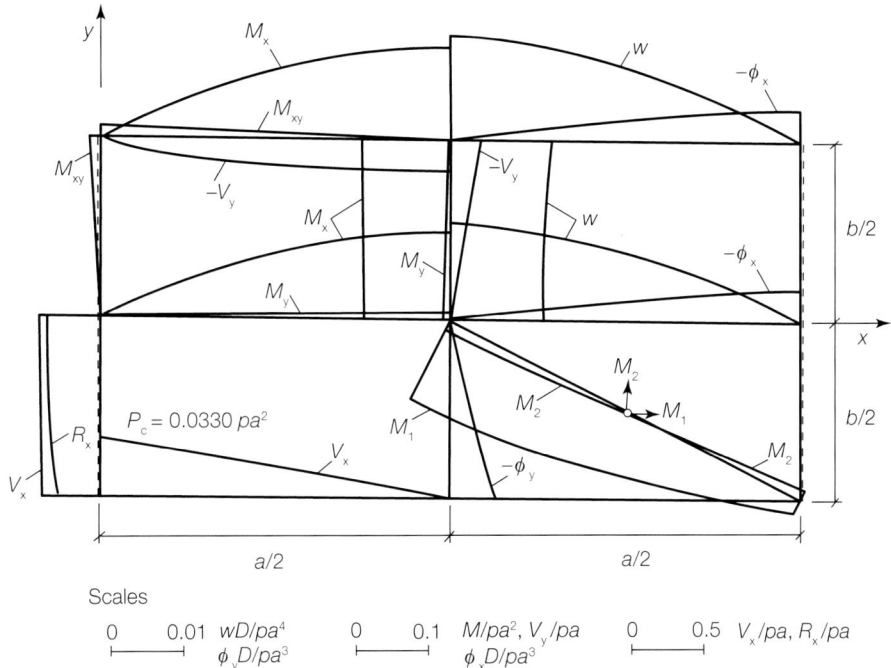

FIGURE A3.13. Elastic flexure of thin rectangular plate (width a, length $b = 0.5a$, Poisson's ratio $v = 0.25$, flexural stiffness D) simply supported on two opposite edges ($x = 0, a$) with two edges 'free', subjected to uniform loading (p)

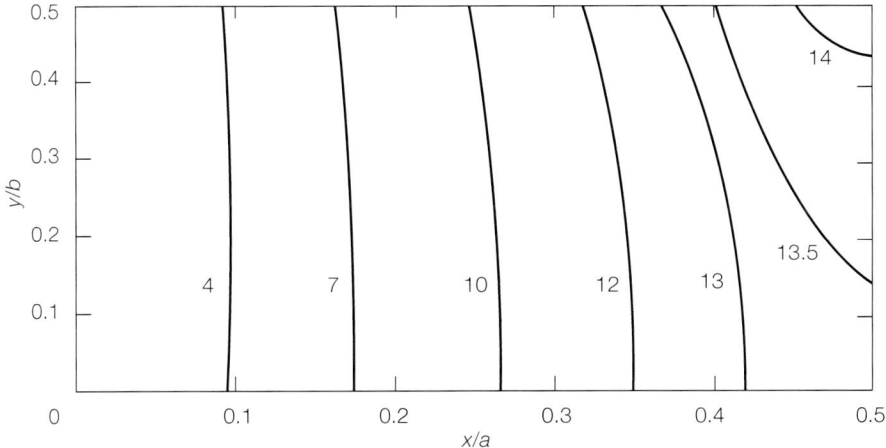

FIGURE A3.14. Deflection contours $w(x, y)D \times 10^3 / pa^4$ within quadrant of thin rectangular plate (width a, length $b = 0.5a$, Poisson's ratio $v = 0.25$, flexural stiffness D) simply supported on two opposite edges ($x = 0, a$) with two edges 'free', subjected to uniform loading (p)

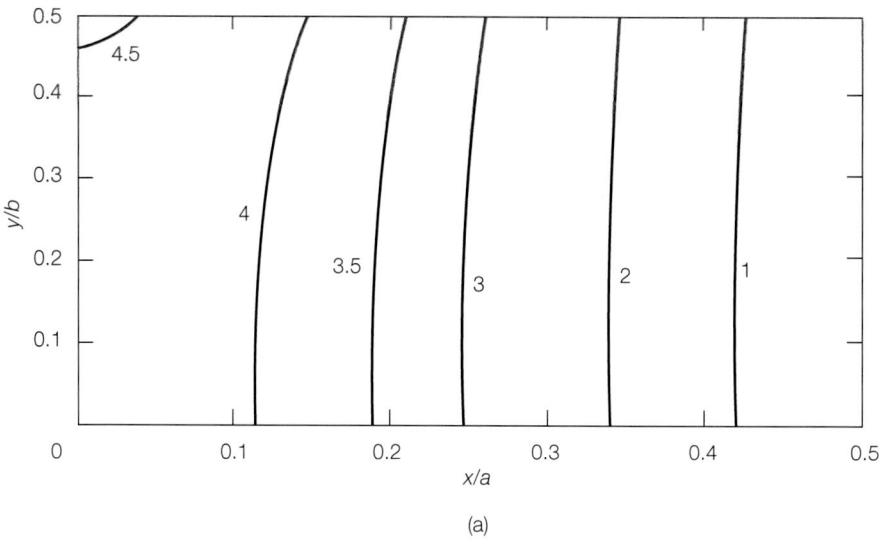

(a)

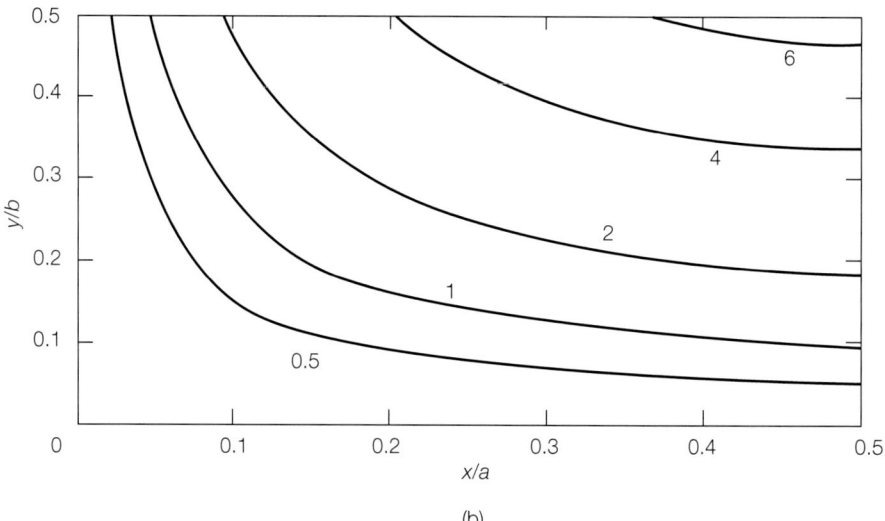

(b)

FIGURE A3.15. Contours of slope within quadrant of thin rectangular plate (width a, length $b = 0.5a$, Poisson's ratio $v = 0.25$, flexural stiffness D) simply supported on two opposite edges ($x = 0, a$) with two edges 'free', subjected to uniform loading (p): (a) $\phi_x(x,y)D \times 10^2/pa^3$; (b) $\phi_y(x,y)D \times 10^3/pa^3$

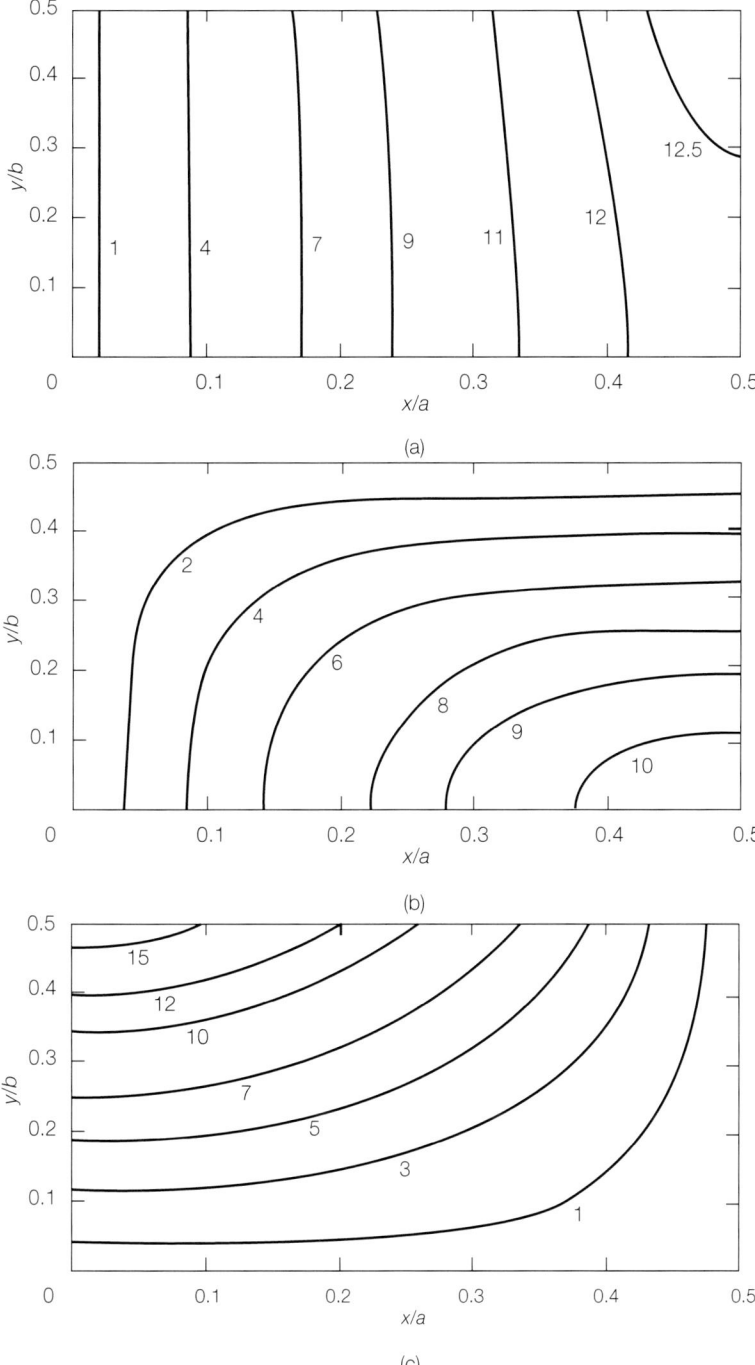

FIGURE A3.16. Contours of bending and twisting moments within quadrant of thin rectangular plate (width a, length $b = 0.5a$, Poisson's ratio $\nu = 0.25$) simply supported on two opposite edges ($x = 0, a$) with two edges 'free', subjected to uniform loading (p): (a) $M_x(x,y) \times 10^2 / pa^2$; (b) $M_y(x,y) \times 10^3 / pa^2$; (c) $M_{xy}(x,y) \times 10^3 / pa^2$

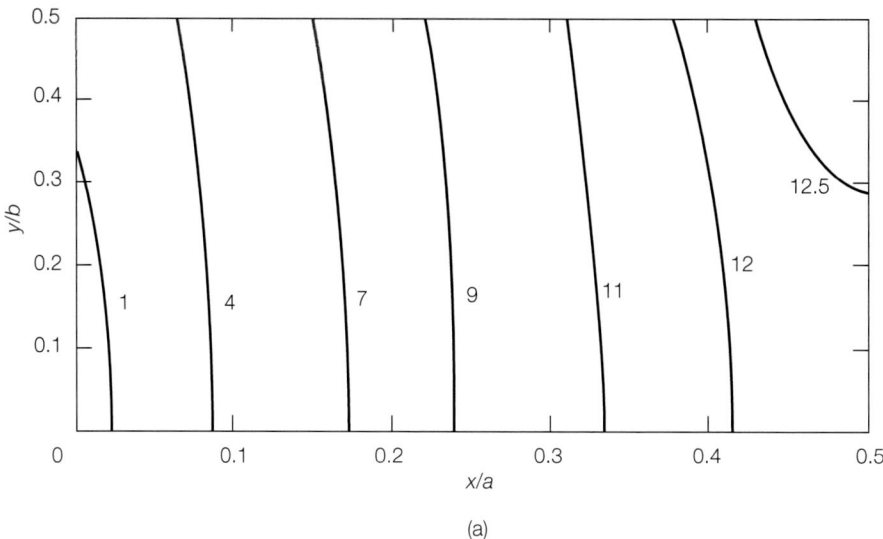

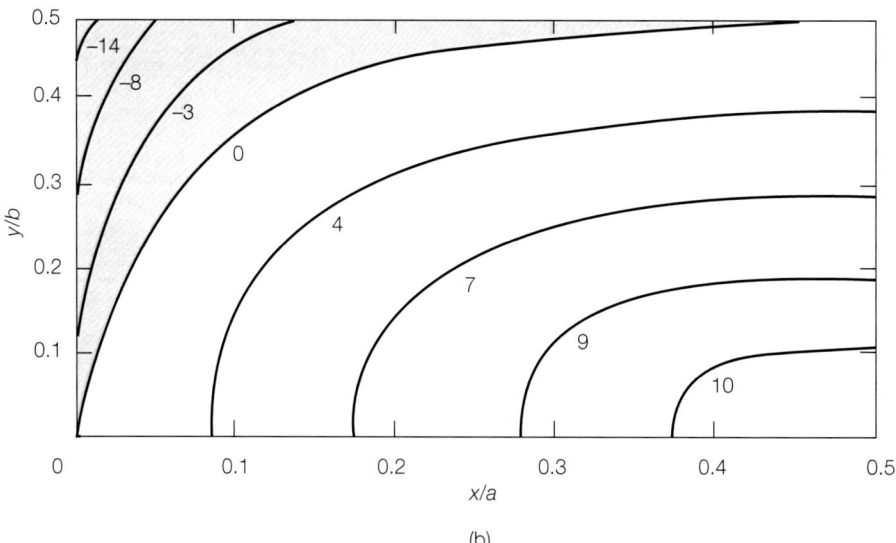

FIGURE A3.17. Contours of major and minor principal moments within quadrant of thin rectangular plate (width a, length $b = 0.5a$, Poisson's ratio $v = 0.25$) simply supported on two opposite edges ($x = 0, a$) with two edges 'free', subjected to uniform loading (p): (a) $M_1(x,y) \times 10^2 / pa^2$; (b) $M_2(x,y) \times 10^3 / pa^2$

ELASTIC FLEXURE OF MONOLITHIC RECTANGULAR PLATE

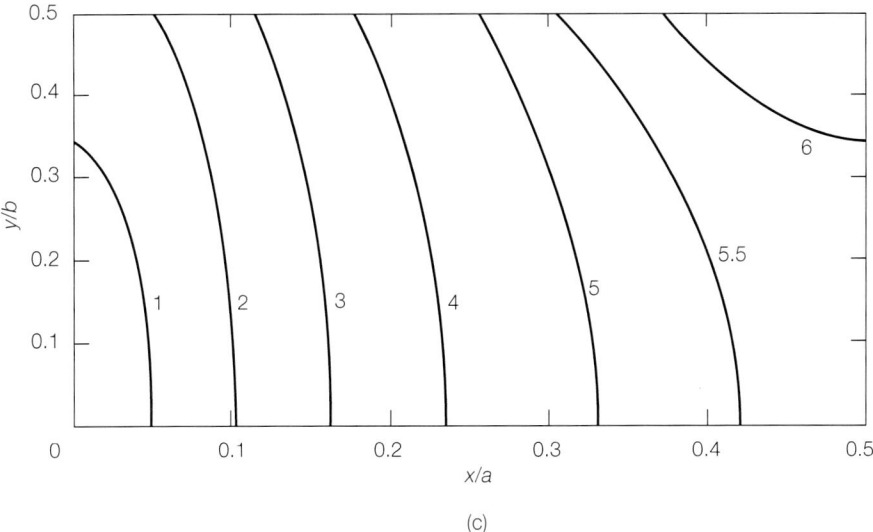

(c)

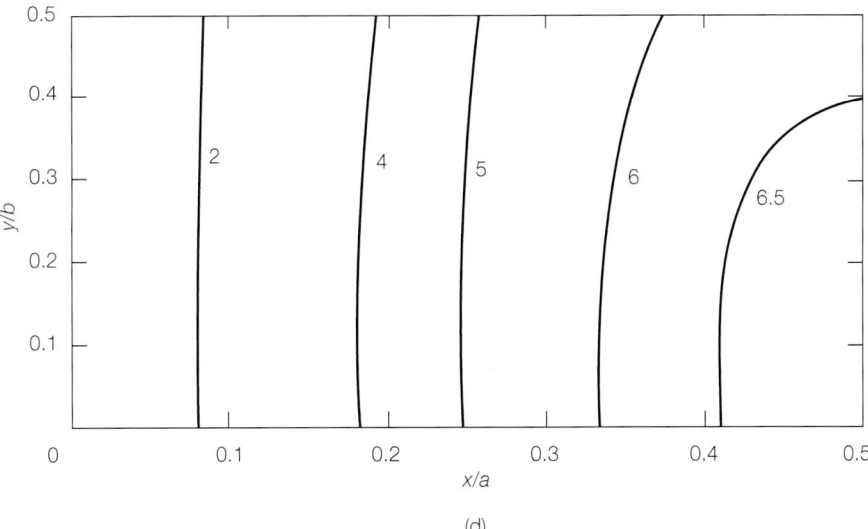

(d)

FIGURE A3.17. (continued) Contours of major and minor principal moments within quadrant of thin rectangular plate (width a, length $b = 0.5a$, Poisson's ratio $\nu = 0.25$) simply supported on two opposite edges ($x = 0, a$) with two edges 'free', subjected to uniform loading (p): (c) $M^*_{xy}(x,y) \times 10^2 / pa^2$; (d) $\bar{M}(x,y) \times 10^2 / pa^2$

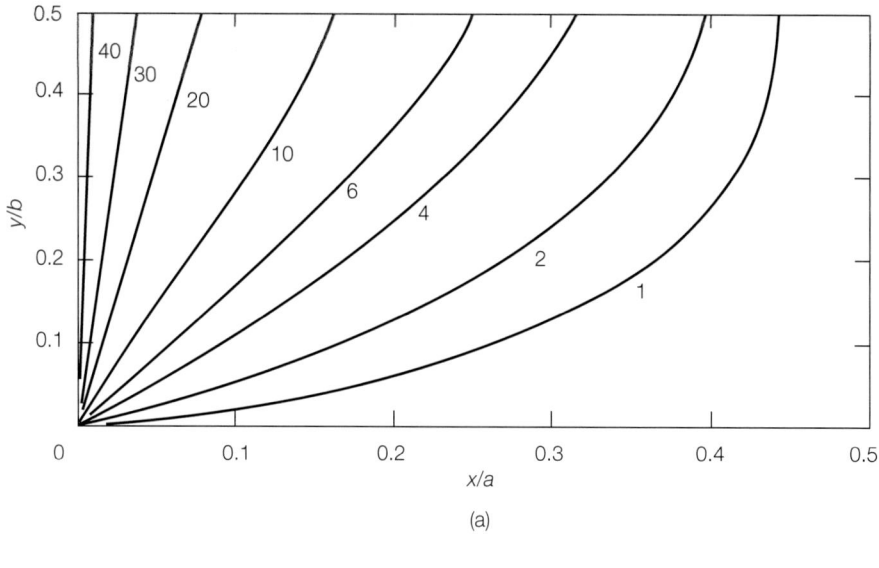

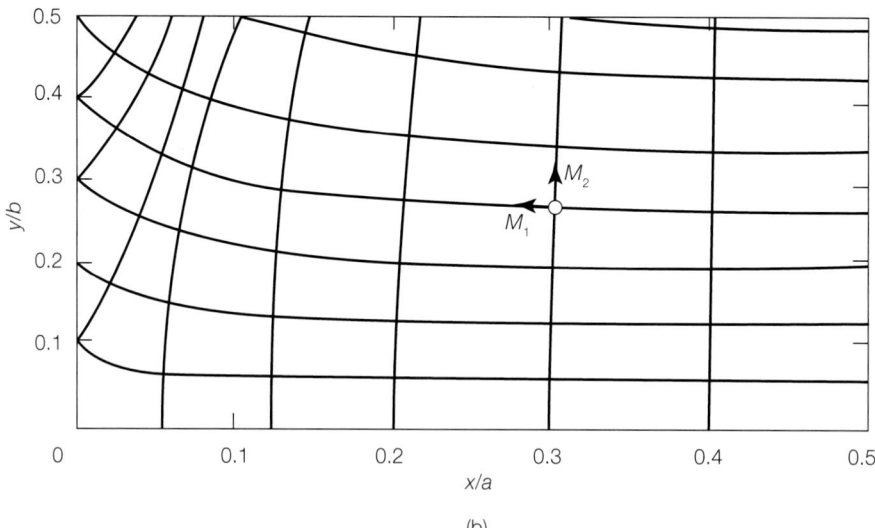

FIGURE A3.18. Orientation of principal bending moments within quadrant of thin rectangular plate (width a, length $b = 0.5a$, Poisson's ratio $v = 0.25$) simply supported on two opposite edges ($x = 0, a$) with two edges 'free', subjected to uniform loading (p): (a) contours of $\theta_1°$; (b) principal moment trajectories

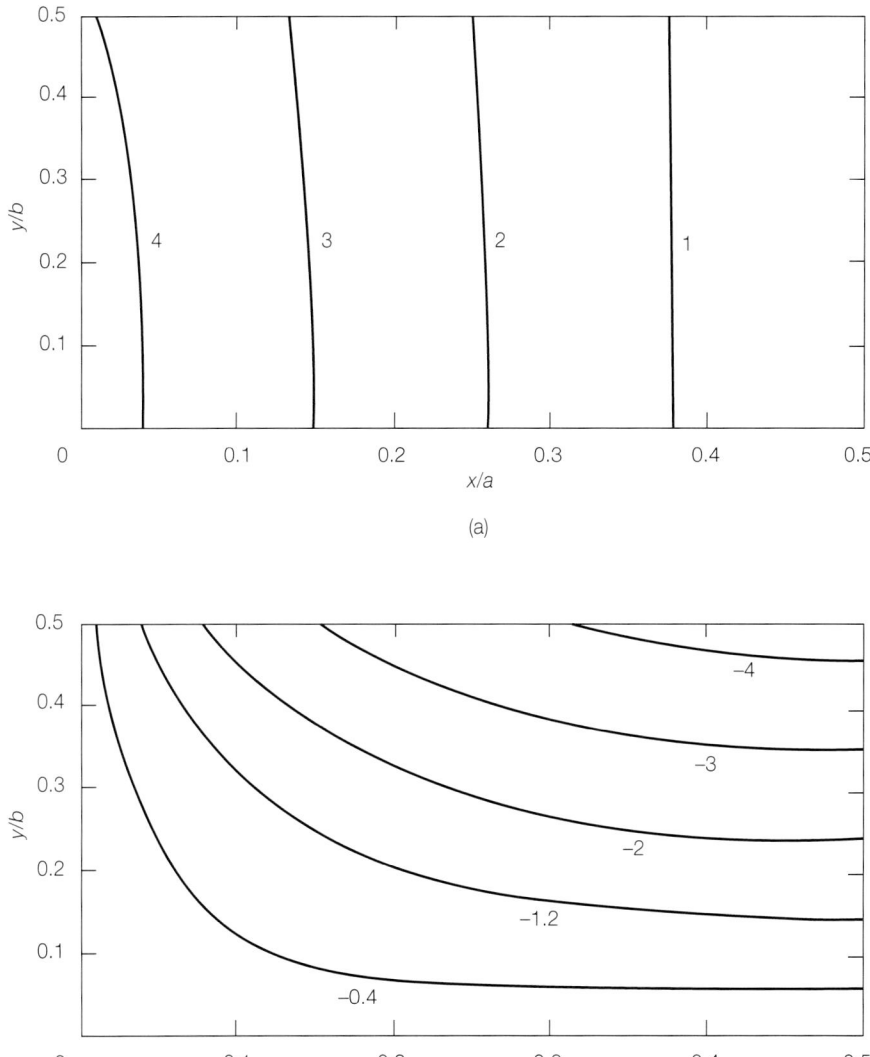

FIGURE A3.19. Contours of vertical shear force within quadrant of thin rectangular plate (width a, length $b = 0.5a$, Poisson's ratio $v = 0.25$) simply supported on two opposite edges ($x = 0, a$) with two edges 'free', subjected to uniform loading (p): (a) $V_x(x,y) \times 10/pa$; (b) $V_y(x,y) \times 10^2/pa$

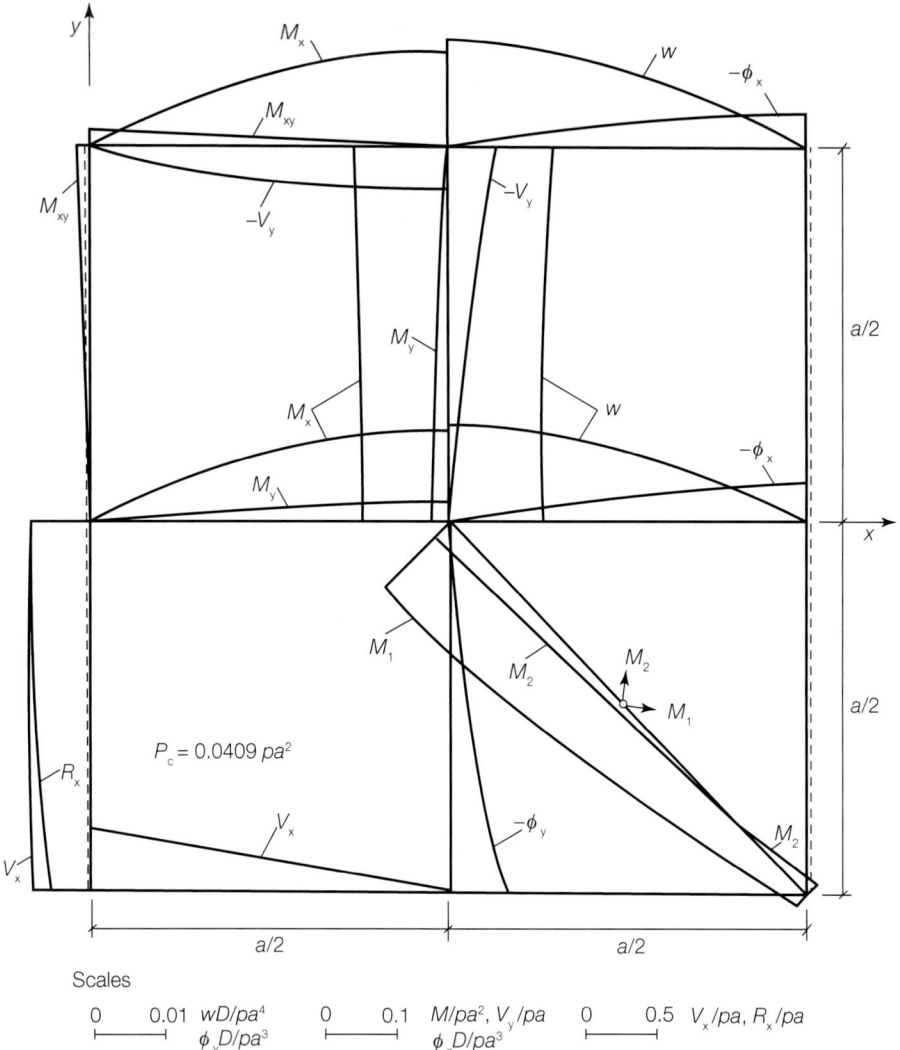

FIGURE A3.20. Elastic flexure of thin square plate (side length a, Poisson's ratio $v = 0.25$, flexural stiffness D) simply supported on two opposite edges ($x = 0, a$) with two edges 'free', subjected to uniform loading (p)

ELASTIC FLEXURE OF MONOLITHIC RECTANGULAR PLATE

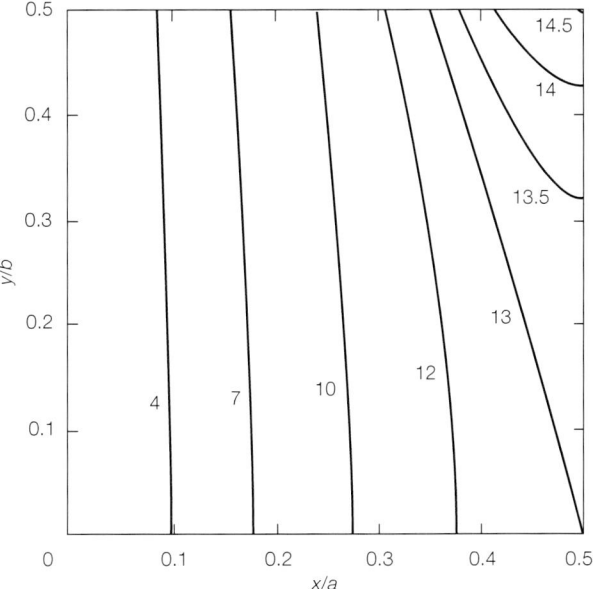

FIGURE A3.21. Deflection contours $w(x,y)D \times 10^3/pa^4$ within quadrant of thin square plate (side length a, Poisson's ratio $\nu = 0.25$, flexural stiffness D) simply supported on two opposite edges ($x = 0, a$) with two edges 'free', subjected to uniform loading (p).

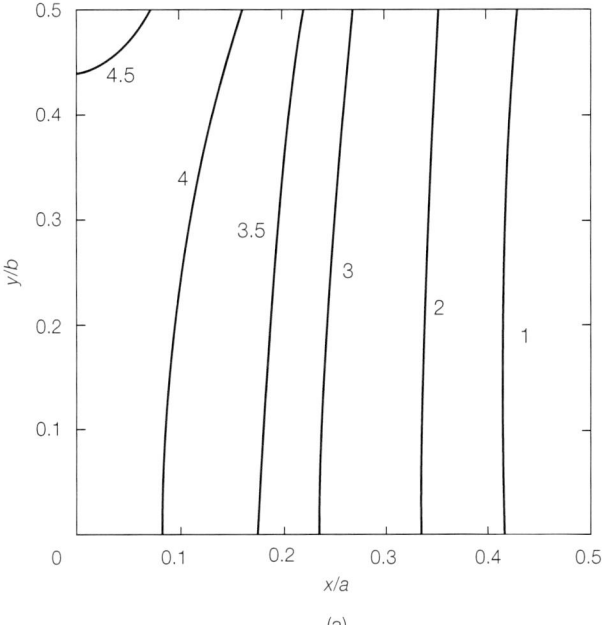

(a)

FIGURE A3.22. Contours of slope within quadrant of thin square plate (side length a, Poisson's ratio $\nu = 0.25$, flexural stiffness D) simply supported on two opposite edges ($x = 0, a$) with two edges 'free', subjected to uniform loading (p): (a) $\phi_x(x,y)D \times 10^2/pa^3$

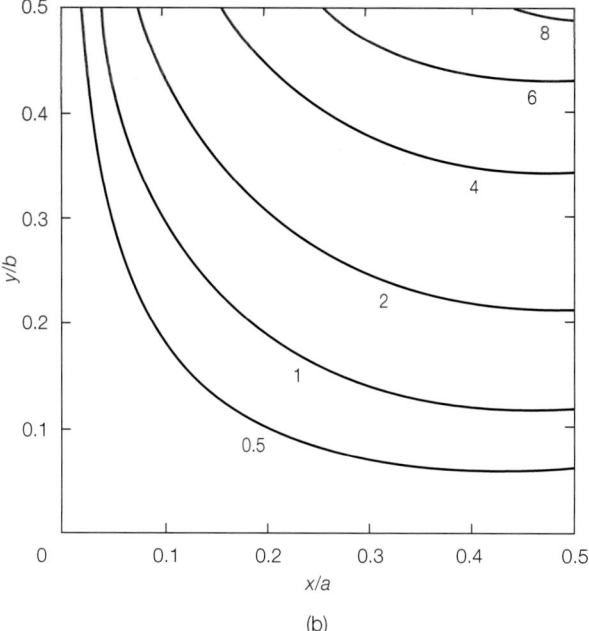

(b)

FIGURE A3.22. (continued) Contours of slope within quadrant of thin square plate (side length a, Poisson's ratio $v = 0.25$, flexural stiffness D) simply supported on two opposite edges ($x = 0, a$) with two edges 'free', subjected to uniform loading (p): (b) $\phi_y(x,y)D \times 10^3/pa^3$

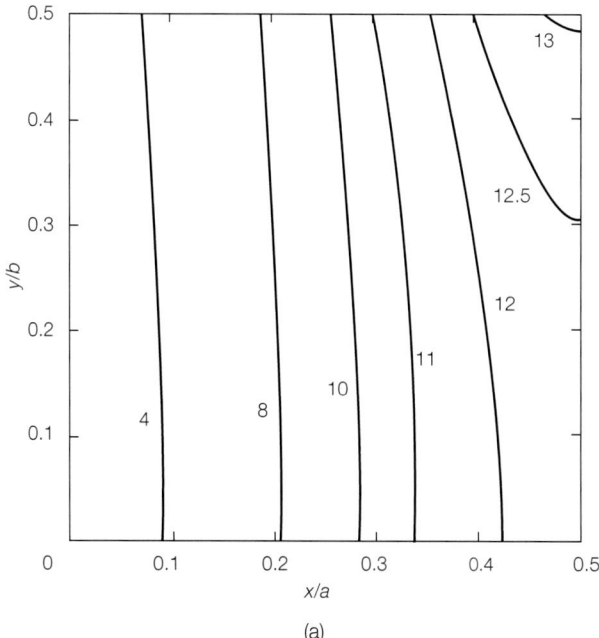

(a)

FIGURE A3.23. Contours of bending and twisting moments within quadrant of thin square plate (side length a, Poisson's ratio $v = 0.25$) simply supported on two opposite edges ($x = 0, a$) with two edges 'free', subjected to uniform loading (p): (a) $M_x(x,y) \times 10^2/pa^2$

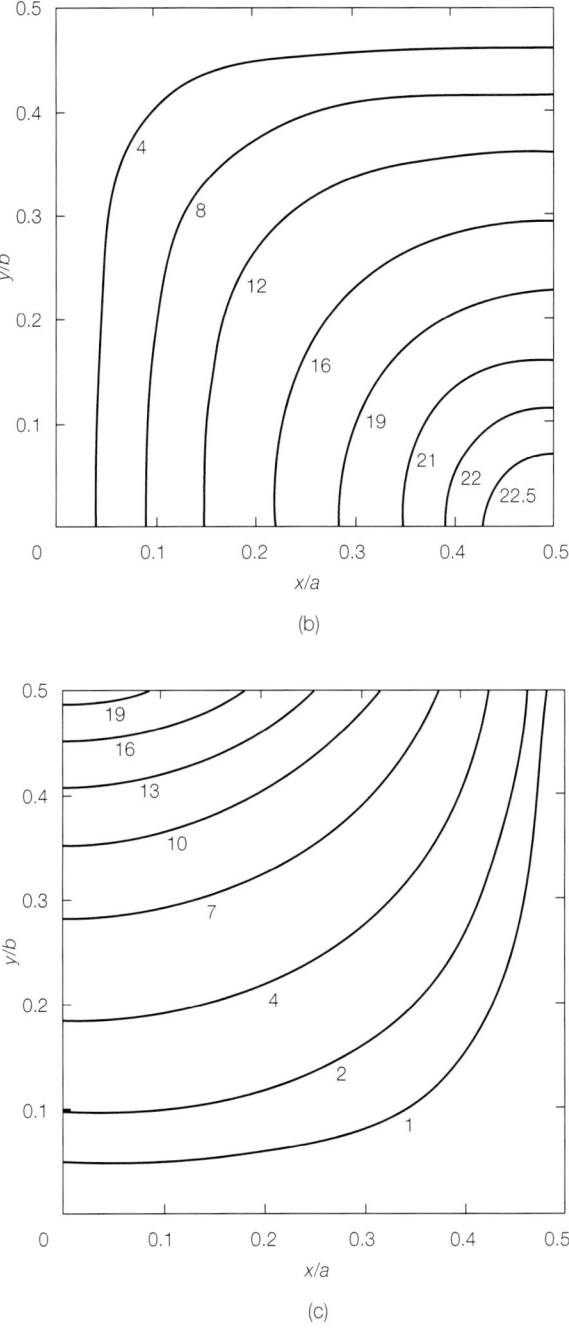

FIGURE A3.23. (continued) Contours of bending and twisting moments within quadrant of thin square plate (side length a, Poisson's ratio $\nu = 0.25$) simply supported on two opposite edges ($x = 0, a$) with two edges 'free', subjected to uniform loading (p): (b) $M_y(x,y) \times 10^3 / pa^2$; (c) $M_{xy}(x,y) \times 10^3 / pa^2$

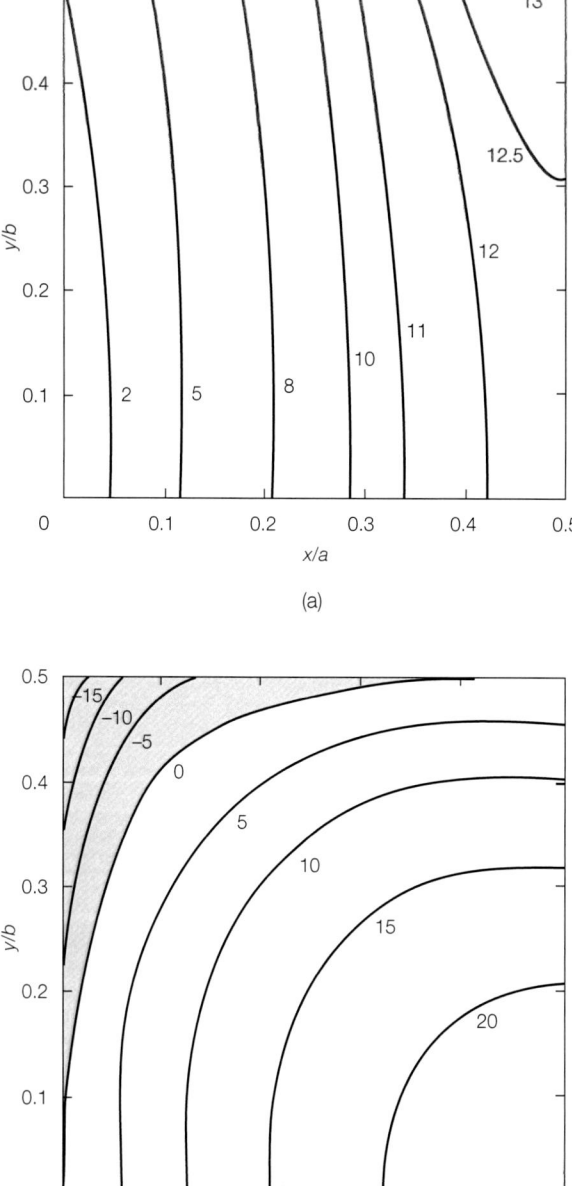

FIGURE A3.24. Contours of major and minor principal moments within quadrant of thin square plate (side length a, Poisson's ratio $v = 0.25$) simply supported on two opposite edges ($x = 0, a$) with two edges 'free', subjected to uniform loading (p): (a) $M_1(x,y) \times 10^2 / pa^2$; (b) $M_2(x,y) \times 10^3 / pa^2$

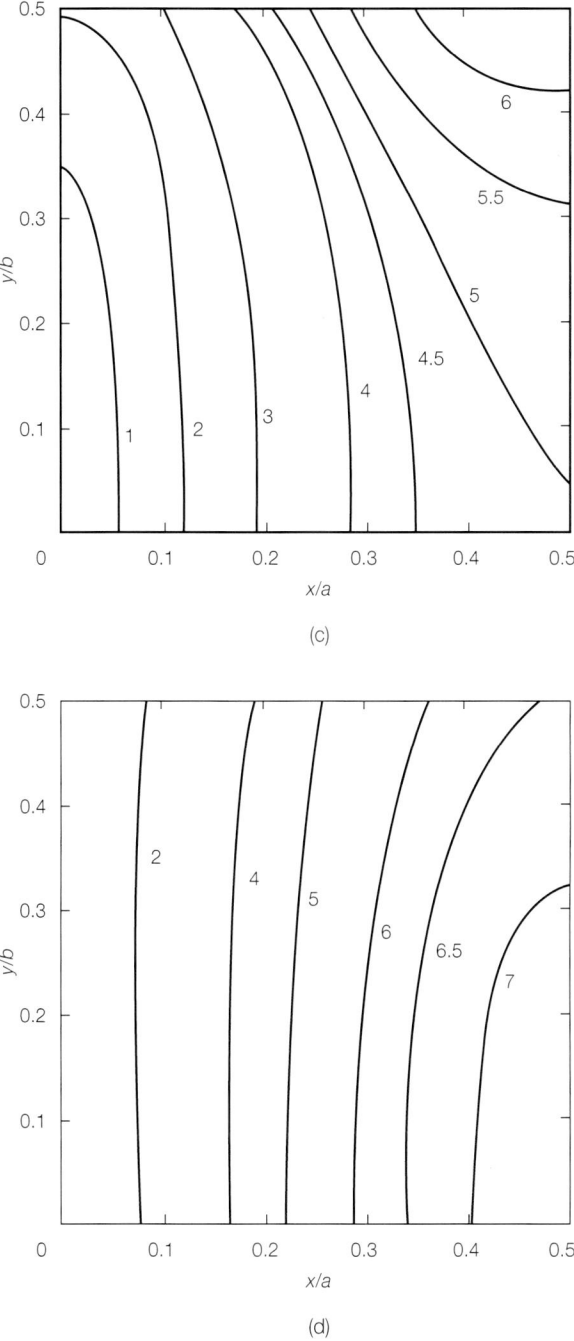

FIGURE A3.24. (continued) Contours of major and minor principal moments within quadrant of thin square plate (side length a, Poisson's ratio $v = 0.25$) simply supported on two opposite edges ($x = 0, a$) with two edges 'free', subjected to uniform loading (p): (c) $M^*_{xy}(x, y) \times 10^2 / pa^2$; (d) $\bar{M}(x, y) \times 10^2 / pa^2$

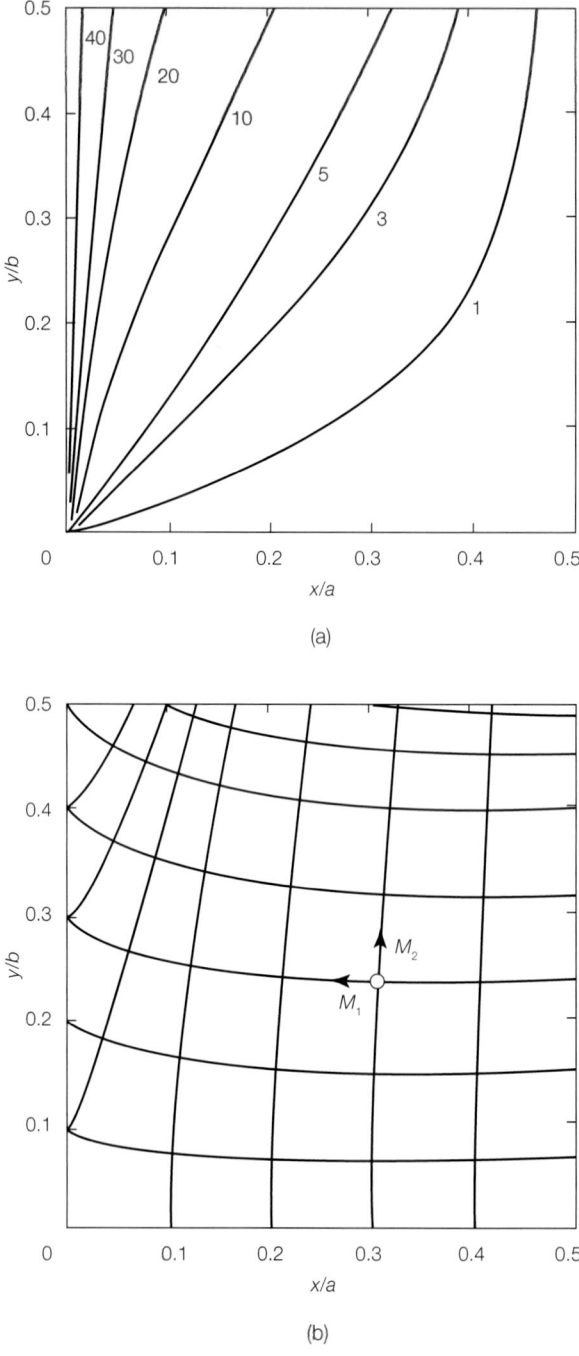

FIGURE A3.25. Orientation of principal bending moments within quadrant of thin square plate (side length a, Poisson's ratio $v = 0.25$) simply supported on two opposite edges ($x = 0, a$) with two edges 'free', subjected to uniform loading (p): (a) contours of $\theta_1°$; (b) principal moment trajectories

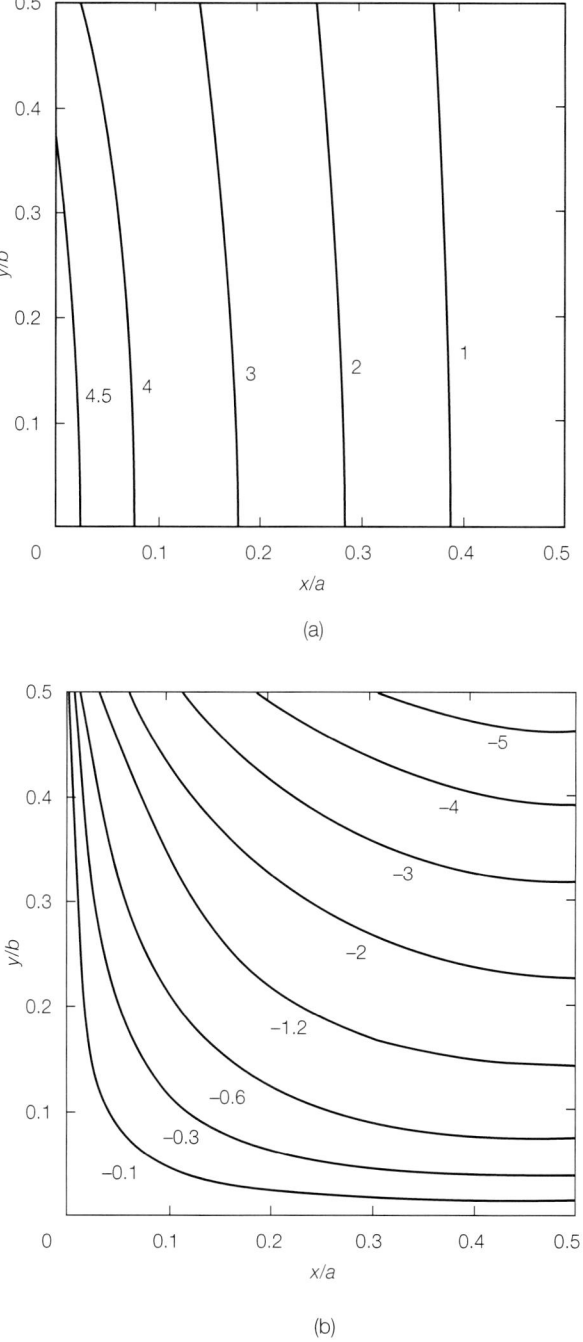

FIGURE A3.26. Contours of vertical shear force within quadrant of thin square plate (side length a, Poisson's ratio $v = 0.25$) simply supported on two opposite edges ($x = 0, a$) with two edges 'free', subjected to uniform loading (p): (a) $V_x(x,y) \times 10/pa$; (b) $V_y(x,y) \times 10^2/pa$

Acknowledgement

The writer is greatly indebted to Chi Hung Chan – a long-standing friend and former research associate – for his most generous computational assistance.

A3.6. References

A3.1. TODHUNTER, I., edited and completed by PEARSON, K. *A history of the theory of elasticity and of the strength of materials from Galilei to Lord Kelvin*. Cambridge Univ. Press, Cambridge, 1886, Vol. I, *Galilei to Saint-Venant 1639–1850*; 1893, Vol. II (Pts I & II), *Saint-Venant to Lord Kelvin*. Constable, London; Dover Publ., New York, 1960.

A3.2. LAGRANGE, J. L. Note communiquée aux Commissaires pour le prix de la surface élastique (décembre 1811).

A3.3. GERMAIN, S. *Recherches sur la théorie des surfaces élastiques*. Mme Ve Courcier, Libraire pour les science, Paris, 1821, 118 pp.

A3.4. GERMAIN, S. *Remarques sur la nature, les bornes et l'étendue de la question des surfaces élastiques, et équation générale de ces surfaces*. Imprimerie Huzard-Courcier, Paris, 1826, 24 pp.

A3.5. GERMAIN, S. Examen des principes qui peuvent conduire á la connaissance des lois de l'équilibre et du mouvement des solides élastiques. *Annales de Chimie et de Physique*, Paris, 1828, Ser. 2, 38, 123–131.

A3.6. NAVIER, C. L. Remarques sur l'Article de M. Poisson, insére dans le Cahier d'août, page 435. *Annales de Chimie et de Physique*, Paris, 1828, Ser. 2, 39, 145–151.

A3.7. BUCCIARELLI, L. L. and DWORSKY, N. *Sophie Germain: an essay in the history of the theory of elasticity*. Reidel, Dordrecht, 1980. (Series: *Studies in the history of modern science*, Vol. 6).

A3.8. LÉVY, M. Sur l'équilibre élastique d'une plaque rectangulaire. *Comptes Rendus des Séances de l'Académie des Sciences*, Paris, 1899, Ser. 2, 129, 15, Oct., 535–539.

A3.9. ESTANAVE, E. Contribution à l'étude de l'équilibre élastique d'une plaque rectangulaire mince. *Annales Scientifiques de l'École Normale Supérieure*, 1900, Ser. 3, 17, Jul.–Aug., 295–358.

A3.10. NÁDAI, Á. *Die Formänderungen und die Spannungen von rechteckigen elastischen Platten*. Julius Springer, Berlin, 1915. (Series: *Forschungsarbeiten auf dem Gebiete des Ingenieurwissens*, Vols 170–171).

A3.11. NÁDAI, Á. *Die elastischen Platten. Die Grundlagen und Verfahren zur Berechnung ihrer Fromänderungen und Spannungen, sowie die Anwendungen der Theorie der ebenen zweidimensionalen elastischen Systeme auf praktische Aufgaben*. Julius Springer, Berlin, 1925. Reprinted 1952; Springer-Verlag, New York, 1968.

A3.12. MARCUS, H. *Die Theorie elastischer Gewebe und ihre Anwendung auf die Berechnung biegsamer Platten: unter besonderer Berücksichtigung der trägerlosen Pilzdecken*. Julius Springer, Berlin, 1924, 1st edn; 1932, 2nd edn.

A3.13. GALERKIN, B. G. *Thin elastic plates*. Gosstrojizdat, Leningrad–Moscow, 1933 (in Russian).

A3.14. HOLL, D. L. *Analysis of thin rectangular plates supported on opposite edges*. Iowa State College of Agriculture & Mechanic Arts, Ames, Iowa, U.S.A., Vol. 35, No. 30, Dec. 1936; Iowa Engineering Experiment Station, Bull. 129, 100 pp.

A3.15. HEMSLEY, J. A. *Glass in engineering science*. Society of Glass Technology, Sheffield, 2015, Vol. 1 (*Optical birefringence in glass*).

A3.16. TIMOSHENKO, S. P. and WOINOWSKY-KRIEGER, S. *Theory of plates and shells*. McGraw-Hill, New York, 1959, 2nd edn.

A3.17. SALERNO, V. L. and GOLDBERG, M. A. Effect of shear deformations on the bending of rectangular plates. *J. Appl. Mech., Amer. Soc. Mech. Engrs*, 1960, 27, 1 (Mar.), 54–58.

A3.18. VOYIADJIS, G. Z., BALUCH, M. H. and CHI, W. K. Effects of shear and normal strain on plate bending. *J. Engng Mech., Amer. Soc. Civ. Engrs*, 1985, 111, 9 (Sept.), 1131–1143. Erratum, 1986, 112, 12 (Dec.), 1391.

A3.19. ZHONG, Y., LI, R., LIU, Y. and TIAN, B. On new symplectic approach for exact bending solutions of moderately thick rectangular plates with two opposite edges simply supported. *Int. J. Solids & Struct.*, 2009, 46, 11–12, Jun., 2505–2513.

A3.20. FRISCH-FAY, R. *Flexible bars*. Butterworths, London, 1962.

A3.21. IYENGAR, K. T. S. R. and RAO, S. K. L. Large deflections of simply supported beams. *J. Franklin Inst.*, 1955, 259, 6 (Jun.), 523–528.

A3.22. WANG, T. M., LEE, S. L. and ZIENKIEWICZ, O. C. A numerical analysis of large deflections of beams. *Int. J. Mech. Sci.*, 1961, 3, 3 (Nov.), 219–228.

A3.23. WANG, T. M. Non-linear bending of beams with uniformly distributed loads. *Int. J. Non-linear Mech.*, 1969, 4, 4, 389–395.

A3.24. STIPPES, M. Large deflections of rectangular plates. *Proc. 1st U.S. Nat. Congr. Appl. Mech., Chicago, Illinois, U.S.A., Jun. 1951*, 339–345. Amer. Soc. Mech. Engrs, New York, 1952 (ed. E. Sternberg).

A3.25. GREAVES, G. N., GREER, A. L., LAKES, R. S. and ROUXEL, T. Poisson's ratio and modern materials. *Nature Mater.*, 2011, 10, 11 (Nov.), 823–837.

A3.26. ABRAMOWITZ, M. and STEGUN, I. A. (eds). *Handbook of mathematical functions, with formulas, graphs, and mathematical tables*. Dover, New York, 1965.

A4. HAILSTONE IMPACT: ILLUSTRATED HISTORICAL PROLOGUE, WITH LITERARY ALLUSIONS IN PROSE AND VERSE

A4.1. Introduction

Although it is widely recognised in the modern world that severe hailstorms may cause widespread damage to crops and property, it may come as a surprise – especially to inhabitants residing in comparatively benign climates, where even hailstones the size of *petits pois* might be disquieting – to learn of the sheer magnitude and scale of such damage, as well as serious injury or worse to living creatures, that can occur in exceptional circumstances. But such destructive natural phenomena have been recorded by various means of human activity during several millennia, and although tangible evidence usually is scarce, the underlying leitmotif changes little throughout the ages. In what follows, selected extracts taken almost entirely from the vast reserves of *non-scientific* world literature are given verbatim, augmented where possible by manuscript artifacts, interpretive artwork and other illustrations. Indeed, long before the invention of writing, it is quite possible that some prehistoric cave-dwellers featured hail amongst primitive drawings on the bare walls or roof of their humble abode.

Much of the literature concerning the maleficent influence of hailstorms has emerged through a long process of cultural osmosis, commonly reflecting a blurred distinction between the natural and the supernatural. The presentation of such material herein is broadly chronological, with occasional thematic interventions or minor detours. These accounts – largely in the form of literary prose and poetry – tend to encompass myths, fables, spells and witchcraft, invariably culminating in a wide variety of local customs and ritualistic characterisations, some of which are more palatable than others; with those of a religious nature often demonstrating the marked contrast between divine strength and human frailty, whilst emphasising basic atavistic instincts. In those cases which ostensibly relate to real events, historical reliability and interpretation are questionable in some instances, as is the connection between cause and effect, given the extent to which extreme weather events are shaped by random natural forces. Moreover, the conversion of archaic written material into modern text can lead to divergencies within such a capacious literary archive.

A4.2. Ancient history (pre 500 A.D.)

Figures of speech encompassing hailstorms appear in ancient Egyptian writings, including the *Pyramid Texts of Unas* (*circa* 2350 B.C.) relating to the afterlife of King Unas (reigned *circa* 2378–2348 B.C.), the ninth and last ruler of the Fifth Dynasty of Egypt. This large corpus of funerary compositions and related literature – possibly the oldest surviving religious texts in the world – was discovered at the *Pyramid of Unas*, one of the royal pyramids built at Saqqara, south of modern-day Cairo. Here it is reckoned that a selection of much older anonymous texts on papyrus (possibly dating up to 3200 B.C., around the time of the invention of writing) have been transposed as hieroglyphs inscribed on the subterranean walls of the tomb. English language translations include those by Mercer [A4.1], Faulkner [A4.2] and Allen [A4.3], wherein noticeable differences are to be found in transliteration and interpretation, as well as in the adopted system of line reference numbering.

Within the burial antechamber, among the lavish array of carved and painted reliefs, it is revealed – according to a recent version of the text produced by the Belgian scholar Wim van den Dungen [A4.4] – that while addressing the deities, the mighty pharaoh remarks (South Wall, part *Utterance 262*):

> '*The Sun-folk have testified concerning me. The hailstorms of the sky have taken me so that they might raise me up to Re.*'

and later, upon sighting the all-powerful Sun God *Re*, implores (North Wall, part *Utterance 311*):

> '*You should take me with You, with You!*
> *He who will drive away storms for You, who will dispel the clouds for You, and who will break up the hail for You.*'

thereby assuring the king's ascent into the sky, to take his place among the gods in the celestial after-world. Figure A4.1 gives a partial view of the antechamber, after Piankoff [A4.5]. Figure A4.1(a) illustrates the gabled West Wall (left) and the North Wall (right), together with a star-patterned inclined roof; Figure A4.1(b) shows a typical closer view of the hieroglyphic text, including four oval cartouches of Unas (topped by the desert hare).

One of the earliest known descriptions of a hailstorm occurs in a Sumerian narrative poem closely related to the *Epic of Gilgamesh*, a Babylonian poem in Akkadian and recognised as a masterpiece in world literature concerning the struggle for eternal life. Poems about Gilgamesh were written in both the Akkadian and Sumerian languages, reflecting the bilingual urban civilisation prevailing in the third millennium B.C. This corpus includes five Sumerian poems of anonymous authorship and great antiquity, which probably were court entertainments sung by minstrels for *King Shulgi of Ur*, whose reign began in the 21st century B.C. They might be regarded as source material for the Babylonian epic.

One such Sumerian poem, *Gilgamesh and the Netherworld*, also known to the ancients as *In those days, in those far-off days*, begins with a message that, shortly after the gods had

HAILSTONE IMPACT: HISTORICAL PROLOGUE 225

(a) (b)

FIGURE A4.1. Inscribed wall texts within *Pyramid of King Unas* at Saqqara, Egypt (*circa* 2350 B.C.): (a) part antechamber with North Wall and gabled West Wall (by courtesy of *Princeton University Press*); (b) part closer view of hieroglyphic text, including four oval cartouches of Unas

divided the universe between them, there occurred a mighty hailstorm. Whereupon the heroic warrior *Gilgamesh* (semi-divine yet mortal, tyrannical ruler of the city of Uruk in southern Babylonia) is addressed by the goddess *Inanna*, brother of *Utu* (*Sun God*), and told a story of a mythical storm that happened when the deities *An* (*Sky God*), *Enlil* (deiform ruler of *Earth*), *Enki* (God of the *Ocean Below*) and *Ereshkigal* (queen of the *Netherworld*, realm of deathly spirits) took up their cosmic abodes. According to a contemporary translation by the English Assyriologist Andrew George [A4.6], [A4.7], this small segment of the mythological composition reads as follows:

> '*Day was dawning, the horizon brightening,*
> *birds were singing in chorus to the dawn,*
> *as the Sun God came forth from his chamber,*
> *and his sister, the holy Inanna,*
> *spoke to the warrior Gilgamesh:*
> *O brother, in those days, after destinies were determined,*
> *when the land flowed with abundance,*
> *when An had made off with the heavens,*
> *Enlil had made off with the earth,*
> *and he had given the Netherworld to Ereshkigal as a dowry-gift,*

after he had set sail, after he had set sail,
after the father had set sail for the Netherworld,
after Enki had set sail for the Netherworld,
on the lord the small ones poured down,
on Enki the big ones poured down –
the small ones were hammer-stones,
the big ones were reed-crushing stones –
into the bottom of Enki's boat
they poured in heaps like surging turtles.'

In the context of a great storm, the stone tools that fell on *Enki* – hammer-stones and reed-crushing stones – are obvious metaphors for giant hailstones. Despite such fearsome tribulations, it appears that *Enki* weathered the storm and reached his final destination, thereupon to take up residence in his enchanted cosmic domain.

Many sources of this epic work have been discovered during the past two centuries, mostly originating within Mesopotamia and revealed in cuneiform text inscribed upon clay tablets dating from at least 1800 B.C; see, for instance, the large number of images compiled as part of the international Cuneiform Digital Library Initiative. In this form of script, each sign comprises a combination of wedge-shaped impressions, while its name derives from *cuneus*, Latin for wedge. Figure A4.2(a) illustrates one such tablet written in the Sumerian language, while Figure A4.2(b) gives a typical example of copied text written in the Akkadian language. Translation of such texts can be problematic, because 'stones' falling from the sky sometimes might refer to either small meteorites or hailstones, as noted by Bjorkman [A4.8] and others.

Another historical account of comparable antiquity (early second millennium B.C.) that refers to hailstorms is found on an almost completely intact cuneiform tablet held at the British Museum in London; see the copied text by King [A4.9] in Figure A4.3, which depicts an anonymous Sumerian hymn of 31 lines honouring *Ishkur* (*Adad* in Akkadian), one of the deities of the Babylonian pantheon and the *God of Storms*. In a setting full of action and divine retribution, a colossal storm progresses from thunder and lightning to flood and finally to a destructive hailstorm. From a modern translation by the American scholar Mark Cohen [A4.10], lines 15–27 become:

15. *Enlil commissioned his son, Ishkur:*
16. *'My young man, …. the storms for yourself! Harness the storms for yourself!*
17. *Ishkur, …. the storms for yourself! Harness the storms for yourself!*
18. *Let the seven storms be harnessed for you like a team (of draft animals)! Harness the storms for yourself!*
19. *Let your howling storm roar for you! Harness the storms for yourself!*
20. *Lightning, your messenger, will precede you. Harness the storms for yourself!*
21. *My young man, joyfully go forth! Go forth! Who is like you when approaching?*
22. *(To) the rebellious land hated by your father, your begettor! Who is like you when approaching?*
23. *Take small …. -stones! Who is like you when approaching?*

HAILSTONE IMPACT: HISTORICAL PROLOGUE

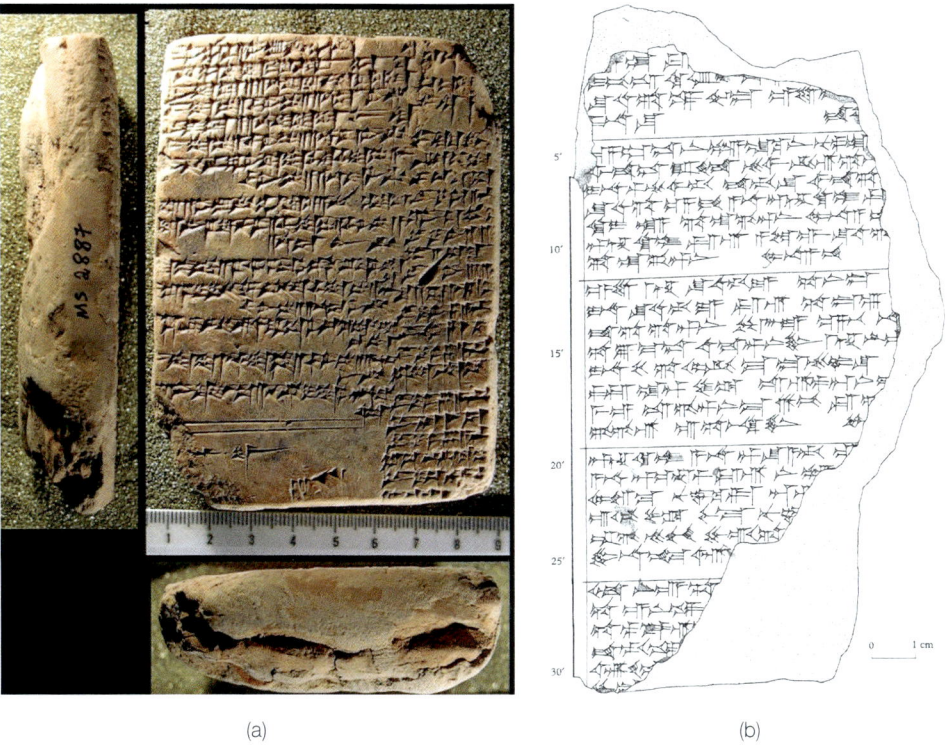

FIGURE A4.2. Fragments of cuneiform text of Sumerian tale *Gilgamesh and the Netherworld*, inscribed around 1800 B.C. or earlier (by courtesy of *Cuneiform Digital Library Initiative*): (a) clay tablet in Sumerian (cm scale); (b) copied text from clay tablet in Akkadian

> 24. *Take large (hail?)-stones! Who is like you when approaching?*
> 25. *Rain down your small (hail?)-stones and your large (hail?)-stones upon it!*
> 26. *In the rebellious land destroy at your right! Overthrow at your left!'*
> 27. *Ishkur gave heed to the spoken words of his father begettor.*

Thereafter *Ishkur* is further dignified as being absolute and all-powerful, a boundless god of heaven and earth.

A second prominent encounter with a hailstorm occurs in a Sumerian poem to King Shulgi, revealed in several copies dating to around 1800–1700 B.C. In the seventh year of his reign, Shulgi announced that he would run from Nippur to Ur and back in a single day to demonstrate his strength and speed; an implausibly long distance to cover between these two ancient cities, but his 'praise poems' are given to exaggeration and boasting. Having ceremoniously bathed and feasted in preparation for the return stage of his challenging peregrination, there follows the passage:

> 60. *Then I arose like an owl,*
> 61. *like a falcon to return to Nibru in my vigour.*

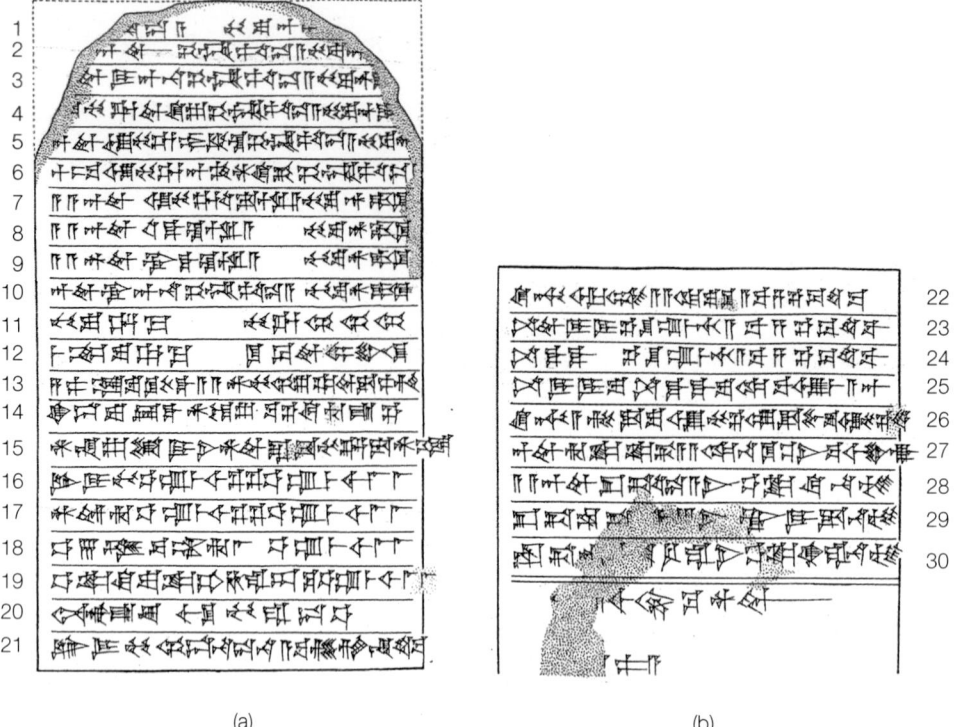

FIGURE A4.3. Copied text from cuneiform tablet (second millennium B.C.) depicting Sumerian *Hymn to Ishkur* that refers to hailstorm (by courtesy of *British Museum, London*): (a) obverse face; (b) reverse face

> 62. *But a storm shrieked,*
> 63. *and the west wind whirled around.*
> 64. *The north wind and the south wind howled at each other.*
> 65. *Lightning together with the seven winds vied with each other in the heavens.*
> 66. *Thundering storms made the earth quake,*
> 67. *and Ishkur roared in the broad heavens.*
> 68. *The rains of heaven mingled with the waters of the earth.*
> 69. *Small and large hailstones drummed on my back.*
> 70. *I, the king, however, did not fear, nor was I terrified.*
> 71. *I rushed forth like a fierce lion.*

being lines 60–71 from an anonymous modern translation [A4.11]. As might be expected, his arrival in Nippur (Nibru) was celebrated at a sumptuous palace banquet, whereupon Shulgi exclaimed that '*singers praised me with songs accompanied by seven tigi drums*'.

A further dramatic reference to hailstorms occurs in the *Book of Exodus* from the *Hebrew Bible* (*Christian Old Testament*), traditionally ascribed to Moses but perhaps more likely to have partially been composed by exiles from the Kingdom of Judea held in Babylonian captivity in the 6th century B.C. Figure A4.4 illustrates a fragment of parchment

(a)   (b)

FIGURE A4.4. Parchment fragment (approximately 5 cm square) from *Dead Sea Scrolls*, Herodian Period (*circa* 30 B.C.–70 A.D.) showing Hebrew text from *Book of Exodus* (by courtesy of *Israel Antiquities Authority*): (a) full spectrum colour image; (b) infrared image

(approximately 5 cm square), found amongst the *Dead Sea Scrolls* discovered in the Qumran caves in the Judaean desert (1946–1956), showing Hebrew text from *Exodus* written during the Herodian Period (*circa* 30 B.C.–70 A.D.).

With the Pharaoh having refused to release the Israelites from enslavement, the *Ten Plagues of Egypt* were unleashed upon the unfortunate inhabitants of this arid land, the seventh of which was *Hail and Fire*. Various English language translations are available, and part of the complete narrative (*Exodus* 9:13–35) from the *King James Version* of the bible [A4.12] – used throughout herein – is:

> 18. *Behold, tomorrow about this time I will cause it to rain a very grievous hail, such as hath not been in Egypt since the foundation thereof even until now.*
>
> 19. *Send therefore now, and gather thy cattle, and all that thou hast in the field; for upon every man and beast which shall be found in the field, and shall not be brought home, the hail shall come down upon them, and they shall die.*
>
> 20. *He that feared the word of the Lord among the servants of Pharaoh made his servants and his cattle flee into the houses:*
>
> 21. *And he that regarded not the word of the Lord left his servants and his cattle in the field.*
>
> 22. *And the Lord said unto Moses: Stretch forth thine hand toward heaven, that there may be hail in all the land of Egypt, upon man, and upon beast, and upon every herb of the field, throughout the land of Egypt.*
>
> 23. *And Moses stretched forth his rod toward heaven: and the Lord sent thunder and hail, and the fire ran along upon the ground; and the Lord rained hail upon the land of Egypt.*

24. *So there was hail, and fire mingled with the hail, very grievous, such as there was none like it in all the land of Egypt since it became a nation.*

25. *And the hail smote throughout all the land of Egypt all that was in the field, both man and beast; and the hail smote every herb of the field, and brake every tree of the field.*

26. *Only in the land of Goshen, where the children of Israel were, was there no hail.*

These traumatic events are also recounted in the *Book of Psalms* (78: 43–51) and have been subjected to numerous artistic interpretations, some of which are illustrated below.

Selected depictions of the *Egyptian Plagues* appear in the illuminated French manuscript (*circa* 1250 A.D.) shown in Figure A4.5 (in Latin, with Persian and Judeo-Persian inscriptions). They include *Hail and Fire* (upper right), where the Pharaoh stubbornly rejects a plea by Moses to release the Israelites. Originating from the same early period, Figure A4.6 shows a miniature from a medieval illuminated manuscript (ink and pigment on parchment, captions in Old French, *circa* 1250 A.D.) by the English artist William de Brailes. Here Moses (horned, to signify his encounter with divinity), followed by his elder brother Aaron (wearing the mitre of a high priest), confronts the Pharaoh (seated and crowned) and his retinue, stretching out a beckoning hand to bring forth hail throughout the land of Egypt.

Figure A4.7(a) is taken from the *Golden Haggadah*, an especially fine version of the prayer book written in Hebrew on vellum – combined with an extravagant use of gold-leaf decoration – and produced in Catalonia, Spain, in the 14th century A.D. This stunning miniature picture shows hailstones (red and white dots) falling on a tree, as a shepherd and his goats shelter underneath, while Moses, on the right, pleads with God to terminate the terrifying storm and thereby appease the Pharaoh, who would then set the people free: alas to no avail, as the promise was broken and the grant of freedom once again refused, signalling an ominous prelude to the *Eighth Plague of Egypt* (*Plague of Locusts*). A similar miniature illustration, Figure A4.7(b), is found in the *Sarajevo Haggadah*, an illuminated Jewish codex that originated around 1350 A.D. in Barcelona, Spain; see also the recent facsimile edition [A4.13]. A different version depicting the horned Moses, taken from among several pre-Lutheran editions of the scriptures, is shown in Figure A4.7(c), being an anonymous woodcut from the *Cologne Bible* of 1478–1480.

Figure A4.8 shows a pair of illustrations centred on Moses. The anonymous woodcut (*Moses stretched forth his rod toward heaven*) in Figure A4.8(a) was published in 1877, after a drawing by the German artist Julius Schnorr von Carolsfeld (1794–1872). Figure A4.8(b) portrays an engraving (*Plaga de granizo y rayos*) by R. Camaron after a drawing by Antonio Martinez.

The treatise *Physica sacra*, written by the Swiss scholar Johann Jakob Scheuchzer (1672–1733), was published concurrently in Latin and German in several volumes from 1731 to 1735 [A4.14], [A4.15]. This *magnus opus* covers a vast field of natural history from a biblical point of reference, and aims to provide scientific explanations of various natural phenomena, as well as presenting a comprehensive taxonomy of the plants and animals mentioned in the scriptures. The text is illustrated with numerous copper plate engravings – thus also being referred to as the *Kupfer-Bibel* (*Copper Bible*) – almost entirely based on drawings by the Swiss artist Johann Melchior Füssli (1677–1736). An engraving from the first volume is

HAILSTONE IMPACT: HISTORICAL PROLOGUE

FIGURE A4.5. Illuminated French manuscript, *circa* 1250 A.D: selected *Egyptian Plagues* including *Hail and Fire* (upper right) in anonymous miniature (by courtesy of *Morgan Library & Museum, New York*, MS M.638, f. 8v)

FIGURE A4.6. Illuminated miniature (*circa* 1250 A.D.) by William de Brailes: Moses (horned) confronts the Pharaoh (seated and crowned) and brings forth hail in Egypt (by courtesy of *Walters Art Museum, Baltimore, Maryland*, MS W.106.7R)

shown in Figure A4.9, entitled *Grando Aegyptiaca* or *Ägyptischer Hagel* (*Egyptian Hail*) and signed by the German artist Jakob Andreas Fridrich (1684–1751); which imposing image conveys a palpable sense of horror at the enduring wholesale destruction. A substantial accompanying commentary in the Latin version begins:

> '*Est Grando ordinarius Naturae effectus, stillarum pluvialium in descensu á Vento Boreali, vel alio frigidiori correptarum conglaciatio. Sed distingui debet Grando á Grandine. Est Aegyptus regio ubi raro, in Mediterraneis scilicet, Pluvia cadit, nunquam Grando.*'

which, although acknowledging hail as being rare in Egypt, appears somewhat optimistic in asserting that hailstorms never occur in the Mediterranean.

HAILSTONE IMPACT: HISTORICAL PROLOGUE

FIGURE A4.7. Anonymous illustrations of *Plague of Hail and Fire*: (a) miniature from *Golden Haggadah* illuminated manuscript, *circa* 1320 A.D. (by courtesy of *British Library, London*); (b) miniature from *Sarajevo Haggadah* illuminated manuscript, *circa* 1350 A.D. (by courtesy of *National Museum of Bosnia and Herzegovina, Sarajevo*); (c) woodcut from *Cologne Bible*, *circa* 1479 A.D.

(a)

(b)

FIGURE A4.8. Biblical illustrations of *Plague of Hail* centred on Moses: (a) anonymous woodcut, 1877 (*And Moses stretched forth his rod toward heaven*), after Julius Schnorr von Carolsfeld; (b) engraving (*Plaga de granizo y rayos*) by R. Camaron after Antonio Martinez (by courtesy of *Wellcome Library, London*)

FIGURE A4.9. Engraving entitled *Egyptian Hail* by Jacob Andreas Fridrich, from 1731 treatise on natural theology by Scheuchzer [A4.14], [A4.15]

The portrayal (*Plaag van Hagel en Vuur over Egipte*) in Figure A4.10(a) is from a book of biblical illustrations [A4.16] published by the Dutch engraver Pieter Mortier (1669–1711) in Amsterdam in 1700. Figure A4.10(b) shows an engraving (*Plague of Hail and Fire*) by the Dutch artist Caspar Luyken (1672–1708), from a similar tome [A4.17] published by the German printer and engraver Christoph Weigel (1654–1725) in Nuremberg in 1708. Two further engravings having related associations are shown in Figure A4.11. The first is again after Pieter Mortier, while Figure A4.11(b) is by the English engraver James Caldwall (1739–1822), published in 1811 and entitled *The Plague of Hail*, after the French artist Clément Pierre Marillier (1740–1808). Other artistic interpretations of this momentous event include the anonymous coloured etching (*Die Siebende Plage in Aegÿptenland*) in Figure A4.12(a), published in Augsburg (*circa* 1777), and the anonymous French woodcut (*Les fleaux de la peste & de la gresle ravagent l'Egypte*) shown in Figure A4.12(b).

(a)

(b)

FIGURE A4.10. Dutch engravings from books of biblical illustrations: (a) *Plaag van Hagel en Vuur over Egipte* after Pieter Mortier, 1700; (b) *Plague of Hail and Fire* by Caspar Luyken, 1708

HAILSTONE IMPACT: HISTORICAL PROLOGUE

(a)

(b)

FIGURE A4.11. Biblical engravings: (a) *The Plague of Hail and Fire*, after Pieter Mortier; (b) *The Plague of Hail* (1811) by James Caldwall

(a)

(b)

FIGURE A4.12. Anonymous illustrations: (a) coloured etching (*Die Siebende Plage in Aegýptenland*), Augsburg, *circa* 1777 (by courtesy of *Wellcome Library, London*, Ref. 6059i); (b) French woodcut (*Les fleaux de la peste & de la gresle ravagent l'Egypte*) (by courtesy of *Wellcome Library, London*, Ref. 6155i)

(a)   (b)

FIGURE A4.13. Extract from *Bristol Psalter*, 11th-century Greek manuscript (*The plagues of Egypt: the plague of locusts; the plague of hail*) (by courtesy of *British Library, London*, Add MS 40731, f. 130r): (a) complete page; (b) anonymous detailed image of plague of hail

An alternative illustrative format is seen in the *Bristol Psalter*, an 11th-century Greek parchment manuscript and one of the few so-called marginal Psalters to survive, with numerous small anonymous images painted beside the text of the Psalms. One complete page (*The plagues of Egypt: the plague of locusts; the plague of hail*) is shown in Figure A4.13(a), while Figure A4.13(b) gives a detailed image of the plague of hail. Another anonymous image illustrating the same two plagues is given in Figure A4.14, sourced from an illuminated French manuscript, *circa* 1250.

Continuing the same religious theme, the imposing oil paintings reproduced in Figure A4.15 vividly express the palpable mood of the prevailing melodrama. Figure A4.15(a), entitled *Landscape with Moses and Aaron calling down the Plague of Hail upon Egypt*, circa 1670, is by the Dutch artist Pieter Mulier the Younger (1637–1701). Figure A4.15(b) is by the English artist Joseph Mallord William Turner (1775–1851), first exhibited at the annual Royal Academy exhibition in London in 1800, and is historically mistitled *The Fifth Plague of Egypt* (as is the similar etching and mezzotint engraving, not shown, published in 1808). Here the emphasis is on the tumultuous sky and turbulent storm cloud formation, taking precedence over the barely discernible figure of Moses (lower right with arms outstretched), shown cast in shadow and dwarfed by the vast ambient scenario. Figure A4.15(c), entitled *Hail and Fire, Seventh Plague of Egypt*, is by the English artist (and later civil engineer) John Martin (1789–1854).

FIGURE A4.14. Anonymous miniature from 13th-century French manuscript (*The Plague of Hail and of Locusts*) (by courtesy of *Granger Historical Picture Archive, New York*)

As a rare example of classical sculptural form, Figure A4.16(a) shows five miniature panels in carved bone (height 10 cm, width 4 cm) from a casket or altarpiece (*circa* 1390–1410 A.D.) made at the workshop of Baldassare Embriachi in Florence or Venice in northern Italy, set in a modern wooden frame. One panel, Figure A4.16(b), depicts the *Plague of Hail*; the remaining panels illustrate another three of the ten biblical plagues of Egypt, and the fifth depicts a shepherd in a rural landscape.

In a musical context, direct references to the *Plagues of Egypt* are found in the biblical oratorio *Israel in Egypt*, composed by George Frederick Handel (1685–1759) and premiered at the King's Theatre in Haymarket, London, in 1739. The libretto, usually attributed to Charles Jennens (1700–1773), has the chorus in part one (section 7) repeatedly giving voice to the words:

'*He gave them hailstones for rain; fire mingled with the hail ran along upon the ground.*'

HAILSTONE IMPACT: HISTORICAL PROLOGUE

(a)

(b)

FIGURE A4.15. Oil paintings depicting *Seventh Plague of Egypt*: (a) *Landscape with Moses and Aaron calling down the Plague of Hail upon Egypt* by Pieter Mulier the Younger, *circa* 1670 (by courtesy of *National Trust, Tatton Park, England*); (b) historically mistitled *The Fifth Plague of Egypt* by J. M. W. Turner, 1800 (by courtesy of *Indianapolis Museum of Art, Newfields, Indiana*)

(c)

FIGURE A4.15. (continued) Oil paintings depicting *Seventh Plague of Egypt*: (c) *Hail and Fire, Seventh Plague of Egypt* by John Martin, 1823 (by courtesy of *Museum of Fine Arts, Boston, Massachusetts*).

A later version occurs in *Redemption: a sacred oratorio*, selected from Handel's works and first performed at the Theatre Royal in Drury Lane, London, in 1786.

A second *Old Testament* example is found in the *Book of Joshua* (10:10–12), where God fulfils his promise to lead the Israelites from Egypt to the *Promised Land*. Following Joshua's initial victory in battle over the Amorites at Gibeon – an ancient hill-city northwest of Jerusalem – and upon pursuing the fleeing warriors towards Azekah, it is stated:

10. *And the Lord discomfited them before Israel, and slew them with a great slaughter at Gibeon, and chased them along the way that goeth up to Beth-horon, and smote them to Azekah, and unto Makkedah.*
11. *And it came to pass, as they fled from before Israel, and were in the going down to Beth-horon, that the Lord cast down great stones from heaven upon them unto Azekah, and they died: they were more which died with hailstones than they whom the children of Israel slew with the sword.*
12. *Then spake Joshua to the Lord in the day when the Lord delivered up the Amorites before the children of Israel, and he said in the sight of Israel, Sun, stand thou still upon Gibeon; and thou, Moon, in the valley of Ajalon.*

while noting in wonder that this miraculous deluge of giant hailstones fell upon only one of the combatant armies.

(a)　　　　　　　　　　　　　　　　(b)

FIGURE A4.16. Miniature carved bone panels (*circa* 1390–1410 A.D.) from workshop of Baldassare Embriachi, northern Italy (by courtesy of *Victoria and Albert Museum, London*, Ref. A.80-1919): (a) framed set of five panels; (b) panel depicting *Plague of Hail*

In the anonymous 5th-century mosaic shown in Figure A4.17(a), the upper part depicts Joshua routing the Amorites, while the lower part represents God's hand casting hailstones upon them. Another evocative interpretation of the unfolding panorama is expressed in an 1816 oil painting (*Joshua commanding the Sun to stand still upon Gibeon*), again by John Martin, shown in Figure A4.17(b). Joshua is standing on a rock outcrop at centre stage in the foreground, arm raised as if beckoning divine intervention, with the two armies stretching along the narrow track from the elevated gates of the city down into the far distance. But whilst the clouds have parted over the Israelites, the storm rages upon the Amorites in the valley below.

A further quotation in a similar vein is found in the *Book of Job* (38: 22–23) where, in reply to Job's overwhelming doubts on theodicy, a lengthy poetic monologue delivered by an omniscient God poses the rhetorical question:

> 22. *Hast thou entered into the treasures of the snow? or hast thou seen the treasures of the hail,*
> 23. *Which I have reserved against the time of trouble, against the day of battle and war?*

once again in the sense of hailstones being deployed by God to overwhelm his foes and inflict defeat upon his enemies. Comparable verses occur in the *Book of Psalms* (18: 12–13):

(a)

(b)

FIGURE A4.17. Joshua's victory in battle over the Amorites at Gibeon: (a) anonymous 5th-century mosaic (upper part, Joshua routing the Amorites; lower part, God's hand casting hailstones upon them); (b) oil painting (*Joshua commanding the Sun to stand still upon Gibeon*) by John Martin, 1816 (by courtesy of *National Gallery of Art, Washington, D.C.*)

12. *At the brightness that was before him his thick clouds passed, hail stones and coals of fire.*

13. *The Lord also thundered in the heavens, and the Highest gave his voice; hail stones and coals of fire.*

while from *Psalm 148* (*Praise ye the Lord*), albeit in a different context, it is stated:

7. *Praise the Lord from the earth, ye dragons, and all deeps:*

8. *Fire, and hail; snow, and vapours; stormy wind fulfilling his word:*

By way of illustrating *Psalm 148*, Figure A4.18 shows a single leaf from the *Utrecht Psalter*, a medieval illuminated codex made near the city of Reims in northern Francia (*circa* 825 A.D.), wherein each psalm is accompanied by a full-width vibrant drawing (pen and ink on parchment, of anonymous origin) comprising an assemblage of individual sketches interpreting the entire text. High above the flooded landscape, three disembodied heads (left side image) emerge from the clouds to generate a violent hailstorm, destroying trees and livestock far below.

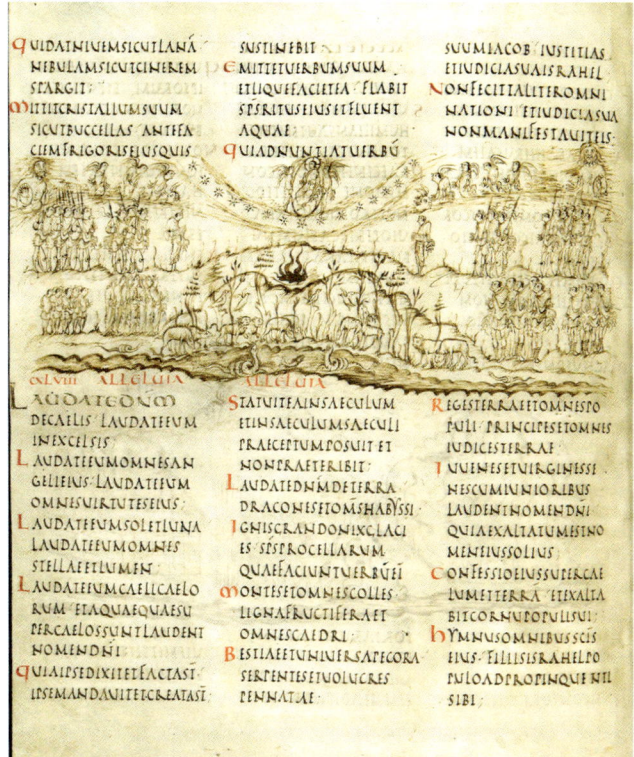

FIGURE A4.18. Leaf from *Utrecht Psalter* (*circa* 825 A.D.) illustrating entire *Psalm 148* (by courtesy of *Universiteitsbibliotheek, Utrecht*, MS Bibl. Rhenotraiectinae I, Nr 32, f. 82v)

It is also noted that the apocryphal *Gospel of St Bartholomew* includes the line:

> 'The angel of the hail is called Mermeōth, and he holdeth the hail upon his head, and my ministers do adjure him and send him whither they will.'

according to a translation [A4.18] by the English scholar Montague Rhodes James (1862–1936).

Another biblical example of God's withering judgement – again with devastating consequences – occurs in the final book of the *New Testament*, where a verse in the *Book of Revelation* (8:7) is written as:

> 7. *The first angel sounded, and there followed hail and fire mingled with blood, and they were cast upon the earth: and the third part of trees was burnt up, and all green grass was burnt up.*

A miniature from an illuminated parchment manuscript [A4.19] shown in Figure A4.19(a) illustrates this seraphic spectacle, with the first of the four angels sounding a trumpet amid the hail and fire. Figure A4.19(b) is from the *Apocalypse Tapestry*, a large medieval set of tapestries woven in Paris between 1377 and 1382, and now housed in a museum at the Château d'Angers, a castle in the city of Angers in the Loire Valley, France. Figure A4.20 displays single leaves of illuminated medieval parchment (*circa* 1180 A.D.) from two of many different *Beatus Manuscripts* based on a *Commentary on the Apocalypse* written by St Beatus (*circa* 730–*circa* 800 A.D.), an Asturian monk who lived and worked in the region of Liébana in northern Spain.

In a quite different setting, early Christian narratives include the story of how the apostle Paul encountered opposition to his ministry at Ephesus, and is sentenced to death by wild beasts. Whereupon, according to Snyder [A4.20]:

> 'A fierce lion is set on him, but it turns out to be the same one Paul had baptized, and it lies down by Paul's feet instead of devouring him. Some other animals are then sent into the arena, but a sudden hailstorm interrupts the proceedings, and both Paul and the lion escape.'

which singular event, against all expectations, portrays a hailstorm in a beneficial light.

The biblical theme is continued – albeit within the more congenial sphere of pastoral care – in a 13th-century illuminated Greek manuscript [A4.21] comprising theological works and orations by Gregory of Nazianzus (*circa* 329–390 A.D.), sometime Archbishop of Constantinople, and eventually elevated to sainthood. Circumstantial details prevailing in the ancient city of Nazianzus (in modern-day Turkey) are sparse, but upon referring to an English language translation by Browne and Swallow [A4.22], their introduction to *Oration 16* begins:

> '*This Oration belongs to the year A.D. 373. A series of disasters had befallen the people of Nazianzus. A deadly cattle plague, which had devastated their herds, had been followed by a prolonged drought, and now their just ripened crops had been ruined by a storm of rain and hail.*'

FIGURE A4.19. Biblical scenes (*circa* 1380 A.D.) from *Book of Revelation*: (a) anonymous miniature (*First angel sounding trumpet amid hail and fire*) in illuminated manuscript (by courtesy of *British Library, London*, MS Yates Thompson 10, f. 13); (b) part of *Apocalypse Tapestry* (by courtesy of *Musée de la Tapisserie, Château d'Angers, France*)

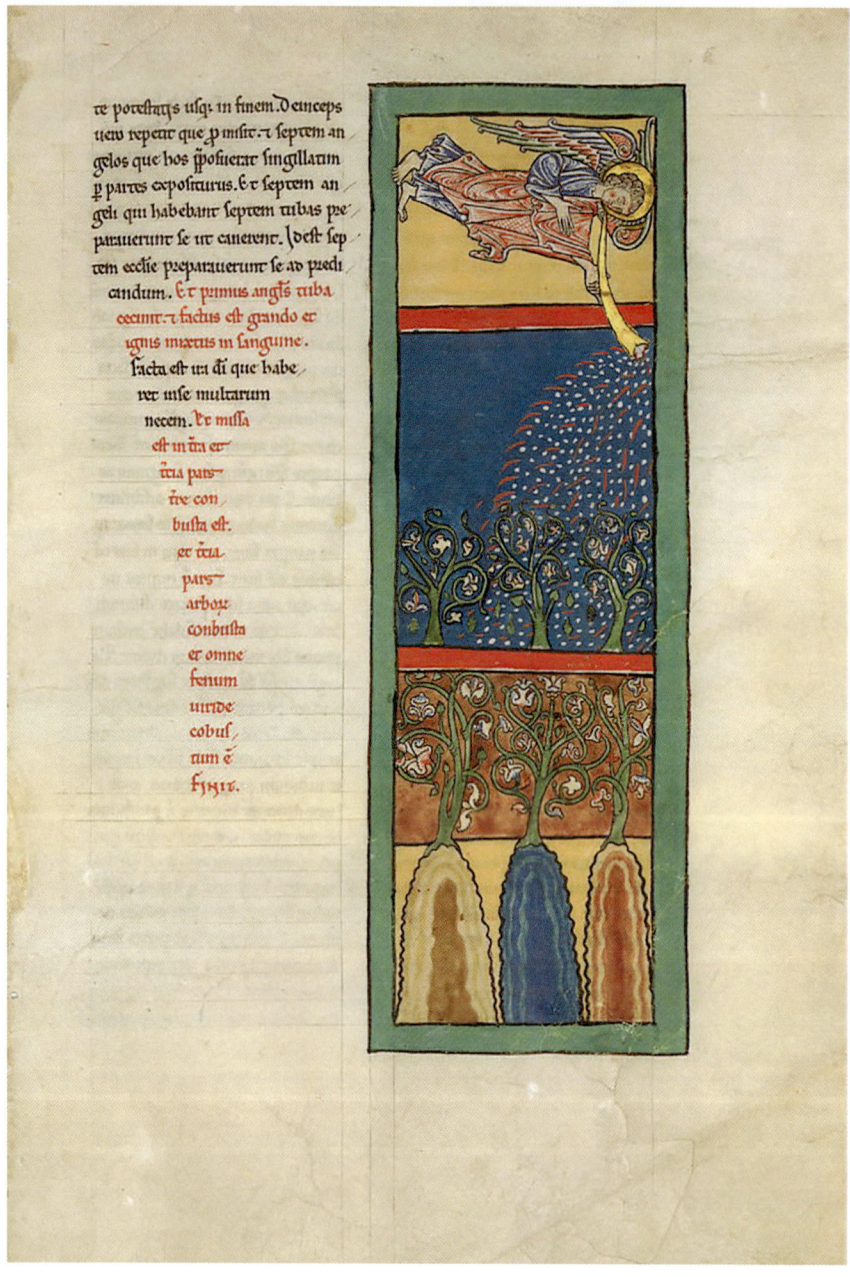

(a)

FIGURE A4.20. Single parchment leaves from two versions (*circa* 1180 A.D.) of *Beatus Manuscript*: (a) *The first angel sounds the trumpet; fire, hailstones, and blood are cast upon the earth* (by courtesy of *Metropolitan Museum of Art, New York*)

FIGURE A4.20. (continued) Single parchment leaves from two versions (*circa* 1180 A.D.) of *Beatus Manuscript*: (b) *An angel with a trumpet; hail falls from the trumpet on to trees below* (by courtesy of *The John Rylands Library, Manchester University*, Latin MS 8, f. 126r)

(a)   (b)

FIGURE A4.21. Anonymous illuminated miniatures from Greek manuscripts: (a) illustration (*circa* 1220 A.D.), *St Gregory comforts people of Nazianzus after disastrous hailstorm* (by courtesy of *Bodleian Library, Oxford*); (b) tempera painting (*circa* 880 A.D.) illustrating homily *On the Plague of Hail* delivered by St Gregory (by courtesy of *Bibliothèque Nationale de France, Paris*, MS Grec 510, f. 78r)

Manifestly real hardships were incurred by real people, and the solemn magnificence expressed in the anonymous large miniature [A4.21] in Figure A4.21(a) is reckoned to show St Gregory comforting the kinsfolk of Nazianzus in the aftermath of this disastrous hailstorm. Returning to an earlier biblical theme, Figure A4.21(b) shows an anonymous tempera painting (*Hailstorm and Gregory of Nazianzus preaching*) illustrating a homily *On the Plague of Hail* delivered by St Gregory, from a Byzantine illuminated manuscript (*Homilies of Gregory of Nazianzus*), *circa* 880 A.D., commissioned in Constantinople by Patriarch Photios I and dedicated to Emperor Basil I (also called *The Macedonian*).

Broadly similar figurative narratives are found in age-old Chinese literature, where exceptional hailstorms oftentimes were interpreted as divine condemnation of a king or emperor, to be carefully recorded for posterity. Wang [A4.23], for example, has listed a series of important meteorological events occurring during the period 2187 B.C.–3 A.D., some originating from engraved bamboo pieces bound together by rope. These include:

1. *King Hsiao, Chou Dynasty: 7th year, winter. Heavy hail killed cattle and horses (948 B.C.).*
2. *King Yee, Chou Dynasty: 7th year, winter. The hailstones were as big as a knife-sharpener stone (918 B.C.).*

3. *Emperor Gin, Han Dynasty:* 2nd year, autumn. Hail in Hen-San, 5 Chun (about 13 cm) in diameter, 2 Che (about 60 cm) in depth (155 B.C.).

4. *Emperor Wu, Han Dynasty:* 3rd year of Yuan-Fong period. Hailstones were as big as horse heads (108 B.C.).

5. *Emperor Shuan, Han Dynasty:* 4th year, 5th month of Ti-Che period. Hail at Chi-Yin of San-Yang area, as big as chicken eggs, 2 che 5 chun (about 74 cm) deep. It killed 20 people. Many birds also died (66 B.C.).

6. *Emperor Chen, Han Dynasty:* 2nd year, 4th month of Ho-Ping period. Hail in Chu, as big as an axe (27 B.C.).

Despite difficulties in translation and the presence of lacunae, together with some overtly colourful allegories, these records provide a modest yet valuable insight into the traditions of another prominent ancient civilisation.

Another interesting source is *The Ramayana*, a prodigiously long Sanskrit epic from ancient India traditionally attributed to the poet Valmiki, describing the life and adventures of the legendary prince *Rama*. This work is dated variously from 500–100 B.C., and with several instances of spears and arrows falling 'as thick as hail', book 3 (canto 34, *Súrpanakhá's Speech*) has the rhyming couplets:

'I looked, I looked, but never saw
His mighty hand the bowstring draw
That sent the deadly arrows out,
While rang through air his battle-shout.
I looked, I looked, and saw too well
How with that hail the giants fell,
As falls to earth the golden grain,
Struck by the blows of Indra's rain.'

from a translation by Ralph Griffith [A4.24], [A4.25], with *Indra* as the Hindu deity of the heavens. Similar phraseology occurs in *The Mahabharata*, an even longer Sanskrit epic (over 100,000 couplets) probably compiled during the period 400 B.C.–400 A.D., with origination attributed to the poet Vyasa, and translated into English prose by Kisari Mohan Ganguli [A4.26].

Historical references to hailstorms also occur in *The Iliad*, an epic poem by the Greek author Homer, possibly composed around 700 B.C., and set during the *Trojan War*. The 24 books were translated into blank verse [A4.27] by the English poet William Cowper (1731–1800), and from early in book 10 is written:

'As when the spouse of beauteous Juno, darts
His frequent fires, designing heavy rain
Immense, or hail-storm, or field-whitening snow,
Or else wide-throated war calamitous,
So frequent were the groans by Atreus' son
Heaved from his inmost heart, trembling with dread.'

citing the goddess *Juno*, wife of *Jupiter*, and *Atreus*, a king of Mycenae and the father of *Agamemnon* and *Menelaus*. Moreover, book 12 contains the passage:

> '*He spake, nor storm-wing'd Iris disobey'd,*
> *But down from the Idaean summit stoop'd*
> *To sacred Ilium. As when snow or hail*
> *Flies drifted by the cloud-dispelling North,*
> *So swiftly, wing'd with readiness of will,*
> *She shot the gulf between, and standing soon*
> *At glorious Neptune's side, him thus address'd.*'

which features the goddess *Iris*, along with *Neptune* (God of the Sea), the venerated *Mount Ida*, and the sacred city of *Ilium* (Troy).

The subsequent Homeric epic poem *The Odyssey*, again written in 24 books, mainly focuses on the Greek hero *Odysseus* and his convoluted journey home to Ithaca after the fall of Troy. With *Odysseus* a captive of the nymph *Calypso* in a foreign land, the despondent *Menelaus* (King of Sparta, husband of *Helen of Troy*) is promised sublime contentment in the *Elysian Fields*, not least marked by the absence of hailstorms; as elucidated in book 4 (*Meeting with Menelaus*):

> '*But oh, beloved by Heaven! reserved to thee*
> *A happier lot the smiling Fates decree:*
> *Free from that law, beneath whose mortal sway*
> *Matter is changed, and varying forms decay;*
> *Elysium shall be thine: the blissful plains*
> *Of utmost earth, where Rhadamanthus reigns.*
> *Joys ever young, unmix'd with pain or fear,*
> *Fill the wide circle of th' eternal year:*
> *Stern winter smiles on that auspicious clime:*
> *The fields are florid with unfading prime;*
> *From the bleak pole no winds inclement blow,*
> *Mould the round hail, or flake the fleecy snow;*
> *But from the breezy deep the blest inhale*
> *The fragrant murmurs of the western gale.*
> *This grace peculiar will the gods afford*
> *To thee, the son of Jove, and beauteous Helen's lord.*'

This extract is based on an original translation [A4.28] by the English poet Alexander Pope (1688–1744), and edited [A4.29] by the English classical scholar Theodore Alois Buckley (1825–1856), with *Jove* being also known as *Jupiter*, and *Rhadamanthus* reining supreme in *Elysium*.

Latent optimism also features in one of the fables (*The bride and the two grooms*) attributed to the legendary Greek storyteller *Aesop* (circa 600 B.C.), where it happened that a hailstorm coincided with a wedding ceremony. As the procession of guests unfolded, the prevailing ambience changed abruptly:

> '*At this moment, Venus, the goddess of love, showed her compassion: the clouds in the sky were tossed by the winds and a crack of thunder shook the heavens. As grim night descended with*

(a) (b)

FIGURE A4.22. Corinthian terracotta globular aryballoi (*circa* 550 B.C.) with hailstorm filling; procession of warriors with circular shields: (a) aryballos (by courtesy of *Harvard Art Museums, Cambridge, Massachusetts*, Ref. 13.1908); (b) aryballos (by courtesy of *Phoebe A. Hearst Museum of Anthropology, Berkeley, California*, Ref. 8-2350)

> *a dense downpour of rain, the light was snatched from everyone's eyes and the terrified party guests were pelted with hail as they scattered in all directions.'*

although eventually there was a gladsome outcome to these proceedings, at least for the successful groom; taken from a modern translation by Laura Gibbs [A4.30]. In a quite different yet contemporary setting, the novel artistic device of a 'hailstorm filling' is illustrated in the two examples of a painted terracotta globular aryballos (small flask or jar) of Corinthian origin (*circa* 550 B.C.) portrayed in Figure A4.22, both of which feature a procession of hoplite soldiers marching with circular shields.

A rudimentary explanation of hail formation was given by the ancient Greek philosopher Anaximenes of Miletus (*circa* 586–*circa* 526 B.C.), who considered air to be a primary source of matter, and is said to have posited that:

> '*Clouds occur when the air is further thickened; when it is compressed further, rain is squeezed out, and hail occurs when the descending water coalesces, snow when some windy portion is included together with the moisture.*'

from historical records compiled by Kirk *et al.* [A4.31].

In *The Daughters of Troy*, one of the tragedies written by the Greek playwright Euripides (*circa* 480–*circa* 406 B.C.) and set in the Greek camp soon after winning the battle of Troy (*Ilium*), the living drama lay in the fate of the conquered women (the men having been slain). With the assured compliance of her father *Zeus* (God of the Sky), the goddess

Athena addresses *Poseidon* (God of the Sea) with dire foreboding for the returning Greeks (*Achaeans*):

> 'When homeward-bound they sail from Ilium.
> Then Zeus shall send down rain unutterable,
> And hail, and blackness of heaven's tempest-breath;
> And to me promiseth his levin-flame
> To smite the Achaeans and burn their ships with fire.'

from a translation into English verse [A4.32], [A4.33] by the English classicist Arthur Sanders Way (1847–1930), with the archaic 'levin-flame' referring to a 'bolt of lightning'. In this play, first produced in 415 B.C., the glories of war become an empty delusion, bringing as much wretchedness to the victors as to the vanquished.

Similar divine intervention occurs in *Oedipus at Kolonus*, one of the three Theban plays by the Greek tragedian Sophocles (*circa* 497–*circa* 406 B.C.), set in the village of Kolonus (near Athens) and first performed in 401 B.C. The blind and banished *Oedipus* (sometime King of Thebes), following a fierce thunderstorm storm which he interprets as a sign of his impending death, is led by his dutiful daughter *Antigone* to meet *Theseus* (mythical King of Athens), who enquires:

> 'What blended clamour echoeth from you?
> Plain was your cry, but his rang clear through all.
> Was Zeus's Levin cause, or burst of hail
> From rain-cloud rack? All things may one forebode
> When God thus sounds the trumpet of the storm.'

being a verse taken from a translation by Arthur Way [A4.34].

In a compendium of bucolic poems known as *The Idylls*, usually attributed to the Greek writer Theocritus (*circa* 310 B.C.–*circa* 250 B.C.), the following dramatic passage occurs in the prelude to hymn 22 in praise of the *Dioscuri* (the twin half-brothers *Castor* and *Pollux* in Roman mythology), especially as saviours of ships about to be lost in storm-tossed seas:

> 'And of ships that, when they defy the stars in the west that set,
> Or that rise in the morning sky, by fierce storm-blasts are met
> Which over the ship's prow curl a huge foam-crested sea,
> Or high o'er the stern, or whirl it as their wild will may be,
> And into the hold they hurl it, and bulwarks to windward and lee
> Are shattered: hang from the mast the tackling and sail, all riven
> In tangled ruin, and fast pour cataract rains from heaven;
> And the night cometh on, and the vast sea rings and roars 'neath the flail
> Of the merciless-scourging blast and the rush of relentless hail.
> Yet despite all this do ye draw from unfathomed abysses to land
> Ships with their crews, when they saw death looming hard at hand.'

from a translation into English verse by Arthur Way [A4.35].

A much later literary sequel takes the form of a fourteen-book epic entitled *The Fall of Troy* (or *Posthomerica*) by Quintus, a Greek poet who flourished in Smyrna (present-day Izmir, Turkey) in the 4th century A.D., but about whom little is known. This mythological work covers the period of the Trojan War from that marked by the completion of Homer's *Iliad* to the final sack of Troy: it ends with the departure of the Greeks, highlighting the extraordinary storm that devastated their fleet. In book 14 (*How the conquerors sailed from Troy unto judgment of tempest and shipwreck*) is the enthralling passage:

> '*Then, when he saw that burg beloved destroyed,*
> *Xanthus, scarce drawing breath from bloody war,*
> *Mourned with his Nymphs for ruin fallen on Troy,*
> *Mourned for the city of Priam blotted out.*
> *As when hail lashes a field of ripened wheat,*
> *And beats it small, and smites off all the ears*
> *With merciless scourge, and levelled with the ground*
> *Are stalks, and on the earth is all the grain*
> *Woefully wasted, and the harvest's lord*
> *Is stricken with deadly grief:*'

once again from a translation [A4.36] by Arthur Way, citing these 'nymphs' as daughters of the river *Xanthus*, and with *Priam* as the legendary King of Troy.

It is of particular interest that the Greek philosopher Aristotle (384–322 B.C.) was minded to make copious remarks on the subject of hail in *Meteorology*, circa 340 B.C. [A4.37]; part of a monumental compendium of work, and generally regarded as the earliest comprehensive treatise on weather science, despite many of its conclusions eventually being shown as incorrect. Whence, for example, the following paragraph (book 1, part 12) from a translation by the young classical scholar Erwin Webster (Captain, King's Royal Rifle Corps, killed in action at Monchy-le-Preux, near Arras, France, 1917, aged 37):

> '*Some think that the cause and origin of hail is this. The cloud is thrust up into the upper atmosphere, which is colder because the reflection of the sun's rays from the earth ceases there, and upon its arrival there the water freezes. They think that this explains why hailstorms are commoner in summer and in warm countries; the heat is greater and it thrusts the clouds further up from the earth. But the fact is that hail does not occur at all at a great height: yet it ought to do so, on their theory, just as we see that snow falls most on high mountains. Again, clouds have often been observed moving with a great noise close to the earth, terrifying those who heard and saw them as portents of some catastrophe.*
>
> *Sometimes, too, when such clouds have been seen, without any noise, there follows a violent hailstorm, and the stones are of incredible size, and angular in shape. This shows that they have not been falling for long and that they were frozen near to the earth, and not as that theory would have it. Moreover, where the hailstones are large, the cause of their freezing must be present in the highest degree: for hail is ice as everyone can see. Now those hailstones are large which are angular in shape. And this shows that they froze close to the earth, for those that fall far are worn away by the length of their fall and become round and smaller in size.*'

The epic poem *Argonautica* by the ancient Greek author Apollonius Rhodius, comprising four books written in the 3rd century B.C., concerns the mythical voyage and heroic adventures of *Jason* and the *Argonauts* in their quest for the *Golden Fleece*. Whence the short passage:

> 'So when great Jove on close-throng'd cities pours
> From hyperborean clouds his haily show'rs;
> Within, the dwellers sit in peace profound,
> Nor heed the rattling storms that rage around;
> In vain the hail descends, the tempests roar,
> Their roofs from harm were well secur'd before:
> Thus on their shields the furies shot their quills,
> The clamouring vanish'd to far distant hills.'

from book 2 of an unfinished translation in rhymed couplets by the English scholar Francis Fawkes (1720–1777), completed posthumously by the English cleric Henry Meen (1744–1817) on behalf of his widow [A4.38].

Occasional allusions to hail are found in records of early Greek astronomy, typically included within a parapegma (star calendar) that gives the time of year based upon observation of the stars; thereby in some respects, following in the tradition of an agricultural almanac embodied in a didactic poem (*Works and Days*) by the ancient Greek writer Hesiod [A4.39] around 700 B.C. Figure A4.23(a) shows an anonymous fragment of one of the earliest known papyrus parapegmas (part of the *Hibeh papyri*) – discovered by Grenfell and Hunt [A4.40] at El-Hibeh on the upper Nile and dating from the 3rd century B.C. – which relates to months of the Egyptian calendar.

A specific literary example relating to hail is from an original manuscript compiled by the Greek scientist Geminus, probably in the 1st century B.C., which combines astronomical and weather lore. Located within the section entitled *The Sun passes through Aries in 31 days*, is the predictive message:

> 'On the 23rd, according to Euktemon, the Hyades hide themselves; and hail follows; and the west wind blows. According to Kallippos, the Claws begin setting; and in many places hail.'

and likewise, within the next section *The Sun passes through Taurus in 32 days*:

> 'On the 1st day, according to Eudoxos, Orion sets at nightfall; rainy weather. According to Kallippos, Aries finishes rising; rainy weather, and in many places hail.
> On the 2nd, according to Euktemon, the Dog hides itself; and there is hail.'

from a modern translation by Evans and Berggren [A4.41].

Rare fragments of inscribed stone parapegmas were discovered at Miletus on the western coast of Anatolia (in modern-day Turkey) around the turn of the 20th century, and described by Diels and Rehm [A4.42] and Rehm [A4.43]. One such example (Nr 456A), dating from the 1st century B.C. and bearing an anonymous inscription referring to hail, is portrayed in Figure A4.23(b). This fragment (54 × 22 × 18 cm) formed part of a much larger parapegma, and represents only around 12 days of the year. The original stone artifact was intended for

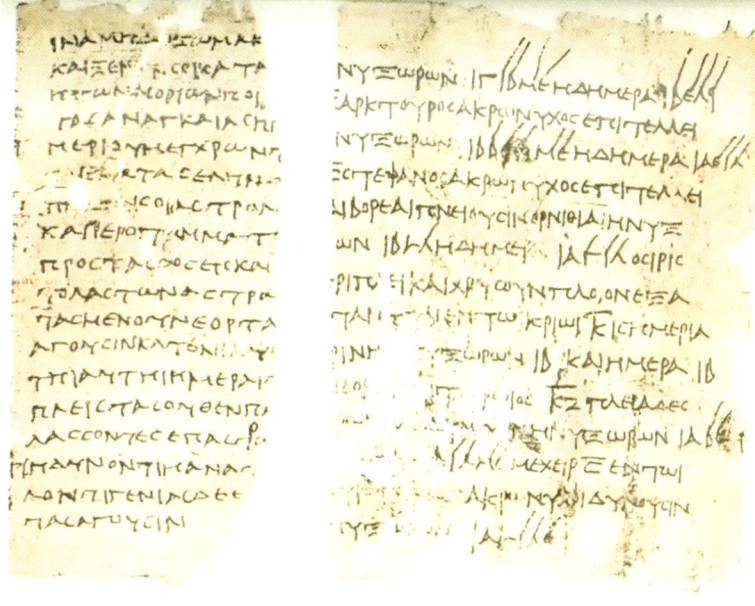

FIGURE A4.23. Fragments of anonymous Greek parapegmas: (a) papyrus parapegma (height 17 cm) discovered at El-Hibeh relating to Egyptian calendar (3rd century B.C.), after Grenfell and Hunt [A4.40]; (b) stone parapegma (54 × 22 × 18 cm) discovered at Miletus (Nr 456A, 1st century B.C.), after Diels and Rehm [A4.42] (by courtesy of *Verlag der Königlich Preussischen Akademie der Wissenschaften, Berlin*)

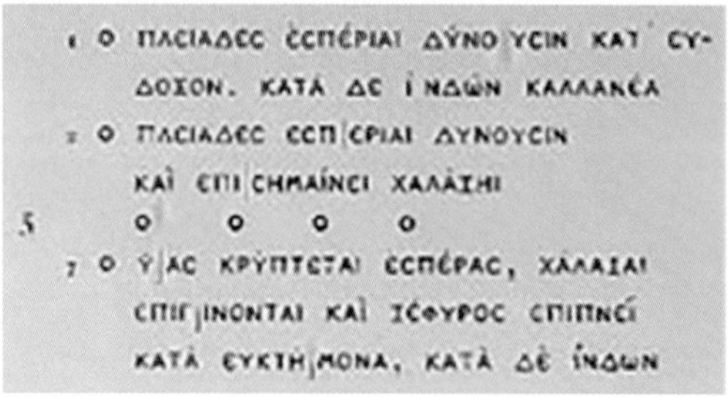

```
ι  Ο  ΠΛCIΛΔCC  ἐCΠἐPIAI  ΔΥΝΟ ΥCIN  ΚΑΤ  CΥ-
       ΔΟΞΟΝ.  ΚΑΤΑ  ΔC  ἰ ΝΔѠΝ  ΚΑΛΛΑΝἐΑ
  ₅  Ο  ΠΛCIΛΔCC  CCΠ CPIAI  ΔΥΝΟΥCIN
       ΚΑἰ  CΠI CHMAἰNCI  ΧΑΛΑΞΗΙ
  5      ο     ο     ο    ο
     7 Ο  ῢ ΑC  ΚΡΥΠΤCΤΑΙ  ἐCΠἐPAC,  ΧΑΛΑΞΙΑΙ
       CΠΙΓ ΙΝΟΝΤΑΙ  ΚΑἰ  ΙCΟΥPΟC  CΠΙΠΝCἰ
       ΚΑΤΑ  CΥΚΤΗ ΜΟΝΑ,  ΚΑΤΑ  ΔἐἰΝΔѠΝ
```

........................]s according to Eu-
....according to the I]ndian Callaneus
..............] sets in the [eve]ning
....there is a ch]ange in the weather,
with hail, [according to ...]   • • •
• ...]ας disappears in the evening. It hails
......]οντα[ι] and Zephyrus blows.
.........]MO[..], and according to the
Indian Callaneus...

(a)   (b)

FIGURE A4.24. Fragment of inscribed stone parapegma (Nr 456A, 1st century B.C.) discovered at Miletus: (a) transcribed Greek text, left column, after Diels and Rehm [A4.42] (by courtesy of *Verlag der Königlich Preussischen Akademie der Wissenschaften, Berlin*); (b) English text, left column, after Lehoux [A4.44] (by courtesy of *Rudolf Habelt GmbH, Bonn*)

public display, acting as a combined celestial and weather calendar, with moveable date pegs used to identify predicted natural occurrences cited on the inscription. Figure A4.24(a) gives the left column of the transcribed text in ancient Greek and its juxtaposition with the circular peg holes [A4.42], while Figure A4.24(b) presents a recent translation by Lehoux [A4.44], noting here that *Zephyrus* was the god of the west wind; see also Lehoux [A4.45].

A striking artefact from Greek antiquity that features hailstones in corporeal form can be seen on the *Horologion of Andronikos Kyrrhestes* (*Tower of the Winds*) within the Roman Agora, Athens, possibly constructed around 50 B.C. or even earlier. The 12 m tall octagonal marble structure, originally functioning as a clocktower and weather station (containing sundials, a water clock and wind vane), is topped by a frieze depicting relief sculptures of eight *Anemoi* – the wind gods of Greek mythology – each facing in a different direction, corresponding to the eight cardinal points of the compass. Figure A4.25(a) illustrates the segment wherein the particular deity *Kaikias*, representing the northeast wind, appears as a winged bearded man wearing a bulky cloak and pouring hailstones from his shield – partly a tribute to the astonishing beauty of the natural world. The philatelic image in Figure A4.25(b) features the same mythical deity on a 50-lepta airmail postage stamp issued in Greece in 1943.

Among many daring exploits of the Carthaginian general Hannibal (247–*circa* 182 B.C.) and his army during the Second Punic War (218–201 B.C.) was the celebrated crossing of the Alps in late 218 B.C. under atrocious weather conditions, leading to great loss of human and animal life, not least the vulnerable war elephants. The oil painting *Snow Storm: Hannibal and his Army Crossing the Alps* by J. M. W. Turner in Figure A4.26(a), first exhibited in 1812, vividly portrays the dark, swirling storm clouds dominating the sky before enveloping the struggling combatants; a forbidding prelude to the oncoming snow and hail, epitomising human vulnerability when exposed to the overwhelming power of nature. Figure A4.26(b) is an illustration from a 1907 book [A4.46] by the English classical scholar Alfred John Church (1829–1912), drawn by the Scottish graphic artist Harold Millar (1869–1942) and entitled *The passage of the Alps was effected under many difficulties*; which modest contention might be regarded by some as an understatement.

HAILSTONE IMPACT: HISTORICAL PROLOGUE

(a)

(b)

FIGURE A4.25. Depictions of *Kaikias*, the Greek mythical deity of northeast wind, pouring hailstones from shield: (a) relief marble sculpture on frieze on *Tower of the Winds* (*circa* 50 B.C.), Roman Agora, Athens; (b) graphic image on Greek airmail postage stamp, issued 1943

(a)

(b)

FIGURE A4.26. Hannibal crossing the Alps in late 218 B.C: (a) oil painting *Snow Storm: Hannibal and his Army Crossing the Alps* by J. M. W. Turner, 1812 (by courtesy of *Tate Gallery, London*); (b) book illustration *The passage of the Alps was effected under many difficulties* drawn by Harold Millar, 1907

Later in the same military campaign, it has been postulated that Rome was saved in 211 B.C. partially on account of prevailing hailstorms. According to Hunt [A4.47], for example, with Hannibal at the city walls:

> 'Twice in the next several days, Hannibal's army and the Romans squared off for full battle, but each time, a fierce torrent of heavy hail prevented it. Was this spring storm from Jupiter or Baal? Both of these gods were storm deities, but it was possibly a worse omen for the invaders. Centuries later, St Augustine even claimed that it was the gods themselves who terrified Hannibal with lightning and tempest. The hailstorm may have shaken Hannibal's resolve, although he knew this city could not be taken even if he slaughtered the army facing him. Hannibal turned around with his troops and headed south again.'

Even the mighty Gaius Julius Caesar (100–44 B.C.) was taken aback by the ferocity of a hailstorm encountered during a military campaign (*De Bello Africo*) in North Africa [A4.48]–[A4.50]. For early in the year 46 B.C., near the Roman town of Ruspina (close to modern-day Monastir in Tunisia), it is stated that:

> '*Per id tempus fere Caesaris exercitui res accidit incredibilis auditu. Namque vergiliarum signo confecto circiter vigilia secunda noctis nimbus cum saxea grandine subito est exortus ingens.*'

Further interesting details of this extraordinary event are given in various English language translations; whence, from the version [A4.48] by the British soldier and politician Martin Bladen (1680–1746):

> '*Much about the same time there happened a most incredible accident to Caesar's army; the Vergiliae being set, about nine a clock at night rose a violent tempest, attended by a dreadful shower of hail. Caesar had not, like other generals, disposed his forces into winter quarters, but decamping every three or four days, lodged himself in a different post nearer the enemy, which rendered the misfortune greater; for the soldiers were so much employed about the works, they had no leisure to take care of themselves. Besides he was so eager to have all his army transported out of Sicily, that he allowed his men to bring nothing over with them but their armour, neither slaves, baggage, nor any thing that might have been serviceable to them. What little provisions they had brought were already consumed, nor could the country, where they were, furnish them with more. Reduced to this necessity, some few lay under tents, others were obliged to erect little huts of reeds, which they covered with their cloaks. Thus surprised by the storm, beaten down by the weight of the hailstones, almost drowned with water, their fires extinguished, and victuals spoiled, they wandered up and down the camp, defending their heads with their shields. This night likewise the tops of the Fifth Legions piles of their own accord took fire.*'

In the first sentence, the Roman *Vergiliae* (the stars of spring) are equivalent to the *Pleiades* from Greek mythology, here referring to the constellation of the seven stars, rising in May and setting in October. In the last sentence, 'pile' means a spear or javelin (*pilum*) comprising an iron shank with a pyramidal head joined to a wooden shaft; while 'of their own accord took fire' describes how these iron shanks shone with a spontaneous light (so-called *St Elmo's fire*) during the storm – a phenomenon nowadays attributed to luminous plasma (ionised gas) created by a localised corona discharge around sharp or pointed objects within an atmospheric electric field.

The Roman poet and philosopher Lucretius (*circa* 99 B.C.–*circa* 55 B.C.) mentions hail on several occasions in his primary work *De rerum natura*, a long poem with a scientific bias, arranged in six untitled books. In book 6, part 2 (*Great meteorological phenomena, etc.*), for example, are the imposing metrical lines:

> '*Oft, too, the multitudinous crash of ice*
> *And down-pour of swift hail gives forth a sound*
> *Among the mighty clouds on high; for when*
> *The wind hath packed them close, each mountain mass*
> *Of rain-cloud, there congealed utterly*
> *And mixed with hail-stones, breaks and booms...*'

from an English language translation by the American scholar William Leonard [A4.51].

With further regard to Roman fact and folklore concerning weather, the Latin poet Virgil (70 B.C.–19 B.C.) gave copious advice on agricultural husbandry in the first book of the *Georgics*, an agricultural poem completed in 29 B.C. Whence the following short extract on aspects of 'weather forecasting', from a translation into English verse by Dryden [A4.52]:

> '*Above the rest, the Sun, who never lies;*
> *Foretels the change of Weather in the Skies:*
> *For if he rise, unwilling to his Race,*
> *Clouds on his Brows, and Spots upon his Face;*
> *Or if thro' Mists he shoots his sullen Beam,*
> *Frugal of Light, in loose and stragling Streams:*
> *Suspect a drisling Day, with Southern Rain,*
> *Fatal to Fruits, and Flocks, and promis'd Grain.*
> *Or if Aurora, with half open'd Eyes,*
> *And a pale sickly Cheek, salute the Skies;*
> *How shall the Vine, with tender Leaves, defend*
> *Her teeming Clusters, when the Storms descend?*
> *When ridgy Roofs and Tiles can scarce avail,*
> *To barr the Ruin of the ratling Hail.*'

Another scene dominated by the vagaries of inclement weather is set by Virgil in his later epic poem *The Aeneid* – one of the pre-eminent works of Latin literature and still unfinished upon his death in 19 B.C. – initially describing the perilous sea journey of the Trojan warrior *Aeneas*, son of the goddess *Venus*, from the ransacked Troy to Italy, eventually to found the city of Rome. A meeting *en route* with *Dido*, Queen of Carthage, initially showed promise but, perhaps in a sense of foreboding engendered by the wrathful goddess *Juno*, the happy pair were overtaken by a violent hailstorm during a convivial hunting expedition. From a modern English language translation [A4.53] by the German-American historian Sabine MacCormack (1941–2012), Virgil wrote (book 4):

> '*The sky darkened with clouds and rumble of thunder,*
> *then rain mixed with hailstones poured down.*
> *Carthaginian nobles and young men from Troy with*
> *the Dardan descendant of Venus all scattered afar*
> *in their fright, seeking shelter. Torrents rush down from the mountains.*
> *Dido and the Trojan commander both found their way*
> *to the very same cave. First the earth and then Juno as guardian of marriage*
> *give the sign. Fires flash in the heavens as witness*
> *to the bridal, and the nymphs give voice from the heights.*'

Such legendary scenes are elegantly portrayed in Figure A4.27, sourced from an illuminated manuscript on vellum [A4.54] produced in the 5th century A.D. Figure A4.27(a) depicts the tempestuous sea voyage with *Aeneas*, arms raised, imploring the gods to grant deliverance during a raging storm sent by the furious *Juno*. In Figure A4.27(b), *Dido* and *Aeneas* shelter safely inside a cave, seated together in perfect accord, while noting that a less fortunate yet seated huntsman – considered by some to be *Aeneas'* son *Ascanius* (the Dardan or Trojan descendant of *Venus*) – is consigned to withstand the hailstorm in nearby woodland by holding a shield above his head. As might be expected in such uncertain times, the liaison did not endure, following the early departure of *Aeneas* and the ultimate demise of the tragic *Dido* soon thereafter.

Broad correlations between personal sentiments and old-world agricultural activity – including the damaging effects of hail – are evident in the works of the Roman lyric poet Horace (65 B.C.–8 B.C.). From a translation (in four volumes) by the Anglo-Irish clergyman Philip Francis (1708–1773) – the English language version preferred by the renowned literary critic and man of words, Samuel Johnson (1709–1784) – it is given [A4.55] in ode 1:

> '*Whether his Vines be smit with Hail,*
> *Whether his promis'd Harvests fail,*
> *Perfidious to his Toil;*
> *Whether his drooping Trees complain*
> *Of angry Winters, chilling Rain,*
> *Or Stars that burn the Soil.*'

while much later [A4.56] in epistle 8 (to *Celsus Albinovanus*):

> '*To Celsus, Muse, my warmest Wishes bear,*
> *And if he kindly ask you how I fare,*
> *Say, tho' I threaten many a vast Design,*
> *Nor Happiness, nor Wisdom yet, are mine.*
> *Not that the driving Hail my Vinyards beat;*
> *Not that my Olives are destroy'd with Heat;*
> *Not that my Cattle pine in foreign Plains –*
> *More in my Mind than Body lie my Pains.*'

Related excerpts occur in the *magnum opus* written by Ovid (43 B.C.–circa 18 A.D.) who, along with his older contemporaries Vergil and Horace, is usually regarded as one of the three canonical poets of Latin literature. Whence in *Metamorphoses*, a mythological narrative in 15 books and first published around 8 A.D., the part entitled *Boreas in love* in book 6 includes:

(a)

(b)

FIGURE A4.27. Illuminated manuscript: Vergil's *Aeneid*, anonymous miniatures, 5th century A.D. (by courtesy of *Biblioteca Apostolica, Vatican City*, MS Vergilius Romanus, Vat. lat. 3867, f. 77r; f. 106r): (a) *Aeneas* entreats deliverance amid tempest and stormy seas; (b) *Dido* and *Aeneas* seek shelter from hailstorm

> *'By force and violence I chiefly live,*
> *By them the louring stormy tempests drive:*
> *In foaming billows raise the hoary deep,*
> *Writhe knotted oaks, and sandy deserts sweep;*
> *Congeal the falling flakes of fleecy snow,*
> *And bruise, with rattling hail, the plains below.'*

Moreover, a segment (*The transformation of Lychas into a rock*) in book 9 contains the passage:

> *'So show'ry drops, when chilly tempests blow,*
> *Thicken at first, then whiten into snow,*
> *In balls congeal'd the rolling fleeces bound,*
> *In solid hail result upon the ground.'*

while a later segment (*The Trojan ships transformed to sea-nymphs*) in book 14 has the lines:

> *'Strait peals of thunder Heav'n's high arches rend,*
> *The hail-stones leap, the show'rs in spouts descend.'*

all from a translated version [A4.57] edited by the English physician and poet Samuel Garth (1661–1719).

The detrimental effects of hail in a rural setting are mentioned by the Roman author and landowner Lucius Columella (4–*circa* 70 A.D.) in his Latin treatise on agriculture (*De Re Rustica*), written in twelve books, with one in blank verse. In book 10 (*Of the Culture of Gardens, in Verse*), for example:

> *'And Gard'ner, free from fear, rejoices in*
> *His adult wares, and, now they're ripe, prepares*
> *The sickle to put in; oft-times fierce Jove*
> *Does dart his grievous show'rs, and with hailstones*
> *The labours both of men and beasts destroys:'*

from an anonymous translation of the complete work [A4.58], including an additional small book on trees (*De Arboribus*), published in 1745.

A somewhat benign effect of hail impact upon roofing tiles was promulgated by the Roman poet Marcus Annaeus Lucanus (39–65 A.D.), better known as Lucan, in his epic work *De Bello Civili* (*On the Civil War*), usually referred to as the *Pharsalia* (after the Battle of Pharsalus in 48 B.C.). Comprising ten books, although probably unfinished upon his premature death by suicide (at the behest of Emperor Nero), the following extract from book 3 (*Massilia*) refers to the defensive measures taken by foot soldiers under attack from 'Grecian bolts', with the aid of their trusty shields, during the lengthy siege of Massilia (present-day Marseille) in 49 B.C:

> *'Long as the shields*
> *Held firm together, like to hail that falls*

> *Harmless upon a roof, so long the stones*
> *Crushed down innocuous; but as the blows*
> *Rained fierce and ceaseless and the Romans tired,*
> *Some here and there sank fainting.'*

from a translation [A4.59] by the English lawyer and politician Edward Ridley (1843–1928). With the enduring resilience of Roman construction having long been widely recognised, perhaps the instinctive confidence of Lucan was justified.

Yet further quotations are to be found in the work of Publius Papinius Statius (*circa* 45–*circa* 96 A.D.), perhaps the foremost Latin poet of the Flavian era of ancient Rome, who influenced several subsequent authors including Dante Alighieri, who included Statius as one of the characters in his *Divine Comedy* (see later). In *The Thebaid*, an epic poem in twelve books based on Theban legend, is the following extract from book 1, set amid travels near Corinth:

> *'The winds arise, and with tumultuous rage*
> *The gath'ring horrors of the storm presage;*
> *And whilst in Heav'n superior sway they claim,*
> *Earth labours, and resounds the starry frame.*
> *But Auster chiefly checks the breaking light,*
> *In clouds incircled, and renews the night;*
> *Then opes the sluices of the pregnant sky,*
> *And bids the tempest from each quarter fly,*
> *Which the fierce north, ere finish'd was its course,*
> *Congeals to show'rs of hail with wond'rous force.'*

in a rhyming couplet version [A4.60] by William Lillington Lewis (*bap.* 1743–1772) first published in 1767, noting that *Auster* represents the god of the south wind in Roman mythology. And from book 6, set among the rival charioteers:

> *'From feet and wheels arise unequal sounds:*
> *Their hands ne'er rest: the driver's lash rebounds*
> *In ecchoing air. — Not thicker in the north*
> *Pale Boreas spreads a spatt'ring Tempest forth*
> *Of noxious hail, nor from the Nurse of Jove*
> *So many show'rs oppress the nodding grove.'*

A much later alternative translation of the above first passage by John Henry Mozley [A4.61] yields:

> *'And now the rocky prisons of Aeolia are smitten and groan, and the coming storm threatens with hoarse bellowing: the winds loud clamouring meet in conflicting currents, and fling loose heaven's vault from its fastened hinges, while each strives for mastery of the sky; but Auster most violent thickens gloom on gloom with whirling eddies of darkness, and pours down rain which keen Boreas with his freezing breath hardens into hail; quivering lightnings gleam, and from the colliding air bursts sudden fire.'*

A second source from the work of Statius is *The Silvae*, a later collection of occasional poems in five books: specifically book 1, part 6 (*The Kalends of December*), being an account of entertainment in the amphitheatre decreed by the emperor during the Saturnalia festivities. Wherefore, upon marvelling at the profusion of delicacies provided at a feast:

> 'Not with such torrents do stormy Hyades o'erwhelm the earth or Pleiades dissolved in rain, as the hail that from a sunny sky lashed the people in the theatre of Rome. Let Jupiter send his tempests through the world and threaten the broad fields, while our own Jove sends us showers like these!'

again due to Mozley [A4.61].

Besides his scientific work, the ancient astronomer Claudius Ptolemy (*circa* 100–*circa* 170 A.D.) also wrote the eclectic treatise *Tetrabiblos*, comprising four books on the history of astrology. Whence in book 2:

> 'Now the sign of Aries as a whole, because it marks the equinox, is characterised by thunder or hail, but, taken part by part, through the variation in degree that is due to the special quality of the fixed stars, its leading portion is rainy and windy, its middle temperate, and the following part hot and pestilential. Its northern parts are hot and destructive, its southern frosty and chilly.'

as translated [A4.62] from the original Greek by the American academic Frank Egleston Robbins (1884–1963); see also an earlier version by Ashmand [A4.63].

The Greek author Nonnos of Panopolis, who flourished in Egypt in the 5th century A.D., wrote *Dionysiaca* in 48 books, perhaps the last of the great epic poems of antiquity. This enormous work relates the story of the Greek god *Dionysos* (son of *Zeus*), also known as *Bacchus* by the Romans, centred on the experiences of godly life and especially an expedition to India. Whence in book 2 (lines 422–435):

> 'Zeus breasting the tempests with his aegis-breastplate swooped down from the air on high, seated in Time's chariot with four winged steeds, for the horses that drew Cronion were the team of the winds. Now he battled with lightnings, now with Levin; now he attacked with thunders, now poured out petrified masses of frozen hail in volleying showers. Waterspouts burst thick upon the Giant's heads with sharp blows, and hands were cut off from the monster by the frozen volleys of the air as by a knife. One hand rolled in the dust, struck off by the icy cut of the hail; it did not drop the crag which it held, but fought on even while it fell, and shot rolling over the ground in self-propelled leaps, a hand gone mad! as if it still wished to strike the vault of Olympos.'

while subsequently observing (lines 540–547):

> 'Now as the son was scourged with frozen volleys of jagged hailstones, his mother the dry Earth was beaten too; and seeing the stone bullets and icy points embedded in the Giant's flesh, the witness of his fate, she prayed to Titan Helios with submissive voice: she begged of him one red hot ray, that with its heating fire she might melt the petrified water of Zeus, by pouring his kindred radiance over frozen Typhon.'

from a translation [A4.64] by the British academic William Rouse (1863–1950).

Hailstorms also feature in the biblical epic *De laudibus Dei* (*Praises of God*), written in three books by the poet Blossius Aemilius Dracontius (*circa* 455–*circa* 505 A.D.) of Carthage. On account of an earlier poem written in honour of a foreign ruler, Dracontius incurred the displeasure of Gunthamund (*circa* 450–496 A.D.), king of the Vandals and tyrannical ruler of Roman Africa, who promptly consigned him to indefinite imprisonment. During this incarceration, he attempted to gain the king's favour by composing two Latin poems, the first of which (*Satisfactio*) failed to achieve its primary purpose, followed by the much longer *De laudibus Dei*, a didactic poem in the nature of a theodicy. Gunthamund's response to the latter work is unknown, although it appears that Dracontius survived this dangerous encounter more or less intact. The following is an engaging excerpt from book 1 (lines 189–193), included in a picturesque description of the Creation:

> '*Non solis anheli*
> *flammatur radiis, quatitur nec flatibus ille,*
> *nec coniuratis furit illic turbo procellis;*
> *non glacies destricta domat, non grandinis ictus*
> *verberat aut gelidis canescunt prata pruinis.*'

which upon translation becomes:

> '*It is not scorched by the rays of the stifling sun,*
> *neither is it shaken by the blasts of the wind,*
> *nor does the whirlwind of conspiring gales there rage against it;*
> *rigid ice does not overpower it, the beating of hail lashes it not,*
> *nor do its meadows whiten with hoarfrost.*'

according to Vollmer [A4.65], [A4.66] and Irwin [A4.67], set within a narrative contrasting the lush surroundings of bountiful *Eden* with the postlapsarian condition of a harsh and frozen landscape. A later excerpt (lines 707–709):

> '*Qui de thesauris ventorum flamina mittit*
> *et frenat rapidas in tempestate procellas,*
> *grandinis atque nivis venientis quae sit origo;*'

which translates as

> '*He sends from His storehouses the blasts of the winds,*
> *and bridles the swift gales in the storm,*
> *which is to be the source of the approaching hail and snow.*'

is from a longer passage extolling the omnipotence of the Almighty creator of heaven and earth.

McCartney [A4.68] has discussed further aspects of Greek and Roman weather lore, with particular reference to hailstorms. More generally, Taub [A4.69] has reviewed a wide variety of Greek and Roman texts relating to ancient meteorology, especially those relating to customary weather phenomena. Chronopoulou and Mavrakis [A4.70] have considered the likely weather conditions experienced during mid-winter – especially the *Halcyon* days – at the open theatre of *Dionysus* in Athens around the 5th century B.C., based upon practical considerations relating to the production of ancient Greek drama. McCormick *et al*. [A4.71] have examined various hypotheses concerning the possible effects of climate change during the rise and fall of the Roman Empire, founded on contemporary written records and later archaeological documentation.

Hailstorms also feature briefly in *Popol Vuh* (*Book of the People*), an indigenous narrative recounting the mythology and history of part of the ancient Maya civilisation in Mesoamerica, originally preserved through oral tradition until recorded in Mayan hieroglyphs in around 1550, and subsequently transcribed and conserved in manuscript form by the Spanish Dominican friar (1666–*circa* 1729). During one episode (*The nations ask for their fire*) centred on the deity *Tohil* is the expressive passage:

> '*Then it began to rain while the fire of the nations burned brightly. Hail fell thickly on the heads of all the nations and their fire went out because of the hail. Thus their fire came to nought. Then Balam Quitze and Balam Acab pleaded again for their fire:*
> '*O Tohil, truly the cold has finished us,' they said to Tohil.*
> '*Very well. Do not mourn,' said Tohil.*
> *Then he brought forth fire. He twist drilled inside his shoe. Thus Balam Quitze, Balam Acab, Mahucutah, and Iqui Balam rejoiced as they warmed themselves.*
> *But the fire of the nations was extinguished and the cold had nearly finished them. Thus they came to plead for fire from Balam Quitze, Balam Acab, Mahucutah, and Iqui Balam. They could no longer bear it because of the cold and the hail. They shuddered and trembled. There was no life in them because of the shivering of their legs and their arms*.'

from a modern translation [A4.72] by the American scholar Allen Christenson, where 'twist drilled' refers to spinning or rotating a stiff wooden stick on a firm wooden platform ('his shoe'), producing sufficient heat *via* friction to ignite kindling.

A4.3. Early Middle Ages (500–1100)

A fascinating reference to hailstorms in Arabic literature occurs in *The Expeditions* (*Kitāb al-Maghāzī*), an early biography of the prophet Muḥammad largely written and compiled by Maʿmar ibn Rāshid (714–770 A.D.), a prominent scholar from Basra in southern Iraq. During the so-called *Battle of the Trench* in early 627 A.D., the last major battle of the Prophet before the conquest of Mecca, the *Messenger of God* and his companions within Medina were besieged by the *Pagans*, one of whose leaders Ḥuyayy ibn Akhṭab requested entry to the Qurayẓah encampment. Whence, from a modern translation by the American author Sean Anthony [A4.73]:

> '*The Qurayẓah clan said, 'By God, we must open our doors to him.' Soon they opened their doors to him, and when he entered he beguiled them. Ḥuyayy exclaimed, 'Sons of Qurayẓah, I have come to you in the nick of time! I come to you with a mighty hailstorm, and nothing can*

> stand in its way!' Their chieftain replied to him, 'Can you promise that this hailstorm will spare us; that you will leave us next to a calm ocean and not abandon us? On the contrary, all you promise is folly.'
>
> Ḥuyayy gave his word to the Qurayẓah clan and made a covenant with them to the effect that if the groups of the united clans dispersed, he would return to join them in their stronghold. When they followed him, they did so in perfidy against the Prophet and the Muslims.'

with the dire consequence that, by command of the Prophet, the unfortunate Ḥuyayy was summarily executed soon thereafter.

Further eastward, in the brief exposition *Encomium for a Horse Painting* written by Du Fu (712–770 A.D.), a prominent Chinese poet of the Tang Dynasty (also known as Tu Fu, depending on transliteration), the divine horse of antiquity is presented as displaying:

> 'Aloof manner, easy and free,
> bounding off, moving at will.
> Four hooves, thunder and hail,
> in one day, across the world.'

from a modern translation of his complete works in six volumes by the American sinologist Stephen Owen [A4.74].

In the realm of linguistic development, it is noted that *Haglaz* is the reconstructed Proto-Germanic name of the *h*-rune ᚺ, sometimes written as the double-barred ᚻ, and is continued in the Anglo-Saxon futhorc as *haegl*, meaning 'hail'. The anonymous Anglo-Saxon parchment (*circa* 9th century) copied in Figure A4.28 illustrates its position within the runic alphabet (with later added names and notes). *Haglaz* is also recorded in the Anglo-Saxon rune poem:

> 'Hægl byþ hwitust corna;   'Hail is the whitest of grains;
> hwyrft hit of heofones lyfte,   it whirls down from the heavens,
> wealcaþ hit windes scura;   and is hurled around by gusts of wind;
> weorþeþ hit to wætere syððan.'   and then it melts into water.'

which incidentally has contemporary Nordic counterparts; this English language version being an amalgam of several similar modern translations.

It is evident from historical accounts by the British historian Reginald Lane Poole (1857–1939) that some medieval church practices concerning the vagaries of weather were on occasion seriously challenged by those from within; one such forthright individual being Agobard of Lyon (*circa* 779–840 A.D.), a Spanish-born priest and archbishop of Lyon during the Carolingian Renaissance. According to Poole [A4.75], from chapter 1 (*Claudius of Turin and Agobard of Lyons*):

> 'It appears that there was a class of impostors who assumed to themselves the office of 'clerk of the weather'. These tempestarii, or weather-wizards, claimed the power not only of controlling the weather, and securing the fields from harm, but also of bringing about hail and thunderstorms, and especially of directing them against their private enemies. Plainly they derived a goodly revenue from a blackmail forced by the double motives of fear and hope.'

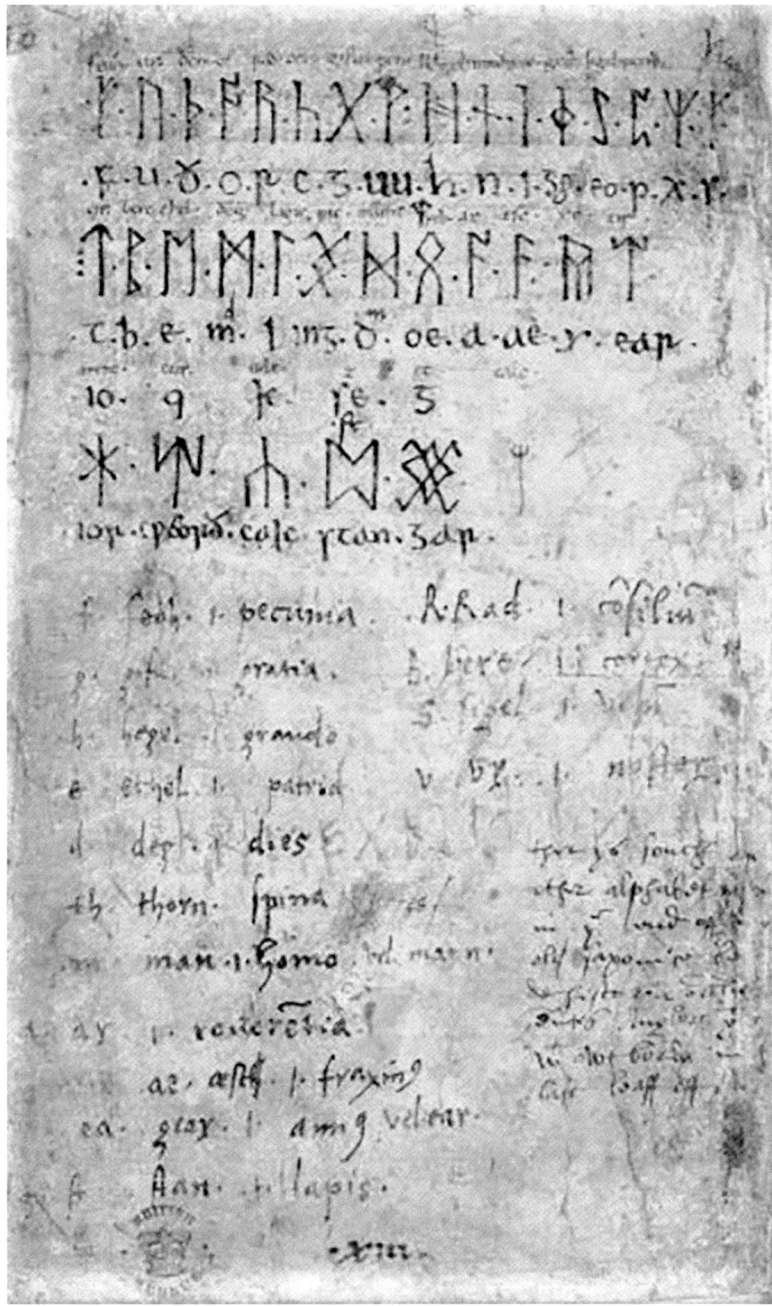

FIGURE A4.28. Anglo-Saxon parchment (*circa* 9th century A.D.) showing runic alphabet (with later added names and notes) including *h*-rune H, continued as *haegl* or hail (by courtesy of *British Library, London*, Cotton MS Domitian A IX, f. 11v)

In an era when much of the population was bound by witchcraft and sorcery, Agobard *'laboured hard for their liberation and attacked unsparingly every form of superstition wherever he found it'*, including that promulgated by the so-called weather-witches (*tempestarii*). A particular example of this spirited resistance is his polemical treatise (*circa* 815 A.D.) *Against the absurd opinion of the vulgar touching hail and thunder* (*Liber contra insulsam vulgi opinionem de grandine et tonitruis*) which *'by argument and ridicule, and at times by a lofty eloquence, he attempted to breast this tide'*, in the words of Poole [A4.75]. One short passage by Agobard reads:

> *'The wretched world lies now under the tyranny of foolishness: things are believed by Christians of such absurdity as no one ever could aforetime induce the heathen, who knew not the Creator of all, to believe.'*

which latent appeal by the great churchman turned out to be in vain, as the tide of superstition continued remorselessly; see also the more recent authoritative study [A4.76] of magical beliefs in early medieval Europe by the English historian Valerie Flint (1936–2009). Indeed, having described some relatively innocuous practices in part 2 (*The magic of the heavens*), chapter 5 (*The magic that persisted: condemned magical agencies*), Flint [A4.76] remarks:

> *'We are some way here from the flesh-and-blood malefactors the secular law codes deal with, but we are certainly concerned with a faith in heavenly magic undesirable to Christians, and with a seemingly persistent faith. Other councils and penitentials return to the simpler 'emissores tempestatum' and levy penances of up to seven years. As with the moon, so in the case of storms, a good deal of noise was sometimes thought to be helpful, and shouts and the blowing of conch shells and trumpets to drive them away attracted the disapproval of pastors.'*

while also citing the treatise *On the nature of things* (*De natura rerum*) by an English Benedictine monk usually known as the Venerable Bede (*circa* 672–735 A.D.), translated into the English language by Kendall and Wallis [A4.77].

At around this period, during the reign of Louis the Pious (778–840 A.D.) and following the death of his father Emperor Charlemagne (747–814), some of the measures taken to implement Carolingian ecclesiastical reform were announced at the Council of Paris in 829. From archived records edited by Pertz [A4.78] and Werminghoff [A4.79]:

> *'Ferunt enim suis maleficiis aera posse conturbare et grandines inmittere, futura praedicere, fructus et lac auferre aliisque dare, et innumera a talibus fieri dicuntur.'*

which, according to the Canadian historian Paul Dutton [A4.80], becomes:

> *'For some say that by their evil deeds they can stir up the air and send down hail, predict the future, and take away the produce and milk of some and give it to others. And countless other things are said to be done by such people.'*

Any man or woman found guilty of these storm-making practices could expect harsh treatment by the church authorities, since they had openly ventured to serve the devil.

In Anglo-Saxon literature, the hardships of ocean life – including exposure to *haegl* (hail) – are recounted in the anonymous poem *The Seafarer*, recorded in the unique 10th-century *Exeter Book* (*Codex Exoniensis*) donated to the library [A4.81] of Exeter Cathedral by Bishop

Leofric in 1072; see also a transcription of the original text alongside a translation [A4.82] by the English scholar Benjamin Thorpe (1782–1870). According to the lonesome mariner:

> '*This he, whose lot happily chances on land, doth not know;*
> *Nor how I on the ice-cold sea passed the winter in exile,*
> *In wretchedness, robbed of my kinsmen, with icicles hung.*
> *The hail flew in showers about me; and there I heard only*
> *The roar of the sea, ice-cold waves, and the song of the swan;*
> *For pastime the gannets' cry served me; the kittiwakes' chatter*
> *For laughter of men; and for mead-drink the call of the sea-mews.*'

from a more liberal translation [A4.83], [A4.84] made in 1901 by the American poet Lola La Motte Iddings (1858–1918), one of several markedly different English language versions. The poem occupies four pages of the original Anglo-Saxon codex, and the first of these manuscript pages is shown in Figure A4.29, from which the above extract is taken.

Elsewhere within the corpus of extant Old English literature is the anonymous Anglo-Saxon religious poem *The Phoenix* (also found in the *Exeter Book*), derived from a 4th-century Latin poem (*Carmen de ave phoenice*), which describes the *Edenic* landscape inhabited by the mythical bird. From a commentary by Appleton [A4.85], with Old English quotations taken from Blake [A4.86], this idyllic distant land is blessed by a temperate climate, protected from inclement weather and natural disasters:

> '*Þæt is wynsum wong, wealdas grene* '*That is a delightful plain, green woods*
> *rume under roderum.* *spacious under the heavens.*
> *Ne mæg þær ren ne snaw,* *Neither rain nor snow,*
> *ne forstes fnæst, ne fyres blæst,* *nor frost's breath, nor fire's blast,*
> *ne hægles hryre, ne hrimes dryre,* *nor hail's destruction, nor rime's fall,*
> *ne sunnan hætu, ne sincaldu,* *nor the sun's heat, nor perpetual cold,*
> *ne wearm weder, ne winterscur,* *nor hot weather, nor winter shower,*
> *wihte gewyrdan, ac se wong seomað,* *may harm anything, but the plain abides,*
> *eadig ond onsund. Is þæt æþele lond* *blessed and sound. That noble land*
> *blostmum geblowen.*' *is blooming with blossoms.*'

whilst in a similar vein:

> '*Þær ne hægl ne hrim hreosað to foldan,* '*There neither hail nor rime falls to earth,*
> *ne windig wolcen, ne þær wæter feallep,* *nor wind-driven cloud, nor does water fall there,*
> *lyfte gebysgad.*' *agitated by wind.*'

which combined imagery is broadly shared with the aforementioned Latin poem (*De laudibus Dei*) by Dracontius.

Several allusions to hail occur in the *Shahnameh* (*The Epic of the Kings*), a very long epic in verse (completed in 1010 A.D.) by the Persian poet Abul-Qâsem Ferdowsi Tusi (*circa* 940–1020 A.D.) – usually referred to as Ferdowsi, although spellings vary – which encompasses both mythical and historical aspects of ancient Persia. For example, in *The Death of Yazdagird*, concerning the untimely fate of the last king of the Sasanian Empire (*Yazdagird III*, 624–651 A.D.), are the astringent lines:

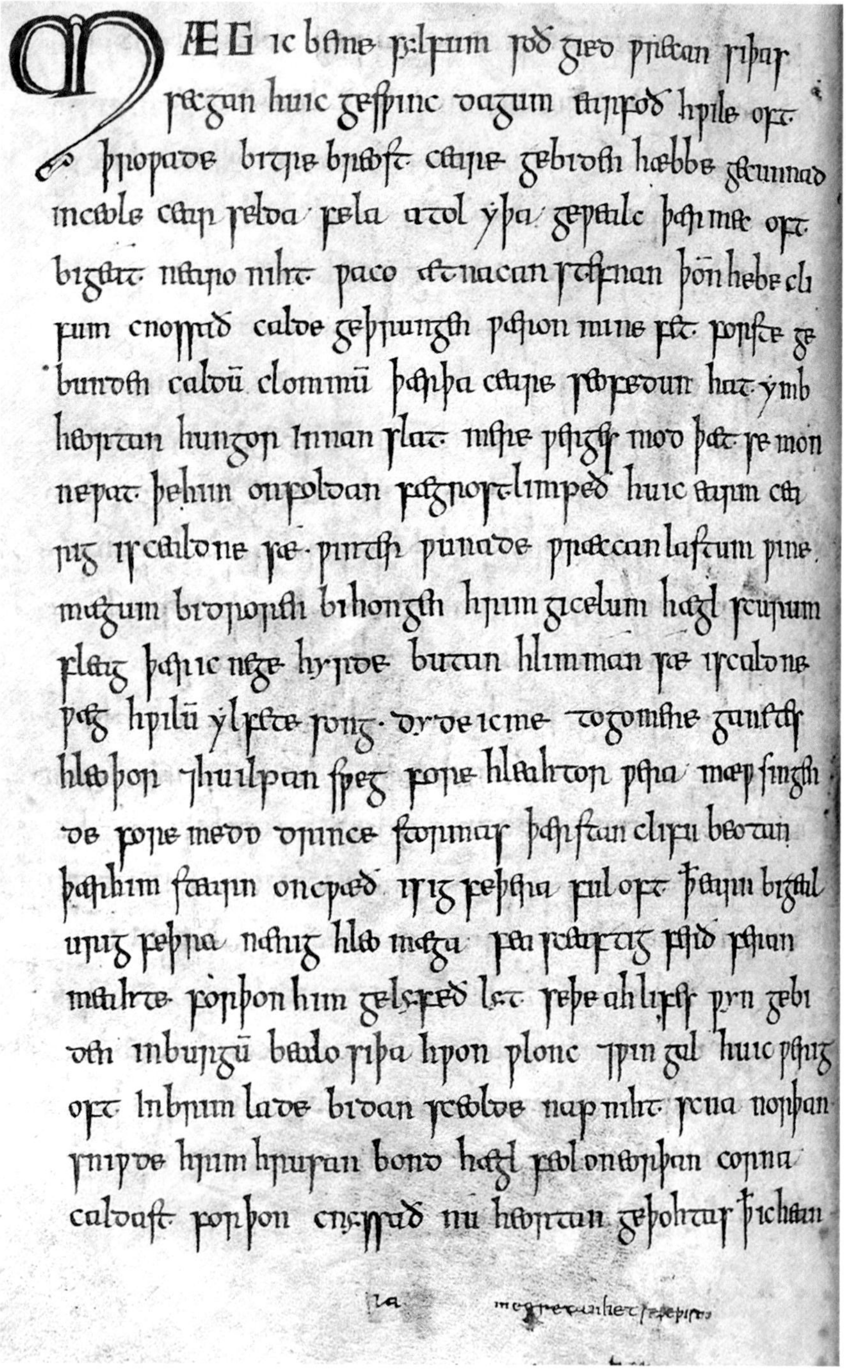

FIGURE A4.29. First page of Anglo-Saxon manuscript (*circa* 970 A.D.) of anonymous poem *The Seafarer*, including reference to hail (by courtesy of *Cathedral Library, Exeter*, MS 3501, f. 81v)

> *'The hail this year like death on me hath come,*
> *Though death itself were better than the hail,*
> *And heaven's lofty, far-extending dome*
> *Hath caused my fuel, wheat, and sheep to fail.'*

from a translation by the Warner brothers [A4.87], based on the first complete Persian text [A4.88] collated and edited by the Irish soldier Turner Macan (1792–1836), Persian interpreter to the Commander-in-Chief of the British Army in India.

Also of special interest is *The Song of Roland* (*La Chanson de Roland*), a medieval narrative in Old French written – possibly by Turoldus – in the 11th century, and based on the *Battle of Roncevaux Pass* in 778 A.D. During a hailstorm on this high mountain pass in the Pyrenees, a strong force of Basques ambushed and slaughtered a rear-guard detachment of Charlemagne's army, including the Frankish commander Roland, duke of the Marches of Brittany. Eventually history transformed itself into legend and then to epic poem, which includes the imposing passage:

> *'But France the while with the turmoil of a marvellous tempest is loud:*
> *There is storm with crashing thunder and wind: from bursting cloud*
> *Cataract rains are pouring, measureless scourging of hail;*
> *Thick and fast descendeth the thunderbolt's flashing flail;*
> *And the very land's foundations with earthquake shiver and sway.'*

from a translation into English verse by Arthur Way [A4.89]. The prevailing ambience of the battle scene is portrayed in an oil painting by the French artist Gustave Doré (1832–1883), shown in Figure A4.30.

A4.4. Late Middle Ages (1100–1500)

Amongst the prodigious literary output of the English writer George Henry Borrow (1803–1881) is the poem *The Hail-Storm (From the Norse)*, published in 1826 [A4.90] and included in volume 8 of a 16-volume compilation of his works [A4.91] edited by the journalist Clement King Shorter (1857–1926). The first half of this short anonymous poem (with anglicised names) is:

> *'When from our ships we bounded,*
> *I heard, with fear astounded,*
> *The storm of Thorgerd's waking,*
> *From Northern vapours breaking;*
> *With flinty masses blended,*
> *Gigantic hail descended,*
> *And thick and fiercely rattled*
> *Against us there embattled.*
>
> *To aid the hostile maces,*
> *It drifted in our faces;*

FIGURE A4.30. Oil painting, *circa* 1870: *Roland à Roncevaux* by Gustave Doré, depicting hailstorm during battle (*private collection, Paris*)

> *It drifted, dealing slaughter,*
> *And blood ran out like water—*
> *Ran reeking, red, and horrid,*
> *From batter'd cheek and forehead;*
> *We plied our swords, but no men*
> *Can stand 'gainst hail and foemen.'*

An alternative version of the poem, translated from Old Norse, appears amidst examples of medieval skaldic poetry [A4.92]; whereupon the first two stanzas become:

> '*For victory as we bounded,*
> *I heard, with fear astounded,*
> *The storm, of Thorgerd's waking,*
> *From Northern vapours breaking.*
> *Sent by the fiend in anger,*
> *With din and stunning clangour,*
> *To crush our might intended,*
> *Gigantic hail descended.*

A pound the smallest pebble
Did weigh, and others treble;
Full dreadful was the slaughter;
And blood ran out like water,
Ran, reeking, red and horrid
From batter'd cheek and forehead.
But though so rudely greeted,
No Jomsborg man retreated.'

which allude to the semi-legendary naval *Battle of Hjorungavag* in 986 A.D. The poem describes the woes of the Jomsviking invaders led by their chieftain *Sigvald*, Earl of Jomsborg, who were repulsed off the coast of Norway by *Hakon Jarl* (*circa* 937–995), with the assistance of the goddess *Thorgerd*, to whom *Hakon* sacrificed his youngest son in order to ensure deliverance from some dire fate.

Figure A4.31(a) illustrates a scene from the battle, as interpreted by the young Norwegian artist Halfdan Egedius (1877–1899) in 1897 and taken from *Heimskringla*, a collection of sagas [A4.93] originally compiled by the Icelandic historian Snorri Sturlaśon (*circa* 1178–1241). Figure A4.31(b) shows a drawing with gouache on paper made in 1899 by another Norwegian artist Gerhard Munthe (1849–1929). Figure A4.31(c) illustrates an anonymous tapestry designed by Munthe in 1908 and made in Norway between 1908 and 1920. Here *Thor* is readily identified by his hammer, fighting alongside a warrior in a chariot pulled by two goats. Whilst accepting a degree of poetic license, it is not entirely clear why the prevailing hailstorm should have benefited only one side in the armed conflict; but see, for example, the detailed translation and commentary by Hollander [A4.94].

At least one reference to hail is found in *The Mabinogion*, a collection of Celtic prose stories or tales in Middle Welsh originating in the 12th–13th centuries, first translated and published in full (in Welsh and English) in separate parts during 1838–45 by the English aristocrat Charlotte Guest (1812–1895), and then in 1877 as a single-volume edition [A4.95]; see also the first book of a later three-volume edition [A4.96]. In *The Lady of the Fountain*, based upon *Arthurian* legend in an era of chivalry, the first of the tales told whilst feasting within the king's chamber is by *Kynon*, a knight eager to demonstrate his superiority over other mortals. Early in the plot, after staying the night at a remote castle and passing through a wood into a sheltered glade, he comes across a *Black Giant* – a man with one foot and one eye, wielding a heavy iron club – who, having displayed his own great strength, directs *Kynon* along a path to a clearing in which stands a tall tree. The giant continues with explicit directions:

'*Under this tree is a fountain, and by the side of the fountain a marble slab, and on the marble slab a silver bowl, attached by a chain of silver, so that it may not be carried away. Take the bowl and throw a bowlful of water upon the slab, and thou wilt hear a mighty peal of thunder, so that thou wilt think that heaven and earth are trembling with its fury. With the thunder there will come a shower so severe that it will be scarce possible for thee to endure it and live. And the shower will be of hailstones; and after the shower, the weather will become fair, but every leaf that was upon the tree will have been carried away by the shower.*'

(a)

Uveiret under Slaget (Hjörungavaag) (übrigt)

(b)

FIGURE A4.31. Depictions of naval *Battle of Hjorungavag* in 986 A.D. during hailstorm: (a) drawing by Halfdan Egedius, 1897; (b) gouache sketch by Gerhard Munthe, 1899 (by courtesy of *National Museum of Art, Architecture and Design, Oslo*)

(c)

FIGURE A4.31. (continued) Depictions of naval *Battle of Hjorungavag* in 986 A.D. during hailstorm: (c) anonymous tapestry designed by Gerhard Munthe, 1908 (by courtesy of *Vesterheim National Norwegian-American Museum, Iowa*)

And in continuing to relate the tale to *Kai* and *Owain*, his fellow knights:

> '*Then I took the bowl, and cast a bowlful of water upon the slab; and thereupon, behold, the thunder came, much more violent than the black man had led me to expect; and after the thunder came the shower; and of a truth I tell thee, Kai, that there is neither man nor beast that can endure that shower and live. For not one of those hailstones would be stopped, either by the flesh or by the skin, until it had reached the bone. I turned my horse's flank towards the shower, and placed the beak of my shield over his head and neck, while I held the upper part of it over my own head. And thus I withstood the shower. When I looked on the tree there was not a single leaf upon it, and then the sky became clear, and with that, behold the birds lighted upon the tree, and sang.*'

Figure A4.32 is a dramatic depiction of the scene where *Kynon* is exposed to the full blast of the hailstorm; from an anonymous illustration in the book [A4.95]. A mysterious *Black Knight* on a black horse who guards the fountain then appears and dismounts *Kynon* in a furious joust, leaving the crestfallen knight to be mocked by the *Black Giant*, and having to return to the castle on foot. The next knight to tell a tale is *Owain*, whose exploits turn out to be more successful.

Cursory observations on the climate of Northern Mongolia were made by the Italian friar, diplomat and explorer Giovanni da Pian del Carpine (*circa* 1185–1252), subsequently Archbishop of Antivari in Dalmatia, during extensive travels to central Asia during 1245–1247, accompanied for most of the journey by a younger Polish friar, Benedict (*circa* 1200–1280), partly to act as an interpreter (on the recommendation of Wenceslaus, King of

FIGURE A4.32. Anonymous illustration relating to ancient Celtic tale *The Lady of the Fountain*, depicting mounted *Arthurian* knight *Kynon* in hailstorm: from 1877 edition of *The Mabinogion* [A4.95]

Bohemia). The two Franciscans reached the 'land of the Mongols' in the summer of 1246, where one of the appointed tasks set by Pope Innocent IV was to meet the future emperor Güyük (grandson of Genghis Khan), whose spectacular enthronement as the new Great Khan (reigned 1246–1248) at a huge tented encampment near Karakoram was delayed for several days in mid-August by a violent hailstorm. Commentaries on a written account of the entire arduous journey compiled by Carpini [A4.97] include those by the French novelist Jules Verne (1828–1905), who wrote of the atmosphere surrounding the imperial pageantry of this momentous occasion [A4.98]:

> 'The climate is very changeable; in summer, storms are very frequent, many fall victims to the vivid lightning, and the wind is often so strong as even to blow over men on horseback: during the winter there is no rain, which all falls in the summer, and then scarcely enough to lay the dust, while the storms of hail are terrible; during Carpini's residence in the country they were so severe that once 140 persons were drowned by the melting of the enormous mass of hailstones that had fallen. It is a very extensive country, but miserable beyond expression.'

Having discharged his formal duty as a papal legate, and despite numerous vicissitudes, Carpini returned home safely the following year, along with his trusted companion.

Another early oriental reference to hail is provided by the Japanese shogun and accomplished *waka* poet Minamoto no Sanetomo (1192–1219). The following piece was composed around 1213 – not long before his premature demise (assassinated by his nephew amid a family feud) – and is included in a sizeable anthology of his work (*Kinkai Wakashu*) compiled posthumously:

> '*Mononofu no*
> *yanami tsukurou*
> *kote no e ni*
> *arare tabashiru*
> *Nasu no shinohara.*'
>
> 'As a warrior reaches back
> to arrows in his quiver,
> upon his raised gauntlet
> hailstones strike down hard,
> in the bamboo field of Nasu.'

where Nasu is located north of present-day Tokyo (originally a fishing village named Edo). Although the above Romanisation of the original script is the same or closely matches those given elsewhere, there are substantial variations in published English language translations, which typically appear as secondary material in studies of later poets [A4.99]–[A4.101].

An intriguing combination of material fact and literary embellishment occurs in the following extract from the *Annals of the Resuli Dynasty of Yemen*, translated by Redhouse [A4.102]:

> 'And in this year (A.H. 695, A.D. 1295–6), in the month of Jumāda-'l-awwal (5th month, about March), there fell in Yemen a rain embracing the whole country. And there was in it a great hailstorm that killed a great number of sheep and goats. There fell then a hailstone as large as a small mountain, with projecting points, each above a cubit (30 inches, about) in length. It fell on a moor between the districts of Sinhān and Rāha. The bulk of it disappeared in the earth, leaving a part visible above the surface. Twenty men could walk round it, who could not see,

some of them, some of the others. Another fell in a place near to the country of Khawtan, the heart of which forty men tried to lift, but were unable.'

Perhaps it is just possible that the aforementioned pair of gargantuan edifices were concentrated accumulations of smaller hailstones, formed under abnormal conditions of local terrain and prevailing wind. By contrast, the more tranquil scene in Figure A4.33 is from a 15th-century Hebrew festival prayer book [A4.103], depicting a man kneeling in prayer upon a hilltop, while two youths find shelter from the heavy hail in a nearby building.

The more malign consequences of hailstorms did not escape the attention of the medieval Italian poet Dante Alighieri (1265–1321) in his epic allegorical work *The Divine Comedy*, comprising three parts (*Hell*, *Purgatory* and *Paradise*). Having initially lost his way but eventually guided by Virgil along the precarious route to salvation, the first part (*Inferno*) tells of the journey by Dante through *Hell*, represented by nine concentric circles of torment located within the *Earth*. In the *Third Circle* (*Gluttony*) of the *Inferno* (canto 6) are written the first four tercets:

FIGURE A4.33. Anonymous miniature (*Man at prayer upon hilltop; two youths shelter from heavy hail*) in illuminated manuscript, 1466 A.D., from Hebrew prayer book (by courtesy of *British Library, London*, MS Harley 5686, Pt 1, f. 60)

> '*At the return of consciousness, that closed*
> *Before the pity of those two relations,*
> *Which utterly with sadness had confused me,*
>
> *New torments I behold, and new tormented*
> *Around me, whichsoever way I move,*
> *And whichsoever way I turn, and gaze.*
>
> *In the third circle am I of the rain*
> *Eternal, maledict, and cold, and heavy;*
> *Its law and quality are never new.*
>
> *Huge hail, and water sombre-hued, and snow,*
> *Athwart the tenebrous air pour down amain;*
> *Noisome the earth is, that receiveth this.*'

from an English language translation of the original Italian text by Henry Longfellow [A4.104].

Figure A4.34 shows two manuscript drawings (*circa* 1587) depicting woeful scenarios in this mythical dystopian world. The first [A4.105], in Figure A4.34(a), was created by the Flemish artist Jan van der Straet, *alias* Johannes Stradanus or Giovanni Stradano (1523–1605); the second [A4.106], in Figure A4.34(b), by the Italian artist Federico Zuccari (*circa* 1540–1609) whilst residing in Spain. Amid violent and incessant hailstorms, the two travellers encounter the fearsome *Cerberus*, a three-headed monster guarding the prostrate gluttons in the freezing mire who did not approve of living mortals entering his realm: whereupon Virgil casts fetid mud into the ravenous mouths of *Cerberus* to assuage its hunger and secure safe passage, free from subjugation. Meanwhile, Dante converses with his former contemporary Florentine, Ciacco, before continuing onward with Virgil; as interpreted in the miniature from an illuminated manuscript [A4.107] shown in Figure A4.35(a) and created by Priamo della Quercia (*circa* 1400–1467), an Italian painter of the early Renaissance. Figure A4.35(b) presents a comparable haunting image from an anonymous earlier illuminated manuscript [A4.108], again evoking doom and destruction, but with a markedly different visualisation of *Cerberus*. All compelling reminders that brutality is invariably the close companion of beauty throughout the literary and artistic cannons.

On a more temporal level, reaching toward the natural world from the supernatural yet often with spiritual undertones, there are several historical accounts of alarming death and destruction caused by hail, some probably apocryphal and others having varying degrees of authenticity. One such disaster occurred during the Hundred Years' War (1337–1453), at a stage where King Edward III of England (1312–1377) was attempting to conquer France, where it has been surmised that a freak thunderstorm on Easter Monday in April 1360 killed an estimated one thousand English soldiers and several thousand horses near the city of Chartres. According to various accounts of this antebellum turmoil – including a modern panoramic narrative by the English historian Jonathan Sumption [A4.109] – a large army (around 10000 combatants, including 4000 men-at-arms and 5000 mounted archers), whilst traversing open fields with little cover, initially was subjected to fierce winds and driving rain, soon turning to sleet and hail as the temperature dropped sharply approaching

(a)

(b)

FIGURE A4.34. Drawings (*circa* 1587) depicting *Third Circle of Hell* in Dante's *Inferno*: (a) *Gourmands lying in the hail, and rain, and snow* by Giovanni Stradano (by courtesy of *Laurentian Medici Library, Florence*, MS Medici Palatino 75, f. 27r); (b) *Terzo cerchio, I golosi, Cerbero* by Federico Zuccari (by courtesy of *Uffizi Gallery, Florence*, MS GDSU inv. 3481F)

HAILSTONE IMPACT: HISTORICAL PROLOGUE

(a)

(b)

FIGURE A4.35. Illuminated manuscripts depicting *Virgil, Dante, Cerberus, and Gluttons under incessant hail, Third Circle of Hell* in Dante's *Divine Comedy* (*Inferno*): (a) miniature by Priamo della Quercia, *circa* 1447 A.D. (by courtesy of *British Library, London*, MS Yates Thompson 36, f. 11); (b) anonymous miniature, *circa* 1370 A.D. (by courtesy of *Bodleian Library, Oxford*, MS Holkham misc. 48, p. 9)

nightfall. Under freezing conditions and without shelter, men and horses died of exposure, while others were killed or injured by enormous hailstones; as rendered in an early sombre vignette:

> 'For an accident befell Edward III and all of his army, who were then before Chartres, that much humbled him and bent his courage. There happened such a storm and violent tempest of thunder and hail, which fell on the English army, that it seemed as if the world was coming to an end. The hailstones were so large as to kill men and beasts, and the boldest were frightened.'

Moreover, in part of a 15th-century manuscript [A4.110], re-printed by Gairdner [A4.111] and attributed to the chronicler William Gregory, Lord Mayor of London (1451–1452), it is recorded – quite literally with icy pessimism – that:

> 'And that day was a foule derke day of myste, rayne, and hayle, and soo bytter colde that men dyde for colde, where fore yet in to thys day hyt ys i-callyd Blacke Monday next aftyr Estyr day.'

while a similar early source [A4.112] quoted by Sumption [A4.109] gives the shorter modern version:

> 'Wherefore, unto this day it is called Black Monday and will be long time hereafter.'

Other writers since have deployed this singular epithet, not least William Shakespeare (1564–1616) in his play [A4.113] *The Merchant of Venice* (Act 2, scene 5), where the young servant and jester *Launcelot Gobbo* refers to 'Black-Monday'.

Casualties occurred across all ranks, and included the young nobleman and military commander Sir Guy de Beauchamp, son and heir to the 11th Earl of Warwick, who was mortally wounded and died soon thereafter. Figure A4.36 depicts two contrasting images of King Edward on the field of battle during the mighty storm. Figure A4.36(a) is an illustration from an early monumental tome [A4.114] by the English scholar Joshua Barnes (1654–1712), showing the resplendent king in full armour amidst the thunder and lightning, with Chartres Cathedral in the far distance. Figure A4.36(b) shows a coloured engraving from a book [A4.115] by the English antiquary and artist James Doyle (1822–1892), portraying a bedraggled king imploring God's forgiveness and reciting a vow for peace, having interpreted the disastrous storm as a grave ill-omen: doubtless a supplication proffered with a generous measure of contrite deliberation, having just laid temporary siege to Paris and pillaged the French countryside, thereby rendering any possible *entente* distinctly less *cordiale*. Mercifully a settlement with France was secured shortly thereafter upon signing the Treaty of Brétigny in May 1360, in which the king renounced his claim to both the French throne and some ancestral possessions, in return for full sovereignty of several French provinces and other concessions. Alas, it was not to be a lasting peace.

Amidst such carnage, it is nonetheless fortunate that this catastrophe did not also claim the life of the eminent English poet Geoffrey Chaucer (*circa* 1344–1400), then a novice man-at-arms in the campaign but captured a few months earlier at Reims; shortly thereafter to be set free following payment of a ransom arranged by the king, and later to become a member of the royal court. Coincidentally, an allusion to hailstorms occurs in his epic poem

(a)

(b)

FIGURE A4.36. Images of King Edward III of England on battlefield near Chartres during Black Monday in April 1360: (a) hailstorm and lightning ravage English army, after Barnes [A4.114]; (b) Edward vows to make peace, after Doyle [A4.115]

The Canterbury Tales, one of the most important early works in English literature. It comprises an unfinished collection of stories written (*circa* 1387) in the archaic vernacular language of Middle English [A4.116], and *The Prologe of the Wyves Tale of Bathe* – from one of several edited versions, Skeat [A4.117] – includes the fragment (lines 460–466):

> '*Metellius, the foule cherl, the swyn,*
> *That with a staf birafte his wyf hir lyf,*
> *For she drank wyn, thogh I hadde been his wyf,*
> *He sholde nat han daunted me fro drinke;*
> *And, after wyn, on Venus moste I thinke:*
> *For al so siker as cold engendreth hayl,*
> *A likerous mouth moste han a likerous tayl.*'

Figure A4.37 shows the last part of the *Prologue* and the opening part of the *Tale* in an anonymous illuminated manuscript on vellum (*Ellesmere Manuscript*) created around 1405.

In a later allegorical poem *The Floure and the Leafe* – originally attributed to Chaucer and first published by Speght [A4.118] in 1598, but probably anonymous – are the lines (365–371):

> '*Save suche as succoured were among the leves*
> *Fro every storme that might hem assaile,*
> *Growing under hegges and thicke greves.*
> *And after that there came a storme of haile*
> *And raine in feare, so that, withouten faile,*
> *The ladies ne the knights nade o threed*
> *Dry on them, so dropping was her weed.*'

given in the modern version edited by Pearsall [A4.119].

Figure A4.38 depicts the manual collection of hailstones in an illustration entitled *baoshui* (hail water) from *Shiwu bencao* (*Materia dietetica*), a classic Chinese dietary compendium in four volumes dating from the Ming period (1368–1644), although the identity of the author and artists is unknown. It is seen that during a hailstorm, one man is holding up a large tray to catch the hailstones, while another man is storing them in a jar. The text states:

> '*Hail water is of indeterminate sapor. 1-2 sheng (1 sheng = circa 1 litre) can be collected at a time and stored in an urn; the original sapor alters during storage.*'

The earlier Yuan dynasty (1271–1368) author Jia Ming also notes in his book *Yinshi xuzhi* (*Essential knowledge for eating and drinking*) that:

> '*If one eats hailstones, it will certainly bring on pestilence (wenyi) and wind madness (fengdian). The liquid is of indeterminate sapor. One or two litres are collected and stored in an urn; its original sapor then changes*'.

which perhaps explains why hailstones are unlikely to feature in any modern cuisine or diet therapy.

FIGURE A4.37. Anonymous illuminated manuscript (*circa* 1405) of Geoffrey Chaucer's *Canterbury Tales* showing opening page of *The Wife of Bath's Prologue and Tale* (by courtesy of *Huntington Library, San Marino, California*, MS EL26C9, f. 72r)

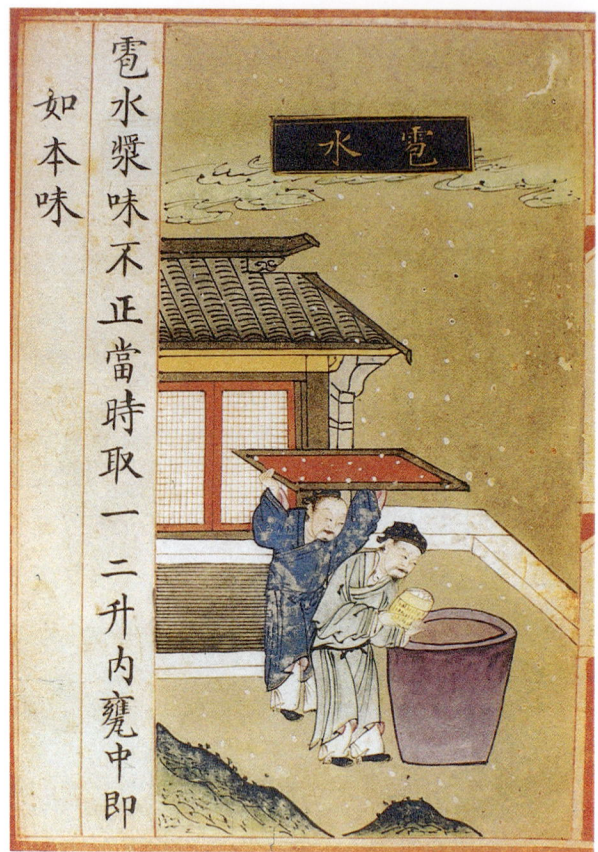

Figure A4.38. Anonymous Chinese illustration entitled *baoshui* (hail water) from Ming period (1368–1644); collection and storage of hailstones for dietary purposes (by courtesy of *Wellcome Library, London*)

In 1411, the Tyrolean author Hans Vintler (died *circa* 1418) completed *Die Pluemen der Tugent* (*Flowers of virtue*), a long poem in rhyming verse, also having the variant title *Das Buch der Tugent* (*The Book of Virtue*) and closely based on the moral didactic poem *Fiore di virtù* by the Italian author Tommaso Gozzadini (1260–1329), written in the early 14th century. This illustrated work [A4.120], with much later commentaries by Zingerle [A4.121] and Zika [A4.122], condemns magic and superstitious practices, and includes the woodcut in Figure A4.39 portraying a female witch producing hail, with a gigantic animal jawbone serving as an essential aid to the nefarious sorcery. And within a military scenario, according to the Greek chronicler Kritovoulos of Imbros (*circa* 1410–*circa* 1470), there occurred a hailstorm of such unprecedented severity in May 1453 during the siege of Constantinople that it was widely regarded as an ill-omen, foreshadowing the imminent fall of the great Byzantine capital.

The medieval treatise on witches entitled *Malleus Maleficarum* (*Hammer of Witches*), first published [A4.123] in Germany in 1487, was written in Latin in three parts by the Dominican friars Heinrich Kramer (*circa* 1430–1505) and Jacob Sprenger (*circa* 1436–1495).

FIGURE A4.39. Anonymous woodcut (*circa* 1486) of female witch producing hail, aided by gigantic animal jawbone, after Vintler [A4.120]

This controversial work endorsed the extermination of witches by elevating sorcery to the criminal status of heresy in the secular courts, almost inevitably resulting in the death penalty for those accused of witchcraft. Here it is noted from the first English language version [A4.124] produced by Montague Summers (1880–1948) and published in 1928, that in part 2, chapter 15 is entitled: *How they raise and stir up hailstorms and tempests, and cause lightening to blast both men and beasts*. One such case begins ominously:

> '*It is better to add an instance which came within our own experience. For in the diocese of Constance, twenty-eight German miles from the town of Ratisbon in the direction of Salzburg, a violent hailstorm destroyed all the fruit, crops and vineyards in a belt one mile wide, so that the vines hardly bore fruit for three years. This was brought to the notice of the Inquisition, since the people clamoured for an inquiry to be held; many beside all the townsmen being of the opinion that it was caused by witchcraft. Accordingly it was agreed after fifteen days' formal deliberation that it was a case of witchcraft for us to consider; and among a large number of suspects, we particularly examined two women, one named Agnes, a bath-woman, and the other Anna von Mindelheim. These two were taken and shut up separately in different prisons, neither of them knowing in the least what had happened to the other.*'

During the following day, upon questioning under interrogation by the chief magistrate, in the presence of a notary, the two unfortunate women were found guilty, and on the third day they were burned at the stake.

Figure A4.40 reproduces anonymous woodcut illustrations from *De lamiis et phitonicis mulieribus* (*Of witches and diviner women*), a treatise on witchcraft by the German legal scholar Ulrich Molitor (*circa* 1442–1507), first published *circa* 1489 in several incunable editions [A4.125]; see also the commentaries by Ashton [A4.126] and Macdonald [A4.127]. The versions in Figures A4.40(a) and (b) depict two witches adding a serpent and a cockerel to the magic brew contained within the flaming cauldron, as part of a devilish scheme to

FIGURE A4.40. Anonymous woodcuts (*circa* 1489) depicting hailstorm witchcraft, after Molitor [A4.125]: (a), (b) two witches brewing up hailstorm by adding serpent and cockerel to flaming cauldron; (c), (d) flying witches mounted on forked stick to summon hailstorm

FIGURE A4.41. Copper engraving (*circa* 1500) by Albrecht Dürer (1471–1528): witch riding backwards on leaping goat, with four putti in attendance, in quest to conjure hailstorm (by courtesy of *National Gallery of Art, Washington, D.C.*)

generate fearful hail and thunder. Figure A4.40(a) is an example of a title page, whereas the in-text version in Figure A4.40(b) includes the caption *An possint puocare demones grandines et tonitrua*. The versions in Figures A4.40(c) and (d) show a leading witch and two followers flying through the air, each with an animal head and mounted on a forked stick, while a hailstorm breaks from the dark clouds to damage the trees and crops far below.

Figure A4.41 shows a copper engraving (*circa* 1500) by the German artist Albrecht Dürer (1471–1528), where a haggard witch clutching a distaff and spindle is riding backwards on a leaping goat – or possibly a horned sea-goat, the legendary astrological symbol for Capricorn, both creatures being long associated with demonic activity – with four engrossed putti in attendance and bearing various emblematic accoutrements; all of which, according to most scholarly interpretations, are configured primarily to instigate a violent hailstorm. Here witchcraft is considered to reverse the natural order, with the long hair of the witch streaming opposite to the direction of travel; a mercurial feature emphasised by the reversed initial within the artist's signed monogram.

A4.5. Early modern period (1500–1750)

Among the innumerable scientific interests of the Italian genius Leonardo da Vinci (1452–1519) were meteorological phenomena, particularly cloud formations and storms, and he made a series of fascinating drawings of some of them [A4.128], [A4.129] towards the end of his life (mostly bequeathed to his former pupil Francesco Melzi). One of an iconic series of *Deluge* drawings, Figure A4.42(a), conveys the vigorous intensity of atmospheric turbulence and vorticity, with a cataclysmic storm engulfing the conurbation far below. Concerning the artistic representation of such scenes, his notebooks also declare:

> '*Let the dark and gloomy air be seen buffeted by the rush of contrary winds and dense from the continued rain mingled with hail and bearing hither and thither an infinite number of branches torn from the trees and mixed with numberless leaves.*'

and with regard to the origins of itinerant storm clouds:

> '*They are often wafted about and borne by the winds from one region to another, where by their density they become so heavy that they fall in thick rain; and if the heat of the sun is added to the power of the element of fire, the clouds are drawn up higher still and find a greater degree of cold, in which they form ice and fall in storms of hail.*'

based on translations [A4.130]–[A4.132] from the original vernacular Italian script (in mirror-writing) by the art historians Jean Paul Richter (1847–1937) and Carlo Pedretti (1928–2018).

The drawing *Storm over a valley* (*circa* 1508) in Figure A4.42(b) depicts a storm breaking over a valley in the foothills of the Alps, with the darkening clouds set against the high peaks in sunlight. Moreover, his handwritten notes (*ricordi*) from this period, contained in the *Codex Leicester* (also briefly re-named *Codex Hammer*), include the passage:

> '*I say that the blueness we see in the atmosphere is not intrinsic colour, but is caused by warm vapour evaporated in minute and insensible atoms on which the solar rays fall, rendering them luminous against the infinite darkness of the fiery sphere which lies beyond and includes it. And this may be seen, as I saw it by going up Monboso (Monte Rosa), a peak of the Alps which divide France from Italy. The base of this mountain gives birth to the four rivers which flow in four different directions through the whole of Europe. And no mountain has its base at so great a height as this, which lifts itself above almost all the clouds; and snow seldom falls there, but only hail in the summer, when the clouds are highest. And this hail lies unmelted there, so that if it were not for the absorption of the rising and falling clouds, which does not happen twice in an age, an enormous mass of ice would be piled up there by the hail, and in the middle of July I found it very considerable. There I saw above me the dark sky, and the sun as it fell on the mountain was far brighter here than in the plains below, because a smaller extent of atmosphere lay between the summit of the mountain and the sun.*'

which description, together with the drawing, clearly are based upon direct personal experience of visiting this alpine region.

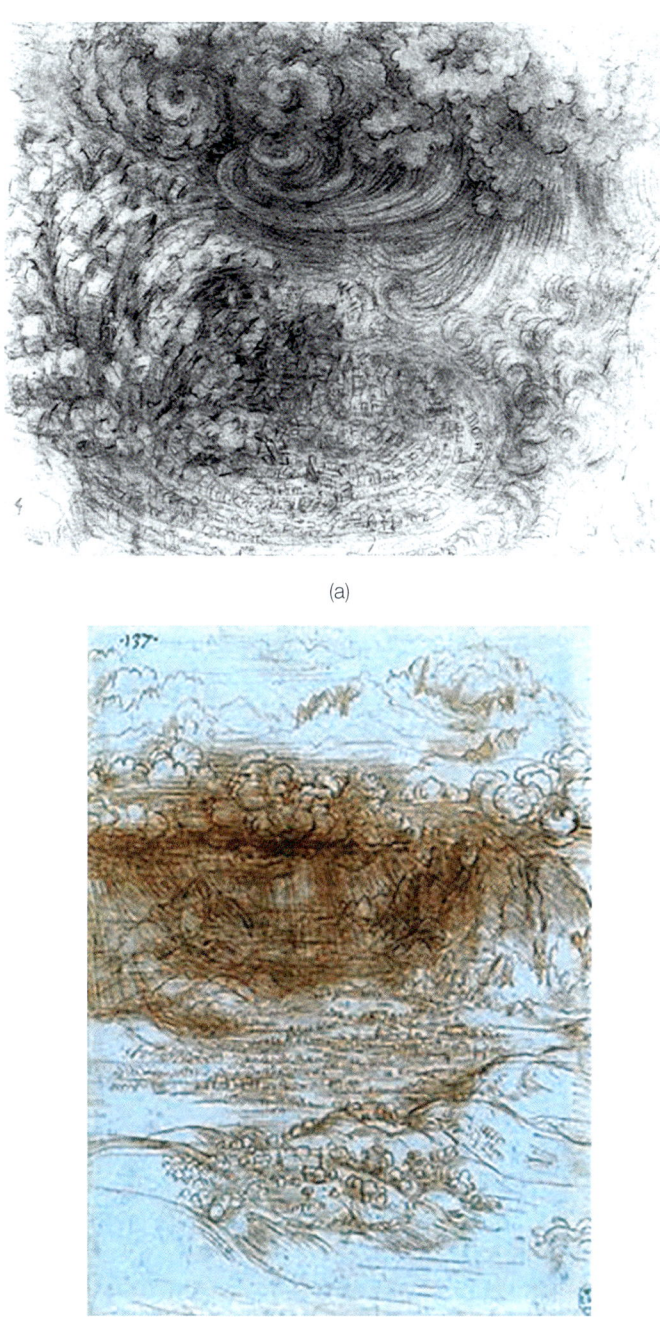

FIGURE A4.42. Drawings by Leonardo da Vinci (by courtesy of *Royal Collection Trust, Windsor*): (a) *Deluge* in black chalk (*circa* 1517) with conurbation engulfed by hailstorm (RCIN 912378, Melzi Nr 145); (b) *Storm over a valley* in red chalk (*circa* 1508) set in alpine foothills (RCIN 912409, Melzi Nr 137)

In a colourful autobiography, the firebrand Italian sculptor Benvenuto Cellini (1500–1571) recalls being caught in a severe hailstorm near Lyons, France, in June 1545:

> 'When we were one day's journey from Lyons, it was nearly twenty-two of the clock, when the heaven commenced to emit certain sharp claps of thunder, but the air was very clear. I was a bolt's distance in front of my companions. After the thunder the heaven emitted a noise so loud and so terrifying that I reckoned to myself that it was the Day of Judgement. And as I stopped for a moment there began to fall a shower of hail without a drop of water. This hail was larger than the pellets of a blow-pipe, and when it struck me, it hurt me very much. And little by little this hail began to grow bigger in such fashion that it was like the bullets of a crossbow. Observing that my horse was very terrified, I turned back at a tremendous gallop until I rejoined my companions, who on account of the same terror had halted within a pinewood. The hailstones increased to the size of large lemons. I intoned a Miserere: and whilst I was communing thus devoutly with God, there came one of those hail-stones so large that it broke in pieces a very thick branch of that pine whereunder I had fancied myself to be safe. Another mass of those hail-stones struck my horse upon the head, which showed signs of falling to the earth: one struck me also, but not with full force, for it would have killed me.'

while also recalling the alarming predicament faced by others nearby, even allowing for possible embellishment:

> 'Likewise one struck that poor old fellow Lionardo Tedaldi in such fashion that, since he was like me upon his knees, it made him set his hands upon the ground. Then I, when I saw that that branch could no longer protect me, and that along with my Misereres it was needful that I should take some action, began to fold my garments around my head: and so I said to Lionardo, who was crying out for help 'Jesu, Jesu,' that He would help him if he helped himself. I had a great deal more trouble to save this man than myself. This business lasted a while, then it ceased, and we, who were all pounded, as best we could remounted our horses: and whilst we went in the direction of our lodging, showing to each other our scratches and bruises, we found, a mile further on, so much greater a destruction than ours as it seems impossible to describe. All the trees were stripped and broken, whilst as many animals as happened to be there were dead: and many shepherds were also dead. We saw a vast quantity of those hailstones which could not be contained in two hands. It seemed to us that we had had a lucky escape, and we knew then that our appeal to God and those Misereres of ours had served us more than we could have been able to do for ourselves. Thus thanking God we journeyed on to Lyons the next day after, and there we remained for eight days.'

from a translation [A4.133] by the English art historian Robert Hobart Cust (1861–1940); see also the similar earlier version [A4.134] by the English author John Addington Symonds (1840–1893).

The poetic work *Theuerdank* is based on material provided by Maximilian I (1459–1519) of the German Habsburg dynasty, who was Holy Roman Emperor from 1508 until his death. Largely written and compiled by his chaplain Melchior Pfintzing (1481–1535), this masterpiece of German literature was first published [A4.135] as a limited edition in 1517, printed either on vellum or paper and accompanied by numerous woodcut illustrations (sometimes hand-coloured) by an accomplished group of artists and engravers; while also noting that its new blackletter typeface was an influential precursor to the German *Fraktur* style of printed *Gothic* script. A larger second edition appeared two years later, followed by several others over many decades; see also the recent first English language edition by Louthan and Green [A4.136]. The allegorical story, written in rhyming couplets, tells of the heroic adventures of Maximilian (cast as the young chivalric knight *Theuerdank*) in 1477 in a quest to claim his wealthy bride Mary of Burgundy (*Ernreich*), daughter and heir to the ruler of Burgundy, Charles the Bold (*Romreich*). He is accompanied by his faithful squire *Ernhold*, while their perilous journey – in part resulting from inclement weather – is frequently harried by three of *Ernreich's* most powerful vassals (*Fürwittig*, *Unfalo* and *Neydlhart*) whose power is threatened by *Teuerdank*, and who wish to prevent the marriage.

The woodcut illustration in Figure A4.43(a) recalls a harrowing episode when the travels are endangered by a severe hailstorm, with *Theuerdank*, *Ernhold* (depicted with the wheel of fortune on his tunic) and two other knights in the boat, and the unscrupulous *Unfalo* standing in the left foreground. Similarly, Figure A4.43(b) depicts a scene where *Theuerdank* is almost struck dead during another storm, which perhaps included hail. Eventually *Theuerdank* reaches his bride, executes his three enemies and finally sets off on a crusade to prove his worthiness as a knight of Christendom; a tidy juxtaposition of religious piety and physical violence.

A nautical framework is central to *The Lusiads*, a Portuguese epic poem by Luís de Camoëns (*circa* 1524–1580), first published in 1572, which celebrates the discovery of a sea route to India by the explorer Vasco da Gama (*circa* 1469–1524). The ten books are written in rhyming stanzas (*ottava rima*), and the following excerpt from book 5 encapsulates the atrocious weather conditions encountered during part of the voyage:

> '*That sage device, whose wonderous use proclaims*
> *Th' immortal honour of its authors' names,*
> *The Sun's height measured, and my compass scann'd,*
> *The painted globe of ocean and of land.*
> *Here we perceived our venturous keels had past*
> *Unharm'd the wintery tropick's howling blast;*
> *And now approach'd dread Neptune's secret reign,*
> *Where the stern Power, as o'er the frozen plain*
> *He rides, wide scatters from the polar star*
> *Hail, ice, and snow, and all the wintery war.*'

based on an original translation [A4.137] by the Scottish poet William Julius Mickle (1734–1788); noting that the 'sage device' refers to an astrolabe, deployed as a navigational instrument aboard ship.

Several references to hail occur in the epic poem *Gerusalemme Liberata* (*Jerusalem Delivered*), arranged in twenty books by the Italian poet Torquato Tasso (1544–1595) and published in 1581 in Parma, Italy. This work is centred upon the conflict between Christians and Muslims towards the end of the *First Crusade*, during the *Siege of Jerusalem* in 1099 A.D. A complete English language translation [A4.138] was made by Edward Fairfax (*circa* 1565–1635), first published in 1600 (under the title *Godfrey of Bouillon*, after the leading

(a)   (b)

FIGURE A4.43. Woodcut illustrations from German poetic work *Theuerdank* by Melchior Pfintzing, 1517: (a) *Theuerdank's* boat engulfed by terrifying hailstorm (by courtesy of *Metropolitan Museum of Art, New York*); (b) *Theuerdank* almost struck dead during storm (by courtesy of *Royal Collection Trust, Windsor*)

French nobleman) and dedicated to Queen Elizabeth I, while the present extracts are from a much later version [A4.139] edited by the English academic Henry Morley (1822–1894). Whence, for example, the following stanza from book 7:

> '*Heaven's glorious lamp, wrapped in an ugly veil*
> *Of shadows dark, was hid from mortal eye,*
> *And hell's grim blackness did bright skies assail;*
> *On every side the fiery lightnings fly,*

> *The thunders roar, the streaming rain and hail*
> *Pour down and make that sea which erst was dry.*
> *The tempests rend the oaks and cedars brake,*
> *And make not trees but rocks and mountains shake.'*

with a further stanza (book 7) in praise of the gallant *Godfrey*:

> '*Thus fled the French, and then pursued in chase*
> *The wicked sprites and all the Syrian train:*
> *But gainst their force and gainst their fell menace*
> *Of hail and wind, of tempest and of rain,*
> *Godfrey alone turned his audacious face,*
> *Blaming his barons for their fear so vain,*
> *Himself the camp gate boldly stood to keep,*
> *And saved his men within his trenches deep.'*

It is also noted that during the *Third Crusade* (1189–1192), the strategic decision by Richard I (1157–1199), King of England, to refrain from besieging Jerusalem in December 1191 was strongly influenced by the appallingly bad weather; being exceedingly cold with heavy rain and hailstorms.

An allusion to hail is found in *The Faerie Queene*, a long poem in six books by Edmund Spenser (*circa* 1552–1599), which examines the different virtues of several knights in the context of Arthurian legend [A4.140]. Whence in book 4, canto 9, there is the stanza:

> '*But they so farre from peace or patience were,*
> *That all at once at him gan fiercely flie,*
> *And lay on load, as they him downe would beare;*
> *Like to a storme which hovers under skie,*
> *Long here and there and round about doth stie,*
> *At length breakes downe in raine, and haile, and sleet,*
> *First from one coast, till nought thereof be drie,*
> *And then another, till that likewise fleet;*
> *And so from side to side till all the world it weet.*'

from the version edited by Wise [A4.141].

Many practices of magic and sorcery, including the raising of tempests, are described in *Compendium Maleficarum*, a notable compendium on witchcraft [A4.142], [A4.143], first published in 1608 in three books, and prepared by the Italian monk Francesco Maria Guazzo of the Ambrosian brotherhood. Whence chapter 8 in book 1 has the Latin title: *Sortilegi, diris imprecationibus, et factis petunt pluuiam, grandines etc.* (*By their terrible deeds and imprecations, witches produce rain and hail, etc.*), with the opening sentence:

> '*It is most clearly proved by experience that witches can control not only the rain and the hail and the wind, but also the lightening when God permits.*'

FIGURE A4.44. Anonymous woodcut (*circa* 1608) of goat-riding witch bringing down hail, from *Compendium Maleficarum* by Francesco Maria Guazzo [A4.142]

These devilish actions are epitomised by the symbolic image in Figure A4.44, which depicts a female witch riding a goat, flying high among the clouds to create a hailstorm.

In parallel with relatively harmless doctrines there evolved more fearful practices, essentially rooted in pagan witchcraft yet frequently sanctioned by the church authorities. From historical accounts of the *Inquisition* by Macdonald [A4.127], one chapter (*The witchcraft delusion*) contains the passage:

> '*These foolish but non-invasive methods of trying to regulate the weather early became associated with others equally foolish, and cruel and murderous to the last degree. The church officially accepted the deduction that men, women and children may act for the devils in bringing evils upon their fellows through storms, hail, floods, and the like. Pope Eugene IV in 1437 and again in 1445 urged the inquisitors to be more relentless in their pursuit of 'weather-makers', and in 1484 Innocent VIII did the same, as already noted. So did Julius II, Alexander X, in 1504, and Adrian VI in 1523. Put to the torture, the accused confessed to anything and everything that their tormentors wanted them to admit. The old pagan laws had not permitted torture to proceed beyond 'human endurance', but under ecclesiastical law there was the principle of 'excepted cases', by which it was meant that those suspected of heresy and witchcraft must confess; they were to be tortured until they did confess. This made suspicion equivalent to confession and conviction.*'

Likewise, Dillinger [A4.144], [A4.145] has discussed magical folk beliefs and their connection to witch hunts in particular European regions, stating in chapter 2 (*Golden goblets and cows' hooves: witchcraft and magic*) that:

> '*The fear of storm damage, particular to vineyards, was one of the most fundamental concerns of ordinary people driving the witch trials in Swabian Austria and the Electorate of Trier. One function of the trials was to prevent such damage in the future. Witch trials accomplished this by identifying and executing the responsible people through a criminal trial.*'

together with observations on hail-festival processions:

> '*In the sixteenth century, parish 'hail festival' processions took place in the Trier region that included prayers for the protection of crops from hail and other storm damage. The decisive difference from the Weingarten cult consisted in the fact that the hail festivals of the Moselle region were always merely local festivals held by individual parishes. These hail festivals lacked church recognition or affiliation with a regional institution, so regionally important cults could not develop. The hail festival processions remained strictly local phenomena, parochial in every sense. Other damage-warding rituals, supported and propagated by church authorities, had degenerated into ad hoc measures. They did not offer a viable alternative to a witchcraft trial. After Urban VIII had endeavoured to limit and standardise festivals with his 1642 bull 'Universa', the archdiocese of Trier even tried to forbid hail processions in 1678. But the populace of the Electorate opposed the restrictive position of their archbishop and clamoured for weather processions. The hail festivals only disappeared in the 1730s, presumably because the Jesuit-supported cult of St Donatus offered an alternative. In Weingarten, as we have seen, the papal bull remained unheeded.*'

Theological aspects of witchcraft are discussed by Midelfort [A4.146], with special reference to hailstorms in 16th-century Europe, as being regularly attributed to the activity of witches (*e.g.* an English translation of the sermon *On hailstorms* preached by the protestant reformer Johann Brenz in 1539). Also Behringer [A4.147] gives a detailed exposition of weather-related witch hunts in Europe, habitually carried out following a perceived maleficium of hail damage to crops, and partly reflecting the importance of climate in an agrarian society; see also Oster [A4.148], Pfister [A4.149], Waite [A4.150] and Kieckhefer [A4.151]. Many innocent lives were lost during these terrifying persecutions.

Literary allusions to hailstorms occur in several of Shakespeare's works [A4.152]. Towards the end of the play *All's Well that Ends Well* (Act 5, scene 3), upon meeting *Count Bertram* once again at his palace at Roussillon, the visiting *King of France* remarks:

> '*I am not a day of season,*
> *For thou mayst see a sunshine and a hail*
> *In me at once: but to the brightest beam*
> *Distracted clouds give way; so stand thou forth;*
> *The time is fair again.*'

Again, in the play *Antony and Cleopatra* (Act 3, scene 13) where *Cleopatra*, splendiferous at her palace in Alexandria, responds pithily to a questioning interjection by *Antony* – and to his satisfaction – with the cogent words:

> '*Ah, dear, if I be so,*
> *From my cold heart let heaven engender hail,*
> *And poison it in the source; and the first stone*
> *Drop in my neck: as it determines, so*
> *Dissolve my life! The next Caesarion smite!*
> *Till by degrees the memory of my womb,*
> *Together with my brave Egyptians all,*
> *By the discandying of this pelleted storm,*
> *Lie graveless, till the flies and gnats of Nile*
> *Have buried them for prey!*'

In other writings, the Bard emphasises the transitory nature of hailstones. In the opening scene of the tragedy *Coriolanus* (Act 1, scene 1), while admonishing a company of mutinous plebeians in Rome, the heroic leader *Caius Marcius* (later surnamed *Coriolanus*) remarks contemptuously to the *First Citizen* that 'you are no surer, no, than is the coal of fire upon the ice, or hailstone in the sun'. And on a lighter note from the comedy *The Merry Wives of Windsor* (Act 1, scene 3), in a room at the Garter Inn, the unscrupulous *Falstaff* promptly dismisses his erstwhile servants *Pistol* and *Nym* upon their refusal to follow his devious instructions: 'Rogues, hence, avaunt! vanish like hailstones, go'.

Earlier battlefield themes encompassing hail continue in Elizabethan verse with the patriotic epic poem *The Battaile of Agincourt* by the English writer Michael Drayton (1563–1631), almost certainly an acquaintance of Shakespeare, expressed [A4.153] in the *ottava rima* stanza:

> '*When all at once the English men assaile,*
> *The French within all valiantly defend,*
> *And in a first assault, if any faile,*
> *They by a second striue it to amend:*
> *Out of the towne come quarries thicke as haile;*
> *As thicke againe their Shafts the English send:*
> *The bellowing Canon from both sides doth rore,*
> *With such a noyse as makes the thunder poore.*'

wherein 'quarries' are crossbow arrows, which featured prominently in the English victory on Saint Crispin's Day in 1415, during the Hundred Years' War.

A prosaic connotation with hail is made by the Spanish writer Cervantes (1547–1616) – a near contemporary of William Shakespeare – in his celebrated novel *Don Quixote*, first published in Madrid in two parts, both in the early 17th century [A4.154]. Whence in chapter 41 of the second part (*Of the arrival of Clavileño and the end of this protracted adventure*), the noble knight-errant addresses his squire *Sancho*:

> '*Don Quixote now, feeling the blast, said, 'Beyond a doubt, Sancho, we must have already reached the second region of the air, where the hail and snow are generated; the thunder, the lightning, and the thunderbolts are engendered in the third region, and if we go on ascending*

at this rate, we shall shortly plunge into the region of fire, and I know not how to regulate this peg, so as not to mount up where we shall be burned."

from the four-volume translation [A4.155] by the Irish-born writer John Ormsby (1829–1895). An illustration of this magical scene, usually attributed to the Spanish artist Ricardo Balaca y Orejas-Canseco (1844–1880) – although possibly created by his compatriot Josep Luís Pellicer (1842–1901) – is shown in Figure A4.45(a), taken from the second volume of an annotated Spanish edition [A4.156] published in 1883. Here Don Quixote and Sancho Panza are depicted as flying blindfolded astride the mythical wooden horse *Clavileño*, which is guided on its perilous airborne adventure by turning a peg at the back of its neck. This iconic scene also is illustrated on postage stamps of various vintages, an example of which is shown in Figure A4.45(b); namely a Spanish 10-peseta airmail stamp issued in 1936.

Quintessential feelings of warmth and admiration towards the daughter (Annie) of a pastor in the local village (Tharaw), near Königsberg (present-day Kaliningrad), are expressed throughout the short lyrical poem *Annie of Tharaw* (*Ännchen von Tharau*), originally written in a Low German dialect by the Prussian writer Simon Dach (1605–1659), as indicated by the extract:

'*As the palm-tree standeth so straight and so tall,
The more the hail beats, and the more the rains fall,—
So love in our hearts shall grow mighty and strong,
Through crosses, through sorrows, through manifold wrong.*'

from a translation made in 1845 by Henry Longfellow [A4.157].

The Welsh author and physician Henry Vaughan (1621–1695) is known chiefly for his religious poetry, and among the anthology in *Silex Scintillans* [A4.158] is *Rules and Lessons*, which includes the stanza:

'*When Seasons change, then lay before thine Eys
His wondrous Method; mark the various Scenes
In heav'n; Hail, Thunder, Rain-bows, Snow, and Ice,
Calmes, Tempests, Light, and darknes by his means;
Thou canst not misse his Praise; Each tree, herb, flowre
Are shadows of his wisedome, and his Pow'r.*'

wherein the biblical allusions are self-evident.

In the epic poem *Paradise Lost* by John Milton (1608–1674), essentially a work of religious mythology initially comprising ten books [A4.159] written in blank verse and published in 1667, the following excerpt is from book 2 (lines 587–595):

'*Beyond this flood a frozen Continent
Lies dark and wilde, beat with perpetual storms*

(a)

(b)

FIGURE A4.45. Illustrations of Don Quixote and Sancho Panza flying blindfolded astride iconic wooden horse *Clavileño*: (a) engraving by Ricardo Balaca, *circa* 1880 (by courtesy of *Biblioteca de la Facultad de Derecho y Ciencias del Trabajo, Universidad de Sevilla*); (b) Spanish airmail postage stamp, issued 1936

> *Of Whirlwind and dire Hail, which on firm land*
> *Thaws not, but gathers heap, and ruin seems*
> *Of ancient pile; all else deep snow and ice,*
> *A gulf profound as that Serbonian Bog*
> *Betwixt Damiata and mount Casius old,*
> *Where Armies whole have sunk: the parching Air*
> *Burns frore, and cold performs th' effect of Fire.'*

whereas much later in book 9 (lines 692–700):

> *'These changes in the Heav'ns, though slow, produc'd*
> *Like change on Sea and Land, sideral blast,*
> *Vapour, and Mist, and Exhalation hot,*
> *Corrupt and Pestilent: Now from the North*
> *Of Norumbega, and the Samoed shoar*
> *Bursting thir brazen Dungeon, armd with ice*
> *And snow and haile and stormie gust and flaw,*
> *Boreas and Caecias and Argestes loud*
> *And Thrascias rend the Woods and Seas upturn;'*

Then again in book 10 (lines 179–186), with clear reference to the aforementioned biblical plagues of Egypt:

> *'His Cattel must of Rot and Murren die,*
> *Botches and blaines must all his flesh imboss,*
> *And all his people; Thunder mixt with Haile,*
> *Haile mixt with fire must rend th' Egyptian Skie*
> *And wheel on th' Earth, devouring where it rouls;*
> *What it devours not, Herb, or Fruit, or Graine,*
> *A darksom Cloud of Locusts swarming down*
> *Must eat, and on the ground leave nothing green:'*

Various references to hail also occur in early descriptions of the *New World* [A4.160], arranged in three books compiled and edited by the Scottish translator and cartographer John Ogilby (1600–1676), largely based on a 1671 publication (*De Nieuwe en Onbekende Weereld*) by the Dutch writer Arnoldus Montanus (*circa* 1625–1683). Regarding hardships at sea, for instance, is an extract from book 1 (*An Accurate Description of America*), chapter 3 (*First Discoverers of America*), section 14 (*Four English Expeditions, under the Conduct of our Famous Sea-Captains Martin Forbisher, Sir Francis Drake, Thomas Candish, and John Smith*):

> *'Captain Forbisher sailing to the Northern Parts of America, Anno 1576, chusing a bad time, the Year being too much spent, and the Ocean so full of Ice, that it forc'd him to return to England: Not many Months after, he renew'd his Voyage, Queen Elizabeth having rigg'd out, and sent under his Command one Frigat and two Ketches, Mann'd*

with a hundred and forty Men: The twenty sixth of May he weigh'd Anchor, and sail'd to the Orkenies, lying to the North of Scotland, where landing, he found the poor Islanders fled out of their Huts, into Caves and Dens among the Rocks. From thence he steer'd North-North-West, through abundance of floating Pieces of Timber, which oftentimes gave him great stops. The fourth of July he made Friezland, where he met with a great Storm of Hail, mix'd with Snow: Before the Shore lay a great Ridge of Ice, which hindred for a while their Landing:'

Elsewhere, in a quest for provisions during a Spanish military expedition to Florida led by Ferdinandus Sottus in the spring of 1539, covered in book 2 (*A description Northern America*), chapter 3 (*Florida*), it happened that:

'In the Village Tolomeco they did the like. But here their Provisions growing scarce, the Army was divided into two Bodies, Balthasar de Gallegos leading one, and Sottus the other; yet the Design of them both was on the Province of Chalaque; whither marching, they were surpris'd by such a violent Storm, that few would have been left to relate their Adventures, had not the Trees bore it off from them; for it not onely Thundred and Lightned as if Heaven and Earth would have met, but also Hail-stones fell down as big as Eggs, which beat down the Boughs of Trees.'

An entirely different cultural thread is centred upon folklore among the indigenous inhabitants of Peru. Whence from book 3 (*A description of Peruana or, Southern America*), chapter 4 (*Peru*):

'This kind of Idolatry was common to the Peruvians, with several other Nations of the antient Heathens, as hath been before observ'd, which made them mock at the Spaniards when they told them of a Crucifi'd Saviour, saying they had a splendid God, who appear'd to them in glory every Morning. In the third Temple near the River Taciquaque in Peru, which exceeded the other two, they worshipp'd Thunder by the Name of Chuquilla, Catuilla, and Intillapa, which the Peruvians believ'd to be a Man that Commanded the Air, and who being Arm'd with a Club and Sling, throws down Rain, Hail, and Snow from the Clouds, and shooting Bullets through the lower Region, causes Thunder and Lightning:'

Substantial hailstorms have even been known to disturb the normal tranquillity of the British countryside; as occurred, for example, in July 1666 over coastal regions of the English county of Suffolk, and briefly recorded by Nathaniel Fairfax [A4.161]. Some hailstones were '*as bigg as turkeys-eggs*', and measured up to '*12 inches about*'. One traveller crossing the heath at Aldeburgh '*had his head broken by the knocks of them through a stiff country-felt*', whilst '*the horses were so pelted, that they hurried away his cart beyond command*'. As for the hailstones:

'They seem'd all white, smooth without, shining within. 'Tis somewhat strange, methinks, that their pillar of air should keep them aloft, if they were not clapt together in the falling; especially at such a time of the year, when the air is less thickned and its spring weaker.'

HAILSTONE IMPACT: HISTORICAL PROLOGUE

An account of a strong hailstorm [A4.162] that occurred in London in May 1680 was given by the English scientist Robert Hooke (1635–1703), which included the drawings reproduced in Figure A4.46. Whence, before mid-day:

> '*It grew very dark, and thunder'd very near; and soon after there began to fall a good quantity of hailstones, some of the bigness of pistol bullets, others as big as pullets eggs, and some above 2½ inches, and near three inches over the broad way; the smaller were pretty round, and white, like chalk, or sugar plums; the other of other shapes; some of the most remarkable were these.*
>
> *Breaking many of them, I found them to be made up of orbs of ice, one encompassing another; some of them transparent, and some white, and opaque; some of these were to the bigness of near an inch in diameter, and were orbicular every way. Some of them had the white spot in the middle, as A; others towards one side, as B; and the variety of white and transparent spots very differing; those, which exceeded these in bigness, were made by an additional accretion of transparent icicles, radiating every way from the surface of the white ball, like the shooting of niter, or toothed sparre. These in some stood, as it were, separate and distinct icicles, which were very clear and transparent, and had no blebs or whiteness in them. Others of them were all concreted into a solid lump, and the interstices filled up with ice, which was not so clear as the stiriae, but whiter; and thereby one side, which, I suppose, was the uppermost, was flat, almost like a turnep; and the radiations appeared to proceed from the ball in the middle, more towards the upper side, and most toward the sides; the edges and top were more rough, and the ends of the stiriae appeared prominent; which the Figures will better express.*'

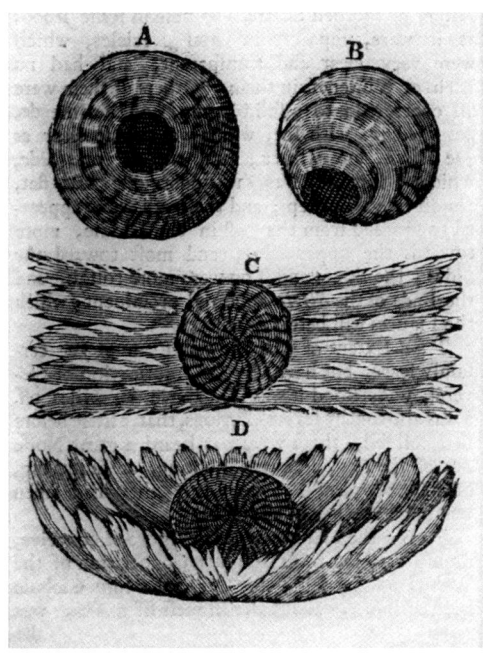
(a)

(b)

FIGURE A4.46. Illustrations by Robert Hooke of hailstones that fell upon London in May 1680 (by courtesy of *Royal Society, London*)

Hooke then made some preceptive remarks on the possible natural processes of hail formation:

> 'From the manner of their figure, I conceive, their accretion was made by a congelation of the water, as they fell; that the small white globule in the middle, about the bigness of a pea, was the first drop that concreted into hail; this, in falling through the clouds beneath, congealed the water thereof into several coats or orbs, till some of them came to the bigness of pigeons eggs, some white, some transparent, according to the several degrees of coldness it passed through, whilst they congealed; that the last accretion was made by a more violent and sudden cold, in the lower part of the cloud, where they passed through almost a continued body of water. Other varieties of their forms, which were very many, I conceive, must be made by their meeting with one another in their passage.'

In the poem *Britannia rediviva* [A4.163] by England's first Poet Laureate, John Dryden (1631–1700), written in 1688 to mark the birth that year of the Prince of Wales, there are hailstorm connotations within a pastoral landscape:

> 'As when a sudden Storm of Hail and Rain
> Beats to the ground the yet unbearded Grain,
> Think not the hopes of Harvest are destroy'd
> On the flat Field, and on the naked void;
> The light, unloaded stem, from tempest free'd,
> Will raise the youthful honours of his head;
> And, soon restor'd by native vigour, bear
> The timely product of the bounteous Year.'

Further notable hailstorms occurred in Britain during the spring and summer of 1697, and are described in several letters to the Royal Society [A4.164]–[A4.169]. Whence the English scientist Edmund Halley (1656–1742) gives an account [A4.164], [A4.165] of a hailstorm that occurred in the Lancashire region of England during the afternoon of 29 April, 1697:

> 'The breadth of the cloud was about two miles; within which compass it did incredible damage, killing all sorts of fowl and small creatures; and scarce leaving any whole panes in any of the windows where it passed; but which is worse, it plowed up the earth, and cut of the blade of the green corn, so as utterly to destroy it, the hailstones burying themselves in the ground; and the bouling-greens where the earth was any thing soft, were quite defaced, so as to be rendered unserviceable for a time. This I had from an eye witness, the hailstones, some of which weighed five ounces, were of differing forms, some round, some half round, some smooth, others embossed and crenulated, like the foot of a drinking glass, the ice very transparent and hard, but a snowy kernel was in the midst of most of them, if not all; the force of their fall argued them to fall from a great height.'

> 'This morning comes our constant butcher, from Ormskirk, and tells Mr Richmond and me, he would not undertake to repair the glass broke by the storm in that town for sixty pounds; and that he was with Dr Tarleton when Gleast's windows were all to the west beaten in, and the tiles off the house. He adds, that Mr Barton the apothecary there, and Mr Ja. Farrer, weighed two hailstones, which came to three quarters of a pound a piece.'

while other witnesses refer to hailstones as being '*as large as duck or goose eggs*', with horses '*knock'd down in the plow*'. And spare a thought for those caught in the hailstorm:

> '*Mrs Mary Clayton, coming this way with her Preston sister and Mrs Langton, were a little while in the shower before they got covert, and were so beaten by it, that they could hardly turn them in their beds next morning; they are now in Liverpool, with their bruises plastred.*'

The prodigious extent of this hailstorm also came as a complete surprise:

> '*What I take to be most extraordinary in this phenomenon is, that such a sort of vapours should continue undispersed for so long a tract, as above sixty miles together, and in all the way of its passage occasion so extraordinary a coagulation and congelation of the watry clouds, as to encrease the hailstones to so vast a bulk in so short a space as that of their fall.*'

Likewise, Robert Tailor [A4.166] describes a hailstorm that occurred at Hitchin, Hertfordshire, on May 4th, 1697 when, following prolonged thunder and lightning:

> '*Then fell a sharp shower, with some hailstones: I sent my man out, and he took up some, which I measured seven and eight inches about; but the extremity of the storm fell about Offley, where there was unhappily a young fellow keeping sheep, who was killed, and one of his eyes stuck out of his head, his body was all over black with bruises; another person nearer to Offley escaped his life, but much bruised; there was in the house of Sir John Spencer, 7000 quarries of glass broke, and there was great damage done to all the neighbouring houses thereabouts; the hail fell in such vast quantities, and so great, that it tore up the ground, split great oaks and other trees, in great numbers; it cut down great fields of rye, as with a scyth, and has destroyed several hundred acres of wheat, barley, etc. insomuch, that they plough it up, and sow it with oats.*'

while also concluding:

> '*The bigness of the hailstones is almost incredible; and truly, were not I a witness to the vastness of the numbers and greatness, I should not have believed, for number it is impossible, to relate some hundred thousand cart-loads, and I see them four days after; and if the beds of hail had not been broke by peoples coming, and trampling of horses, it might have lain till Michaelmas. They have been measured from one, to thirteen and fourteen inches certain; some people talk largely of it, seventeen and eighteen inches, but the other is certain truth: the figures of them are various, some oval, others round, others picked, some flat, we were not so curious to weigh them.*'

Brief comments on hailstorms occurring in the same year (1697) relate to those on 9th June in the English county of Herefordshire [A4.167], where much damage was sustained in the parish of Westhide and '*many of the stones were measured above nine inches in compas*'; and on 6th June at Pontypool in the Welsh county of Monmouthshire [A4.168], where Edward Lhwyd observed that '*some of the hail were eight inches about, their figure very irregular and unconstant, several of the hailstones being compounded*'.

Prompted by these singular events, some deliberations on the natural formation of such large hailstones are given by the English scholar John Wallis (1616–1703) towards the end of his letter [A4.169]:

> '*Tis very possible, that, though their first concretion, upon their suddain congelation, might be but moderately great, as in other hail; yet, in their long descent, if the medium through which they fall were alike inclined to congelation, they might receive a great accession to their bulk, and divers of them incorporate into one.*'

while noting more generally that these hailstorms occurred during the 'Little Ice Age' in Europe. On occasions, for example, ice covering parts of the River Thames in London was thick enough to sustain extensive 'Frost Fairs' on the frozen surface, providing every possible form of food and entertainment over several days or weeks, including spit-roasted oxen, coach racing, skating lessons and bowling matches among the carnival activities. The anonymous engraving in Figure A4.47 depicts one such scene during a 'Frost Fair' held in 1608, from the title page of a pamphlet [A4.170] attributed to the English writer Thomas Dekker (*circa* 1572–1632). During this period, however, parts of the river were much wider that at present (prior to the construction of massive embankments) which, coupled with the additional resistance offered by numerous support piers of Old London Bridge (partly shown on the left in Figure A4.47), resulted in the water flowing comparatively slowly.

The aforementioned hailstorm in Lancashire is also recalled by the English physician and naturalist Charles Leigh (1662–*circa* 1701) in an early book on natural history [A4.171]:

> '*The wind blowing high at north-west, happened a violent storm of hail, several stones were nine inches in circumference, others were six, seven and eight. In this storm several rooks were killed in their nests; some hares upon their seats; vast quantities of glass broke, and all kinds of cattle in a general consternation.*'

Figure A4.48 is an engraving (by Mr Burgher of Oxford) from this book depicting the battered landscape, including '*the flashes of lightning, the largeness of the hailstones that fell in that storm and the hares and birds that were killed by them*'.

Other early accounts of English hailstorms include those occurring at Everdon in Northamptonshire in July 1691, recorded by Wallis [A4.172]; at Oundle in Northamptonshire in March 1693, described by Rutty [A4.173]; and near Rotherham in Yorkshire in June 1711, recalled by Thoresby [A4.174], where it was observed:

> '*The hail-stones were from 3 to 5 inches in circumference, and some say larger, which killed several pidgeons; but the chief damage done here was in the glass windows, which cost forty pounds in repairing. In Washfield, about two miles from thence, it did vast damage. This field is generally computed to be worth a thousand pounds when in white corn (to use the countryman's expression). Some part of it escaped, and the barley received no damage; but the generality of the wheat was cut off, about half a yard from the ground, and the rye about two foot.*'

There are also anonymous records [A4.175] of hailstorms in Flanders in May 1686.

FIGURE A4.47. Anonymous engraving depicting 'Frost Fair' on frozen River Thames, London in 1608 (by courtesy of *Houghton Library, Harvard University*, STC 11403)

Occasionally, the circumstances of real storms have been addressed by prominent literary figures. Whence descriptive material [A4.176] compiled by the English writer Daniel Defoe (*circa* 1660–1731) concerning '*the greatest storm that ever we had in England before*' includes an anonymous abstract that commences:

> '*Upon the 18th of February, 1661, being Tuesday, very early in the Morning, there began a very great and dreadful Storm of Wind (accompanied with Thunder, Lightning, Hail, and Rain,*

FIGURE A4.48. Engraving (by Mr Burgher) illustrating dreadful hailstorm in Lancashire in 1697, from book by Charles Leigh [A4.171]

> which in many Places were as salt as Brine) which continued with a strange and unusual Violence till almost Night: the sad Effects whereof throughout the Nation are so many, that a very great Volume is not sufficient to contain the Narrative of them.'

taken from chapter 4 (*Of the Extent of this Storm, and from what Parts it was suppos'd to come; with some Circumstances as to the Time of it*). This is a preamble to the substance of the book – a precursor to modern journalistic reportage – which relates to the *Great Storm* of November 1703, where the colossal destruction and loss of life resulted almost entirely from hurricane strength winds; see also a contemporary anonymous account [A4.177].

The consummate brevity and simplicity of Japanese *haiku* poetry – with particular reference to hail – is demonstrated by the following three compositions (selected among others) by Matsuo Bashō (1644–1694), the foremost poet of the Edo period (1603–1868). Either alone or with a companion, he undertook several long and perilous journeys through the Japanese countryside, relishing the changing scenery and the seasons, and some of the resulting poetic diaries were published. Towards the end of his life, he compiled a substantial anthology as part of his major work *The Narrow Road to the Interior* (*Oku no Hosomichi*), a combination of prose and *haiku*, first published posthumously in 1702. In each case herein, the transliteration is written on a single line, while noting that Japanese *haiku* poems generally are written or printed in a single vertical column of characters. The English language translation of each poem is based on several similar versions recorded in the literature, with an emphasis on perceived meaning rather than syllable count: see, for instance, references [A4.178]–[A4.182], where all translations differ from each other.

> '*Arare kiku ya kono mi wa moto no furu gashiwa.*' '*The sound of hail –*
> *I remain as before*
> *like an aging oak.*'

> '*Ikameshiki oto ya arare no hinoki-gasa.*' '*How harsh –*
> *the sound of hail*
> *on my cypress hat.*'

> '*Ishiyama no ishi ni tabashiru arare kana.*' '*Showering down*
> *on the rocks of Mount Ishiyama*
> *hailstones.*'

It was customary for authors to inscribe particular compositions on a *tanzaku* (poem-card, typically 6 × 36 cm) to present to acquaintances, and one such example by Bashō (*Old Pond*, probably his most famous *haiku*) is illustrated in Figure A4.49(a); after Addiss [A4.183], although the original source is unspecified.

Likewise, some two centuries later, is the following example from the author and ascetic Zen priest Taneda Shōichi, *alias* Taneda Santōka (1882–1940), included among a collection of *haiku* poems first published in 1933 (*Somokuto: Grass and Tree Stupa*):

> '*Teppatsu no naka e mo arare.*' '*Even into my iron begging bowl –*
> *Hailstones.*'

while similar versions are given in references [A4.184]–[A4.188]. The poem was composed by the wandering mendicant on a wintry day during a long solitary trek along deserted pathways, with the dull metallic sound of hail striking the iron alms bowl emphasising his isolation amid the raw and pervasive coldness of the season. Here the poet refers to the small

(a)　　　　(b)

FIGURE A4.49. Original Japanese calligraphy showing *haiku* poems inscribed by authors on *tanzaku* (ink on decorated paper, typically 6 × 36 cm), after Addiss [A4.183]: (a) *Old Pond* by Matsuo Bashō, *circa* 1686; (b) *Hailstones* by Taneda Santōka, *circa* 1933

hailstones of winter (*arare*), in contrast to the larger ones that usually fall in summer. This is one of Santōka's most admired poems that he would occasionally write out on a *tanzaku*, as illustrated in Figure A4.49(b), adding his signature below in a smaller size.

A picturesque encounter with hailstones occurs in the fantastical narrative of *Gulliver's Travels* (or *Travels into several remote nations of the world*) by the Anglo-Irish satirist Jonathan Swift (1667–1745), originally published in 1726 (two volumes, four parts) under the pseudonym Lemuel Gulliver [A4.189]. Whence in chapter 5, part 2 (*A voyage to Brobdingnag*), with Gulliver in the company of his 'little nurse' *Glumdalclitch* (the farmer's daughter) whilst visiting the remote fictional land of *Brobdingnag* (occupied by giants), is the paragraph (with modern spelling):

> 'Another day Glumdalclitch left me on a smooth grass-plot to divert myself while she walked at some distance with her governess. In the meantime there suddenly fell such a violent shower of hail, that I was immediately by the force of it struck to the ground: and when I was down, the hailstones gave me such cruel bangs all over the body, as if I had been pelted with tennis balls; however I made a shift to creep on all fours, and shelter myself by lying flat on my face on the

lee-side of a border of lemon thyme, but so bruised from head to foot that I could not go abroad in ten days. Neither is this at all to be wondered at, because nature in that country observing the same proportion through all her operations, a hailstone is near eighteen hundred times as large as one in Europe; which I can assert upon experience, having been so curious as to weigh and measure them.'

Hailstone metrics can be challenging at the best of times, even more so in the distorted parallel universe of giants; as exemplified by Figure A4.50, which shows an oil painting entitled *Gulliver exhibited to the Brobdingnag farmer* by the English artist Richard Redgrave (1804–1888).

Various indications of oncoming hail were discussed some three centuries ago in a book [A4.190] by the English cleric John Pointer (1668–1754), first published in 1723, and the early signs were deemed to be:

'If the sun at its rising cast a glittering light as if it reflected on some lucid matter, tho' few or no clouds appear at that time; the vapours are condens'd in the cold region, and forerunning into clouds that will scatter violent hail.

If (in the morning) the Eastern skies, before the sun-rising, look pale; and refracted rays appear in thick clouds, great storms of hail will ensue to the great damage of corn and fruits.'

FIGURE A4.50. *Gulliver exhibited to the Brobdingnag farmer*, painted in 1836 by Richard Redgrave (by courtesy of *Victoria and Albert Museum, London*)

with hail clouds recognised as follows:

> 'If the clouds look fleecy, dusky, white inclining to yellow, and move but heavily, tho' the wind be pretty rough, the vapours composing them are engender'd and frozen, and ratling hail ensues. If the clouds appear of a whitish blue, and expand much, it will be small hail or drisling (i.e. frozen mists); for that happens in the winter or spring, when it cannot be carried high enough to be condens'd by a greater quantity of cold, because the refracted rays of the sun are but weak, and this appears by a curdling in the clouds as they rise, and in appearance expand themselves. White clouds in the summer-time are a sign of hail, but in winter-time they are a sign of snow, especially when we perceive the air to be a little warm, occasion'd by some warm eruptions out of the clouds.'

Further early comments on the atmospheric conditions necessary for hail formation are found in a 1730 publication [A4.191] written by the English scholar Edward Saul (1677–1754):

> 'When the vapours are beginning to run into small drops, and are precipitated by the cold above, before they are completely form'd, they then fall down in mizzling rains; or, if frozen, in sleet. And as the ordinary drops of rain, freezing in their descent, form hail; so whenever they happen afterwards in their farther descent downwards, to pass through a cloud of snow, they encrease in bulk, and gathering, in a literal sence, as a snow ball, form those larger hail-stones (of six, or more inches in circumference) of which we often read, and sometimes with surprize admire and observe.'

More generally, Oberholzner [A4.192] has studied hailstorms that occurred during the Early Modern period and beyond, set against the prevailing background of theological and scientific opinion.

A4.6. Late modern period (1750–1900)

Whilst Fauquier [A4.193] described '*an extraordinary storm of hail*' that occurred in Virginia in July 1758, and Dewey [A4.194], [A4.195] reported a severe hailstorm that raged in Connecticut in July 1799, the return to a nautical theme is signalled by the *The Shipwreck*, an epic poem in three cantos by the Scottish sailor William Falconer (1732–1769), whose wide maritime experience enabled realistic descriptions of the many dangers and privations encountered at sea. The fictional yet semi-autobiographical work – first published in 1762, and revised in three further editions by 1778 (the last, posthumously) – recounts the fateful last voyage of a merchant ship bound from Alexandria to Venice [A4.196]. Whence amidst the impending doom (canto 2):

> 'But here the Queen of shade around them threw
> Her dragon wing, disastrous to the view!
> Dire was the Scene with whirlwind, hail, and shower;
> Black Melancholy ruled the fearful hour:
> Beneath, tremendous rolled the flashing Tide,
> Where Fate on every billow seemed to ride.'

HAILSTONE IMPACT: HISTORICAL PROLOGUE

With passing time, the prodigious tempest only worsened (canto 3):

> '*Her joints unhinged in palsied languors play,*
> *As ice-flakes part beneath the noon-tide ray:*
> *The Gale howls doleful through the blocks and shrouds,*
> *And big Rain pours a deluge from the clouds;*
> *From wintery magazines that sweep the sky,*
> *Descending globes of Hail impetuous fly;*'

It is especially tragic to note that, in real life, the poet was among all hands lost upon the sinking of the frigate *H.M.S. Aurora*, by some unknown and fatal mischance towards the end of 1769, having rounded the Cape of Good Hope *en route* to India. This maritime catastrophe also claimed the life of Robert Pitcairn (1752–1769) who, as a young midshipman high in the crow's nest aboard *H.M.S. Swallow* in 1767, had first sighted the land of what thereafter was named Pitcairn Island.

Much of the literary work of William Cowper (1731–1800) reveals an affinity and empathy with capricious nature in all its guises, as exemplified by the following extract of blank verse from *The Task*, a long descriptive poem in six books [A4.197] first published in 1785. Whence from book 5 (*The winter morning walk*):

> '*In such a palace Poetry might place*
> *The armoury of Winter; where his troops,*
> *The gloomy clouds, find weapons, arrowy sleet,*
> *Skin-piercing volley, blossom-bruising hail,*
> *And snow that often blinds the traveller's course,*
> *And wraps him in an unexpected tomb.*'

In stark reality, a notable hailstorm occurred in the greater Paris region on 13 July 1788, causing enormous losses of agricultural produce, particularly vines and cereal crops, with consequential hardship among the general population resulting from acute shortages of staple foods. Some days later a descriptive note of this singular event was sent by the British Ambassador to France (Lord Dorset) to the Foreign Secretary (Lord Carmarthen) in London, which included the passage [A4.198]:

> '*A storm of thunder, lightning and hail which was experienced in the environs of Paris last Sunday morning, was so incommonly violent and has done so much mischief in those parts where its full force was felt, that I am induced to give your Lordship such particulars of it as have come to my knowledge.*
>
> *About 9 o'clock in the morning the darkness at Paris was very great and the appearance of the heavens seemed to threaten a dreadful storm, the clouds however dispersed in a short time, having, by what I have since learned, wasted their force in other parts, the accounts from where give a melancholy description of the effects of the hurricane which appears to have commenced in the Forest of Rambouillet: His Majesty on his return to Versailles (having hunted the preceding day in that country) was obliged to stop on the road and to take shelter in a farmhouse.*

The hailstones that fell were of a size and weight never heard of before in this country, some of them measuring sixteen inches in circumference and in some places, it is said, they were even much larger. Not far from St Germains two men were found dead upon the road, and a horse so much bruised that it was determined to kill him from a motive of humanity to put an end to his misery: it is impossible to give description in detail of the damage that has been done: some of the largest trees were torn up by the roots; all the corn and vines destroyed, windows broken and even some houses beaten down.'

This natural catastrophe, combined with the dire effects of a preceding drought and a subsequent gruelling winter, have been viewed as significant contributary factors to widespread civil unrest throughout the country, culminating in the French Revolution during the following summer: as discussed, for example, by Neumann [A4.199], Neumann and Dettwiller [A4.200] and Cashman [A4.201].

On a more tranquil note, hail showers feature in the short poem *A Hail-storm* written by the English poet William Wordsworth (1770–1850) in 1798 [A4.202], also entitled *A Whirl-blast from Behind the Hill* in the first volume of his edited poetical works [A4.203]:

'A whirl-blast from behind the hill
Rushed o'er the wood with startling sound;
Then—all at once the air was still,
And showers of hailstones pattered round.
Where leafless oaks towered high above,
I sat within an undergrove
Of tallest hollies, tall and green;
A fairer bower was never seen.
From year to year the spacious floor
With withered leaves is covered o'er,
And all the year the bower is green.
But see! where'er the hailstones drop
The withered leaves all skip and hop;
There's not a breeze—no breath of air—
Yet here, and there, and every where
Along the floor, beneath the shade
By those embowering hollies made,
The leaves in myriads jump and spring,
As if with pipes and music rare
Some Robin Good-fellow were there,
And all those leaves, in festive glee,
Were dancing to the minstrelsy.'

while recalling the following diary entry [A4.204] by his sister Dorothy Wordsworth (1771–1855) on 18 March 1798, when residing at Alfoxden manor house in Somerset:

'The Coleridges left us. A cold, windy morning. Walked with them half way. On our return, sheltered under the hollies, during a hail-shower. The withered leaves danced with the hailstones. William wrote a description of the storm.'

upon bidding farewell to their close friend and near neighbour Samuel Taylor Coleridge.

As might be expected from such an accomplished observer of nature, Wordsworth's later poem *The Small Celandine*, composed in 1804, reflects a fondness of this yellow-flowering plant within the buttercup family (also known as the lesser celandine or pilewort), usually emerging in late winter and regarded by many as a harbinger of spring. From the third edited volume [A4.203], the first half of the poem is:

> '*There is a Flower, the lesser Celandine,*
> *That shrinks, like many more, from cold and rain;*
> *And, the first moment that the sun may shine,*
> *Bright as the sun himself, 'tis out again!*
>
> *When hailstones have been falling, swarm on swarm,*
> *Or blasts the green field and the trees distrest,*
> *Oft have I seen it muffled up from harm,*
> *In close self-shelter, like a Thing at rest.*
>
> *But lately, one rough day, this Flower I passed*
> *And recognised it, though an altered form,*
> *Now standing forth an offering to the blast,*
> *And buffeted at will by rain and storm.*'

Evocative accounts of hailstorms are given in *The Columbiad*, a patriotic epic poem by the American diplomat and politician Joel Barlow (1754–1812), embracing the expeditions of Christopher Columbus and the history of the Americas. The work was first published in 1807 as a single volume [A4.205] comprising ten books, and a forthright extract from the first book is:

> '*Indignant Frost, to hold his captive, plies*
> *His hosted fiends that vex the polar skies,*
> *Unlocks his magazines of nitric stores,*
> *Azotic charms and muriatic powers;*
> *Hail, with its glassy globes, and brume congeal'd,*
> *Rime's fleecy flakes, and storm that heaps the field*
> *Strike thro the sullen Stream with numbing force.*'

where 'brume' refers to mist or fog. And in book 5, with the leader of the indigenous militia sailing on the St Laurence River and first catching sight of '*Quebec's dread walls, and Wolfe's unclouded height*', there is the following mordant passage:

> '*With skilful glance he views the fortress round,*
> *Bristled with pikes, with dark artillery crown'd;*
> *Resolves with naked steel to scale the towers,*
> *And snatch a realm from Britain's hostile powers.*
> *Now drear December's boreal blasts arise,*
> *A roaring hailstorm sweeps the shuddering skies,*
> *Night with condensing horror mantles all,*
> *And trembling watch-lights glimmer from the wall.*'

FIGURE A4.51. Oil painting *Spring: A Mill on a Common; Hail Squalls; East Bergholt, Suffolk* by John Constable, 1814 (by courtesy of *Victoria and Albert Museum, London*, Ref. 144-1888)

History records on this occasion that the British forces eventually emerged victorious, but with appalling loss of life, including that of their illustrious young commander.

Ever at one with local nature, Figure A4.51 shows a variously titled oil painting *Spring: A Mill on a Common; Hail Squalls; East Bergholt, Suffolk* by the English landscape artist John Constable (1776–1837), first exhibited at the Royal Academy, London, in 1814. The scene is centred around the family-owned Flatford Mill near the Suffolk county village of East Bergholt, situated within the fertile valley of the River Stour, where Constable was born and worked as a boy, and which stimulating surroundings he remembered thereafter with abiding affection. According to a biography [A4.206] written by his friend and fellow artist Charles Leslie (1794–1859), which includes a mezzotint version by the English engraver David Lucas (1802–1881), Constable said of this painting:

> '*It may perhaps give some idea of one of those bright and silvery days in the spring, when at noon large garish clouds surcharged with hail or sleet sweep with their broad shadows the fields, woods, and hills; and by their depths enhance the value of the vivid greens and yellows so peculiar to the season. The 'natural history', if the expression may be used, of the skies, which are so particularly marked in the hail squalls at this time of year, is this: —*
>
> *The clouds accumulate in very large masses, and from their loftiness seem to move but slowly: immediately upon these large clouds appear numerous opaque patches, which are only small clouds passing rapidly before them, and consisting of isolated portions detached probably from the larger cloud. These floating much nearer the earth may perhaps fall in with a stronger current of wind, which as well as their comparative lightness causes them to move with greater rapidity; hence they are called by wind-millers and sailors 'messengers', and always portend bad weather. They float midway in what may be termed the lanes of the clouds; and from being*

so situated, are almost uniformly in shadow, receiving a reflected light only, from the clear blue sky immediately above them. In passing over the bright parts of the large clouds they appear as darks; but in passing the shadowed parts, they assume a grey, a pale, or a lurid hue.'

Details of a memorable hailstorm in Orkney off the north-eastern coast of Scotland in July 1818 are given by Neill [A4.207], where the localised nature of the storm is clearly illustrated by the narrow tracking path sketched in Figure A4.52. Of particular interest in the present context are early comments on the weather patterns typically experienced in the urban settlement at Liverpool in New South Wales (now an outer suburb of Sydney) that appear in an 1819 report by Wentworth [A4.208], including:

'In November the weather may be again called hot. Dry parching winds prevail as the month advances, and squalls of thunder and lightning with rain or hail. The thermometer at daylight is seldom under 65 degrees, and frequently at noon rises to 80 degrees, 84 degrees, and even 90 degrees.'

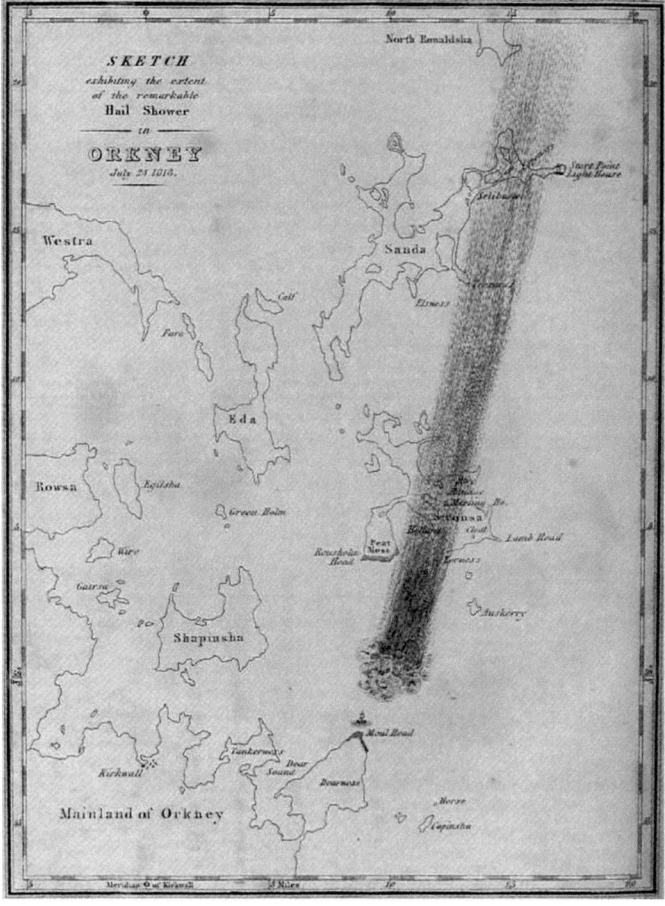

FIGURE A4.52. Narrow tracking path taken by severe hailstorm over Orkney Islands in July 1818 (by courtesy of *Royal Society of Edinburgh*)

while noting that, because the primary objective of this study was to promote and encourage increased emigration to Australia, its success or otherwise in this regard remains unknown.

Contrasting references to hail are evident in two poems by the Scottish writer Walter Scott (1771–1832). In *Marmion*, a descriptive tale in six cantos [A4.209], the pastoral theme continues in the following beguiling excerpt (*Introduction to canto 4*):

> '*When red hath set the beamless sun,*
> *Through heavy vapours dark and dun;*
> *When the tired ploughman, dry and warm,*
> *Hears, half-asleep, the rising storm*
> *Hurling the hail, and sleeted rain,*
> *Against the casement's tinkling pane;*'

while in a later collection of poetical works [A4.210], the first two verses from *The Orphan Maid* are:

> '*November's hail-cloud drifts away,*
> *November's sun-beam wan*
> *Looks coldly on the castle grey,*
> *When forth comes Lady Anne.*
>
> *The orphan by the oak was set,*
> *Her arms, her feet, were bare;*
> *The hail-drops had not melted yet,*
> *Amid her raven hair.*'

An apposite quotation can even be found in juvenile literature, namely *Beauty and the Beast* by the English writer Charles Lamb (1775–1834), a poem in rhyming verse first published [A4.211] in 1811. This magical tale concerns the adventures of the beautiful youngest daughter (*Beauty*) of a once-prosperous merchant, and their encounters with a terrifying beast. Whereupon, during one of the merchant's treks:

> '*A short day's travel from his Cot,*
> *New misadventures were his lot;*
> *Dark grew the air, the wind blew high,*
> *And spoke the gathering tempest nigh;*
> *Hail, snow, and night-fog join'd their force,*
> *Bewildering rider and his horse.*
> *Dismay'd, perplext, the road they crost,*
> *And in the dubious maze were lost.*'

where *cot* is taken to refer to a country cottage (occupied by *cotters*).

Hail forms part of the compelling imagery encapsulated in the lyric poem *Kubla Khan* by the English writer Samuel Taylor Coleridge (1772–1834), completed in 1797 and published

[A4.212] with minor revisions (and other verse) in 1816 at the suggestion of Lord Byron. In scenes by the sacred river *Alph* in *Xanadu*, the summer capital of the 13th-century Mongol ruler, the second stanza includes the lines:

> '*And from this chasm, with ceaseless turmoil seething,*
> *As if this earth in fast thick pants were breathing,*
> *A mighty fountain momently was forced:*
> *Amid whose swift half-intermitted burst*
> *Huge fragments vaulted like rebounding hail,*
> *Or chaffy grain beneath the thresher's flail:*
> *And 'mid these dancing rocks at once and ever*
> *It flung up momently the sacred river.*'

which captures the ambience of the erupting geyser and the power of its showering debris. Figure A4.53 shows a late draft of the first page of the two-page handwritten manuscript.

In the lengthy work *Endymion* by the English Romantic poet John Keats (1795–1821), divided into four books [A4.213] and based on an ancient Greek myth of the titular *Aeolian* shepherd, ruler of *Elis*, there is (book 4) the rhyming couplet:

> '*The kings of Inde their jewel-sceptres vail,*
> *And from their treasures scatter pearled hail;*'

with *Endymion* having just met a captivating Indian maiden, followed much later by the words:

> '*Foot-feather'd Mercury appear'd sublime*
> *Beyond the tall tree tops; and in less time*
> *Than shoots the slanted hail-storm, down he dropt*
> *Towards the ground; but rested not, nor stopt*
> *One moment from his home:*'

marking progress towards his ardent quest for heavenly beauty and eternal life.

Contemporarily, in the first stanza of his joyous poem *The Cloud*, published with other verse [A4.214] in 1820, Percy Bysshe Shelley (1792–1822) – also one of the major English Romantic poets – writes:

> '*I wield the flail of the lashing hail,*
> *And whiten the green plains under;*
> *And then again I dissolve it in rain,*
> *And laugh as I pass in thunder.*'

as part of an emotive portrayal of clouds being an unending cyclical facet of nature, coupled metaphorically with creative and dynamic elements that greatly benefit the human condition, even serving as a harbinger of social change and revolution.

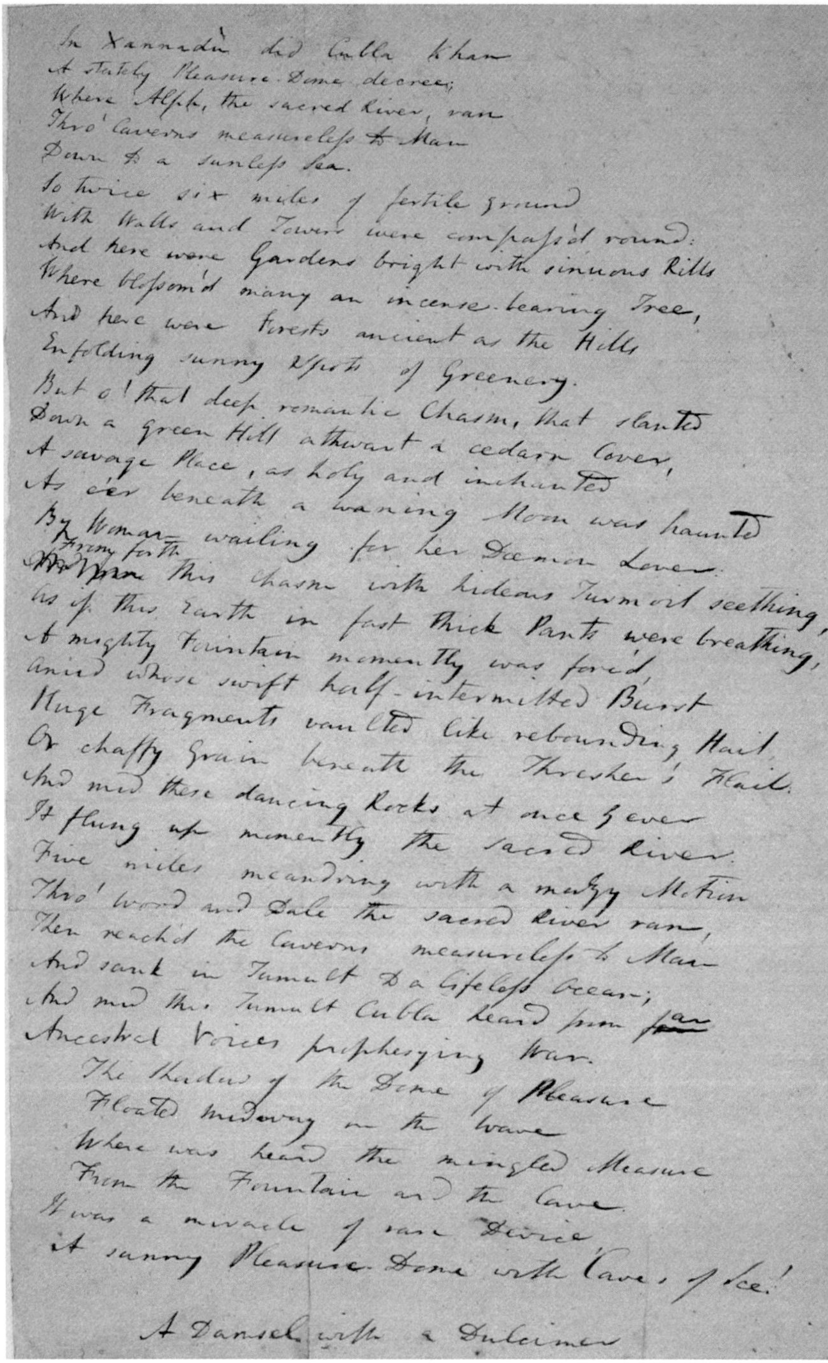
In Xannadù did Cubla Khan
A stately Pleasure-Dome decree;
Where Alph, the sacred River, ran
Thro' Caverns measureless to Man
Down to a sunless Sea.
So twice six miles of fertile ground
With Walls and Towers were compass'd round:
And here were Gardens bright with sinuous Rills
Where blossom'd many an incense-bearing Tree,
And here were Forests ancient as the Hills
Enfolding sunny Spots of Greenery.
But o! that deep romantic Chasm, that slanted
Down a green Hill athwart a cedarn Cover,
A savage Place, as holy and inchanted
As e'er beneath a waning Moon was haunted
By Woman wailing for her Dæmon Lover.
~~Amid~~ Forth from This Chasm with hideous Turmoil seething,
As if this Earth in fast thick Pants were breathing,
A mighty Fountain momently was forc'd,
Amid whose swift half-intermitted Burst
Huge Fragments vaulted like rebounding Hail,
Or chaffy Grain beneath the Thresher's Flail.
And mid these dancing Rocks at once & ever
It flung up momently the sacred River.
Five miles meandring with a mazy Motion
Thro' Wood and Dale the sacred River ran,
Then reach'd the Caverns measureless to Man
And sank in Tumult to a lifeless Ocean;
And mid this Tumult Cubla heard from far
Ancestral Voices prophesying War.
The Shadow of the Dome of Pleasure
Floated midway on the Wave
Where was heard the mingled Measure
From the Fountain and the Cave.
It was a miracle of rare Device,
A sunny Pleasure-Dome with Caves of Ice!
A Damsel with a Dulcimer

FIGURE A4.53. Part handwritten draft manuscript of poem *Kubla Khan* by Samuel Taylor Coleridge shortly before publication in 1816 (by courtesy of *British Library, London*, MS 50847)

Occasional allusions to hail are found in *Grimms' Fairy Tales*, compiled by the German scholars Jacob Grimm (1785–1863) and his brother Wilhelm Grimm (1786–1859), appearing originally as *Children's and Household Tales* (*Kinder- und Hausmärchen*), but with seven editions published between 1812 and 1857. One such tale, usually entitled *The Wolf and the Man* (*Der Wolf und der Mensch*), begins with a fox talking to a wolf about the strength of men who, the fox maintains, are feared by all beasts. The wolf demurs, boasting that he could easily '*make him run*', yet comes to grief when confronted by a huntsman armed with a shotgun:

> '*Well, brother wolf*', said the fox. '*How did you come out in your fight with the man?*'
>
> '*Oh,*' groaned the wolf. *He is, as you said, much stronger than I am. When I ran at him, he took a stick from his shoulder. Out of this he blew lightning in my face and hailstones against my nose*'.

from a version [A4.215] edited by the American author Edna Turpin (1867–1952), and re-titled *The Boasting Wolf*.

Likewise, the Russian ethnographer Alexander Afanasyev (1826–1871) assembled a large collection of national folk and fairy tales, initially published in eight volumes from 1855 to 1867. The tale *Elijah the Prophet and St Nicholas* is centred on a peasant farmer who always observed St Nicholas' Day, but never that of St Elias, as conveyed in a translation [A4.216] by the British scholar Leonard Magnus (1879–1924):

> '*He used to say a Te Deum to Nicholas, and burn a taper, but never gave as much as a thought to the Prophet Elijah.*
>
> *One day Elijah and Nicholas were walking through this peasant's fields, going along and surveying; and the ears were so large, so full, that it warmed one's heart to look at them!*
>
> '*What a fine crop this will be!*' said Nicolas. '*Yes, and he's a fine fellow, a good, brave peasant, pious; he remembers God, and reveres the Holy Saints. Whatever he turns his hand to shall prosper.*'
>
> '*Ha, let's have a look, brother,*' Elijah demurred. '*Will there be so much over? My lightnings shall glint and my hail beat his field down; then your peasant shall learn right, and regard my name-day.*'
>
> *So they wrangled and argued, and at last agreed to go each his own way.*'

Then follows a somewhat convoluted account of how the resourceful peasant overcame this damaging hailstorm, which is described thus:

> '*A heavy thunderous cloud gathered and, with frightsome lightning and hail, played on the peasant's field, cut through his crops like a scythe, and left not one blade to tell the tale.*'

and, with *Elijah* eventually reconciled, lived a merry life thereafter and honoured both name-days equally.

The English soldier and diplomat William Miller (1795–1861), who fulfilled a prominent military role in several South American revolutions, encountered powerful hailstorms during a campaign in the Andes in the summer of 1824:

> 'During certain months of the year, tremendous hailstorms occur. They have fallen with such violence that the army has been obliged to halt, and the men, being compelled to hold up their knapsacks to protect their faces, have had their hands so severely bruised and cut by large hailstones, as to bleed copiously.'

from an account in the second tome of a two-volume biography [A4.217] written by his brother (John Miller).

An engaging allusion to hail occurs in the epic poem *Don Juan* by the inimitable Englishman George Gordon Byron (1788–1824), better known as Lord Byron [A4.218]. Named after the legendary Spanish libertine, this *magnus opus* comprises 17 cantos (in *ottava rima*) written during the period 1818–1823, and published intermittently (the last canto being unfinished and published posthumously). In canto 4, contemplation of the possible early demise of the young *Juan* and his lady companion *Haidée* (the beautiful daughter of a Greek pirate, smuggler and vagabond) engenders the moving stanza:

> 'Their faces were not made for wrinkles, their
> Pure blood to stagnate, their great hearts to fail;
> The blank grey was not made to blast their hair,
> But like the climes that know nor snow nor hail
> They were all summer: lightning might assail
> And shiver them to ashes, but to trail
> A long and snake-like life of dull decay
> Was not for them – they had too little day.'

The lesser-known epic narrative *King Alfred* written by the English poet John Fitchett (1776–1838) contemplates the dramatic life and times of Alfred the Great (*circa* 849–899 A.D.), particularly his battles against the Danish invaders. The work was first published privately in five volumes at intervals between 1808 and 1834, followed by a version in six volumes [A4.219] edited by his former pupil Robert Roscoe, which includes the following extract from a passage explaining how the devilish *Nicka* (*Elf of the waters*) can invoke supernatural powers to track down the covert location of Alfred and his liegemen:

> 'Or by some opening river's mouth at times
> Wait where along the bleak and shelly beach
> The cormorant stalks, or on the neighbour cliffs
> Sits, below which some bark with flagging sail
> Interprets secret watch and deem'd escape.
> There amid drenching rains and drifted hail
> The skulking michers, from their haunts beguiled,
> Gallow'd with sudden terror, we may snare
> And in the trammels of our watch surprise.'

Firmly in the Victorian mode, *Paracelsus* is a five-part epic poem written by Robert Browning (1812–1889) and published [A4.220] in 1835. In an early scene (part 1, *Paracelsus aspires*) set in a garden near Würzburg in 1512, whilst in conversation with his friends *Festus* and *Michal*, the 16th-century European savant (*Aureolus Paracelsus*) declares his personal beliefs (upon prompting by *Festus*), concluding with the incisive passage:

> '*I see my way as birds their trackless way –*
> *I shall arrive! what time, what circuit first,*
> *I ask not: but unless God send his hail*
> *Or blinding fire-balls, sleet, or stifling snow,*
> *In some time – his good time – I shall arrive:*
> *He guides me and the bird. In his good time!*'

at which juncture *Michal* comments '*Vex him no further, Festus; it is so!*'.

An example among the litany of praises ascribed to nature by the English writer Elizabeth Barrett Browning (1806–1861) – the wife of Robert Browning – occurs in her early poem *Earth and Her Praisers* [A4.221], which also reveals the inner conflict between her religious and poetic visions:

> '*Earth, we Christians praise thee thus,*
> *Even for the change that comes,*
> *With a grief, from thee to us!*
> *For thy cradles and thy tombs;*
> *For the pleasant corn and wine,*
> *And summer-heat; and also for*
> *The frost upon the sycamore,*
> *And hail upon the vine!*'

Countless other references to hail are found in English prose literature, not least those by the incomparable English writer Charles Dickens (1812–1870). Towards the end of his third novel [A4.222] *The life and adventures of Nicholas Nickleby*, the roguish and malevolent uncle *Ralph Nickleby*, shortly before his demise and looking out from the garret of his London home amidst quotidian drabness, observed with foreboding the dark, threatening sky and closed the window:

> '*The rain and hail pattered against the glass; the chimneys quaked and rocked; the crazy casement rattled with the wind, as though an impatient hand inside were striving to burst it open. But no hand was there, and it opened no more.*'

Subsequently, in his novella [A4.223] *A Christmas Carol*, part of the opening stave (*Marley's Ghost*) describes the elderly miser *Ebenezer Scrooge* in the following excoriating manner:

> '*Oh! But he was a tight-fisted hand at the grindstone, Scrooge! A squeezing, wrenching, grasping, scraping, clutching, covetous old sinner! Hard and sharp as flint, from which no steel had ever struck out generous fire; secret and self contained and solitary as an oyster. The cold within him froze his old features, nipped his pointed nose, shrivelled his cheek, stiffened his gait; made his eyes red, his thin lips blue; and spoke out shrewdly in his grating voice. A*

frosty rime was on his head, and on his eyebrows, and his wiry chin. He carried his own low temperature always about him; he iced his office in the dogdays; and didn't thaw it one degree at Christmas.

External heat and cold had little influence on Scrooge. No warmth could warm, nor wintry weather chill him. No wind that blew was bitterer than he, no falling snow was more intent upon its purpose, no pelting rain less open to entreaty. Foul weather didn't know where to have him. The heaviest rain, and snow, and hail, and sleet could boast of the advantage over him in only one respect. They often 'came down' handsomely; and Scrooge never did.'

These words are a transcription of the first two complete paragraphs from the original manuscript page shown in Figure A4.54. In contrast to the grim fate of so many predecessors, it appears that the redoubtable *Scrooge* was able to withstand the ordeals of extreme weather, including hail.

Among several references to hail by the German writer Johann Wolfgang von Goethe (1749–1832) – one of the leading figures in Western literature – occurs in the fourth stanza of his ode *The Godlike*, which takes the simple yet acerbic form:

'*Tempest and torrent,*
Thunder and hail,
Roar on their path,
Seizing the while,
As they haste onward,
One after another.'

A brighter note is sounded in *The bequest of the ancient Persian faith*, from a collection of his poems in twelve books comprising the *West-Eastern divan*, written in the Persian manner. Whence in book 11 (*Parsi Nameh, Book of the Parsees*):

'*When we oft have seen the monarch ride,*
Gold upon him, gold on ev'ry side;
Jewels on him, on his courtiers all,
Thickly strew'd as hailstones when they fall,'

A different style is followed in *Hermann and Dorothea*, an elegy in nine cantos, where in canto 8 (*Melpomene*):

'*So tow'rd the sun, now fast sinking to rest, the two walk'd together, whilst he veil'd himself deep in clouds which thunder portended. Out-of his veil now here, now there, with fiery glances beaming over the plain with rays foreboding and lurid. 'May this threatening weather,' said Hermann, 'not bring to us shortly hail and violent rain, for well does the harvest now promise.' And they both rejoiced in the corn so lofty and waving, well nigh reaching the heads of the two tall figures that walk'd there.*'

All three translations [A4.224] are by the English civil servant and politician Edgar Bowring (1826–1911), the latter composition appearing only in the second edition of the collected

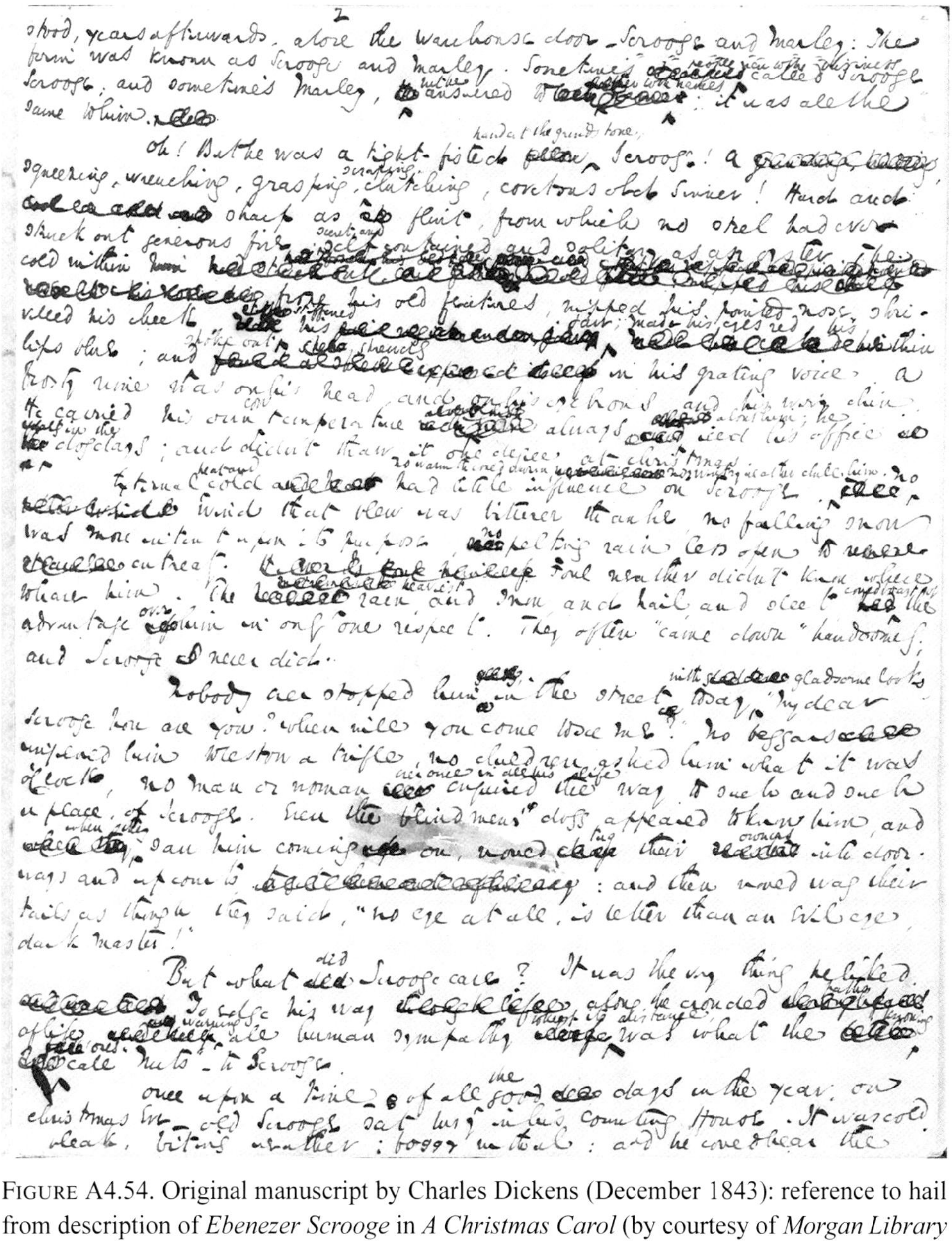

FIGURE A4.54. Original manuscript by Charles Dickens (December 1843): reference to hail from description of *Ebenezer Scrooge* in *A Christmas Carol* (by courtesy of *Morgan Library & Museum, New York*, MS MA97, p. 2)

works. Remarkably, Goethe was also an early pioneer in the science of colour measurement and perception (see also section 6 of the main *Glass Walls* text), with his original book published in 1810 (*Zur Farbenlehre*) and translated into English by Charles Eastlake [A4.225].

Factual accounts of hailstorms were expounded by the English naturalist Charles Darwin (1809–1882), following information gathered [A4.226] during the second voyage of *H.M.S. Beagle* (1831–1836), which circumnavigated the globe. The extract below – from a later edition of his records [A4.227] – is part of a journal entry for 16 September 1833 when, having set out on horseback for the Argentinian city of Buenos Aires some days earlier, his small party reached the foot of the Sierra Tapalguen:

> *'We were here told a fact, which I would not have credited, if I had not had partly ocular proof of it; namely, that, during the previous night, hail as large as small apples, and extremely hard, had fallen with such violence, as to kill the greater number of the wild animals. One of the men had already found thirteen deer (Cervus campestris) lying dead, and I saw their fresh hides; another of the party, a few minutes after my arrival, brought in seven more. Now I well know, that one man without dogs could hardly have killed seven deer in a week. The men believed they had seen about fifteen dead ostriches (part of one of which we had for dinner); and they said that several were running about evidently blind in one eye. Numbers of smaller birds, as ducks, hawks, and partridges, were killed. I saw one of the latter with a black mark on its back, as if it had been struck with a paving-stone. A fence of thistle stalks round the hovel was nearly broken down, and my informer, putting his head out to see what was the matter, received a severe cut, and now wore a bandage.*
>
> *The storm was said to have been of limited extent: we certainly saw from our last night's bivouac a dense cloud and lightning in this direction. It is marvellous how such strong animals as deer could thus have been killed; but I have no doubt, from the evidence I have given, that the story is not in the least exaggerated. I am glad, however, to have its credibility supported by the Jesuit Drobrizhoffer, who, speaking of a country much to the northward, says, hail fell of an enormous size and killed vast numbers of cattle: the Indians hence called the place Lalegraicavalca, meaning 'the little white things'. Dr Malcolmson, also, informs me that he witnessed in 1831 in India, a hailstorm, which killed numbers of large birds and much injured the cattle. These hailstones were flat, and one was ten inches in circumference, and another weighed two ounces. They ploughed up a gravel walk like musket balls, and passed through glass windows, making round holes, but not cracking them.'*

the last sentence of which includes an interesting observation on dynamic glass fracture.

More generally, within the *pampas* of South America, predominantly a vast lowland region under a temperate climate, severe thunderstorms are common in the spring and summer, sometimes leading to intense precipitation of hail. With specific reference to the Argentinian *pampa*, Simpich [A4.228] has commented:

> *'But it is hail – not wind, lightning, or torrential rain – which wreaks havoc. Big hailstones, pounding the farms like artillery fire, not only beat crops, fruits, and vegetables to pieces, but have been known to kill sheep, young cattle, and even horses.'*

Many other historical accounts of real hailstorms in various parts of the world are summarised by the English meteorologist Rollo Russell (1849–1914) – son of a British Prime Minister (Lord John Russell) and uncle of a celebrated philosopher (Bertrand Russell) – in a monograph [A4.229] that also includes an extensive review of related scientific aspects developed by the end of the 19th century.

Passing references to hail occur in a few of the celebrated fairy tales [A4.230] created by the Danish author Hans Christian Andersen (1805–1875). In *The garden of paradise* (*Paradisets Have*), dating from 1839, a prince is walking alone in a wood at night and seeks shelter in a large cavern (*Cavern of the Winds*), which turns out to be occupied by a fearsome elderly woman (*Mother of the Winds*) slowly turning a roasting stag on a spit over an immense fire. Her absent sons are the four *Winds of Heaven*, but the first is soon to arrive:

> '*It was the North Wind who came in, bringing with him a cold, piercing blast; large hailstones rattled on the floor, and snowflakes were scattered around in all directions. He wore a bearskin dress and cloak. His sealskin cap was drawn over his ears, long icicles hung from his beard, and one hailstone after another rolled from the collar of his jacket.*'

The drawing in Figure A4.55 is by Andersen's fellow countryman Vilhelm Pedersen (1820–1859), taken from an 1849 illustrated Danish publication and showing the young prince meeting the *Mother of the Winds* in the cavern.

Following the much earlier Jomsviking adventure, the Nordic tradition continues with *The Kalevala*, a Finnish national epic created by Elias Lönnrot (1802–1884), first published in 1835 but enlarged to the definitive version of fifty runes in 1849. The work is based on the written and oral records of rural storytellers over the ages, mostly short ballads and lyric poems, gathered during extensive travels throughout his native countryside. In an early episode, one of the principal heroes (*Lemminkainen*) invokes the forest deities when set the task of hunting an evil fire-breathing steed (*Hisi*). On appeal to the sky-god *Ukko*, akin to the Greek *Zeus*, it is declared in rune 14 (*Death of Lemminkainen*):

> '*Spake the daring Lemminkainen,*
> *This the hero's supplication:*
> *'Ukko, thou O God above me,*
> *Thou that rulest all the storm-clouds,*
> *Open thou the vault of heaven,*
> *Open windows through the ether,*
> *Let the icy rain come falling,*
> *Let the heavy hailstones shower*
> *On the flaming horse of Hisi,*
> *On the fire-expiring stallion.*'

having been translated into blank verse [A4.231] – with anglicised names – by the American scholar John Martin Crawford (1845–1916). Another adventure occurs much later when *Louhi* (*Hostess of Pohyola*), in her efforts to regain the *Sampo* (a priceless magic jewel) from the heroes of *Kalevala*, threatens to cause almighty disruption throughout the land, including the scourge of iron-hard hailstones:

FIGURE A4.55. Illustration by Vilhelm Pedersen, *circa* 1849, from Hans Christian Andersen's fairy tale *The garden of paradise*, showing young prince meeting *Mother of the Winds* in cavern

> '*In the rocks I'll sink the moonbeams,*
> *Hide the sun within the mountain,*
> *Let the frost destroy thy sowings,*
> *Freeze the crops on all thy corn-fields;*
> *Iron-hail I'll send from heaven,*
> *On the richness of thine acres,*
> *On the barley of thy planting;*'

taken from rune 43 (*The Sampo lost in the sea*): an especially daunting prospect in such a cold northern climate, with a short summer growing season for crops.

A closely related work of literature is the Estonian epic *Kalevipoeg*, comprising twenty cantos created from ancient folklore material by Friedrich Kreutzwald (1803–1882), first published in 1857, and translated into German in 1861. In canto 14 (*The palace of Sarvik*), the hero *Kalevide* is implored to flee upon hearing the thunderous approach of *Sarvik* (Prince of *Põrgu*, or *Hades*), but instead stands firm:

> '*Like the oak-tree in the tempest,*
> *Or the red glow 'mid the cloudlets,*
> *Or the rock amid a hailstorm,*
> *Or a tower in windy weather.*'

from a translation and commentary [A4.232] by the English entomologist and folklorist William Forsell Kirby (1844–1912). A marked contrast in general ambience is discernible in the Estonian ballad *The Blue Bird*, wherein the mythical blue bird *Siuru* (daughter of *Taara*, or *Ukko*), in response to her divine father's probing on her latest flight, declares:

> '*Long I flew on path of thunder,*
> *On the roadway of the rainbow,*
> *And the hailstone's toilsome pathway;*'

Turning to the folklore of another of the Baltic states, the misfortune of being lost at sea is encountered in the epic work *Lacplesis* (*Bearslayer*) by the Latvian writer Andrejs Pumpurs (1841–1902), first published in 1888 in six cantos, and recently translated into English verse [A4.233] by the Australian academic Arthur Cropley. Whence in canto 4 (scene 3: *In the realm of the North Wind*), the hero *Lacplesis* sets sail for Germany to rescue his heroine *Laimdota*:

> '*But battered by the force*
> *Of wind and storm that blew,*
> *And lost, far from the course,*
> *The way no more they knew.*
>
> *It seemed that evil powers,*
> *Sea ghosts, were ever near,*
> *In day and night-time hours,*
> *They filled the crew with fear.*
>
> *Dank mists and deepest gloom*
> *The light blocked as they swirled;*
> *While hail and snow-filled spume*
> *Were by the North Wind hurled.*'

The fabled pair were soon reunited, yet following a brief period together in peaceful harmony, both were to meet a prematurely tragic end.

Composed in a contrasting jocular vein is the poem *Peg of Limavaddy* by the Indian-born English writer William Makepeace Thackeray (1811–1863), better known in the arena of Victorian literature for his satirical novels. It was written during his travels in Ireland in 1842 – and published the following year under the *nom de plume* M. A. Titmarsh [A4.234] – featuring the enchanting maid (*Peggy*) who served him ale on a stormy October day whilst staying at an inn located in the market town of Limavaddy. Upon being warmly welcomed by the landlord and his wife, it is observed in one of the early stanzas:

> '*Up and down the stair*
> *Two more young ones patter*
> *(Twins were never seen*
> *Dirtier nor fatter);*
> *Both have mottled legs,*

Both have snubby noses,
Both have—Here the host
Kindly interposes:
'Sure you must be froze
With the sleet and hail, sir:
So will you have some punch,
Or will you have some ale, sir?'

A plaintive note is struck by the American writer Henry Wadsworth Longfellow (1807–1882) in his epic poem *Evangeline: a Tale of Acadie*, first published [A4.235] in 1847, and loosely based upon the tribulations of the Acadians (early settlers in north-eastern North America) being driven into exile during the preceding century, with a fictional maiden *Evangeline Bellefontaine* as the title character. Whence the lines (canto 4, part 1):

'As, when the air is serene in the sultry solstice of summer,
Suddenly gathers a storm, and the deadly sling of the hailstones
Beats down the farmer's corn in the field and shatters his windows,
Hiding the sun, and strewing the ground with thatch from the house-roofs,
Bellowing fly the herds, and seek to break their enclosures;'

Also related to historical events is Longfellow's much later piece [A4.236] entitled *A Ballad of the French Fleet*, subtitled *October, 1746, Mr Thomas Prince loquitur*, concerning the providential deliverance of the coastal town of Boston, Massachusetts, from a French attack in 1746. Based upon a sermon by Thomas Prince (1687–1758), pastor at South Church in Boston, the poem emphasises the crucial part played by the ferocious weather, as exemplified by the verse:

'The lightning suddenly
Unsheathed its flaming sword,
And I cried: 'Stand still, and see
The salvation of the Lord!'
The heavens were black with cloud,
The sea was white with hail,
And ever more fierce and loud
Blew the October gale.'

Whence the fleet was scattered far and wide by the storm, with many ships wrecked along the coast, yet leaving the township intact. In a different setting, the harsh reality of nature also is portrayed in Figure A4.56, with a pen and ink drawing entitled *Gleaners in the Hail* by the French artist Charles-François Daubigny (1817–1878), dated 1862.

Many allegorical references to hailstorms are found in Victorian poetry, including several by the English author Alfred Tennyson (1809–1892), sometime Poet Laureate. In *Morte*

FIGURE A4.56. Pen and ink drawing *Gleaners in the Hail* by Charles-François Daubigny, 1862 (by courtesy of *Metropolitan Museum of Art, New York*)

d'Arthur, for example [A4.237], the mortally wounded *King Arthur* tells the last surviving knight *Sir Bedivere* of his impending mystical journey to the tranquil pastures of *Avilion*:

> '*Where falls not hail, or rain, or any snow,*
> *Nor ever wind blows loudly; but it lies*
> *Deep-meadow'd, happy, fair with orchard-lawns*
> *And bowery hollows crown'd with summer sea,*
> *Where I will heal me of my grievous wound.*'

while continuing the *Arthurian* legend with his poem *Sir Galahad*, in which the pious *Knight of the Round Table* contemplates a vision of the *Holy Grail*, and where the following lines of blank verse also strike a sanguine note:

> '*The tempest crackles on the leads,*
> *And, ringing, spins from brand and mail;*
> *But o'er the dark a glory spreads,*
> *And gilds the driving hail.*'

A broadly similar thread is echoed in the poem *The Defence of Guenevere* by William Morris (1834–1896), a leading *Pre-Raphaelite* better known in the modern era for his close association with the British Arts and Crafts Movement: see the first volume of his poetry [A4.238] published in 1858. Following an initial *mea culpa* by the beautiful *Queen*

Guenevere, it chanced that *Launcelot* came to dwell at *Arthur's* court at Christmas time, after which she declares (in *terza rima* verse):

> '*Christmas and whitened winter passed away,*
> *And over me the April sunshine came,*
> *Made very awful with black hail-clouds, yea*
>
> *And in the Summer I grew white with flame,*
> *And bowed my head down: Autumn, and the sick*
> *Sure knowledge things would never be the same,*'

Likewise rooted in the Victorian era is the short poem *Heaven–Haven* (sub-titled *A nun takes the veil*) by the English poet and priest Gerard Manley Hopkins (1844–1889), written as an undergraduate in 1864, in which the opening stanza (as spoken by a nun seeking the spiritual joys of a heavenly world) is:

> '*I have desired to go*
> *Where springs not fail,*
> *To fields where flies no sharp and sided hail*
> *And a few lilies blow.*'

where *blow* (from the Old English *blōwan*) means to bloom or blossom; a melancholy description comparable to that of Tennyson's *Avilion*. A contrasting narrative occurs in *Strike, Churl*, composed (perhaps as an unfinished sonnet) in May 1885 whilst abhorring the tardy progression from a harsh winter to more spring-like weather. The entire poem comprises four lines:

> '*Strike, churl; hurl, cheerless wind, then; heltering hail*
> *May's beauty massacre and wispèd wild clouds grow*
> *Out on the giant air; tell Summer No,*
> *Bid joy back, have at the harvest, keep Hope pale.*'

and, along with *Heaven–Haven*, is included in an anthology first published posthumously [A4.239].

As epitomised in some previous quotations, the English *Pre-Raphaelite* poet Algernon Charles Swinburne (1837–1909) also encompasses the pastoral world in *The Masque of Queen Bersabe: a Miracle-Play*, first published [A4.240] in 1866, where in the words of *King David*:

> '*Lord God, alas, what shall I sain?*
> *Lo, thou art as an hundred men*
> *Both to break and build again:*
> *The wild ways thou makest plain,*
> *Thine hands hold the hail and rain,*
> *And thy fingers both grape and grain;*
> *Of their largess we be all well fain,*
> *And of their great pity:*'

Also of special interest are extracts from two poems from the collected works [A4.241] of the French writer Victor Hugo (1802–1885). From *La légende des siècles*, completed in 1859, is the poem *Eviradnus: the knight errant* (*Qu'est-ce que Sigismond et Ladislas ont dit?*), translated by the English writer Mrs Newton Crosland (born Camilla Dufour Toulmin, 1812–1895). In part 3 (*In the forest*) are the trenchant lines:

> '*Winter, the savage warrior, pleases well,*
> *With its storm clouds, the mighty citadel,—*
> *Restoring it to life. The lightning flash*
> *Strikes like a thief and flies; the winds that crash*
> *Sound like a clarion, for the Tempest bluff*
> *Is Battle's sister. And when wild and rough,*
> *The north wind blows, the tower exultant cries*
> '*Behold me!' When hail-hurling gales arise*
> *Of blustering Equinox, to fan the strife,*
> *It stands erect, with martial ardour rife,*
> *A joyous soldier!*'

In Hugo's work *L'art d'être grandpère*, dated 1877, the dramatic poem *The epic of the lion* (*Un lion avait pris un enfant*), translated by the English poet and journalist Edwin Arnold (1832–1904), describes the tribulations and eventual deliverance of a child (the princely son of the king) taken by a raging lion:

> '*Frightful! – they saw the Lion! Not one pace*
> *Further stirred any man; the very trees*
> *Grew blacker with his presence, and the breeze*
> *Blew shudders into all hearts present there:*
> *Yet, whether 'twas from valour or wild fear,*
> *The archers drew – and arrow, bolt and dart*
> *Made target of the Beast. He, on his part –*
> *As calm as Pelion in the rain or hail –*
> *Bristled majestic from the nose to tail,*
> *And shook full fifty missiles from his hide;*'

An early scene from this epic story – at once playful and profound – is depicted in Figure A4.57, based on a painting by Georges Moreau de Tours (1848–1901) and produced as a photogravure image by Goupil et Cie, Paris. Having made short work of a brave knight intent on rescue, the lion dismisses a passing hermit with a thunderous roar – his life grudgingly spared – before confronting and soon dispersing an armed group gathered at his cavernous den. Then pacing towards the town and entering the king's palace, its terrified occupants hidden from view, the merciful lion finally leaves the unharmed boy with his intrepid younger sister, following her determined remonstration with the mighty beast.

In the natural world, large hailstones of unusual form have been described by Abich [A4.242], relating to storms that took place near Tiflis (also known as Tbilisi) in Georgia

FIGURE A4.57. Early scene from *The epic of the lion* by Victor Hugo; photogravure image based on painting by Georges Moreau de Tours, *circa* 1890 (by courtesy of *Goupil et Cie, Paris*)

during May and June 1869, causing enormous devastation, with '*strong branches struck down as if cut by some sharp implement*'. Drawings of two of the hailstones are shown in Figure A4.58, drawn to natural scale in the original paper, and probably quite close in size to the present images. The author comments that:

> '*In the two cases under notice, personal observation sets more or less at defiance any theory of the formation of hail hitherto established.*'

and further states that:

> '*These aggregations may have had a long stay within a medium of highly refrigerated aqueous vapour before they fell to the ground. It must be remarked, to fully understand the drawings, that the shaded portion of the flattened spheroidal fundamental form of the groups is not always opaque in the original. Only the circle round the centre has a milky aspect, due to the air bubbles enclosed in it, as also the nucleus of the greater number; in other specimens the nucleus is transparent, especially when reduced by melting away into disks of ¾ to 1 inch in diameter, sometimes affecting the form of a perfect regular hexagon. In this case the milky circle around the centre appeared distinctly as an intricate tissue of minute lengthened pores and of capillary fissures filled with air. The shadow next to the margin of the larger peripherical circle is only intended to indicate the rounded and flattened spheroidal form of the chief body, on whose broader margin the crystals themselves adhere parasitically, or are inserted, as in*

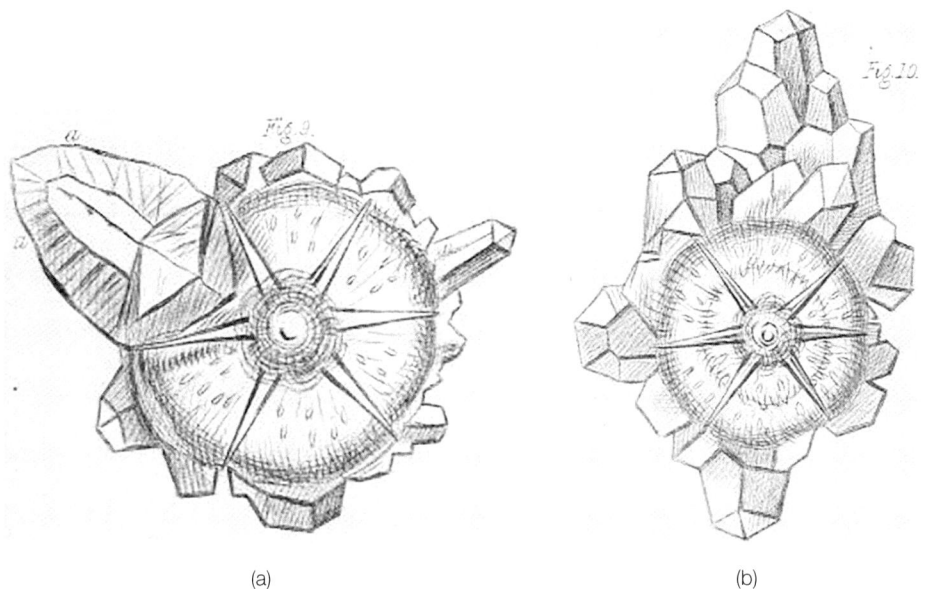

FIGURE A4.58. Drawings of two hailstones that fell in Georgia during May and June 1869, after Abich [A4.242]

an alveole, made visible by the commencement of fusion (see 'a' in fig. 9). All the specimens presented lengthened vermiform and pyriform pores filled with air, extending radially from the centre to the circumference.'

Curiously – but not entirely unexpectedly from such a Promethean author – the pragmatic world of commercial insurance and hail damage is mentioned in verse by the English mathematician and writer of children's fiction Charles Dodgson, *alias* Lewis Carroll (1832–1898), in his humorous poem *The Hunting of the Snark: an Agony, in eight Fits*; with the following two consecutive stanzas [A4.243] from *Fit the First; The Landing*:

*'The Beaver's best course was, no doubt, to procure
A second-hand dagger-proof coat –
So the Baker advised it – and next, to insure
Its life in some Office of note:*

*This the Banker suggested, and offered for hire
(On moderate terms), or for sale,
Two excellent Policies, one Against Fire,
And one Against Damage From Hail.'*

Contemporaneously, the English novelist and horticulturist Richard Doddridge Blackmore (1825–1900) – usually known as R. D. Blackmore – suffered substantial flower and fruit crop losses, coupled with a major financial setback, resulting from a hailstorm in the summer of 1879 at his market garden at Teddington (then a country village, now part of south-west

FIGURE A4.59. Italian lithograph (*circa* 1870) entitled *Grandine*, after Giulio Gorra; glasshouse destroyed by hail (by courtesy of *Wellcome Library, London*, Ref. 573190i)

London), including extensive glass damage to greenhouses from hailstones '*as large as hens eggs, and some even larger*'; see Webber and Ching [A4.244]. Similarly, Figure A4.59 shows a lithograph entitled *Grandine*, after the Italian painter Giulio Gorra (1832–1884), where a forlorn onlooker is standing in front of a glasshouse destroyed by hail.

Much further afield in South Africa, the engraving in Figure A4.60 depicts a Zulu warrior sheltering from a hailstorm under his shield near an ant-heap during the Anglo-Zulu War, and appeared in the British weekly illustrated magazine *The Graphic* in April 1879, based on an original work by the English artist Samuel Edmund Waller (1850–1903). The related commentary notes that:

> '*The stones are often as large as walnuts, and hurt excessively. They weigh four or five ounces, and going before a high wind at great speed often break tiles, and have been known to pierce iron sheet-roofing.*'

Even in peacetime, hailstorm conditions can be challenging to travellers on horseback, as epitomised in the novel *Anna Karenina* by the Russian writer Leo Tolstoy (1828–1910), first published in book form in 1878, wherein a wealthy country landowner (*Konstantin Levin*) sets off alone on a homeward journey in late September:

> '*The weather had become worse than ever towards evening; the hail lashed the drenched mare so cruelly that she went along sideways, shaking her head and ears; but Levin was all right*

FIGURE A4.60. Zulu warrior sheltering from hailstorm during Anglo-Zulu War, South Africa, 1879; engraving after Samuel Waller

> *under his hood, and he looked cheerfully about him at the muddy streams running under the wheels, at the drops hanging on every bare twig, at the whiteness of the patch of unmelted hailstones on the planks of the bridge, at the thick layer of still juicy, fleshy leaves that lay heaped up about the stripped elm-tree.'*

being an extract from a translation [A4.245] by the English linguist Constance Garnett (1861–1946).

The Selfish Giant is a short fantasy story for children by the versatile Irish writer Oscar Wilde (1854–1900), written long before his bleak final years. It tells of a selfish giant who builds a high wall to keep the village children out of his beautiful castle garden, but whose heart is changed by an innocent child to one of compassion [A4.246]. The poor children had nowhere to play, so instead of the garden blossoming in spring, it remained firmly in the grip of winter:

> *'The Snow covered up the grass with her great white cloak, and the Frost painted all the trees silver. Then they invited the North Wind to stay with them, and he came. He was wrapped in furs, and he roared all day about the garden, and blew the chimney-pots down. 'This is a delightful spot,' he said, 'we must ask the Hail on a visit.' So the Hail came. Every day for three hours he rattled on the roof of the castle till he broke most of the slates, and then he ran round and round the garden as fast as he could go. He was dressed in grey, and his breath was like ice.*
>
> *'I cannot understand why the Spring is so late in coming,' said the Selfish Giant, as he sat at the window and looked out at his cold white garden; 'I hope there will be a change in the weather.'*
>
> *But the Spring never came, nor the Summer. The Autumn gave golden fruit to every garden, but to the Giant's garden she gave none. 'He is too selfish,' she said. So it was always Winter there, and the North Wind, and the Hail, and the Frost, and the Snow danced about through the trees.'*

The seasons were partially restored only when the children gained access to the garden through a hole in the wall, and were able to enjoy themselves as before. But in the farthest corner of the garden it was still winter, and the recalcitrant giant lifted a small boy into the nearest tree – which immediately broke into blossom – and resolved to demolish the abhorrent wall. This aphoristic tale has a strong moral message, with a moving ending that resonates with redemption and forgiveness; blessings alas not bestowed upon the renowned author towards the tragic end of his own short life. Figure A4.61 is a poignant illustration of a notable scene from this wistful tale, created by the British artist Walter Crane (1845–1915).

With a light-hearted touch firmly embedded in the Victorian era, the majestic queen – by then also titled Empress of India – is mentioned directly by the Indian-born English writer Rudyard Kipling (1865–1936) in the short poem *The Overland Mail* (sub-titled *Foot-Service to the Hills*), first published [A4.247] in 1886. Employed by the Indian postal service to deliver mail in the more remote regions of the country, the so-called 'dak runners' formed a hardy band of indigenous men whose daily task was arduous and often dangerous. The poem may be regarded as a tribute to their splendid service amidst grinding poverty and oppression, and the final verse is:

FIGURE A4.61. Illustration from Oscar Wilde's story *The Selfish Giant*, 1888, by Walter Crane

> '*There's a speck on the hillside, a dot on the road –*
> *A jingle of bells on the foot-path below –*
> *There's a scuffle above in the monkey's abode –*
> *The world is awake, and the clouds are aglow.*
> *For the great Sun himself must attend to the hail:*
> *'In the name of the Empress, the Overland Mail!'*'

The mail runners in Bengal portrayed in Figure A4.62(a) – published by the *Illustrated London News* in 1858 – were following a traditional role that already had existed for several hundred years. The runner shown in Figure A4.62(b), armed only with a spear for self-defence, is from a later photogravure image based on a drawing (*circa* 1899) by the American artist Victor Searles, and included in an edition of collected works [A4.248].

A brief return to nautical matters is prompted by Kipling's narrative poem *McAndrews' Hymn*, first published [A4.249] in 1894. The rhyming verse forms a reflective monologue by a Scottish marine engineer on night-watch aboard a passenger steamship circumnavigating the globe, written in a strong Glaswegian dialect, and includes the characteristic wording:

> '*An' home again, the Rio run: it's no child's play to go*
> *Steamin' to bell for fourteen days o' snow an' floe an' blow –*
> *The bergs like kelpies overside that girn an' turn an' shift*
> *Whaur, grindin' like the Mills o' God, goes by the big South drift.*
> *(Hail, snow an' ice that praise the Lord: I've met them at their work,*
> *An' wished we had anither route or they anither kirk.)*'

wherein 'kelpies' are mythical water creatures of Celtic legends, and 'girn' means grumble or growl.

A quizzical reference to hail was made by Gertrude Bell (1868–1926), an English writer and political administrator who, prior to embarking upon an influential career as a diplomat in the Middle East, travelled to Persia in May 1892 to visit her uncle (British minister at Tehran) and explore the local countryside. She described this journey in her book *Persian Pictures* [A4.250], and in one chapter (*Three noble ladies*) recalled visiting a princess and her two daughters at their home in Tehran, on the eve of departure for their summer camp in the mountains. And soon thereafter:

> '*We went to see the three ladies again when we were in the mountains. Their camp was pitched about a mile lower down the river than ours, on a grassy plateau, from which they had a magnificent view down the long bare valley and across mountains crowned by the white peak of Demavend. No sooner had we forded the river in front of our tents than a storm of wind and rain and hail broke upon us, but we continued dauntlessly on our way, for the day of our visit had been fixed some time before, and it was almost pleasant after the summer's drought to feel the rain beating on our faces.*'

Upon reaching the Persian camp they dismounted, and were taken into a large tent to meet the Princess:

(a)

(b)

FIGURE A4.62. Portrayals of Indian mail runners referred to in poem *The Overland Mail* by Rudyard Kipling: (a) set in Bengal, 1858 (by courtesy of *Illustrated London News*); (b) from original drawing by Victor Searles, *circa* 1899 (by courtesy of *H. M. Caldwell Co., New York*)

'We greeted her with chattering teeth and sat down on some wooden chairs round her, carrying on a laboured conversation in the French tongue, while our wet clothes grew ever colder upon us. We remembered the steaming cups of tea of our former visit, and prayed that they might speedily make their appearance, but, alas! on this occasion they were omitted, and lemon ices alone were offered to us. It is not to be denied that lemon ices have their merit on a hot summer afternoon, but the Persian's one idea of hospitality is to give you lemon ices—lemon ices in hailstorms, lemon ices when you are drenched with rain, lemon ices when a biting wind is blowing through the tent door—it was more than the best regulated constitution could stand. We politely refused them.'

However, a most genial and agreeable conversation followed, after which the visitors returned to their own camp in bright sunshine, the stormy weather having cleared as rapidly as it first appeared.

In the science fiction novella *The Time Machine* [A4.251] by the English writer Herbert Wells (1866–1946), usually known as H. G. Wells, it occurs early in the journey of the *Time Traveller* that:

'There was the sound of a clap of thunder in my ears. I may have been stunned for a moment. A pitiless hail was hissing round me, and I was sitting on soft turf in front of the overset machine. Everything still seemed grey, but presently I remarked that the confusion in my ears was gone. I looked round me. I was on what seemed to be a little lawn in a garden, surrounded by rhododendron bushes, and I noticed that their mauve and purple blossoms were dropping in a shower under the beating of the hailstones. The rebounding, dancing hail hung in a cloud over the machine, and drove along the ground like smoke. In a moment I was wet to the skin.'

An acclaimed work of maritime fiction [A4.252] by the Polish-born writer Joseph Conrad (1857–1924), following an extensive seafaring career, is the novella *The Nigger of the 'Narcissus'* (sub-titled *A Tale of the Sea*) – a main title from a bygone era that nowadays would be considered inappropriate, despite the author's proclaimed affinity with *all* seamen, irrespective of class, colour or creed. The narrative follows the declining fortunes of an increasingly frail black sailor aboard a merchant ship (*Narcissus*) sailing from Bombay to London, which runs into frightful difficulties when rounding the Cape of Good Hope. Whence from chapter 3 is the evocative passage:

'The ship tossed about, shaken furiously, like a toy in the hand of a lunatic. Just at sunset there was a rush to shorten sail before the menace of a sombre hail cloud. The hard gust of wind came brutal like the blow of a fist. The ship relieved of her canvas in time received it pluckily: she yielded reluctantly to the violent onset; then coming up with a stately and irresistible motion, brought her spars to windward in the teeth of the screeching squall. Out of the abysmal darkness of the black cloud overhead white hail streamed on her, rattled on the rigging, leaped in handfuls off the yards, rebounded on the deck – round and gleaming in the murky turmoil like a shower of pearls.'

In contrast to the pitiful sailor, who was buried at sea, the ship eventually returned safely to its home port.

In a short quotation by the Indian writer Rabindranath Tagore (1861–1941) – see, for example, Seth [A4.253] – his intense dislike of traditional early schooling, largely based on personal experience, is expressed in a forthright manner:

'We had to sit inert, like dead specimens of some museum, whilst lessons were pelted at us from on high, like hailstones on flowers.'

which oppressive and stultifying conditions he was eager to avoid in various pedagogic missions throughout his long life, especially in creating independent educational establishments having an alternative approach to learning.

Journeys into the remote land of Tibet around the turn of the 20th century by the Japanese Buddhist monk Ekai Kawaguchi (1866–1945) are recorded in considerable detail in his own English translation [A4.254], and include several encounters with hailstorms, together with tales of mystical preventive measures (see section A4.8). Regarding the former, the following abstract from chapter 26 (*Sacred Manasarovara and its legends*) describes an awesome panorama viewed in early August 1900:

'After proceeding about ten miles over an undulating range of mountains we came in sight of Man-ri, a peak of perpetual snow, which has an altitude of 25,600 feet above the sea-level. The view of Man-ri, rising majestically high above the surrounding mountains (themselves of great elevation) was sublimely grand. While standing absorbed in the severe magnificence of the scenery, I was treated to another experience, which was as soul-stirring as any earthly phenomenon could be. A magical change in the weather was heralded by a sudden flash of lightning, followed by another, yet another and another, new accompanied by rolling thunder. Heavy pelting hailstones then joined in the war of elements, which literally shook the mighty mountains to their very foundations, and filled the air with the utmost confusion of terrific noises and lurid tongues of fire. Standing almost alone upon a great height, I saw black clouds with fearful suddenness envelope the world of vision in frightful darkness, made doubly dark by the contrasts produced by the momentary glare of pale, penetrating lightning, which, in the same instant, revealed the glittering snow on the grand peaks of the Himalayas, and the deepest chasms, thousands of fathoms below!'

In yet another facet of Far Eastern culture, Chinese paper gods were created for worshipping the heaven and gods in various rituals, and an example in the form of a 19th-century woodblock print (width 28 cm) entitled *God of Hail* from the Beijing region is displayed in Figure A4.63.

A4.7. Near-contemporary history (post 1900)

A more traditional quotation is found in the narrative poem *Baile and Aillinn* by the Irish writer William Butler Yeats (1865–1939), characteristically steeped in mythology and folklore, namely:

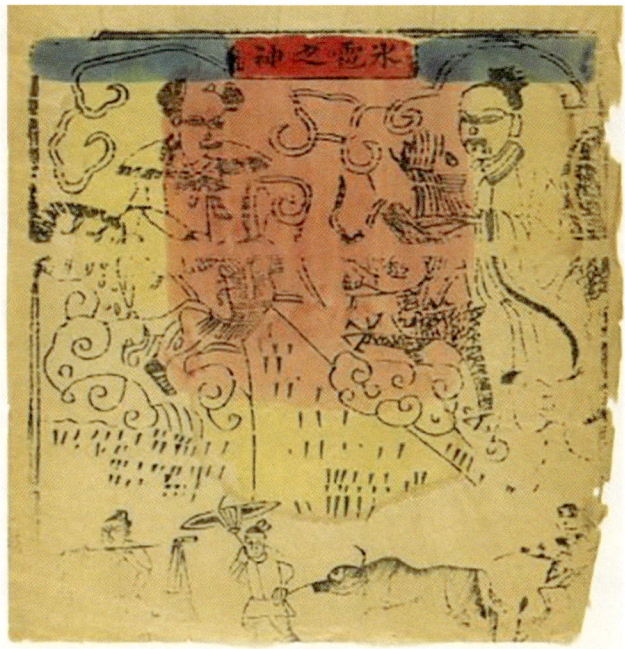

FIGURE A4.63. *God of Hail*, 19th-century woodblock print on paper, Beijing, China (by courtesy of *Royal Ontario Museum, Toronto*)

> '*Who was it put so great a scorn*
> *In the grey reeds that night and morn*
> *Are trodden and broken by the herds,*
> *And in the light bodies of birds*
> *That north wind tumbles to and fro*
> *And pinches among hail and snow?*'

first published at the turn of the 20th century [A4.255].

An empathy with the natural countryside continues in the poem *Ryton Firs*, written around 1919 by the British scholar Lascelles Abercrombie (1881–1938), who recalls the splendid woods near his former homestead in the English hamlet of Ryton, near Dymock, Gloucestershire, where trees had been felled and used as pit-props in Welsh mines:

> '*And only yesterday it was I saw*
> *Veil'd in streamers of grey wavering smoke*
> *My shapely Malvern Hills.*
> *That was the last hailstorm to trouble spring:*
> *He came in gloomy haste,*
> *Pusht in front of the white clouds quietly basking,*
> *In such a hurry he tript against the hills*
> *And stumbling forward spilt over his shoulders*

All his black baggage held,
Streaking downpour of hail.
Then fled dismayed, and the sun in golden glee
And the high white clouds laught down his dusky ghost.'

from a contribution [A4.256] to the last of five anthologies (*Georgian Poetry*) edited by the British civil servant Edward Marsh (1872–1953).

Hailstorms generated only a little further north in Derbyshire are mentioned by the English writer David Lawrence (1885–1930), better known as D. H. Lawrence, in the novel *Kangaroo*, first published [A4.257] in 1923 and set in Australia, following the author's three-month visit to the Sydney region in 1922. But chapter 12 (*The Nightmare*) includes the experiences of one of the principal characters (*Richard Lovat Somers*) and his wife in wartime England:

'They had been living in this remote cottage in the Derbyshire hills: and they must leave at half-past seven in the morning, to complete their journey in a day. It was a black morning, with a slow dawn. Somers had the trunks ready. He stood looking at the dark gulf of the valley below. Meanwhile heavy clouds sank over the bare, Derbyshire hills, and the dawn was blotted out before it came. Then broke a terrific thunderstorm, and hail lashed down with a noise like insanity. He stood at the big window over the valley, and watched. Come hail, come rain, he would go: forever.'

Much further afield in the countryside east of Naples, from the final chapter (*Suspense*) of Lawrence's earlier novel *The Lost Girl* [A4.258], the title character *Alvina* becomes reflective in springtime:

'Ciccio and Pancrazio were busy with the vines. As she watched them hoeing, crouching, tying, tending, grafting, mindless and utterly absorbed, hour after hour, day after day, thinking vines, living vines, she wondered they didn't begin to sprout vine-buds and vine stems from their own elbows and neck-joints. There was something to her unnatural in the quality of the attention the men gave to the wine. It was a sort of worship, almost a degradation again. And heaven knows, Pancrazio's wine was poor enough, his grapes almost invariably bruised with hailstones, and half-rotten instead of ripe.'

in keeping with the recognised susceptibility of Italian vineyards to hailstorm damage. Moreover, a passing reference to weather science occurs in *Fantasia of the Unconscious*, the second of a pair of Lawrentian works on aspects of psychoanalysis [A4.259]:

'But wait. Existence is truly a matter of propagation between the two infinites. But it needs a third presence. Sun-principle and moon-principle, embracing through the aeons, could never by themselves propagate one molecule of matter. The hailstone needs a grain of dust for its core. So does the universe. Midway between the two cosmic infinites lies the third, which is more than infinite. This is the Holy Ghost Life, individual life.'

being a paragraph from chapter 13 (*Cosmological*).

The Scottish geologist and writer Archibald Geikie (1835–1924) also has remarked [A4.260] on the prevalence of hailstorms during Italian summers throughout recorded history:

> *'In respect to such sudden and destructive storms, when vegetation which has been parched in a torrid heat is the next hour beaten down by lumps of ice, the present time fares no better than did the Augustan age. Thus on 22nd July 1911, when the thermometer had reached 93° Fahr. in the shade, a heavy hail-storm and furious wind fell on Rome and the surrounding country. Part of the zinc roof of the railway station was torn off, the column of Victory in the grounds of the Exhibition fell, and the wall of the Sardinian pavilion was much injured. When the storm passed away, Monte Gennaro, above Tivoli, the highest summit of the Sabine Hills and so conspicuous a landmark from Rome, presented an extraordinary appearance, being covered with hail and as white as in mid-winter.'*

while more general aspects of hail damage are discussed by Talman [A4.261].

A particular series of hailstorms in the Middle East is mentioned by the British soldier and diplomat Thomas Lawrence (1888–1935), usually known as T. E. Lawrence, in the book *Seven Pillars of Wisdom* [A4.262], first printed privately in 1926, describing his experiences while serving as a liaison officer with rebel forces during the Arab Revolt against the Ottoman Turks (1916–1918). In book 7 (*The Dead Sea Campaign*), following the capture of Jerusalem, progress towards the next objective started well, but upon reaching the Red Sea, their advance over high terrain early in 1918 was seriously blunted by bad weather:

> *'The air seemed cold enough to freeze anything, but did not: the wind, which had changed during the night, swept into us from the west in hindering blizzards. Our cloaks bellied out and dragged like sails, against us. At last we skinned them off, and went easier, our bare shirts wrapped tightly about us to restrain their slapping tails. The whirling direction of the squalls was shown to our eyes by the white mist they carried across hill and dale. Our hands were numbed into insensibility, so that we knew the cut on them only by red stains in their plastered mud: but our bodies were not so chill, and for hours quivered under the hailstones of each storm. We twisted ourselves to get the sharpness on an unhurt side, and held our shirts free from the skin, to shield us momentarily.'*

Despite this and other setbacks, Lawrence eventually reached the next staging post alone and exhausted, not least by virtue of his own fortitude and that of his sturdy camel.

The short poem *To the Song* (*An den Gesang*) was written in 1932 by the German-Israeli scholar Werner Kraft (1896–1991), shortly before fleeing his native land to avoid further persecution. In a distinctly Kafkaesque mode [A4.263], serving as a heartfelt lamentation and mourning all that is lost in a devastated and desolate world:

> *'Lost and forgotten song! / If you'd only return! / If you'd only stay long! // Peaceful and blessed songs! / Too grievous the world. / Hailstorm laid waste the land. // Everything declines. / I fearfully listen / On fields of lament // To the chant of longing / That rings in my ears. / Oh that for a long time // The Horae would grant me / The dreary remnant / Of the sacred choir // At a mournful feast!'*

being a modern translation from the original German by Caroline Jessen [A4.264].

Figurative sketches of hailstorms by the Scottish writer and politician John Buchan (1875–1940) occur in book 3 of *The Island of Sheep*, the last novel [A4.265] in the *Richard Hannay* series of buccaneering adventures, set partly among the remote so-called *Norland Isles* in the North Atlantic Ocean. Whence:

> '*Before he reached the Bird Marsh the weather had changed with a vengeance. The purple cloud had crossed the Channel from Halder, and the afternoon had grown as dark as a winter's gloaming. There was no lightning, but the gloom suddenly burst in a tornado of hail. So violent was the fall that the boy was beaten to the ground, where he lay with his back humped, protecting every inch of exposed skin from that blistering bastinado.*'

from chapter 14 (*The ways of the Pink-Foot*), and

> '*I did not see her anchor and lower her boats. For she was no sooner off the mouth of the voe than the gloom which had been brooding over the Channel burst in the father and mother of a storm. I would have been beaten off my perch if I had not found some shelter from the chimney stack. In a minute or two the grass of the roof was white with hailstones the size of a sparrow's eggs. The garden, the terrace, the hillside looked deep in snow. And with the hail came a wind that cut like a knife. It must have been the better part of half an hour before the tornado passed, and I could look seaward at anything but a blinding scurry.*'

from chapter 15 (*Transformation by Fire*).

One of several allusions to hail by the Welsh poet Dylan Thomas (1914–1953) occurs in *Before I Knocked*, an early example of his stylistic composition [A4.266] written as a newspaper competition entry in 1933 (contest sponsored by the *Sunday Referee*). The third stanza is:

> '*I knew the message of the winter,*
> *The darted hail, the childish snow,*
> *And the wind was my sister suitor;*
> *Wind in me leaped, the hellborn dew;*
> *My veins flowed with the Eastern weather;*
> *Ungotten I knew night and day.*'

A further example of his impassioned verse is from the longer poem *In Country Sleep*, written in 1947 and first published in the monthly magazine *Horizon* [A4.267], which has the stanza:

> '*As the rain falls, hail on the fleece, as the vale mist rides*
> *Through the haygold stalls, as the dew falls on the wind-*
> *Milled dust of the apple tree and the pounded islands*
> *Of the morning leaves, as the star falls, as the winged*
> *Apple seed glides,*
> *And falls, and flowers in the yawning wound at our sides,*
> *As the world falls, silent as the cyclone of silence.*'

while noting that both poems are included in a collection [A4.268] selected by the author and published in 1952, shortly before his premature demise.

On a much lighter note from comedic literature, a suitably far-fetched and amusing commentary on the formation and effects of hailstorms is given in the children's novel *James and the Giant Peach* by Roald Dahl (1916–1990), first published [A4.269] in 1961. With *James* and his companions (giant insects) floating high in the sky aboard the magical *Giant Peach*, an extraordinary scenario came to pass:

> 'Once, as they drifted silently past a massive white cloud, they saw on the top of it a group of strange, tall, wispy-looking things that were about twice the height of ordinary men. They were not easy to see at first because they were almost as white as the cloud itself, but as the peach sailed closer, it became obvious that these 'things' were actually living creatures – tall, wispy, wraithlike, shadowy, white creatures who looked as though they were made out of a mixture of cotton-wool and candyfloss and thin white hairs.'

Amid the ensuing trepidation, *James* announced with impressive insouciance that these mysterious sightings were of demonic *Cloud-Men*, who were busily occupied in their primary task:

> 'The Cloud-Men were all standing in a group, and they were doing something peculiar with their hands. First, they would reach out (all of them at once) and grab handfuls of cloud. Then they would roll these handfuls of cloud in their fingers until they turned into what looked like large white marbles. Then they would toss the marbles to one side and quickly grab more bits of cloud and start over again.'

And after some further querulous remarks by the fearful passengers:

> 'Then the watchers on the peach saw one of the Cloud-Men raising his long wispy arms above his head and they heard him shouting, 'All right, boys! That's enough! Get the shovels!' And all the other Cloud-Men immediately let out a strange high-pitched whoop of joy and started jumping up and down and waving their arms in the air. Then they picked up enormous shovels and rushed over to the pile of marbles and began shovelling them as fast as they could over the side of the cloud, into space. 'Down they go!' they chanted as they worked.
>
> 'Down they go!
>
> Hail and snow!
>
> Freezes and sneezes and noses will blow!'
>
> 'It's hailstones!' whispered James excitedly. 'They've been making hailstones and now they are showering them down onto the people in the world below!'
>
> 'Hailstones?' the Centipede said. 'That's ridiculous! This is summertime. You don't have hailstones in summertime.'
>
> 'They are practicing for the winter,' James told him.
>
> 'I don't believe it!' shouted the Centipede, raising his voice.'

Here the *Centipede* had made what appeared to be a reasonable assertion, despite being then admonished by *James* and slated as a 'loathsome pest' by the quarrelsome

Earthworm, as hailstorms can indeed occur in summertime. Unfortunately, the raucous *Centipede* attracted the attention of the *Cloud-Men*, who immediately launched a ferocious attack. The hailstones smashed into the *Giant Peach*, broke *Glow-worm's* light, hit *Ladybird* on her shell and *Centipede* on the nose; yet the barrage eventually subsided and the plucky travellers survived to face another day. Amid such magic and enchantment, Figure A4.64 offers an engaging depiction of the furtive activities of the *Cloud-Men* painted by Quentin Blake, included in a later illustrated version of the book [A4.270].

Patently ensconced in the modern era, the short poem *For a space prober* by the American scholar Thomas Bergin (1904–1987) is reported to be the first literary work [A4.271] to leave the confines of Earth, being inscribed upon an instrumentation panel within the U.S. Navy's

FIGURE A4.64. Hailstorm generation by *Cloud-Men* in Roald Dahl's *James and the Giant Peach*, illustrated by Quentin Blake in 1995 (by courtesy of *Quentin Blake, London*)

Transit Research and Attitude Control (TRAAC) satellite launched in November 1961. The first half of the poem is:

> '*From Time's obscure beginning, the Olympians*
> *Have, moved by pity, anger, sometimes mirth,*
> *Poured an abundant store of missiles down*
> *On the resigned, defenceless sons of Earth.*
> *Hailstones and chiding thunderclaps of Jove,*
> *Remote directives from the constellations:*
> *Aye, the celestials have swooped down themselves,*
> *Grim bent on miracles or incarnations.*'

The satellite continues to orbit the earth, and is expected to do so for several hundred years. In this connection, based on scientific calculations explained by Graves *et al.* [A4.272], future space travellers landing on Titan – the largest moon of Saturn – could well be subjected to solid methane hail, although any such precipitation might be the least of their concerns as the surface temperature of this natural satellite is around $-180\ °C$.

Progressing even further along the path from the spiritual to the secular, there is little doubt that extreme hailstorms have caused considerable fatalities amongst the population of various lands, even allowing for inevitable inaccuracies and embellishments. Comprehensive documentary evidence summarised by Cerveny *et al.* [A4.273] indicates that the highest human mortality substantially verified in comparatively recent times was 246 killed during a hailstorm near Moradabad in Uttar Pradesh, India, in April, 1888, coupled with the demise of some 1600 head of cattle, sheep and goats; apparently the hailstones were '*as large as goose eggs and oranges and cricket balls*'; the latter analogy revealing a somewhat tenuous relationship with sporting activity in the bygone days of the British Raj. A comparable loss of life was recorded in Nanking, China, in 1932, where some 200 people were killed and thousands injured during a hailstorm that struck Honan Province. More recently, a storm in Gopalganj, Bangladesh, in April 1986 is reported to have killed 92 people, with hailstones weighing up to 1 kg. Singh *et al.* [A4.274] have used remote sensing data to assess crop damage in the major wheat-growing areas of India, while Bal *et al.* [A4.275] have examined various post-hail measures to aid crop recovery in the Deccan Plateau region.

A4.8. Hail prevention measures since antiquity

Efforts have been made by mankind over the ages to circumvent or alleviate the damage wrought by hailstorms, ranging from conjuration and appeals to various deities through to modern-day suppression techniques. Examples of the former are given in a seven-book work *Natural questions* (*Quaestiones naturales*) by the Roman Stoic philosopher and statesman Lucius Annaeus Seneca (*circa* 4 B.C.–65 A.D.), usually known as Seneca the Younger, who discussed the subject of hail at some length. Whence in book 4 (*Containing a discussion of snow, hail, and rain*), having covered various scientific aspects, Seneca is drawn to consider the magical arts:

> '*I cannot refrain from trotting out all the silly fancies of our Stoic friends. The assertion in question is that there are some people skilled in observing the clouds who foretell when a hail shower is coming on. They gather this just from experience by marking the colour of the clouds and noting which was on previous occasions followed by hail. It seems incredible that at Cleonae there were hail-guards appointed by the state to look out for the approach of hail. When they had given the signal that the hail was close at hand, what do you think? That people ran off to get their overcoats or cloaks? Nay, they each offered a sacrifice as fast as they could, one a lamb, another a chicken.*'

from a translation [A4.276] by the British academic John Clarke (1853–1939). This great work was completed shortly before Seneca was ordered to commit suicide by the despotic emperor Nero, to whom he had once acted as tutor and advisor; by any measure, a lamentable recompense for services rendered unto Caesar.

It is also recalled in a book [A4.277] by the English classical scholar Arthur Bernard Cook (1868–1952) that it was Zeus the weather god who sent both rain and hail, and that within the ancient kingdom of Phrygia in Anatolia he was worshipped as the '*God of Hail who gives Deliverance*'. By way of illustration, the white marble Greek stele (62 × 82 cm) depicted in Figure A4.65 and found near the coastal town of Panderma (now Bandirma in north-western Turkey, on the Sea of Marmara), has an inscribed relief possibly dating from the 1st century B.C. Figure A4.65(a) is a photograph from a paper [A4.278] by the English archaeologist Frederick William Hasluck (1878–1920), while the drawing in Figure A4.65(b) includes a transcription of the barely decipherable Greek text. According to Cook [A4.277]:

> '*A sunk panel between pilasters shows Zeus, in chitón and himátion, standing with a phiále in his right hand, a long sceptre in his left. Beside him is his eagle. Beyond it, a small altar decorated with a bull sinking on its knees and held by a young attendant. A draped worshipper approaches the altar from the left. The background is occupied by a sacred tree, presumably an oak. On the architrave above the pilasters is inscribed:*
>
> > *Zeus Chalázios Sózon*
> > *In the time of Dionysios –*
>
> *Then below the relief the inscription runs on:*
>
> > *the Thrakiokometai consecrated this stéle to the god*
> > *to secure good crops and the safety of their fruits*
> > *and the health and preservation of the land-lessees and*
> > *those who repair to the god and reside in*
> > *Thrakia Kome.*
> > *Meidias, son of Straton, as first mayor handed over the stéle*
> > *to the god and to the villagers at his own charges*
> > *as a free-will offering.*
>
> *It will be noticed that, in the matter of hail, Greek religion like Greek magic was throughout concerned to avoid damage, not to cause it. Things were otherwise with the vindictive witchcraft of the Middle Ages.*'

Further textural interpretations of the original Greek inscription are given by Hasluck [A4.278], whilst many other historical examples of weather lore featuring hail are cited in

(a) (b)

FIGURE A4.65. White marble Greek stele (62 × 82 cm), *circa* 1st century B.C., found near Panderma (north-western Turkey): (a) photograph; (b) drawing with transcribed text

both the aforementioned tome [A4.277], and in the commentaries by Bellucci [A4.279], Fehrle [A4.280], Stegemann [A4.281] and Rodgers [A4.282].

Hail is mentioned on several occasions in the encyclopaedic *Natural History* (*Naturalis Historia*) by the Roman author, military commander and natural philosopher Gaius Plinius Secundus (*circa* 23–79 A.D.) – usually called Pliny the Elder – which massive work (comprising 37 books) he completed shortly before perishing during the eruption of Mount Vesuvius while attempting to rescue others. He observed, for example, that:

> '*Hail is produced from frozen rain and snow from the same fluid less solidly condensed, but hoar frost from cold dew; that snow falls during winter but not hail, and hail itself falls more often in the daytime than at night, and melts much faster than snow.*'

from a translation [A4.283] by the English classical scholar Harris Rackham (1868–1944). However, he appears to have been noticeably circumspect on the subject of magical charms:

> '*There are in existence, also, certain charms against hailstorms, diseases of various kinds, and burns, some of which have been proved, by actual experience, to be effectual; but so great is the diversity of opinion upon them, that I am precluded by a feeling of extreme diffidence from entering into further particulars, and must therefore leave each to form his own conclusions as he may feel inclined.*'

from a translation by Bostock and Riley [A4.284]. Indeed, when his nephew Gaius Plinius Caecilius Secundus (*circa* 61–*circa* 113 A.D.) – better known as Pliny the Younger – experienced crop damage caused by severe hail on his farmland in Tuscany, as recorded in book 4 (*Letter to Julius Naso*) of the *Epistulae* [A4.285], [A4.286], there is no mention of sorcery.

However, a contentious religious backdrop within the farming community of this geographical region appears to have lasted until comparatively recent times. Whence, for example, the aftermath of a hailstorm in the Italian village of Campia in Tuscany was recalled by Cyriax [A4.287] in all its stark reality:

> '*I saw the devastated terraces of Campia and walked down the road past the fontana to San Lorenza. Everywhere the same miserable spectacle. Half-naked vines with withered leaves scattered round about, grapes on the ground, often whole bunches of them. The edges of the torn and perforated leaves, still left on the plants, were turning brown and every grape that had been hit by a hailstone showed its bruise. The maize plants were broken, and the long leaves hung in ribbons, as if they had been combed. All the plants had the most bedraggled appearance. In places the road was thick with olive leaves and little black olives.*'

while the local inhabitants questioned why God had sent such storms, despite the Madonna having been carried through the streets, thereupon blaming the church, the saints and the priests. Further examples of ancient defensive measures taken against hail throughout Italy are summarised by Malossini [A4.288].

According to Kotansky [A4.289], the bronze amulet or phylactery (80 × 134 mm) shown in Figure A4.66(a) dates from ancient Greece and was deployed by its owner to protect crops from hail and snow. It was found (before 1873) at Bouchet near Avignon in southern France, and may have been affixed to a stone or stake in a field or vineyard, since the centre of the plate is pierced. The English text of the magical incantation is:

> '*Thôsouderkyô vineyard oumixonthei, divert from this property all hail and all snow, and whatever might injure the land. The god, Ôamouoa, orders it, and you Abrasax, assist!*'

This apotropaic artifact is broadly similar to engraved Christian crosses from a later period found in the provinces of Gaul and North Africa. One such example described by Audollent [A4.290] takes the form of a lead cross (340 × 350 mm) from the Roman era found in Ain-Fourna, Tunisia; see Figure A4.66(b), although the Latin text of the inscribed prayer against hail is barely decipherable in this image.

Nieto [A4.291] has given a detailed interpretation of the Latin text inscribed upon a slate amulet from the western part of the Roman Empire (Hispania), dating from the 8th century. The Visigothic artifact was found near Carrió, within the province of Asturias in northern Spain, and paraphrasing his introductory remarks:

> '*This is clearly a charm devised to protect agricultural land against hail, one of the commonest severe weather phenomena in the Mediterranean area, and perhaps the one most dreaded by those dependent for their livelihoods on agrarian production. That it is directed against hail is*

(a)

(b)

FIGURE A4.66. Ancient hailstorm phylacteries: (a) Greek bronze amulet (80 × 134 mm) from Avignon, France (by courtesy of *Westdeutscher Verlag, Opladen*); (b) Roman lead cross (340 × 350 mm) from Ain-Fourna, Tunisia (by courtesy of *Institut national de France*)

> clear from the words *reuersus est grando in pluuia in alia parte monte cimeteri* (and the hail turned into rain on the other part of the cemetery mount). The text is thus an example of a genre of charms known as φυλακτήρια, κωλυτήρια or κωλύματα, designed to protect their owner from a potential threat.'

Further related studies include those by Grégoire [A4.292], Delatte [A4.293], Manganaro [A4.294], Ferchiou and Gabillon [A4.295] and Faraone [A4.296].

Examples of weather lore were outlined by the Greek traveller and geographer *Pausanias* (*circa* 110–180 A.D.) in his *Description of Greece*, where an emphasis on the mystical nature of hailstorms is clear from the outset. In volume 1, book 2 (*Corinth*), from an English language translation [A4.297] entitled *Pausanias's description of Greece* by the Scottish anthropologist James Frazer (1854–1941), Pausanias recalls having '*seen folk before now trying to keep off hail by sacrifices and spells*' whilst visiting the town of Methana on the Peloponnese peninsula. Further such tales are recorded in an extensive commentary in volume 3. Whence a paraphrased account originating from the aforementioned Roman sage Seneca relates to the well-fortified citadel at Cleonae, located on high ground between the cities of Corinth and Argos:

> 'Seneca describes a curious custom which prevailed at Cleonae. Watchmen were maintained at the public expense to look out for hailstorms. When these watchmen saw a hail-cloud approaching they made a signal, where upon the farmers turned out and sacrificed lambs or fowls. It was thought that when the clouds had tasted the blood, they would turn aside and go somewhere else. People who were too poor to offer a lamb or a fowl pricked their fingers and offered their own blood to the clouds to induce them to go away. If the vines and crops suffered from a hailstorm, the watchmen were brought before the magistrates and punished for neglect of duty.'

And in a second quotation:

> 'Another Greek way of keeping hail from the vines was to tie a strap round one of the vines; it was supposed that this would save all the rest (Philostratus). Many other equally absurd modes of averting a hailstorm (by brandishing bloody axes in a threatening manner at the sky, holding up a mirror to the clouds, rubbing the pruning knives with bear's grease, etc.) are gravely recorded by ancient writers (Palladius, Pliny).'

It is also recalled from biblical history that one of the specific responsibilities of Barnabas, Patron Saint of Cyprus (*See of Antioch*), was to protect against hailstorms.

Other primitive measures taken to ward off hail during past centuries have included shooting arrows and later cannon balls or gunshot into advancing storm clouds, and the ringing of church bells; the results being generally unknown or inconclusive, except for tangible collateral damage that included the occasional serious injury or fatality within the local peasantry, and bell ringers being killed by lightning. The illustration in Figure A4.67 is purported to represent Nordic archers shooting arrows towards a thunderous sky, from a book [A4.298] written in Latin by the Swedish author Olaus Magnus (1490–1557), first published in 1555.

FIGURE A4.67. Illustration from book by Olaus Magnus published in 1555; Nordic archers shooting arrows towards thunderous sky

Numerous historical references to hail within the context of Christian theology are given in a comprehensive tome [A4.299] by the American historian and politician Andrew Dickson White (1832–1918). In chapter 11 (*From 'The Prince of the Power of the air' to meteorology*) he states:

> '*The first and most natural means taken against this work of Satan in the air was prayer; and various petitions are to be found scattered through the Christian liturgies – some very beautiful and touching. This means of escape has been relied upon, with greater or less faith, from those days to these. Various medieval saints and reformers, and devoted men in all centuries, from St Giles to John Wesley, have used it with results claimed to be miraculous. Whatever theory any thinking man may hold in the matter, he will certainly not venture a reproachful word: such prayers have been in all ages a natural outcome of the mind of man in trouble.*'

while other means included those of processions, exorcism and specific liturgical prayer (*Agnus Dei*). It is further stated, for example, that in Protestant Swabia during the 17th century:

> '*Pastor Georg Nuber issued a volume of 'weather-sermons', in which he discusses nearly every sort of elemental disturbances—storms, floods, droughts, lightning, and hail. These, he says, come direct from God for human sins, yet no doubt with discrimination, for there are five sins which God especially punishes with lightning and hail—namely, impenitence, incredulity, neglect of the repair of churches, fraud in the payment of tithes to the clergy, and oppression of subordinates, each of which points he supports with a mass of scriptural texts.*'

Even during comparatively recent times in rural France, it was reckoned that:

> 'Against storms St Barbara is very generally considered the most powerful protectress; but, in the French diocese of Limoges, Notre Dame de Crocq has proved a most powerful rival, for when, a few years since, all the neighbouring parishes were ravaged by storms, not a hailstone fell in the canton which she protected. In the diocese of Tarbes, St Exupere is especially invoked against hail, peasants flocking from all the surrounding country to his shrine.'

However, it was the ringing of church bells that became the most widely used means of '*baffling the powers of the air*'. The ancient tradition of ringing bells to ward off hailstorms lasted several centuries, and sometimes was formally endorsed by the ecclesiastical authorities. According to Oddie [A4.300], for example, Pope Urban VIII (1568–1644) authorised the following prayer for use by bishops when consecrating church bells:

> 'Grant O Lord, that the sound of this bell may drive away the harmful storms, hail and strong winds, and that the evil spirits that dwell in the air may by Thy Almighty power be struck to the ground.'

While discussing the meteorological use of bell ringing in Slovenia, Kovačič [A4.301] remarks:

> 'People believed that demonic forces caused storms, lightning, and hail, and they tried to drive them away by ringing bells. The fact that the custom was already practiced in medieval times is testified to by the frequent inscription on medieval bells 'Fulgura frango' (I break the lightning). Today this type of inscription is still one of the most frequent on bells. The custom of ringing bells during a storm started to be seriously questioned in the 18th century, and in some places the practice was forbidden and people were advised to seek shelter instead. In 1786 the Parliament of Paris signed an order that forbade bell ringing during storms because of the many victims among bell ringers.'

However, the current Roman Ritual (*Rituale Romanum*) of the Catholic Church [A4.302] for the sacred blessing of a church bell includes the passage:

> 'At its sound let all evil spirits be driven afar; let thunder and lightning, hail and storm be banished; let the power of your hand put down the evil powers of the air, causing them to tremble at the sound of this bell, and to flee at the sight of the holy cross engraved thereon.'

In the aforementioned treks around Tibet described by the Japanese monk Ekai Kawaguchi [A4.254], it is stated in chapter 43 (*Manners and customs*):

> 'I will now relate the strange method which the Tibetans have for keeping off hailstones, which they dread exceedingly, especially in summer, for then the crops of wheat and barley, which they can reap only once in a year or two, may be entirely destroyed. So they naturally try to find some means to keep off the hailstones, and the method they have discovered is certainly curious enough.'

Every village had at least one priest (*Ngak-pa*), and following various religious ceremonies and customs, the energetic priest moved valiantly into action:

> '*As early as March or April the ploughing of the fields and sowing of wheat begins, and then the Ngak-pa proceeds to the Hail-Subduing Temple, erected on the top of one of the high mountains. This kind of temple is always built on the most elevated place in the whole district, for the reason that the greatest advantage is thus obtained for ascertaining the direction from which the clouds containing hail issue forth.*'

> '*When it happens that big masses of clouds are gathering overhead, the Ngak-pa first assumes a solemn and stern aspect, drawing himself up on the brink of the precipice as firm as the rock itself, and then pronounces an enchantment with many flourishes of his rosary much in the same manner as our warrior of old did with his baton. In a wild attempt to drive away the hail clouds, he fights against the mountain, but it often happens that the overwhelming host comes gloomily upon him with thunders roaring and flashes of lightning that seem to shake the ground under him and rend the sky above, and the volleys of big hailstones follow, pouring down thick and fast, like arrows flying in the thick of battle. The priest then, all in a frenzy, dances in fight against the air, displaying a fury quite like a madman in a rage.*'

If the outcome of such rituals was beneficial, the priest received a 'hail-prevention tax', substantially enhancing his annual salary; while an unsuccessful outcome could lead to dire consequences for the unfortunate holy man. The original Japanese illustration [A4.254] in Figure A4.68 depicts a Tibetan priest fighting with hail, characteristic of this forceful and mildly disturbing story.

From the commentary by Frazer [A4.297], other customs based upon other sources were followed in other lands:

> '*Among the Aztecs of Mexico there were sorcerers who by their spells endeavoured to charm away hail from the maize and divert it to wastelands. There are villages in India at the present day in which a professional charmer is kept for the sole purpose of repeating incantations to drive away the hail from the growing crops. Roman and Greek writers record the belief that if you carry the skin of a hyaena, a crocodile, or a seal round your land, and then hang up the skin over the door of the house, no hail will fall on your land. It is said to be an Austrian custom, on the approach of a hailstorm, to bury an egg at each of the four corners of the field. When the people of Car Nicobar (the most northerly of the Nicobar Islands) see signs of an approaching storm, the people of every village march round their own boundaries, and fix up at different distances small sticks split at the top, into which split they put a piece of cocoa-nut, a wisp of tobacco, and the leaf of a certain plant. Among the Esthonians it used to be customary for a farmer to go to his fields on the day of the Annunciation and let fall three drops of blood from the ring finger of his left hand at each of the four corners of all his fields; this was to make the crops thrive.*'

The eventual outcome of these rituals and incantations is unknown, as is the ultimate fate of the quixotic petitioners.

It is also noted that in the Aztec religion, the rain god *Tlāloc* was at once revered as the beneficent giver of life and sustenance, yet also feared for his ability to unleash hail, thunder and lightning upon his earthly subjects. Other examples of a similar complexion are to be found in a wide range of historical literature. According to Andrews [A4.303], for instance:

FIGURE A4.68. Anonymous Japanese illustration depicting Tibetan priest fighting with hail (by courtesy of *Theosophical Publishing Society, Benares and London*)

> 'The Maori had many gods of hail, including Au-Whatu, the god of rain that turned into hail. Vritra, the drought demon from India also caused hail, and so did the storm deity Mundur-Tengri in Siberia and a malevolent spirit who lived on an island in Lake Titicaca in Peru. The Quechua of Peru prayed to their hail god, Santiago, for mercy and protection, and the Chinese prayed to Hu-shen, who was one of several deities who could avert the hail and protect the fields.'

Numerous other customs and festivals centred on hailstorms are described in a separate monumental work (*The Golden Bough*) by Frazer [A4.304], some of which are paraphrased below:

> 'In neighbouring villages of Hesse, between the Rhön and the Vogel Mountains, it is thought that wherever the burning wheels roll, the fields will be safe from hail and storm.'

> '*In many parts of Germany a bonfire is also kindled, by means of the new fire, on some open space near the church. It is consecrated, and the people bring sticks of oak, walnut, and beech, which they char in the fire, and then take home with them. Some of these charred sticks are thereupon burned at home in a newly-kindled fire, with a prayer that God will preserve the homestead from fire, lightning, and hail. Others are placed in the fields, gardens, and meadows, with a prayer that God will keep them from blight and hail.*'

> '*To this day the ritual of bringing in the Yule log is observed with much solemnity among the Southern Slavs, especially the Serbians. The log is usually a block of oak, but sometimes of olive or beech. Some people carry a piece of the log out to the fields to protect them against hail.*'

> '*Again, the bonfires are often supposed to protect the fields against hail and the homestead against thunder and lightning. But both hail and thunderstorms are frequently thought to be caused by witches; hence the fire which bans the witches necessarily serves at the same time as a talisman against hail, thunder, and lightning.*'

In a supplementary narrative, Frazer [A4.305] also writes of the lands occupied by the Kikuyu tribe in Kenya, where some of the hills were crowned with sacred groves, which served as a place of sacrifice to remedy the perceived evil of famine or drought. Elected priests ascended the chosen hill with a sheep, which was slaughtered and roasted upon a fire, then partially eaten by tribal elders, the residue being cast into the fire for consumption by God or Ngai.

> '*It is said that no sooner was this sacrifice completed than thunder rolled up and hail rolled down with such force that the old men had to wrap their garments around their heads and run for their huts.*'

Many of the more bizarre and colourful descriptions of such witchcraft and wizardry are omitted herein, but despite some extravagant rhetoric, they doubtless achieved multiple levels of ethereal resonance with the local inhabitants.

The sorcerer's horn illustrated in Figure A4.69 is from Tibet, and typically was used by itinerant magicians in the 19th century to avert any hailstorms that might destroy crops. The horn would have contained seeds or small pebbles which, during a ritual, were magically transformed into weapons, to be hurled at the hail-creating demons to frighten them away. The bone horn is carved on its exterior surface with conches, two scorpions, a stupa with syllables engraved upon it, and the astrological symbol held by a tortoise consisting of the nine planets at its centre, with eleven trigrams further encircled by trigrams. The narrow end is carved in the form of a makara. In a similar vein, the photograph in Figure A4.70 – taken at the Trappist Mission, Mariann Hill, Natal, South Africa in the 19th century – shows a Zulu medicine-man or shaman performing a ritual to fend off a hailstorm.

The widespread deployment of specialised cannon as a hail prevention measure in several areas of Europe continued until comparatively recent times, followed shortly thereafter by primitive rockets to initiate explosions within the storm clouds; see, for instance, Leschevin [A4.306], Anon. [A4.307], Murray [A4.308], Anon. [A4.309], [A4.310], Rowell [A4.311], Hilgard [A4.312], Sigaux [A4.313], Ward [A4.314], Abbe [A4.315]–[A4.320], Pernter

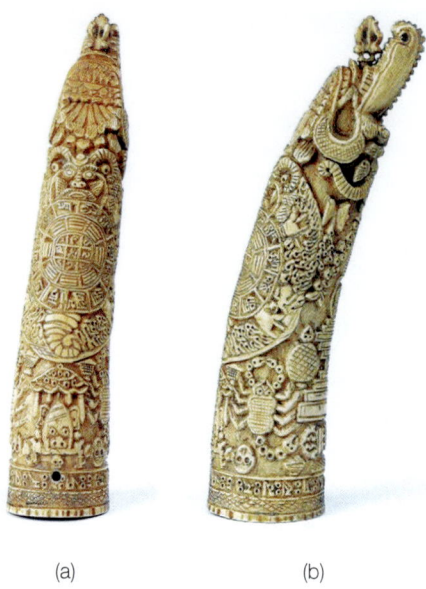

(a)   (b)

FIGURE A4.69. Tibetan sorcerer's horn from 19th century containing magic seeds or pebbles used to fend off hailstorms (by courtesy of *Victoria and Albert Museum, London*, IS 9-1947)

FIGURE A4.70. Zulu medicine-man fighting coming hailstorm, 19th century (photo by *Trappist Mission, Mariann Hill, Natal, South Africa*; by courtesy of *Wellcome Library, London*, Ref. 21325i)

[A4.321], [A4.322], Shaw [A4.323], [A4.324], Plumandon [A4.325]–[A4.327], Anon. [A4.328], Dastre [A4.329], Angot [A4.330], Anon. [A4.331], Ludlam [A4.332], Heninger [A4.333], Roberts [A4.334], Morgan [A4.335], Atals [A4.336], Changnon *et al.* [A4.337], Changnon and Ivens [A4.338], Steiner [A4.339], Wieringa and Holleman [A4.340], Morgan [A4.341] and Baker [A4.342]. Here it was anticipated that the propagation of sonic pressure waves would inhibit the formation of hail, and many thousands of such cannons were in use around the turn of the 20th century. The grainy photograph reproduced in Figure A4.71 illustrates various types of hail cannon on public display in Lyons, France, as part of an Exposition for the *Troisième Congrès international de défense contre la grêle* (*Third International Congress on Hail Shooting*) held there in November 1901.

A spectacular illustration of a discharging hail cannon is portrayed in Figure A4.72(a), set among the vineyards of the Beaujolais region of France in 1900; see the weekly colour supplement [A4.343] accompanying the newspaper *Le Petit Parisien*, including an explanatory covering note (*Le canon contre la grêle: les expériences dans le Beaujolais*). Whilst the wine grower – turned temporary gunner – displays a certain *sang-froid*, retaining both his composure and even his sombrero amid the tumultuous explosion, the same cannot be said of the decorous female spectators; which is hardly surprising in view of their alarmingly close proximity to such a deafening mechanical device, somewhat reminiscent of a smaller version of an ancient siege gun.

An informative short article that appeared in a Sunday colour supplement [A4.344] of the French daily newspaper *Le Petit Journal* in 1901, entitled *L'artillerie contre la grêle*, is reproduced here:

> '*On sait les terribles désastres que cause la grêle, fauchant les moissons, abattant les fruits, ruinant en un mot les contrées sur lesquelles elle s'abat. Depuis longtemps, on essaye de la combattre par un moyen original dont l'idée appartient à Benvenuto Cellini, le célèbre ciseleur du seizième siècle, dont le génie, comme celui de Michel Ange, était universel. Grand*

FIGURE A4.71. Various types of hail cannon on public display in Lyons, France, November 1901, after Plumandon [A4.325]

homme de guerre, une légende veut que ce soit lui qui ait tué d'un coup d'arquebuse le traître connétable de Bourbon, il rêva d'opposer l'artillerie à la grêle. Mais ses expériences ne furent pas trouvées concluantes et l'on y renonça. Ce n'est que dans ces derniers temps, il y a cinq ans environ, que les habitants de Trieste les reprirent pour leur compte. Leurs premiers engins d'abord très primitifs se perfectionnèrent et leur usage se répandit en Italie.

Mais leur application définitive devait être réglée par un de nos compatriotes, M. Guinand, de Denicé-en-Beaujolais près de Villefranche-sur-Saône. Ses canons contre la grêle ressemblent à d'immenses flûtes à champagne, de trois mètres de hauteur, dirigés vers le ciel; au fond se trouve la chambre où l'on met la cartouche de 80 grammes de poudre qu'enflammera un percuteur. Il dispose ses pièces à 500 mètres l'une de l'autre, 44 suffisent à garantir 25 hectares. Quand les nuages s'amoncellent, quand la grêle est dans l'air, sur un signal de lui, ses canonniers, pris parmi d'anciens artilleurs de l'armée, font feu, et le ciel, au lieu d'envoyer la grêle meurtrière, ne répand plus qu'une ondée fine et douce, qui vivifie la récolte au lieu de la détruire. Les essais tentés jusqu'ici ont presque toujours réussi, et si l'on songe que, par hectare, la dépense nécessitée est à peu près de 4 francs par an, on reconnaîtra que M. Guinand et ses collaborateurs ont rendu un bien grand service à l'agriculture.'

The illustration accompanying this article, reproduced in Figure A4.72(b), portrays one of several cannons (vertical conical funnel) set in the wine producing region of Denicé-en-Beaujolais, with the two *canonniers* sportingly attired, including straw boaters.

Similarly, from a later anonymous article [A4.345] entitled *Canonnade pacifique*, Figure A4.72(c) depicts rather more workmanlike horticulturists firing an array of vertical tubular cannons, with the intention of protecting their flower beds and fruit crops at Bagnolet, near Paris, in 1910. Figure A4.72(d), in the form of a postcard (*circa* 1904), shows a hail cannon (vertical conical funnel) located in a vineyard at Saint-Émilion in the *Bordeaux* wine region of France, probably one of many spaced out on a local geographical grid to provide wider coverage.

Very few of these early hail cannons are extant, but Figure A4.73 illustrates one such specimen (inclined conical funnel), currently on display at the Old Castle at Banská Štiavnica in Slovakia. Perhaps surprisingly, the manufacture of specialised hail cannon for horticultural protection continues to the present day, typically incorporating a weather radar that can predict approaching hailstorms. Figure A4.74(a), for instance, illustrates one such installation located in the New Zealand countryside. According to the manufacturer, a charge of acetylene gas is ignited in the lower blast chamber every few seconds before and during the storm to generate high energy shockwaves emitting from the vertical conical funnel, thereby disrupting the formation of hailstone embryos and changing the precipitation from destructive hail to more benign sleet or rain over the protected area.

Nowadays, in conjunction with weather radars, hail damage can also be reduced by chemical thundercloud seeding using shells charged with small crystalizing particles such as silver or lead iodide, which change the cloud microstructure and inhibit the growth of large hailstones. These explosive charges can be launched using modified artillery guns or rockets positioned either on the ground or in an aircraft. An example of an automated ground installation sited in Russia is described by Abshaev [A4.346] and Abshaev et al. [A4.347], and illustrated schematically in Figure A4.74(b). Other methods include the deployment of

HAILSTONE IMPACT: HISTORICAL PROLOGUE

(a)

(b)

(c)

FIGURE A4.72. Firing cannon into clouds to prevent hailstorm damage in French countryside (by courtesy of *Bibliothèque nationale de France*): (a) Beaujolais region, 1900; (b) Denicé-en-Beaujolais, near Villefranche-sur-Saône, 1901; (c) Bagnolet, near Paris, 1910

(d)

FIGURE A4.72. (continued) Firing cannon into clouds to prevent hailstorm damage in French countryside (by courtesy of *Bibliothèque nationale de France*): (d) Saint-Émilion, *circa* 1904

FIGURE A4.73. Vintage hail cannon on display at Old Castle, Banská Štiavnica, Slovakia (by courtesy of photographer *Etan Tal*)

(a)

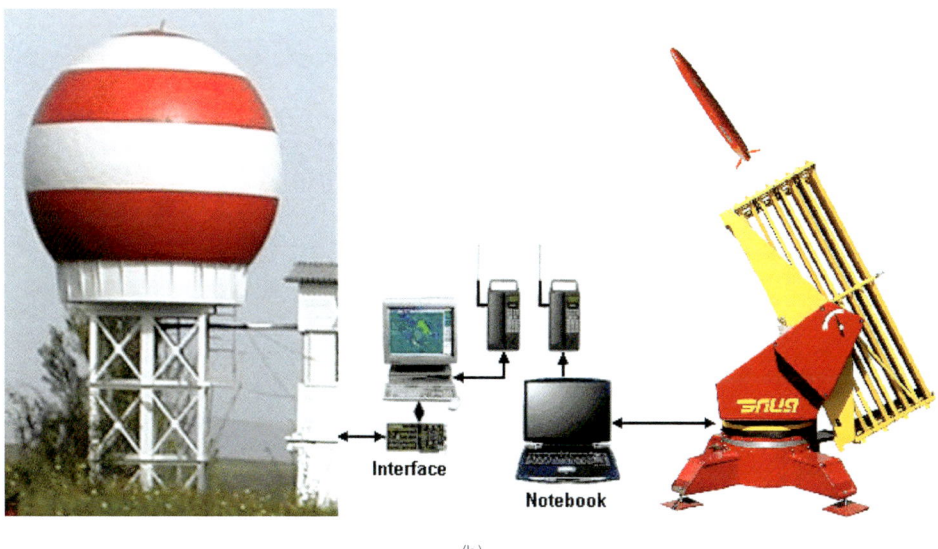

(b)

FIGURE A4.74. Examples of 21st-century techniques of hail suppression: (a) hail cannon in horticultural use in New Zealand (by courtesy of *Eggers Ltd, New Zealand*); (b) schematic arrangement of automated anti-hail rocket complex in Russia incorporating computerised radar signals, after Abshaev *et al*. [A4.347] (by courtesy of *Ministry for Presidential Affairs, Abu Dhabi, U.A.E.*)

ground-based burners to vaporise the seeding solution, and the use of netting to prevent large hailstones from reaching vulnerable crops.

Many studies of hail suppression have been reported over several decades, including the more recent accounts by Favreau and Goyer [A4.348], Schleusener [A4.349], Summers *et al.* [A4.350], Marwitz [A4.351], Battan [A4.352], Browning and Atlas [A4.353], Foote and Knight [A4.354], Paluch [A4.355], Swanson *et al.* [A4.356], Knight *et al.* [A4.357], Federer *et al.* [A4.358], Cheng [A4.359], Zotov *et al.* [A4.360], Mesinger and Mesinger [A4.361], Morgan [A4.362], Simeonov [A4.363], [A4.364], Ćurić *et al.* [A4.365], Dessens [A4.366], Zheng [A4.367], Počakal and Štalec [A4.368], Krauss and Santos [A4.369], Guo *et al.* [A4.370], Makitov [A4.371], Tsagalidis and Georgiou [A4.372], Gavrilov *et al.* [A4.373], Chen *et al.* [A4.374], Yang *et al.* [A4.375], Abshaev *et al.* [A4.376], Javanmard and Pirhayati [A4.377], Najafi *et al.* [A4.378], Bosco *et al.* [A4.379], Dessens *et al.* [A4.380], Tani *et al.* [A4.381], Gandorfer *et al.* [A4.382], Geresdi *et al.* [A4.383], Vujović and Protić [A4.384], Arakelyan [A4.385], [A4.386], Birsan *et al.* [A4.387], Kovačević [A4.388], Bozic [A4.389], Rivera *et al.* [A4.390] and Auf der Maur and Germann [A4.391].

Far more could be written on these fascinating topics, and further literary allusions identified, but to preserve a semblance of brevity in accordance with practical constraints, here endeth this epigrammatic prologue.

Natura nihil frustra facit.

A4.9. References

A4.1.   Mercer, S. A. B. *The Pyramid Texts*. Longmans, Green & Co., London, 1952, 4 vols.

A4.2.   Faulkner, R. O. *The ancient Egyptian Pyramid Texts*. Clarendon Press, Oxford, 1969, 2 vols.

A4.3.   Allen, J. P. *The ancient Egyptian Pyramid Texts*. Society of Biblical Literature, Atlanta, 2005 (ed. P. Der Manuelian), 1st edn; 2015, 2nd edn. (Series: *Writings from the ancient world*, Vols 23 & 38).

A4.4.   Dungen, W. van den. *Ancient Egyptian readings*. Taurus Press, Brasschaat, 2018, 2nd edn.

A4.5.   Piankoff, A. *The Pyramid of Unas*. Princeton Univ. Press, Princeton, 1968, Plate 4. (Bollingen Series 40: *Egyptian religious texts and representations*, Vol. 5).

A4.6.   George, A. R. *The Epic of Gilgamesh: a new translation*. Penguin Books Ltd, London, 2020, 2nd edn.

A4.7.   George, A. R. *The Babylonian Gilgamesh Epic; introduction, critical edition and cuneiform texts*. Oxford Univ. Press, Oxford, 2003, 2 vols.

A4.8.   Bjorkman, J. K. Meteors and meteorites in the ancient Near East. *Meteoritics*, 1973, 8, 2 (Jun.), 91–130.

A4.9.   King, L. W. *Cuneiform texts from Babylonian tablets, etc., in the British Museum*. Trustees of British Museum, London, 1902 (*Dept Egyptian & Assyrian Antiquities*), Pt 15, Plates 15 & 16 (Ref. 29631).

A4.10. Cohen, M. E. *Sumerian hymnology: the eršemma*. Hebrew Union College, Cincinnati, 1981 (*Eršemma of Iškur*, 23.1); Pt 2A, 52–54 (text), Pt 3, 151–152 (commentary). (Series: *Hebrew Union College Annual Suppl.*, 2).

A4.11. ANONYMOUS. *A praise poem of Šulgi (Šulgi A)*. Electronic Text Corpus of Sumerian Literature, Oxford Univ. (Faculty of Oriental Studies), 2006, ETCSL transl: t.2.4.2.01.

A4.12. THE HOLY BIBLE, *Conteyning the Old Testament, and the New: Newly Translated out of the Originall tongues: & with the former Translations diligently compared and revised, by his Majesties speciall Comandement. Appointed to be read in Churches*. Imprinted at London by Robert Barker, Printer to the Kings most Excellent Majestie, 1611.

A4.13. SABAR, S. (Ed.). *The Sarajevo Haggadah: History & Art*. National Museum of Bosnia and Herzegovina, Sarajevo, 2018.

A4.14. SCHEUCHZER, J. J. *Physica sacra*. Johann Andreas Pfeffel, Augustae Vindelicorum & Ulmae (Augsburg & Ulm), 1731, Vol. 1; 1732, Vol. 2; 1733, Vol. 3; 1735, Vol. 4.

A4.15. SCHEUCHZER, J. J. *Kupfer-Bibel: in welcher Die Physica Sacra, Oder Beheiligte Naturwissenschafft Derer in Heil. Schrifft Vorkommenden Natürlichen Sachen, Deutlich Erklärt und Bewährt*. Christian Ulrich Wagner, Augsburg & Ulm, 1731–1735, 5 vols.

A4.16. MORTIER, P. *Histoire des Ouden en Nieuwen Testaments, Verrykt met meer Dan Vierhonderd Printverbeeldingen in Koper Gesneeden*. Pieter Mortier, Amsterdam, 1700.

A4.17. WEIGEL, C. (Ed.). *Historiae celebriores Veteris Testamenti iconibus repraesentatae et ad excitandas bonas meditationes selectis epigrammatibus exornatae*. Christoph Weigel, Nuremberg, 1708.

A4.18. JAMES, M. R. *The apocryphal New Testament: being the Apocryphal Gospels, Acts, Epistles, and Apocalypses, with other narratives and fragments; newly translated by Montague Rhodes James*. Clarendon Press, Oxford, 1924, Ch. *Gospel of Bartholomew*, 166–181.

A4.19. APOCALYPSE (REVELATION). Illuminated French parchment codex, Paris, France, *circa* 1380: anonymous miniature; *First angel sounding trumpet amid hail and fire*. British Library, London (MS Yates Thompson 10, f. 13).

A4.20. SNYDER, J. *Paul as a character in early Christian narratives*. In: *The Oxford Handbook on Pauline studies*. Oxford Univ. Press, Oxford, 2022 (eds Matthew V. Novenson and R. Barry Matlock), Ch. 6.

A4.21. SAINT GREGORY OF NAZIANZEN. *Homilies*. Illuminated Greek parchment codex, Byzantine Empire (possibly Nicaea), *circa* 1220: anonymous large miniature; *St Gregory comforts people of Nazianzus after disastrous hailstorm*. Bodleian Library, Oxford (MS Roe 6, f. 175v).

A4.22. SAINT GREGORY OF NAZIANZEN. *Oration 16: On his father's silence, because of the plague of hail* (transl. C. G. Browne and J. E. Swallow). In: *A select library of Nicene and post-Nicene fathers of the Christian church*. Christian Literature Co., New York, 1894, 2nd Ser., Vol. 7 (eds P. Schaff and H. Wace), 247–254.

A4.23. WANG, P-K. Meteorological records from ancient chronicles of China. *Bull. Amer. Meteorological Soc.*, 1979, 60, 4 (Apr.), 313–318.

A4.24. VALMIKI. *The Rámáyan of Válmíki. Translated into English verse by Ralph T. H. Griffith*. Trübner & Co. Ltd, London; E. J. Lazarus & Co., Benares; 1870–1874, 5 vols.

A4.25. VALMIKI. *The Rámáyan of Válmíki. Translated into English verse by Ralph T. H. Griffith. Complete in one volume*. E. J. Lazarus & Co., Benares, 1895.

A4.26. VYASA. *The Mahabharata of Krishna-Dwaipayana Vyasa. Translated into English prose, from the original Sanskrit text by Kisari Mohan Ganguli*. Bharata Press, Calcutta, 1883–1896, 18 vols.

A4.27. HOMER. *The Iliad and Odyssey of Homer, translated into English blank verse, by W. Cowper*. J. Johnson, London, 1791; Vol. 1 (*Iliad*), Vol. 2 (*Odyssey*).

A4.28. HOMER. *The Odyssey of Homer. Translated from the Greek by Alexander Pope, assisted by Elijah Fenton and William Broome.* Printed for Bernard Lintot, London, 1725–1726, 5 vols.

A4.29. HOMER. *The Odyssey of Homer, translated by Alexander Pope. With notes, by Theodore Alois Buckley, with Flaxman's designs, and other engravings.* Ingram, Cooke & Co., Strand, London, 1853.

A4.30. AESOP. *Aesop's Fables. Translated with an introduction and notes by Laura Gibbs.* Oxford Univ. Press, Oxford, 2002. (Series: *Oxford World's Classics*).

A4.31. KIRK, G. S., RAVEN, J. E. and SCHOFIELD, M. *The Presocratic Philosophers: a critical history with a selection of texts.* Cambridge Univ. Press, Cambridge, 1983, 2nd edn, Ch. 4 (*Anaximenes of Miletus*), 143–162.

A4.32. EURIPIDES. *The tragedies of Euripides in English verse. By Arthur S. Way. In three volumes.* Macmillan & Co., London, 1896, Vol. 2.

A4.33. EURIPIDES. *Euripides: with an English translation by Arthur S. Way. In four volumes.* William Heinemann Ltd, London, 1912, Vol. 1 (Greek and English texts).

A4.34. SOPHOCLES. *Sophocles. In English verse. By Arthur S. Way.* Macmillan & Co. Ltd, London, 1909, Pt 1 (*Oedipus the King; Oedipus at Kolonus; Antigone*).

A4.35. THEOCRITUS. *Theocritus, Bion and Moschus. Translated into English verse. By Arthur S. Way.* University Press, Cambridge, 1913.

A4.36. QUINTUS. *Quintus Smyrnaeus. The Fall of Troy. Translated by A. S. Way.* William Heinemann, Ltd, London, 1913. (Series: *Loeb Classical Library*, Vol. 19).

A4.37. ARISTOTLE. *Meteorology.* Circa 340 B.C. Greek language version in: *Aristotelis, edidit Academia Regia Borussica.* Reimer, Berlin, 1831–1870, 5 vols (ed. A. I. Bekker *et al.*). Greek and Latin language versions in: *Aristotelis Meteorologicorum libri quattuor.* Harvard Univ. Press, Cambridge, 1919 (transl. F. H. Fobes). English language versions. *Meteorologica.* Clarendon Press, Oxford, 1923 (transl. E. W. Webster). Also in: *The works of Aristotle.* Oxford Univ. Press, Oxford, 1931, Vol. 3 (ed. W. D. Ross); *The complete works of Aristotle.* Princeton Univ. Press, Princeton, 1984, Vol. 1 (ed. J. Barnes), 555–625; *Meteorologica.* William Heinemann Ltd, London & Harvard Univ. Press, Cambridge, 1952 (transl. H. D. P. Lee).

A4.38. APOLLONIUS RHODIUS. *The Argonautics of Apollonius Rhodius, in four books, by Francis Fawkes.* Printed for J. Dodsley, in Pall-Mall, London, 1780.

A4.39. EVELYN-WHITE, H. G. *Hesiod: the Homeric Hymns and Homerica.* William Heinemann Ltd, London, 1914, 2–65 (*Works and Days*).

A4.40. GRENFELL, B. P. and HUNT, A. S. *The Hibeh papyri. Part 1.* Egypt Exploration Fund, London, 1906, Pt 3 (*Calendar*), 138–157 (& Plate).

A4.41. EVANS, J. and BERGGREN, J. L. *Geminos's Introduction to the Phenomena: a translation and study of a Hellenistic survey of astronomy.* Princeton Univ. Press, Princeton, 2006.

A4.42. DIELS, H. and REHM, A. Parapegmenfragmente aus Milet. *Sitzungsberichte der Königlich Preussischen Akademie der Wissenschaften*, 1904, 3, Jan., 92–111 (& Plate).

A4.43. REHM, A. Weiteres zu den milesischen Parapegmen. *Sitzungsberichte der Königlich Preussischen Akademie der Wissenschaften*, 1904, 23, Apr., 752–759.

A4.44. LEHOUX, D. The parapegma fragments from Miletus. *Zeitschrift für Papyrologie und Epigraphik*, 2005, 152, 125–140.

A4.45. LEHOUX, D. *Astronomy, weather, and calendars in the ancient world: parapegmata and related texts in classical and Near Eastern societies.* Cambridge Univ. Press, Cambridge, 2007.

A4.46. CHURCH, A. J. *Stories from Ancient Rome*. Cassell & Co., London, 1907.

A4.47. HUNT, P. N. *Hannibal*. Simon & Schuster, New York, 2017.

A4.48. GAIUS JULIUS CAESAR. *C. Julius Caesar's Commentaries of his wars in Gaul, and civil war with Pompey. To which is added Aulus Hirtius or Oppius's Supplement of the Alexandrian, African and Spanish wars. With the author's life. Adorn'd with sculptures from the designs of the famous Palladio. Made English from the original Latin, by Martin Bladen, Gent.* Printed for Richard Smith, at the Angel and Bible without Temple-Bar, London, 1705, 325–326.

A4.49. GAIUS JULIUS CAESAR. *Caesar's commentaries on the Gallic and Civil Wars: with the supplementary books attributed to Hirtius; including the Alexandrian, African, and Spanish Wars. Literally translated, with notes and a very elaborate index*. Henry G. Bohn, Covent Garden, London, 1851 (transl. W. A. McDevitte and W. S. Bohn), Pt 47, 447.

A4.50. GAIUS JULIUS CAESAR. *The Works of Julius Caesar. Parallel Latin and English texts*. Harper & Bros, New York, 1869 (transl. W. A. McDevitte and W. S. Bohn).

A4.51. LUCRETIUS (TITUS LUCRETIUS CARUS). *De rerum natura. Circa* 60 B.C. *Of the nature of things; a metrical translation by William Ellery Leonard*. J. M. Dent & Sons Ltd, London; E. P. Dutton & Co., New York; 1916.

A4.52. VIRGIL (PUBLIUS VERGILIUS MARO). *The Georgics*. 29 B.C. *The works of Virgil; containing his Pastorals, Georgics, and Aeneis: adorn'd with a hundred sculptures; translated into English verse, by Mr Dryden*. Jacob Tonson, at the Judges-Head in Fleetstreet, near the Inner-Temple-Gate, London, 1697.

A4.53. VIRGIL (PUBLIUS VERGILIUS MARO). *The Aeneid*. 19 B.C., 12 Books. Part English language translation in: *The shadows of poetry: Vergil in the mind of Augustine by Sabine MacCormack*. California Univ. Press, Berkeley, 1998.

A4.54. VIRGIL (PUBLIUS VERGILIUS MARO). *Aeneid, Georgics, and Eclogues (part)*. Illuminated Latin vellum manuscript, 5th century A.D; anonymous drawings from *Aeneid*: (a) *Aeneas* implores the gods to grant deliverance amid tempest and stormy seas; (b) *Dido* and *Aeneas* seek shelter from hailstorm. Biblioteca Apostolica Vaticana (MS Vergilius Romanus, Vat. lat. 3867, f. 77r; f. 106r).

A4.55. HORACE (QUINTUS HORATIUS FLACCUS). *The Odes, Epodes and Carmen Seculare of Horace. In Latin and English. With critical notes collected from his best Latin and French commentators. By Philip Francis*. Printed for A. Millar, at Buchanan's Head, opposite to Katharine-street in the Strand, London, 1743, Vol. 1.

A4.56. HORACE (QUINTUS HORATIUS FLACCUS). *The Epistles and Art of Poetry of Horace. In Latin and English. With critical notes collected from his best Latin and French commentators. By Philip Francis*. Printed for A. Millar, at Buchanan's Head, opposite to Katharine-Street, in the Strand, London, 1746, Vol. 4.

A4.57. OVID (PUBLIUS OVIDIUS NASO). *Ovid's Metamorphoses in fifteen books. Translated by the most Eminent Hands. Adorn'd with Sculptures*. Printed for Jacob Tonson at Shakespear's-Head over-against Katharine-Street in the Strand, London, 1717 (ed. Samuel Garth).

A4.58. COLUMELLA, L. J. M. *L. Junius Moderatus Columella. Of Husbandry. In Twelve Books: and His Book concerning Trees. Translated into English, with several illustrations from Pliny, Cato, Varro, Palladius, and other antient and modern authors*. Printed for A. Millar, opposite to Catharine-street in the Strand, London, 1745.

A4.59. LUCAN (MARCUS ANNAEUS LUCANUS). *The Pharsalia of Lucan. Translated into blank verse, by Edward Ridley.* Longmans, Green & Co., London, 1896.

A4.60. STATIUS (PUBLIUS PAPINIUS STATIUS). *The Thebaid of Statius, translated into English Verse, with Notes and Observations; and a Dissertation upon the Whole by Way of Preface*. Printed for T. Becket, in the Strand, London, 1773, 2nd edn (transl. William Lillington Lewis).

A4.61. STATIUS (PUBLIUS PAPINIUS STATIUS). *Statius, with an English translation by J. H. Mozley, in two volumes*. William Heinemann Ltd, London, 1928, Vol. 1 (*Silvae, Thebaid I–IV*).

A4.62. PTOLEMY. *Tetrabiblos. Translated by F. E. Robbins*. Harvard Univ. Press, Cambridge; William Heinemann Ltd, London; 1940, Bk 2, Ch. 11 (*Of the nature of the signs, part by part, and their effect upon the weather*), 200–205. (Series: *Loeb Classical Library*, Vol. 435).

A4.63. PTOLEMY. *Ptolemy's Tetrabiblos, or Quadripartite: being four books of the influence of the stars. Newly translated from the Greek paraphrase of Proclus. With a preface, explanatory notes, and an appendix, containing extracts from the Almagest of Ptolemy and the whole of his Centiloquy; together with a short notice of Mr Ranger's Zodiacal planisphere, and an explanatory plate. By J. M. Ashmand*. Davis and Dickson, London, 1822, Bk 2, Ch. 12 (*The particular natures of the signs by which the different constitutions of the atmosphere are produced*), 94–96.

A4.64. NONNOS OF PANOPOLIS. *Nonnos Dionysiaca. With an English translation by W. H. D. Rouse, mythological introduction and notes by H. J. Rose, notes on text criticism by L. R. Lind*. Harvard University Press, Cambridge; William Heinemann Ltd, London; 1940, Vol. 1 (*Books 1–15*). (Series: *Loeb Classical Library*, Vol. 344).

A4.65. VOLLMER, F. K. (Ed.). *Dracontius: de laudibus Dei*. In: *Flavius Merobaudis Reliquiae, Blossii Aemilii Dracontii Carmina, Eugenii Toletani Episcopi Carmina et Epistulae; cum appendicula carminum spuriorum*. Weidmann, Berlin, 1905 (ed. Frederick Vollmer). (Series: *Monumenta Germaniae Historica, Auctores Antiquissimi*, Vol. 14).

A4.66. VOLLMER, F. K. (Ed.). *Dracontii de laudibus Dei, Satisfactio, Romulea, Orestis tragoedia, Fragmenta incerti Aegritudo Perdicae*. B. G. Teubner, Leipzig, 1914, 2nd edn. (Series: *Poetae Latini Minores*, Vol. 5, eds Emil Baehrens and Frederick Vollmer).

A4.67. IRWIN, J. F. *Liber I: Dracontii de laudibus Dei. With introduction, text, translation and commentary*. Ph.D. Dissertation, University of Pennsylvania, Philadelphia, 1942, 133 pp.

A4.68. McCARTNEY, E. S. Greek and Roman weather lore of two destructive agents, hail and drought. *Classical Weekly*, 1934, 28, 1, 745, 1 Oct., 1–7; 2, 746, 8 Oct., 9–12; 3, 747, 15 Oct., 17–23; 4, 748, 22 Oct., 25–31.

A4.69. TAUB, L. C. *Ancient meteorology*. Routledge, London, 2003. (Series: *Sciences of antiquity*).

A4.70. CHRONOPOULOU, C. and MAVRAKIS, A. Ancient Greek drama as an eyewitness of a specific meteorological phenomenon: indication of stability of the Halcyon days. *Weather*, 2014, 69, 3 (Mar.), 66–69.

A4.71. McCORMICK, M., BÜNTGEN, U., CANE, M. A., COOK, E. R., HARPER, K., HUYBERS, P., LITT, T., MANNING, S. W., MAYEWSKI, P. A., MORE, A. F. M., NICOLUSSI, K. and TEGEL, W. Climate change during and after the Roman Empire: reconstructing the past from scientific and historical evidence. *J. Interdiscip. Hist.*, 2012, 43, 2, 169–220.

A4.72. CHRISTENSON, A. J. *Popol Vuh: the Sacred Book of the Maya*. Oklahoma Univ. Press, Norman, 2007, 2nd edn.

A4.73. MA MAR IBN R SHID. *The Expeditions: an early biography of Muhammad*. New York Univ. Press, New York, 2014 (transl. Sean W. Anthony), Ch. 8 (*The incident involving the United Clans and the Qurayzah Clan*), 83–89. (Series: *Library of Arabic literature*).

A4.74. OWEN, S. *The poetry of Du Fu: translated and edited by Stephen Owen*. De Gruyter, Boston, 2016, Vol. 6 (eds P. W. Kroll and D. X. Warner).

A4.75. POOLE, R. L. *Illustrations of the history of medieval thought in the departments of theology and ecclesiastical politics*. Williams & Norgate, London, 1884, Ch. 1 (*Claudius of Turin and Agobard of Lyons*), 38–42.

A4.76. FLINT, V. I. J. *The rise of magic in early medieval Europe.* Princeton Univ. Press, Princeton, 1991.

A4.77. BEDE. *On the Nature of Things and On Times.* Liverpool Univ. Press, Liverpool, 2010 (transl. Calvin B. Kendall and Faith Wallis).

A4.78. PERTZ, G. H. (Ed.). *Monumenta Germaniae Historica: inde ab anno Christi quingentesimo usque ad annum millesimum et quingentesimum.* Impensis Bibliopolii Aulici Hahniani, Hanover, 1837, Vol. 2, Pt 2 (*Capitularia Spuria; Canones Ecclesiastici; Bullae Pontificum*), 136.

A4.79 WERMINGHOFF, A. (Ed.). *Monumenta Germaniae Historica. Concilia 2.2.* Impensis Bibliopolii Hahniani, Hanover, 1908, Pt 50D (*Concilium Parisiense a. 829 m. Iunio*), 605–679.

A4.80. DUTTON, P. E. *Charlemagne's Mustache and other cultural clusters of a Dark Age.* Palgrave Macmillan, New York, 2004, Ch. 7 (*Thunder and hail over the Carolingian countryside*), 169–188.

A4.81. ANONYMOUS. *The Seafarer.* Circa 970 A.D. In: *The Exeter Book* (*Codex Exoniensis*), Exeter Cathedral library (MS 3501, f. 81v–f. 83r).

A4.82. ANONYMOUS. *The Seafarer.* Circa 970 A.D. In: *Codex Exoniensis. A collection of Anglo-Saxon poetry, from a manuscript in the library of the Dean and Chapter of Exeter, with an English translation, notes, and indexes. By Benjamin Thorpe.* Published for the Society of Antiquaries of London, by William Pickering, London, 1842, 306–313.

A4.83. ANONYMOUS. *The Seafarer* (La Motte Iddings version). In: *Select translations from Old English poetry.* Gin & Co., Boston, 1902 (eds A. S. Cook and C. B. Tinker), 44–49.

A4.84. ANONYMOUS. *The Seafarer: a translation from Old English lyric, 1901.* In: *Poems by Lola La Motte Iddings.* Yale Univ. Press, New Haven, 1920, 109–116.

A4.85. APPLETON, H. The insular landscape of the Old English poem *The Phoenix*. *Neophilologus*, 2017, 101, 4 (Oct.), 585–602.

A4.86. BLAKE, N. F. (Ed.). *The Phoenix.* Exeter Univ. Press, Exeter, 1990, 2nd edn. (Series: *Exeter medieval English texts and studies*).

A4.87. FERDOWSI (ABUL-QÂSEM FERDOWSI TUSI). *The Sháhnáma of Firdausí, done into English by Arthur George Warner and Edmond Warner.* Kegan Paul, Trench, Trübner & Co. Ltd, London, 1905–1925, 9 vols. (Series: *Trübner's Oriental*, Vols 4–12).

A4.88. FERDOWSI (ABUL-QÂSEM FERDOWSI TUSI). *The Shah nameh: an heroic poem. Containing the history of Persia from Kioomurs to Yesdejird; that is, from the earliest times to the conquest of that empire by the Arabs. By Abool Kasim Firdousee. Carefully collated with a number of the oldest and best manuscripts, and illustrated by a copious glossary of obsolete words and obscure idioms: with an introduction and life of the author, in English and Persian; and an appendix, containing the interpolated episodes, etc. found in different manuscripts. By Turner Macan, Persian interpreter to the Commander in Chief, and member of the Asiatic Society of Calcutta.* Baptist Mission Press, Calcutta, 1829, 4 vols.

A4.89. TUROLDUS. *The Song of Roland. Translated into English verse. By Arthur S. Way.* University Press, Cambridge, 1913.

A4.90. BORROW, G. H. *Romantic Ballads, translated from the Danish; and miscellaneous pieces.* John Taylor, London, 1826.

A4.91. BORROW, G. H. *The Works of George Borrow.* Constable & Co. Ltd, London, 1923 (ed. Clement Shorter), 16 vols (*Norwich Edition*); Vol. 8 (*The songs of Scandinavia and other poems and ballads*). Reprinted by AMS Press Inc., New York, 1967.

A4.92. BORROW, G. H. *Targum: or metrical translations from thirty languages and dialects.* Schulz & Beneze, St Petersburg, 1835.

A4.93. STURLAŚON, S. *Heimskringla*. J. M. Stenersen & Co., Oslo, 1899.

A4.94. HOLLANDER, L. M. *The Saga of the Jómsvikings*. Texas Univ. Press, Austin, 1955.

A4.95. ANONYMOUS. *The Mabinogion. From the Welsh of the Llyfr Coch o Hergest (The Red Book of Hergest). In the library of Jesus College, Oxford. Translated, with notes, by Lady Charlotte Guest*. Bernard Quaritch, London, 1877.

A4.96. ANONYMOUS. *The Mabinogion. Translated from The Red Book of Hergest by Lady Charlotte Guest*. T. Fisher Unwin, London, 1902, Vol. 1 (ed. Owen M. Edwards).

A4.97. GIOVANNI DA PIAN DEL CARPINE (JOHN OF PLANO CARPINI). *Historia Mongalorum quos nos Tartaros appellamus*. Italy, 13th-century Latin manuscript. English language version: *The story of the Mongols whom we call the Tartars, by Friar Giovanni DiPlano Carpini*. Branden Publ. Co., Boston, 1996 (transl. Erik Hildinger).

A4.98. VERNE, J. *Celebrated travels and travellers: the exploration of the world*. Sampson Low, Marston, Searle, & Rivington, London, 1882, Pt 1 (transl. Dora Leigh).

A4.99. RAMIREZ-CHRISTENSEN, E. U. *Heart's Flower: the life and poetry of Shinkei*. Stanford Univ. Press, Stanford, 1994.

A4.100. BEICHMAN, J. *Masaoka Shiki: his life and works*. Cheng & Tsui Co., Boston, 2002.

A4.101. KEENE, D. *The winter sun shines in: a life of Masaoka Shiki*. Columbia Univ. Press, New York, 2013.

A4.102. REDHOUSE, J. W. Remarkable hailstorm. *Quart. J. Royal Met. Soc.*, 1884, 10, 52 (Oct.), 303.

A4.103. FESTIVAL PRAYER BOOK. Illuminated Hebrew parchment codex, Reggio nell'Emilia, Italy, 1466: miniature, probably by Giorgio d'Alemagna (workshop artist for Duke Borso d'Este); *Man at prayer upon hilltop; two youths shelter from heavy hail*. British Library, London (MS Harley 5686, Pt 1, f. 60).

A4.104. DANTE ALIGHIERI. *La Divina Commedia: Inferno, Purgatorio, Paradiso*. Circa 1320. English language translation: *The Divine Comedy of Dante Alighieri*. Ticknor & Fields, Boston, 1867 (transl. H. W. Longfellow), Vol. 1 (*Inferno*).

A4.105. STRADANO, G. *Dante, Commedia, Inferno, Cerchio III (Canto VI): Golosi distesi a la grandine, e pioggia, e neve*. Manuscript drawing, Florence, Italy, circa 1587, Biblioteca Medicea Laurenziana, Firenze (MS Medici Palatino 75, f. 27r).

A4.106. ZUCCARI, F. *Dante, Inferno, Canti V–VI: Terzo cerchio, I golosi, Cerbero*. Manuscript drawing, Spain, circa 1587, Uffizi Gallery, Florence (MS GDSU inv. 3481F).

A4.107. DANTE ALIGHIERI. *Divina Commedia*. Illuminated Italian parchment codex, Italy (possibly Siena), circa 1447: miniature by Priamo della Quercia; *Virgil, Dante, and Cerberus*. British Museum, London (MS Yates Thompson 36, f. 11).

A4.108. DANTE ALIGHIERI. *Divine Comedy*. Illuminated Italian parchment codex, Italy (possibly Genoa), circa 1370: anonymous miniature; *Inferno: Third Circle of Hell*. Bodleian Library, Oxford (MS Holkham misc. 48, p. 9).

A4.109. SUMPTION, J. P. C. *Trial by fire. The Hundred Years War II*. Faber & Faber, London, 1999.

A4.110. MISCELLANEOUS. *Historical and other pieces in prose and verse*. 15th century. British Library, London (MS Egerton 1995).

A4.111. GAIRDNER, J. (Ed.). *The historical collections of a citizen of London in the fifteenth century*. Camden Soc., London, 1876 (*Gregory's Chronicle*: 1250–1367).

A4.112. KINGSFORD, C. L. (Ed.). *Chronicles of London: with introduction and notes*. Clarendon Press, Oxford, 1905, 13.

A4.113. SHAKESPEARE, W. *The most excellent Historie of the Merchant of Venice. With the extreame crueltie of Shylocke the Jewe towards the sayd Merchant, in cutting a just pound of his flesh: and the obtayning of Portia by the choyse of three chests. As it hath beene divers times acted by the Lord Chamberlaine his servants. Written by William Shakespeare.* Printed by James Roberts for Thomas Heyes, and are to be sold in Paules Church-yard, at the signe of the Greene Dragon, London, 1600.

A4.114. BARNES, J. *The history of that most victorius monarch, Edward IIId, King of England and France, and Lord of Ireland, and first founder of the most noble Order of the Garter being a full and exact account of the life and death of the said king, together with that of his most renowned son, Edward, Prince of Wales and of Aquitain, sirnamed the Black-Prince: faithfully and carefully collected from the best and most antient authors, domestick and foreign, printed books, manuscripts and records.* Printed by John Hayes for the author, Cambridge, 1688.

A4.115. DOYLE, J. W. E. *A chronicle of England: B.C. 55–A.D. 1485.* Longman, London, 1864.

A4.116. CHAUCER, G. *The Canterbury Tales.* William Caxton, Westminster, London, 1476, 1st printed edn (sourced from diverse much earlier manuscripts).

A4.117. SKEAT, W. W. (Ed.). *The complete works of Geoffrey Chaucer.* Clarendon Press, Oxford, 1894, Vol. 4 (*The Canterbury Tales: text*), 1st edn; 1900, 2nd edn.

A4.118. SPEGHT, T. (Ed.). *The Workes of our Antient and lerned English Poet, Geffrey Chaucer, newly Printed. In this Impression you shall find these Additions: 1. His Portraiture and Progenie shewed. 2. His Life collected. 3. Arguments to every Booke gathered. 4. Old and obscure Words explained. 5. Authors by him cited, declared. 6. Difficulties opened. 7. Two Bookes of his never before printed.* Impensis Geor. Bishop, Londini, 1598.

A4.119. PEARSALL, D. A. (Ed.). *The Floure and the Leafe, and The Assembly of Ladies.* Thomas Nelson & Sons, London, 1962.

A4.120. VINTLER, H. *Das Buch der Tugent.* Johann Blaubirer, Augsburg, 1486.

A4.121. ZINGERLE, I. V. (Ed.). *Die pluemen der tugent des Hans Vintler.* Wagner, Innsbruck, 1874.

A4.122. ZIKA, C. *Exorcising our demons: magic, witchcraft, and visual culture in early modern Europe.* Brill, Leiden, 2003.

A4.123. HENRICUS INSTITORIS and JACOBUS SPRENGER. *Malleus Maleficarum.* Peter Drach, Speyer, 1487.

A4.124. KRAMER, H. and SPRENGER, J. *Malleus Maleficarum. Translated with an introduction, bibliography and notes by the Rev. Montague Summers.* John Rodker, London, 1928, Pt 2, Ch. 15 (*How they raise and stir up hailstorms and tempests, and cause lightening to blast both men and beasts*).

A4.125. MOLITOR, U. *De lamiis et phitonicis mulieribus.* Michael Furter, Basel; Cornelis de Zierikzee, Cologne; Johann Prüss, Strassburg; Johann Otmar, Reutlingen; *circa* 1489.

A4.126. ASHTON, J. *The devil in Britain and America.* Ward & Downey Ltd, London, 1896, Ch. 11, 151.

A4.127. MACDONALD, E. M. *A short history of the Inquisition. What it was and what it did. To which is appended an account of persecutions by Protestants, persecutions of witches, the war between religion and science, and the attitude of the American churches toward African slavery.* The Truth Seeker Co., New York, 1907, Ch. (*The witchcraft delusion*), 373–409 (& Plate).

A4.128. LEONARDO DA VINCI. *The drawings and miscellaneous papers of Leonardo da Vinci in the collection of Her Majesty the Queen at Windsor Castle.* Royal Collection Trust, London (ed. Carlo Pedretti): 1982, Vol. 1 (*Landscapes, plants and water studies*); 1987, Vol. 2 (*Horses and other animals*).

A4.129. CLAYTON, M. *Leonardo da Vinci: a life in drawing.* Royal Collection Trust, London, 2018.

A4.130. LEONARDO DA VINCI. *The literary works of Leonardo da Vinci, compiled and edited from the original manuscripts by Jean Paul Richter.* Sampson Low, Marston, Searle & Rivington, London, 1883, 2 vols, 1st edn; Oxford Univ. Press, London, 1939, 2nd edn; Phaidon Press, London, 1970, 3rd edn.

A4.131. LEONARDO DA VINCI. *The literary works of Leonardo da Vinci. A commentary to Jean Paul Richter's edition by Carlo Pedretti.* Phaidon Press, London, 1977, 2 vols.

A4.132. LEONARDO DA VINCI. *The Codex Hammer of Leonardo da Vinci. Translated into English and annotated by Carlo Pedretti.* Giunti-Barbèra, Florence, 1987.

A4.133. CELLINI, B. *The life of Benvenuto Cellini: a new version by Robert H. Hobart Cust.* G. Bell & Sons Ltd, London, 1910, Vol. 2, Bk 2, Ch. 8 (*1545*), 232–257.

A4.134. CELLINI, B. *The life of Benvenuto Cellini: newly translated into English by John Addington Symonds. With engraved portrait and eight etchings by F. Laguillermie; also eighteen reproductions of the works of the Master.* John C. Nimmo, London, 1888, 2 vols.

A4.135. PFINZING, M. *Theuerdank* (*Die geuerlicheiten und einsteils der geschichten des loblichen streytparen und hochberümbten helds und Ritters herr Tewrdannckhs*). Johann Schönsperger, Nürnberg, 1517.

A4.136. LOUTHAN, H. (Ed.). *Theuerdank. The Illustrated Epic of a Renaissance Knight.* Routledge, London, 2022 (transl. Jonathan Green).

A4.137. CAMOËNS, L. DE. *The Lusiad; or, the discovery of India. An epic poem. Translated from the original Portuguese of Luís de Camoëns. By William Julius Mickle.* Printed by Jackson and Lister, Oxford; and sold by Cadell, in the Strand; Dilly, in the Poultry; Bew, Pater-noster-Row; Flexney, Holborn; Evans, near York-Buildings; Richardson and Urquhart, under the Royal-Exchange; and Goodman, near Charing-Cross; London, 1776.

A4.138. TASSO, T. *Godfrey of Bulloigne or The Recoverie of Jerusalem. Done into English heroicall verse, by Edward Fairefax, Gent.* Imprinted by Ar. Hatfield, for J. Jaggard and M. Lownes, London, 1600.

A4.139. TASSO, T. *Jerusalem Delivered. A poem. By Torquato Tasso. Translated by Edward Fairfax. Edited by Henry Morley.* George Routledge & Sons Ltd, London, 1890.

A4.140. SPENSER, E. *The Faerie Queene.* Printed for William Ponsonbie, London, 1590 (Books I–III), 1596 (Books I–VI).

A4.141. WISE, T. J. (Ed.). *Spenser's Faerie Queene.* George Allen, London, 6 vols (illus. W. Crane), 1894–1897.

A4.142. GUAZZO, F. M. *Compendivm Maleficarvm. Ex quo nefandissima in genus humanum opera venefica, ac ad illa vitanda remedia conspiciuntur. Per Fratrem Franciscum Mariam Guaccium Ord. S. Ambrosij ad Nemus Mediolani compilatum.* Ex Collegij Ambrosiani typographia, Milan, 1626, Primo libro, Cap. VIII (*Sortilegi, diris imprecationibus, et factis petunt pluuiam, grandines etc.*), 46–51.

A4.143. GUAZZO, F. M. *Compendium Maleficarum. Collected in 3 books from many sources by Brother Francesco Maria Guazzo of the Order of St Ambrose ad Nemus, showing the iniquitous and execrable operations of witches against the human race, and the divine remedies by which they may be frustrated. Edited with notes by the Rev. Montague Summers. Translated by E. A. Ashwin.* John Rodker, London, 1929, Bk 1, Ch. 7 (*By their terrible deeds and imprecations, witches produce rain and hail, etc.*), 19–22.

A4.144. DILLINGER, J. *'Böse Leute'. Hexenverfolgungen in Schwäbisch-Österreich und Kurtrier im Vergleich.* Paulinus Verlag, Trier, 1999.

A4.145. DILLINGER, J. *'Evil people'. A comparative study of witch hunts in Swabian Austria and the Electorate of Trier*. Univ. Virginia Press, Charlottesville & London, 2009 (transl. Laura Stokes), Ch. 2 (*Golden goblets and cows' hooves: witchcraft and magic*), 41–73. (Series: *Studies in early modern German history*, ed. H. C. Erik Midelfort).

A4.146. MIDELFORT, H. C. E. Were there really witches? In: *Transition and revolution: problems and issues of European renaissance and reformation history*. Burgess Publ. Co., Minneapolis, 1974 (ed. R. M. Kingdon), 213–231.

A4.147. BEHRINGER, W. Weather, hunger and fear: origins of the European witch-hunts in climate, society and mentality. *German History*, 1995, 13, 1 (Jan.), 1–27.

A4.148. OSTER, E. Witchcraft, weather and economic growth in renaissance Europe. *J. Economic Perspectives*, 2004, 18, 1, 215–228.

A4.149. PFISTER, C. Climatic extremes, recurrent crises and witch hunts: strategies of European societies in coping with exogenous shocks in the late sixteenth and early seventeenth centuries. *Medieval History J.*, 2007, 10, 1 & 2, 33–73.

A4.150. WAITE, G. K. *Eradicating the Devil's Minions: Anabaptists and Witches in Reformation Europe, 1535–1600*. Univ. Toronto Press, Toronto, 2007, Ch. 5 (*The Devil's Sabbat: nocturnal Anabaptist meetings, hailstorms, and witchcraft in southern Germany*), 130–165.

A4.151. KIECKHEFER, R. Constructing narratives of witchcraft. In: *Inquisition and knowledge, 1200–1700*. York Medieval Press, York, 2022 (eds P. Biller and L. J. Sackville), Pt 1 (*Medieval*), Ch. 8, 195–208. (Series: *Heresy and Inquisition in the Middle Ages*, Vol. 10).

A4.152. SHAKESPEARE, W. *Mr William Shakespeares Comedies, Histories, & Tragedies. Published according to the True Originall Copies*. Printed by Isaac Jaggard and Edward Blount, London, 1623.

A4.153. DRAYTON, M. *The Battaile of Agincourt. Fought by Henry the fift of that name, King of England, against the whole power of the French: under the raigne of their Charles the sixt, Anno Dom. 1415. The miseries of Queene Margarite, the infortunate wife, of that most infortunate King Henry the sixt. Nimphidia, the court of Fayrie. The quest of Cinthia. The shepheards Sirena. The moone-calfe. Elegies upon sundry occasions*. Printed for William Lee, at the Turkes Head in Fleete-Streete, next to the Miter and Phaenix, London, 1627, 1st edn; 1631, 2nd edn.

A4.154. CERVANTES (MIGUEL DE CERVANTES SAAVEDRA). *El ingenioso hidalgo (cavallero) Don Quixote de la Mancha, compuesto por Miguel de Cervantes Saavedra*. Juan de la Cuesta, Madrid, 1605, Pt 1; 1615, Pt 2 (titular *hidalgo* replaced by *cavallero*).

A4.155. CERVANTES (MIGUEL DE CERVANTES SAAVEDRA). *The ingenious gentleman Don Quixote of La Mancha by Miguel de Cervantes Saavedra. A translation, with introduction and notes, by John Ormsby, in 4 vols*. Smith, Elder & Co., London, 1885, Vol. 4, 15–29.

A4.156. CERVANTES (MIGUEL DE CERVANTES SAAVEDRA). *El ingenioso hidalgo Don Quijote de la Mancha, compuesto por Miguel de Cervantes Saavedra; edición anotada por don Nicolás Diaz de Benjumea é ilustrada por don Ricardo Balaca*. Montaner y Simón, Barcelona, 1880, Vol.1; 1883, Vol. 2 (*ilustrada por don Ricardo Balaca y don J. Luís Pellicer*), 324.

A4.157. DACH, S. *Annie of Tharau (Ännchen von Tharau)*. In: *The Poetical Works of Henry Wadsworth Longfellow. With illustrations by John Gilbert*. George Routledge & Co., London, 1855, 297–298.

A4.158. VAUGHAN, H. *Silex Scintillans: or sacred poems and priuate eiaculations. By Henry Vaughan, Silurist*. Printed by T. W. for H. Blunden at ye Castle in Cornehill, London, 1650.

A4.159. MILTON, J. *Paradise Lost. A poem written in ten books by John Milton*. Printed, and are to be sold by Peter Parker, under Creed Church neer Aldgate; and by Robert Boulter at the Turks

Head in Bishopsgate-street; and Matthias Walker, under St Dunstons Church in Fleet-street; London, 1667.

A4.160. OGILBY, J. (Ed.). *America: being the Latest, and Most Accurate Description of the New World; containing the Original of the Inhabitants, and the Remarkable Voyages thither. The Conquest of the Vast Empires of Mexico and Peru, and other large Provinces and Territories, with the several European Plantations in those Parts. Also their Cities, Fortresses, Towns, Temples, Mountains, and Rivers. Their Habits, Customs, Manners, and Religions. Their Plants, Beasts, Birds, and Serpents. With an Appendix, containing, besides several other considerable Additions, a brief Survey of what hath been discover'd of the Unknown South-Land and the Arctick Region. Collected from most Authentick Authors, Augmented with later Observations, and Adorn'd with Maps and Sculptures, by John Ogilby Esq; His Majesty's Cosmographer, Geographick Printer, and Master of the Revels in the Kingdom of Ireland.* Printed by the Author, and are to be had at his House in White Fryers, London, 1671.

A4.161. FAIRFAX, N. An account of hail-stones of an unusual bigness, communicated by D. Nath. Fairfax, with his reflections on them. *Phil. Trans Royal Soc. (Lond.)*, 1667, 2, 26 (Jun.), 481–482.

A4.162. HOOKE, R. *Dr Hook's account of the great hailstones that fell in London, on May 18, 1680.* In: *Philosophical experiments and observations of the late eminent Dr Robert Hooke, S.R.S. and Geom. Prof. Gresh., and other eminent virtuosos in his time. With copper plates. Publish'd by W. Derham, F.R.S.* Printed by W. and J. Innys, London, 1726; Printers to the Royal Society, at the West End of St Paul's, 49–53.

A4.163. DRYDEN, J. *Britannia rediviva: a poem on the birth of the Prince.* Printed for J. Tonson, at the Judges-Head in Chancery-Lane, near Fleet-street, London, 1688.

A4.164. HALLEY, E. A letter from Mr Halley at Chester, giving an account of an extraordinary hail in these parts, on the 29th of April last. *Phil. Trans Royal Soc. (Lond.)*, 1697, 19, 229 (Jun.), 570–572.

A4.165. HALLEY, E. Part of another letter, dated May 1, giving a larger account of the same hailstorm. *Phil. Trans Royal Soc. (Lond.)*, 1697, 19, 229 (Jun.), 572–576.

A4.166. TAILOR, R. Part of a letter from Mr Robert Tailor, apothecary at Hitchin in Hartfordshire, to Hans Sloan, giving account of a great hail storm there, May 4th, 1697. *Phil. Trans Royal Soc. (Lond.)*, 1697, 19, 229 (Jun.), 577–578.

A4.167. ANONYMOUS. Part of a letter, dated June the 9th, 1697, from Herefordshire, giving a relation of the effects of a great hailstorm there, June 1697. *Phil. Trans Royal Soc. (Lond.)*, 1697, 19, 229 (Jun.), 579.

A4.168. LHWYD, E. A note concerning an extraordinary hail in Monmouthshire, extracted out of a letter sent from Mr Edward Lhwyd to Dr Tancred Robinson, Fell. of Coll. of Phys. & R. S. Dat. Usk in Monmouthshire, June 15, 1697. *Phil. Trans Royal Soc. (Lond.)*, 1697, 19, 229 (Jun.), 579–580.

A4.169. WALLIS, J. A letter of Dr Wallis to Dr Sloane, concerning the generation of hail, and of thunder and lightning, and effects thereof. Oxford, July 26, 1697. *Phil. Trans Royal Soc. (Lond.)*, 1697, 19, 231 (Aug.), 653–658.

A4.170. DEKKER, T. *The Great Frost. Cold doings in London, except it be at the lotterie. With newes out of the country. A familiar talke betwene a country-man and a citizen touching this terrible frost and the great lotterie, and the effects of them.* Printed at London for Henry Gosson, 1608.

A4.171. LEIGH, C. *The Natural History of Lancashire, Cheshire, and the Peak, in Derbyshire: with an account of the British, Phoenician, Armenian, Gr. and Rom. Antiquities in Those Parts.* Printed for the Author, and to be had at Mr George West's, and Mr Henry Clement's,

Booksellers, Oxford; Mr Edward Evet's, at the Green-Dragon, in St Paul's Church-yard; and Mr John Nicholson, at the King's-Arms, in Little-Britain; London, 1700, Ch. 1 (*Of the ancient inhabitants, and of the air in those counties*), 9.

A4.172. WALLIS, J. A letter from Dr Wallis of Jan. 11, 1697/98, to Dr Sloane, concerning the effects of a great storm of thunder and lightning at Everdon in Northamptonshire (wherein divers persons were killed) on July 27, 1691. *Phil. Trans Royal Soc. (Lond.)*, 1698, 20, 236 (Jan.), 5–11.

A4.173. RUTTY, W. The relation of a storm of thunder, lightning and hail at Oundle in Northamptonshire on the 20th of March 1692/3. *Phil. Trans Royal Soc. (Lond.)*, 1693, 17, 199 (Apr.), 710–711.

A4.174. THORESBY, R. A letter from Mr Ralph Thoresby, FRS to D. Hans Sloane, RS Secr. giving an account of the damage done by a storm of hail, which happen'd near Rotherham in Yorkshire, on June 7, 1711. *Phil. Trans Royal Soc. (Lond.)*, 1712, 27, 335 (Jul.–Sept.), 514–516.

A4.175. ANONYMOUS. The extract of a letter from Lisle in Flanders, May 25 N. S. 1686, giving an account of an unusual storm of hail which fell there. *Phil. Trans Royal Soc. (Lond.)*, 1693, 17, 203 (Sept.), 858.

A4.176. DEFOE, D. *The Storm: or, a Collection of the most Remarkable Casualties and Disasters which happen'd in the Late Dreadful Tempest, both by Sea and Land*. Printed for G. Sawbridge in Little Britain, and sold by J. Nutt near Stationers-Hall, London, 1704.

A4.177. ANONYMOUS. *An Exact Relation of the Late Dreadful Tempest: or, a Faithful Account of the most Remarkable Disasters which hapned on that occasion: the places where, and Persons Names who suffer'd by the same, in City and Countrey; the Number of Ships, Men and Guns, that were lost, the miraculous Escapes of several Persons from the Dangers of that Calamity both by Sea and Land. Faithfully collected by an Ingenious Hand, to Preserve the Memory of so Terrible a Judgment*. Printed and sold by A. Baldwin at the Oxford-Arms in Warwick-Lane, London, 1704, 24 pp.

A4.178. MATSUO BASHŌ. *On Love and Barley: Haiku of Basho. Translated from the Japanese with an introduction by Lucien Stryk*. Penguin Books Ltd, London, 1985.

A4.179. MATSUO BASHŌ. *Basho's Haiku. Literal translations for those who wish to read the original Japanese text, with grammatical analysis and explanatory notes. Toshiharu Oseko*. Privately published with Maruzen, Tokyo, 1990, Vol. 1; 1996, Vol. 2.

A4.180. MATSUO BASHŌ. *Bashō and his interpreters. Selected Hokku with commentary. Compiled, translated, and with an introduction by Makoto Ueda*. Stanford Univ. Press, Stanford, 1992.

A4.181. MATSUO BASHŌ. *Bashō's Haiku. Selected poems of Matsuo Bashō. Translated and with an introduction by David Landis Barnhill*. State Univ. New York Press, Albany, 2004.

A4.182. MATSUO BASHŌ. *Basho: the complete Haiku. Translated with an introduction, biography and notes by Jane Reichhold. Original artwork by Shiro Tsujimura*. Kodansha Int., Tokyo, 2008.

A4.183. ADDISS, S. L. *The art of Haiku: its history through poems and paintings by Japanese masters*. Shambhala Publications Inc., Boston, 2012.

A4.184. BLYTH, R. H. *A history of Haiku*. Hokuseido Press, Tokyo, 1964, Vol. 2 (*From Issa up to the Present*), *Santōka*, 173–188.

A4.185. TANEDA SHŌICHI (*alias* TANEDA SANTŌKA). *Mountain tasting. Zen Haiku by Santōka Taneda. Translated and introduced by John Stevens*. John Weatherhill Inc., New York, 1980.

A4.186. TANEDA SHŌICHI (*alias* TANEDA SANTŌKA). *Santoka: Grass and Tree Cairn. Translations by Hiroaki Sato, illustrations by Stephen Addiss*. Red Moon Press, Winchester, 2002.

A4.187. TANEDA SHŌICHI (*alias* TANEDA SANTŌKA). *Taneda Santōka translated by Burton Watson. For all my walking: free-verse Haiku of Taneda Santôka, with excerpts from his diary*. Columbia Univ. Press, New York, 2003.

A4.188. TANEDA SHŌICHI (alias TANEDA SANTŌKA). *Ziarna gradu. Hailstones: haiku by Taneda Santoka, edited and illustrated by Lidia Rozmus. English translations by John Stevens, Polish translations by Wioletta Laskowska and Lidia Rozmus, Japanese calligraphy by Masanobu Hoshikawa.* Deep North Press, Evanston, 2006.

A4.189. SWIFT, J. *Travels into several remote nations of the world. In four parts. By Lemuel Gulliver, first a surgeon, and then a captain of several ships.* Printed for Benj. Motte, at the Middle Temple-Gate in Fleet-street, London, 1726, Vol. 1, 227–228.

A4.190. POINTER, J. *A rational account of the weather.* Printed for Aaron Ward, at the King's Arms in Little-Britain, London, 1738, 2nd edn, Pt 32 (Signs of hail), 108–112.

A4.191. SAUL, E. *An historical and philosophical account of the barometer, or weather-glass. Wherein the reason and use of that instrument, the theory of the atmosphere, the causes of its different gravitation are assign'd and explain'd. And a modest attempt from thence made towards a rational account and probable judgement of the weather.* Printed for A. Bettesworth and C. Hitch, at the Red-Lyon in Pater-Noster-Row, London, 1730.

A4.192. OBERHOLZNER, F. From an Act of God to an Insurable Risk: the change in the perception of hailstorms and thunderstorms since the Early Modern period. *Environment and History*, 2011, 17, 1 (Feb.), 133–152.

A4.193. FAUQUIER, W. An account of an extraordinary storm of hail in Virginia. *Phil. Trans Royal Soc. (Lond.)*, 1758, 50, 2, 746–747.

A4.194. DEWEY, S. *Account of a hail storm, which fell on part of the towns of Lebanon, Bozrah and Franklin, on the 15th of July, 1799; perhaps never equalled by any other ever known, not even in Egypt.* Thomas & Thomas Press, Walpole, Newhampshire, 1799.

A4.195. DEWEY, S. *An impartial relation of the hail-storm on the fifteenth of July and the tornado on the second of August 1799. Which appeared in the towns of Bozrah, Lebanon, and Franklin, in the state of Connecticut. To which is annexed an estimate of the damages done by the storm, made by a committee, from said towns. The whole published under their direction for the information of the public.* John Trumbull, Norwich, Connecticut, 1799.

A4.196. FALCONER, W. *The Shipwreck, a Poem, by William Falconer, a Sailor. The text illustrated by additional notes, and corrected from the first and second editions; with the life of the author, by J. S. Clarke.* Printed for William Miller, Albemarle-street; by W. Bulmer & Co., Cleveland-row; London, 1811.

A4.197. COWPER, W. *The Task, a poem, in six books. By William Cowper, of the Inner Temple, Esq. To which are added, by the same author, An epistle to Joseph Hill, Esq. Tirocinium, or a Review of Schools, and the History of John Gilpin.* Printed for J. Johnson, St Paul's Church-yard, London, 1785.

A4.198. BROWNING, O. (Ed.). *Despatches from Paris 1784–1790. Selected and edited from the Foreign Office correspondence by Oscar Browning. Volume II (1788–1790). Camden third series Vol. XIX.* Camden Soc., London, 1910, 75–76.

A4.199. NEUMANN, J. Great historical events that were significantly affected by the weather: 2, the year leading to the Revolution of 1789 in France. *Bull. Amer. Meteorological Soc.*, 1977, 58, 2 (Feb.), 163–168.

A4.200. NEUMANN, J. and DETTWILLER, J. Great historical events that were significantly affected by the weather: Part 9, the year leading to the Revolution of 1789 in France (II). *Bull. Amer. Meteorological Soc.*, 1990, 71, 1 (Jan.), 33–41.

A4.201. CASHMAN, L. Killer hailstones: weather and the French Revolution. *Agora*, 2010, 45, 3, 18–23.

A4.202. WORDSWORTH, W. *Select Pieces from the Poems of William Wordsworth.* James Burns, London, 1843.

A4.203. WORDSWORTH, W. *The poetical works of William Wordsworth*. Macmillan & Co. Ltd, London, 1896 (ed. William Knight), 8 vols.

A4.204. WORDSWORTH, W. *Journals of Dorothy Wordsworth*. Macmillan & Co. Ltd, London, 1897 (ed. William Knight), Vol. 1, Ch. 1.

A4.205. BARLOW, J. *The Columbiad: a poem. By Joel Barlow*. Printed by Fry & Kammerer for C. & A. Conrad & Co., Philadelphia; Conrad, Lucas & Co., Baltimore; 1807.

A4.206. LESLIE, C. R. *Memoirs of the Life of John Constable, Esq., R.A. Composed Chiefly of His Letters. By C. R. Leslie, R.A.* Longman, Brown, Green, and Longmans, Paternoster Row, London, 1845, 2nd edn, Ch. 1 (1776–1810), 4–5.

A4.207. NEILL, P. Notice respecting a remarkable shower of hail which fell in Orkney on the 24th of July 1818. *Trans Royal Soc. Edin.*, 1823, 9, 1, 187–199 (& Plate).

A4.208. WENTWORTH, W. C. *Statistical, historical, and political description of the colony of New South Wales, and its dependent settlements in Van Diemen's Land: with a particular enumeration of the advantages which these colonies offer for emigration, and their superiority in many respects over those possessed by the United States of America. By William Charles Wentworth, a native of the colony*. G. & W. B. Whittaker, London, 1819.

A4.209. SCOTT, W. *Marmion; a Tale of Flodden Field*. Archibald Constable & Co., Edinburgh; William Miller & John Murray, London; 1808.

A4.210. SCOTT, W. *The Poetical Works of Sir Walter Scott, Baronet. In ten volumes*. Archibald Constable & Co., Edinburgh: Longman, Hurst, Rees, Orme & Brown; J. Murray; and Hurst, Robinson & Co; London: 1821, Vol. 10.

A4.211. LAMB, C. *Beauty and the Beast. Or a Rough Outside with Gentle Heart. A Poetical Version of an Ancient Tale. Illustrated with a series of elegant engravings. And Beauty's Song at Her Spinning Wheel, set to music by Mr Whitaker*. Printed for M. J. Godwin, at the Juvenile Library, Skinner Street, London, 1811.

A4.212. COLERIDGE, S. T. *Christabel: Kubla Khan, a vision: The Pains of Sleep*. Printed for John Murray, Albemarle-Street, London, 1816.

A4.213. KEATS, J. *Endymion: a poetic romance*. Printed for Taylor & Hessey, Fleet Street, London, 1818.

A4.214. SHELLEY, P. B. *Prometheus Unbound, a lyrical drama in four acts, with other poems*. C. & J. Ollier, Bond Street, London, 1820.

A4.215. GRIMM, J. L. K. and GRIMM, W. C. *Grimm's Fairy Tales. Selected and edited for primary reader grades by Edna Henry Lee Turpin*. Maynard, Merrill & Co., New York, 1903.

A4.216. AFANASYEV, A. N. *Russian folk-tales (translated from the Russian). With introduction and notes by Leonard A. Magnus*. Kegan Paul, Trench, Trubner & Co. Ltd, London; E. P. Dutton & Co., New York; 1916.

A4.217. MILLER, W. *Memoirs of General Miller, in the Service of the Republic of Peru. By John Miller. In two volumes*. Longman, Rees, Orme, Brown & Green, Paternoster-Row, London; 1828, Vol. 1; 1829, Vol. 2, Ch. 27, 249.

A4.218. BYRON, G. G. *Don Juan: in sixteen cantos, with notes. By Lord Byron*. Milner & Sowerby, Halifax, 1837.

A4.219. FITCHETT, J. *King Alfred. A poem by John Fitchett. Edited by Robert Roscoe*. William Pickering, London, 1842, Vol. 4, Bk 26, 74.

A4.220. BROWNING, R. *Paracelsus. By Robert Browning*. Effingham Wilson, London, 1835, Pt 1 (*Paracelsus aspires*), 28.

A4.221. BROWNING, E. B. *The poems of Elizabeth Barrett Browning*. Frederick Warne & Co., London, 1850.

A4.222. DICKENS, C. *The life and adventures of Nicholas Nickleby. With illustrations by Phiz*. Chapman & Hall, Strand, London, 1839 (published in serial form, Mar. 1838–Sept. 1839).

A4.223. DICKENS, C. *A Christmas Carol. In Prose. Being a Ghost Story of Christmas. With illustrations by John Leech*. Chapman & Hall, Strand, London, 1843.

A4.224. GOETHE, J. W. VON. *The Poems of Goethe: translated in the Original Metres, with a Sketch of Goethe's life, by Edgar Alfred Bowring*. John W. Parker & Son, London, 1853, 1st edn; George Bell & Sons, London, 1874, 2nd edn.

A4.225. GOETHE, J. W. VON. *Goethe's Theory of Colours; translated from the German: with notes by Charles Lock Eastlake*. John Murray, London, 1840.

A4.226. DARWIN, C. R. *Journal of researches into the geology and natural history of the various countries visited by H.M.S. Beagle, under the command of Captain Fitzroy R.N. from 1832 to 1836*. Henry Colburn, London, 1839.

A4.227. DARWIN, C. R. *Journal of researches into the natural history and geology of the countries visited during the voyage of H.M.S. Beagle round the world, under the command of Capt. Fitz Roy R.N.* John Murray, London, 1860, Ch. 6, 115–116.

A4.228. SIMPICH, F. Life on the Argentine pampa. *The National Geographic Magazine*, 1933. 64, 4 (Oct.), 449–492.

A4.229. RUSSELL, F. A. R. *On hail*. Edward Stanford, London, 1893, 224 pp.

A4.230. ANDERSEN, H. C. *Hans Andersen's Fairy Tales. A new translation by Mrs H. B. Paull. With a special adaption and arrangement for young people*. Frederick Warne & Co., London & New York, 1887. (Series: *The Chandos Classics*).

A4.231. LÖNNROT, E. *The Kalevala. The Epic Poem of Finland into English. By John Martin Crawford. In two volumes*. John A. Berry & Co., New York, 1888.

A4.232. KREUTZWALD, F. R. *The Hero of Esthonia and other studies in the romantic literature of that country. Compiled from Esthonian and German sources by W. F. Kirby. In two volumes*. John C. Nimmo, London, 1895.

A4.233. PUMPURS, A. *Bearslayer by Andrejs Pumpurs. A free translation from the unrhymed Latvian into English heroic verse, by Arthur Cropley*. Private Publn (A. J. Cropley), Adelaide, 2005.

A4.234. THACKERAY, W. M. (alias M. A. TITMARSH). *The Irish Sketch-Book. By Mr M. A. Titmarsh. With numerous engravings on wood, drawn by the author*. Chapman & Hall, Strand, London, 1843, 2 vols.

A4.235. LONGFELLOW, H. W. *Evangeline, a Tale of Acadie. By Henry Wadsworth Longfellow*. William D. Ticknor & Co., Boston, 1847.

A4.236. LONGFELLOW, H. W. *Kéramos and Other Poems by Henry Wadsworth Longfellow*. Houghton, Osgood & Co., Boston, 1878.

A4.237. TENNYSON, A. (later 1ST BARON TENNYSON). *Poems*. Edward Moxon, Dover Street, London, 1842, Vol. 2.

A4.238. MORRIS, W. *The Defence of Guenevere, and other Poems. By William Morris*. Bell & Daldy, Fleet Street, London, 1858.

A4.239. HOPKINS, G. M. *Poems of Gerard Manley Hopkins, now first published. Edited with notes by Robert Bridges, Poet Laureate*. Humphrey Milford, London, 1918.

A4.240. SWINBURNE, A. C. *Poems and Ballads*. Edward Moxon & Co., London, 1866.

A4.241. HUGO, V. M. *The works of Victor Hugo.* Jefferson Press, Boston, 1900, Vol. 9 (*Poems, Dramas. Poems of Victor Hugo, translated into English by various authors, now first collected by Henry Llewellyn Williams. Dramas of Victor Hugo, translated by Frederick L. Slous and Mrs Newton Crosland*).

A4.242. ABICH, M. Hailstorms in Russian Georgia. *Phil. Mag.*, 1869, Ser. 4, 38, 257 (Dec.), 440–441 (& Plate).

A4.243. DODGSON, C. L. (*alias* LEWIS CARROLL). *The Hunting of the Snark: an Agony, in eight Fits. By Lewis Carroll, with nine illustrations by Henry Holiday.* Macmillan & Co., London, 1876.

A4.244. WEBBER, R. (with additional material by P. A. CHING). *R. D. Blackmore: author and horticulturist of Teddington.* Borough of Twickenham Local History Society, 1980, Paper 44, 29 pp.

A4.245. TOLSTOY, L. N. *Anna Karenin. A novel by Leo Tolstoy. A new and complete translation from the Russian, by Constance Garnett. In two volumes.* William Heinemann Ltd, London, 1901, Vol. 1, Pt 3, Ch. 30.

A4.246. WILDE, O. F. O'F. W. *The Happy Prince and Other Tales by Oscar Wilde. Illustrated by Walter Crane and Jacomb Hood.* David Nutt, London, 1888.

A4.247. KIPLING, J. R. *The Overland Mail.* In: *Departmental Ditties and Other Verses.* Civil & Military Gazette, Lahore, Sept. 1886, 2nd edn.

A4.248. KIPLING, J. R. *Poems and Ballads. By Rudyard Kipling. With original illustrations by Victor A. Searles.* H. M. Caldwell Co., New York, 1899 (Oriental edn), 129–130 (& Plate).

A4.249. KIPLING, J. R. McAndrews' Hymn. By Rudyard Kipling. Illustrations by Howard Pyle. *Scribner's Magazine, New York*, 1894, 16, 6 (Dec.), 667–674.

A4.250. BELL, G. M. L. *Safar Nameh: Persian Pictures. A book of travel.* Richard Bentley & Son, London, 1894.

A4.251. WELLS, H. G. *The Time Machine. An Invention.* William Heinemann Ltd, London, 1895.

A4.252. CONRAD, J. *The Nigger of the 'Narcissus'. A Tale of the Sea. By Joseph Conrad.* William Heinemann Ltd, London, 1897.

A4.253. SETH, S. *Subject lessons: the Western education of colonial India.* Duke Univ. Press, Durham, 2007, 163.

A4.254. KAWAGUCHI, E. *Three years in Tibet, with the original Japanese illustrations. By the Shramana Ekai Kawaguchi.* The Theosophist Office, Adyar, Madras; Theosophical Publishing Society, Benares and London; 1909.

A4.255. YEATS, W. B. Baile and Aillinn. *The Monthly Review* (ed. Henry Newbolt), 1902, 8, 22, 1 (Jul.), 156–164.

A4.256. ABERCROMBIE, L. *Ryton Firs.* In: *Georgian Poetry 1920-1922.* The Poetry Bookshop, London, 1922 (ed. E. H. Marsh), 3–7.

A4.257. LAWRENCE, D. H. *Kangaroo.* Martin Secker Ltd, London, 1923.

A4.258. LAWRENCE, D. H. *The Lost Girl.* Martin Secker Ltd, London, 1920.

A4.259. LAWRENCE, D. H. *Fantasia of the Unconscious.* Martin Secker Ltd, London, 1923.

A4.260. GEIKIE, A. *The Love of Nature among the Romans during the later decades of the Republic and the first century of the Empire.* John Murray, London, 1912, Ch. 11 (*The seasons*), 247–248.

A4.261. TALMAN, C. F. *The realm of the air: a book about weather.* Bobbs-Merrill Co., Indianapolis, 1931, Ch. 7 (*Hail and the damage it does*), 75–82.

A4.262. LAWRENCE, T. E. *Seven Pillars of Wisdom: a Triumph.* Jonathan Cape, London, 1935.

A4.263. KRAFT, W. *Figur der Hoffnung. Ausgewählte Gedichte 1925–1953*. Lambert Schneider, Heidelberg, 1955, 7–8.

A4.264. JESSEN, C. Tradition of loss: Werner Kraft on Franz Kafka. In: *Kafka after Kafka: Dialogical engagement with his works from the Holocaust to Postmodernism*. Camden House, New York, 2019 (eds Iris Bruce and Mark H. Gelber), Pt 1 (*Philosophical and literary hermeneutics after the Holocaust*), Ch. 1, 11–28. (Series: *Studies in German literature, linguistics, and culture*).

A4.265. BUCHAN, J. *The Island of Sheep*. Hodder and Stoughton Ltd, London, 1936.

A4.266. THOMAS, D. M. *18 Poems. Dylan Thomas*. Sunday Referee & Parton Bookshop, London, 1934.

A4.267. THOMAS, D. M. In Country Sleep. *Horizon (Lond.)*, 1947, 16, 96 (Dec.), 302–305 (with misprints).

A4.268. THOMAS, D. M. *Collected Poems 1934–1952. Dylan Thomas*. J. M. Dent & Sons, London, 1952.

A4.269. DAHL, R. *James and the Giant Peach. Illustrated by Nancy Ekholm Burkert*. Alfred A. Knopf, Inc., New York, 1961.

A4.270. DAHL, R. *James and the Giant Peach. Illustrated by Quentin Blake*. Viking Penguin Books Ltd, London, 1995.

A4.271. BERGIN, T. G. For a space prober. In: *Advances in Geophysics*. Academic Press, New York, 1971 (eds H. E. Landsberg and J. Van Mieghem), Vol. 15, 136.

A4.272. GRAVES, S. D. B., MCKAY, C.P., GRIFFITH, C. A., FERRI, F. and FULCHIGNONI, M. Rain and hail can reach the surface of Titan. *Planetary & Space Sci.*, 2008, 56, 3–4 (Mar.), 346–357.

A4.273. CERVENY, R. S., BESSEMOULIN, P., BURT, C. C., COOPER, M. A., DEWAN, A., FINCH, J., HOLLE, R. L., KALKSTEIN, L., KRUGER, A., LEE, T-C., MARTÍNEZ, R., MOHAPATRA, M., PATTANAIK, D. R., PETERSON, T. C., SHERIDAN, S., TREWIN, B., TAIT, A., WAHAB, M. M. A. and ZHANG, C. *WMO assessment of weather and climate mortality extremes: lightning, tropical cyclones, tornadoes, and hail (World Meteorological Organization)*. Weather, Climate & Society (*Amer. Meteorological Soc.*), 2017, 9, Jul., 487–497.

A4.274. SINGH, S. K., SAXENA, R., PORWAL, A., RAY, N. and RAY, S. S. Assessment of hailstorm damage in wheat crop using remote sensing. *Current Science*, 2017, 112, 10, May, 2095–2100.

A4.275. BAL, S. K., MINHAS, P. S., SINGH, Y., KUMAR, M., PATEL, D. P., RANE, J., SURESH KUMAR, P., RATNAKUMAR, P., CHOUDHURY, B. U. and SINGH, N. P. Coping with hailstorm in vulnerable Deccan Plateau region of India: technological interventions for crop recovery. *Current Science*, 2017, 113, 10, Nov., 2021–2027.

A4.276. LUCIUS ANNAEUS SENECA (SENECA THE YOUNGER). *Physical science in the time of Nero. Being a translation of the Quaestiones Naturales of Seneca. By John Clarke, with notes on the treatise by Archibald Geikie*. Macmillan & Co. Ltd, London, 1910, Bk 4 (*Containing a discussion of snow, hail, and rain*), 181–182.

A4.277. COOK, A. B. *Zeus: a Study in Ancient Religion, by Arthur Bernard Cook*. University Press, Cambridge, 1940, Vol. 3 (*Zeus God of the Dark Sky; earthquakes, clouds, wind, dew, rain, meteorites*), Pt 1 (*Text and notes*), 875–881 (*Zeus and the Hail*).

A4.278. HASLUCK, F. W. Unpublished inscriptions from Cyzicus neighbourhood. *J. Hellenic Studies*, 1904, 24, Nov., 20–40.

A4.279. BELLUCCI, G. *La grandine nell'Umbria, con note esplicative e comparative e con illustrazioni*. Unione Tipografica Cooperativa Editrice, Perugia, 1903. (Series: *Tradizioni popolari Italiane*, Nr 1).

A4.280. FEHRLE, E. *Studien zu den griechischen Geoponikern*. B. G. Teubner, Leipzig, 1920. (Series: *Studien zur Geschichte des antiken Weltbildes und der griechischen Wissenschaft*, Vol. 3).

A4.281. STEGEMANN, V. *Hagel, Hagelzauber*. In: *Handwörterbuch des deutschen Aberglaubens*. Walter de Gruyter, Berlin (eds Hanns Bächtold-Stäubli and Eduard Hoffmann-Krayer), Vol. 3, 1931, 1304–1320.

A4.282. RODGERS, R. H. Hail, frosts, and pests in the vineyard: Anatolius of Berytus as a source for the Nabataean Agriculture. *J. Amer. Oriental Soc.*, 1980, 100, 1 (Jan.–Mar.), 1–11.

A4.283. GAIUS PLINIUS SECUNDUS (PLINY THE ELDER). *Pliny: Natural History. In ten volumes, with an English translation by H. Rackham*. Harvard Univ. Press, Cambridge; William Heinemann Ltd, London; 1938, 1st edn, 1949 revised edn, Vol. 1, Bk 2, 289. (Series: *Loeb Classical Library*, Vol. 330).

A4.284. GAIUS PLINIUS SECUNDUS (PLINY THE ELDER). *The Natural History of Pliny*. Henry G. Bohn, London, 1856 (transl. J. Bostock and H. T. Riley), Vol. 5, Bk 28, Ch. 5.

A4.285. GAIUS PLINIUS CAECILIUS SECUNDUS (PLINY THE YOUNGER). *The letters of the Younger Pliny, literally translated by John Delaware Lewis*. Kegan Paul, Trench, Trübner & Co. Ltd, London, 1890.

A4.286. GAIUS PLINIUS CAECILIUS SECUNDUS (PLINY THE YOUNGER). *The letters of the Younger Pliny: translated by John B. Firth*. Walter Scott Ltd, London, 1900, 2 vols (ed. Heinrich Keil).

A4.287. CYRIAX, T. *Among Italian peasants: written and illustrated by Tony Cyriax, with an introduction by Muirhead Bone*. W. Collins Sons & Co. Ltd, London, 1919, Ch. 14 (*Hailstones*), 187–193.

A4.288. MALOSSINI, A. *50 modi (non troppo sicuri) per difendersi dalla grandine*. Associazione Regionale dei Consorzi di Difesa dell'Emilia-Romagna, Bologna, 1992, 75 pp.

A4.289. KOTANSKY, R. D. Two amulets against hailstorm. In: *Greek magical amulets. The inscribed gold, silver, copper, and bronze lamellae. Part I. Published texts of known provenance. Text and Commentary by Roy Kotansky*. Westdeutscher Verlag, Opladen, 1994, 46–53. (Series: *Papyrologica Coloniensia*, Vol. 22-1).

A4.290. AUDOLLENT, A. Double inscription prophylactique contre la grêle, sur une croix de plomb trouvée en Tunisie. *Mémoires de l'Institut national de France*, 1951, 43, 2, 45–76.

A4.291. NIETO, F. J. F. A Visigothic charm from Asturias and the classical tradition of phylacteries against hail. *Proc. Int. Conf. Magical Practice in Latin West, Univ. Zaragoza, Spain, Sept. 2005*. In: *Magical practice in the Latin West*. Brill, Leiden, 2010 (eds R. L. Gordon and F. M. Simón), Ch. 16, 551–599.

A4.292. GRÉGOIRE, H. *Recueil des inscriptions grecques-chrétiennes d'Asie mineure*. E. Leroux, Paris, 1922; reprint, A. M. Hakkert, Amsterdam, 1968.

A4.293. DELATTE, A. *Anecdota Atheniensia*. H. Vaillant-Carmanne, Liége, É. Champion, Paris, 1927, Vol. 1 (*Textes grecs inédits relatifs à l'histoire des religions*). (Series: *Bibliothèque de la Faculté de philosophie et lettres de l'Université de Liége*, Fasc. 36).

A4.294. MANGANARO, G. Nuovi documenti magici della Sicilia orientale. *Rendiconti della Classe di Scienze morali, storiche e filologiche dell'Accademia dei Lincei*, 1963, Ser. 8, 18, 57–74.

A4.295. FERCHIOU, N. and GABILLON, A. Une inscription grecque magique de la région de Bou Arada (Tunisie), ou les quatre plaies de l'agriculture antique en Proconsulaire. *Bulletin archéologique du Comité des travaux historiques et scientifiques*, 1983, 19b, 109–123.

A4.296. FARAONE, C. A. *The transformation of Greek amulets in Roman imperial times*. Univ. Pennsylvania Press, Philadelphia, 2018.

A4.297. PAUSANIAS. *Description of Greece*. 2nd century A.D. English language versions. *Pausanias's description of Greece*. Macmillan, London, 1898, Vol. I (transl. J. G. Frazer) & Vol. III (commentary on Books II–V by J. G. Frazer). *Pausanias: description of Greece*. William Heinemann Ltd, London, 1918, Vol. 1 (Books I & II), with Greek text (transl. W. H. S. Jones).

A4.298. OLAUS MAGNUS. *Historia de Gentibus Septentrionalibus*. Rome, 1555. English language version: *A description of the Northern Peoples, 1555*. Hakluyt Society, London, 1996–1998, 3 vols (transl. P. Fisher and H. Higgens; ed. P. Foote).

A4.299. WHITE, A. D. *A history of the warfare of science with theology in Christendom. Two volumes combined*. D. Appleton & Co., New York, 1898.

A4.300. ODDIE, B. C. V. The hail cannon: an early attempt at weather control. *Weather*, 1965, 20, 5 (May), 154–156.

A4.301. KOVAČIČ, M. The bell and its symbolic role in Slovenia. *Proc. 16th Int. Mtg ICTM Study Group on Folk Musical Instruments, Vilnius, Lithuania, Apr. 2006*, 105–116.

A4.302. ROMAN RITUAL (RITUALE ROMANUM). Bruce Publ. Co., Milwaukee, 1964 (ed. Philip T. Weller).

A4.303. ANDREWS, T. *Dictionary of nature myths: legends of the earth, sea, and sky*. Oxford Univ. Press, Oxford, 1998.

A4.304. FRAZER, J. G. *The Golden Bough: a study in comparative religion*. Macmillan, London, 1890, 1st edn, 2 vols; *The Golden Bough: a study of magic and religion*. Macmillan, London, 1900, 2nd edn, 3 vols; 1906–1915, 3rd edn, 12 vols; 1922, abridged edn, 1 vol.

A4.305. FRAZER, J. G. *Aftermath: a supplement to The Golden Bough*. Macmillan, London, 1936.

A4.306. LESCHEVIN, M. Memoir upon a process employed in the ci-devant Mâconnais of France, to avert showers of hail, and to dissipate storms. *Phil. Mag.*, 1806, 1st Ser., 26, 103, 212–218.

A4.307. ANONYMOUS. Hail-rod. *Amer. J. Sci. & Arts*, 1825, 1st Ser., 10, 1, Oct., 196–198.

A4.308. MURRAY, J. On the paragrêle or protector from hail. *Edinburgh New Phil. J.*, 1827, 3, 103–107.

A4.309. ANONYMOUS. On the efficacy of paragrêles. *Amer. J. Sci. & Arts*, 1828, 1st Ser., 14, 1, Jul., 37–40.

A4.310. ANONYMOUS. Protection against hail. *Amer. J. Sci. & Arts*, 1854, 2nd Ser., 18, 54, Nov., 432–433.

A4.311. ROWELL, G. A. Forests and hailstorms. *The Builder (Lond.)*, 1882, 43, Dec., 730.

A4.312. HILGARD, E. W. Prevention of hail. *Science*, 1900, New Ser., 11, 265, Jan., 153.

A4.313. SIGAUX, J. La défense des vignes contre la grêle par le tir du canon. *Revue Scientifique*, 1900, 4th Ser., 14, Oct., 461–464.

A4.314. WARD, R. DEC. Current notes on meteorology. Hail prevention by cannonading. *Science*, 1901, New Ser., 14, 363, Dec., 938.

A4.315. ABBE, C. (Ed.). Prevention of hail by cannonading. *Monthly Weather Rev.*, 1900, 28, 6 (Jun.), 251–252.

A4.316. ABBE, C. (Ed.). Bombarding hail clouds. *Monthly Weather Rev.*, 1900, 28, 12 (Dec.), 542–543.

A4.317. ABBE, C. (Ed.). The third international congress on hail shooting. *Monthly Weather Rev.*, 1902, 30, 1 (Jan.), 33–35.

A4.318. ABBE, C. (Ed.). Bombarding against hail. *Monthly Weather Rev.*, 1903, 31, 1 (Jan.), 30.

A4.319. ABBE, C. (Ed.). Cannonading against hail. *Monthly Weather Rev.*, 1904, 32, 7 (Jul.), 328–329.

A4.320. ABBE, C. (Ed.). Hail shooting in Italy. *Monthly Weather Rev.*, 1907, 35, 8 (Aug.), 358.

A4.321. PERNTER, J. M. Damage by hail in spite of cannonading. *Monthly Weather Rev.*, 1901, 29, 3 (Mar.), 117.

A4.322. PERNTER, J. M. Das Ende des Wetterschiessens. *Meteorologische Zeitschrift*, 1907, 24, 3 (Mar.), 97–102.

A4.323. SHAW, W. N. Hailstorm artillery. *Nature*, 1901, 64, 1650, Jun., 159–161.

A4.324. SHAW, W. N. *Manual of meteorology*. Cambridge Univ. Press, Cambridge, 1926, Vol. 1 (*Meteorology in history*).

A4.325. PLUMANDON, J. R. Les orages et la grêle. *Encyclopédie Scientifique des Aide-Mémoire, Paris, 1901*, 1–26.

A4.326. PLUMANDON, J. R. General Report on hail shooting presented to the Congress at Lyons (transl. R. S. Hotze). *Monthly Weather Rev.*, 1902, 30, 1 (Jan.), 35–38.

A4.327. PLUMANDON, J. R. Cannon and hail. *Monthly Weather Rev.*, 1902, 30, 13 (Dec.), 604–607.

A4.328. ANONYMOUS. International Conference on Weather-Shooting. *Nature*, 1903, 67, 1731, Jan., 213.

A4.329. DASTRE, A. Revue scientifique: la lutte contre la grêle. *Revue des Deux Mondes*, 1905, 5th Ser., 27, 697–708.

A4.330. ANGOT, A. Electric paragrêles. *Monthly Weather Rev.*, 1914, 42, 3 (Mar.), 166–167 (transl. R. E. Edwards).

A4.331. ANONYMOUS. Guns that protect crops from the ravages of hailstorms. *Scientific American*, 1916, 114, 22, May, 557.

A4.332. LUDLAM, F. H. The hail problem. *Nubila* (Verona, Italy), 1958, 1, 1, 12–96.

A4.333. HENINGER, S. K. *A handbook of Renaissance meteorology, with particular reference to Elizabethan and Jacobean literature*. Duke Univ. Press, Durham, 1960.

A4.334. ROBERTS, W. O. Russian hail-suppression experiments. *Science*, 1967, 156, 3782, Jun., 1580.

A4.335. MORGAN, G. M. A general description of the hail problem in the Po Valley of northern Italy. *J. Appl. Meteorology*, 1973, 12, 2 (Mar.), 338–353. Discussion: 1974, 13, 1 (Feb.), 182–184.

A4.336. ATLAS, D. The paradox of hail suppression. *Science*, 1977, 195, 4274, Jan., 139–145.

A4.337. CHANGNON, S. A., FARHAR, B. C. and SWANSON, E. R. Hail suppression and society. *Science*, 1978, 200, 4340, Apr., 387–394.

A4.338. CHANGNON, S. A. and IVENS, J. L. History repeated: the forgotten hail cannons of Europe. *Bull. Amer. Meteorological Soc.*, 1981, 62, 3 (Mar.), 368–375.

A4.339. STEINER, J. T. Can we reduce the hail problem? *Weather & Climate*, 1988, 8, 1 (Feb.), 23–32.

A4.340. WIERINGA, J. and HOLLEMAN, I. If cannons cannot fight hail, what else? *Meteorologische Zeitschrift*, 2006, 15, 6 (Dec.), 659–669.

A4.341. MORGAN, G. M. The return of the anti-hail cannons. *Weatherwise*, 2008, 61, 4 (Jul.–Aug), 14–19.

A4.342. BAKER, A. R. H. Hail as hazard: changing attitudes to crop protection against hail damage in France, 1815–1914. *Agricultural History Rev.*, 2012, 60, 1, 19–36.

A4.343. ANONYMOUS. Le canon contre la grêle: les expériences dans le Beaujolais. *Le Petit Parisien: Supplément littéraire illustré*, 1900, Nr 607, Dimanche 23 Septembre 1900, 303–304.

A4.344. ANONYMOUS. L'artillerie contre la grêle. *Le Petit Journal: Supplément illustré*, 1901, Nr 555, Dimanche 7 Juillet, 215–216.

A4.345. ANONYMOUS. Canonnade pacifique. *Le Petit Journal: Supplément illustré*, 1910, Nr 1020, Dimanche 5 Juin, 177–178.

A4.346. ABSHAEV, A. M. Crystallizing agent dispersion at rocket and artillery seeding of hailstorms. *Proc. 8th W.M.O. Sci. Conf. Weather Modification, Casablanca, Morocco, Apr. 2003*. World Meteorological Organisation, Geneva, Switzerland, 2003, WMO/TD 1146, 357–360.

A4.347. ABSHAEV, M. T., SULAKVELIDZE, G. K., BURTSEV, I. I., FEDCHENKO, L. M., JEKAMUKHOV, M. K., ABSHAEV, A. M., KUZNETSOV, B. K., MALKAROVA, A. M., TEBUEV, A. D., NESMEYANOV, P. A., SHAKIROV, I. N. and SHEVELA, G. F. Development of rocket and artillery technology for hail suppression. In: *Achievements in weather modification*. Department of Atmospheric Studies, Ministry for Presidential Affairs, Abu Dhabi, U.A.E., Feb. 2006 (ed. Rumen D. Bojkov), 109–127.

A4.348. FAVREAU, R. F. and GOYER, G. G. The effect of shock waves on a hailstone model. *J. Appl. Meteorology*, 1967, 6, 2 (Apr.), 326–335.

A4.349. SCHLEUSENER, R. A. Hailfall damage suppression by cloud seeding; a review of the evidence. *J. Appl. Meteorology*, 1968, 7, 6 (Dec.), 1004–1011.

A4.350. SUMMERS, P. W., MATHER, G. K. and TREDDENICK, D. S. The development and testing of an airborne droppable pyrotechnic flare system for seeding Alberta hailstorms. *J. Appl. Meteorology*, 1972, 11, 4 (Jun.), 695–703.

A4.351. MARWITZ, J. D. Hailstorms and hail suppression techniques in the U.S.S.R. – 1972. *Bull. Amer. Meteorological Soc.*, 1973, 54, 4 (Apr.), 317–325.

A4.352. BATTAN, L. J. Weather modification in the Soviet Union – 1976. *Bull. Amer. Meteorological Soc.*, 1977, 58, 1 (Jan.), 4–19.

A4.353. BROWNING, K. A. and ATLAS, D. Some new approaches in hail suppression experiments. *J. Appl. Meteorology*, 1977, 16, 4 (Apr.), 327–332.

A4.354. FOOTE, G. B. and KNIGHT, C. A. (Eds). *Hail: a review of hail science and hail suppression*. American Meteorological Society, Boston, Meteorological Monograph Vol. 16, Nr 38, Dec. 1977, 295 pp.

A4.355. PALUCH, I. R. Size sorting of hail in a three-dimensional updraft and implications for hail suppression. *J. Appl. Meteorology*, 1978, 17, 6 (Jun.), 763–777.

A4.356. SWANSON, E. R., SONKA, S. T., TAYLOR, C. R. and BLOKLAND, P. J. VAN. An economic analysis of hail suppression. *J. Appl. Meteorology*, 1978, 17, 10 (Oct.), 1432–1440.

A4.357. KNIGHT, C. A., FOOTE, G. B. and SUMMERS, P. W. Results of a randomised hail suppression experiment in northeast Colorado. Part IX: overall discussion and summary in the context of physical research. *J. Appl. Meteorology*, 1979, 18, 12 (Dec.), 1629–1639.

A4.358. FEDERER, B., WALDVOGEL, A., SCHMID, W., SCHIESSER, H. H., HAMPEL, F., SCHWEINGRUBER, M., STAHEL, W., BADER, J., MEZEIX, J. F., DORAS, N., D'AUBIGNY, G., DERMEGREDITCHIAN, G. and VENTO, D. Main results of Grossversuch IV. *J. Appl. Meteorology & Climatology*, 1986, 25, 7 (Jul.), 917–957.

A4.359. CHENG, L. An attempt to normalise the hailstorm variability for the evaluation of cloud seeding. *J. Climate & Appl. Meteorology*, 1987, 26, 4 (Apr.), 443–456.

A4.360. ZOTOV, YE. I., NIKORICH, T. D., NIKORICH, V. D., PLAUDE, N. O., POTAPOV, YE. I. and UTKINA, G. V. Some characteristics of the atmospheric aerosol in Moldavia. *J. Aerosol Sci.*, 1991, 22, Suppl. 1, S649–S652.

A4.361. MESINGER, F. and MESINGER, N. Has hail suppression in eastern Yugoslavia led to a reduction in the frequency of hail? *J. Appl. Meteorology & Climatology*, 1992, 31, 1 (Jan.), 104–111.

A4.362. MORGAN, G. M. Some results of aircraft investigation of internal properties of thunderstorm cloud systems in northeastern Italy with an interpretation for hail prevention. *Atmospheric Res.*, 1992, 28, 3–4 (Dec.), 259–269.

A4.363. SIMEONOV, P. Comparative study of the hail suppression efficiency in Bulgaria and in France. *Atmospheric Res.*, 1992, 28, 3–4 (Dec.), 227–235.

A4.364. SIMEONOV, P. An overview of crop hail damage and evaluation of hail suppression efficiency in Bulgaria. *J. Appl. Meteorology*, 1996, 35, 9 (Sept.), 1574–1581.

A4.365. ĆURIĆ, M., JANC, D. and VUČKOVIĆ, V. The influence of cloud drip size distribution on simulation seeding effects of hail-bearing cloud. *J. Weather Modification*, 1997, 29, 1 (Apr.), 70–73.

A4.366. DESSENS, J. A physical evaluation of a hail suppression project with silver iodide ground burners in southwestern France. *J. Appl. Meteorology*, 1998, 37, 12 (Dec.), 1588–1599.

A4.367. ZHENG, G. An overview of weather modification activities in China. *Proc. 8th W.M.O. Sci. Conf. Weather Modification, Casablanca, Morocco, Apr. 2003*. World Meteorological Organisation, Geneva, Switzerland, 2003, Rep. WMO/TD 1146, 25–29.

A4.368. POČAKAL, D. and ŠTALEC, J. Statistical analysis of hail characteristics in the hail-protected western part of Croatia using data from hail suppression stations. *Atmospheric Res.*, 2003, 67–68 (Jul.–Sept.), 533–540.

A4.369. KRAUSS, T. W. and SANTOS, J. R. Exploratory analysis of the effect of hail suppression operations on precipitation in Alberta. *Atmospheric Res.*, 2004, 71, 1–2 (Jul.), 35–50.

A4.370. GUO, X., ZHENG, G. and JIN, D. A numerical comparison study of cloud seeding by silver iodide and liquid carbon dioxide. *Atmospheric Res.*, 2006, 79, 3–4 (Mar.), 183–226.

A4.371. MAKITOV, V. Radar measurements of integral parameters of hailstorms used on hail suppression projects. *Atmospheric Res.*, 2007, 83, 2–4 (Feb.), 380–388.

A4.372. TSAGALIDIS, E. G. and GEORGIOU, A. C. Using simulation modelling to support decisions in hail suppression programmes with airborne means. *J. Operational Res. Soc.*, 2009, 60, 1 (Jan.), 14–22.

A4.373. GAVRILOV, M. B., LASIĆ, L., PEŠIC, A., MILUTINOVIĆ, M., MARKOVIĆ, D., STANKOVIĆ, A. and GAVRILOV, M. M. Influence of hail suppression on the hail trend in Serbia. *Physical Geography*, 2010, 31, 5, May, 441–454.

A4.374. CHEN, B. and XIAO, H. Silver iodide seeding impact on the microphysics and dynamics of convective clouds in the high plains. *Atmospheric Res.*, 2010, 96, 2–3 (May), 186–207.

A4.375. YANG, H-L., XIAO, H. and HONG, Y-C. A numerical study of aerosol effects on cloud microphysical processes of hailstorm clouds. *Atmospheric Res.*, 2011, 102, 4 (Dec.), 432–443.

A4.376. ABSHAEV, M. T., ABSHAEV, A. M. and MALKAROVA, A. M. Estimation of antihail projects efficiency considering the tendency of hail climatology change. *Proc. 10th W.M.O. Sci. Conf. Weather Modification, Bali, Indonesia, Oct. 2011*. World Meteorological Organisation, Geneva, Switzerland, Apr. 2012, WWRP 2012-2, 1–4.

A4.377. JAVANMARD, S. and PIRHAYATI, M. K. AgI cloud seeding modelling for hail suppression of cold clouds. *J. Geography & Geology*, 2012, 4, 2, 81–93.

A4.378. NAJAFI, M., JAVANMARD, S. and MOHAMMAD-HOSSEINZADEH, F. Comparative study of liquid carbon dioxide and silver iodide seeding effects on cumulonimbus clouds rainfall enhancement and hail suppression. *Int. J. Environ. Sci. Technol.*, 2015, 12, 1 (Jan.), 87–104.

A4.379. BOSCO, L. C., BERGAMASCHI, H., CARDOSO, L. S., DE PAULA, V. A., MARODIN, G. A. B. and NACHTIGALL, G. R. Apple production and quality when cultivated under anti-hail cover in Southern Brazil. *Int. J. Biometeorology*, 2015, 59, 7 (Jul.), 773–782.

A4.380. DESSENS, J., SÁNCHEZ, J. L., BERTHET, C., HERMIDA, L. and MERINO, A. Hail prevention by ground-based silver iodide generators: results of historical and modern field projects. *Atmospheric Res.*, 2016, 170, Mar., 98–111.

A4.381. Tani, S., Paulitsch, H., Teschl, R. and Süsser-Rechberger, B. Evaluation of hail suppression programme effectiveness using radar derived parameters. *European Geophysical Society General Assembly, Vienna, Austria, April 2016*, EPSC2016-13356.

A4.382. Gandorfer, M., Hartwich, A. and Bitsch, V. Hail risk management in fruit production: antihail net versus hail insurance in Germany. *Proc. 18th Int. Symp. Horticultural Economics & Management, Alnarp, Sweden, Apr. 2016 (Int. Soc. Horticultural Sci.).* In: *Acta Horticulturae*, 2016, Vol. 1132 (eds L. E. Axelson and F. Fernqvist), 19, 141–146.

A4.383. Geresdi, I., Xue, L. and Rasmussen, R. Evaluation of orographic cloud seeding using a bin microphysics scheme: two-dimensional approach. *J. Appl. Meteorology & Climatology*, 2017, 56, 5 (May), 1443–1462.

A4.384. Vujović, D. and Protić, M. The behaviour of the radar parameters of cumulonimbus clouds during cloud seeding with AgI. *Atmospheric Res.*, 2017, 189, Jun., 33–46.

A4.385. Arakelyan, A. K. A new approach in hail prevention technique for a locally restricted area. *Agricultural Sciences*, 2017, 8, 7 (Jul.), 559–571.

A4.386. Arakelyan, A. K. Flaws and advantages of current and prospective anti-hail protection methods, stations and networks (Review). *Global Adv. Res. J. Agricultural Sci.*, 2018, 7, 12 (Dec.), 366–376.

A4.387. Birsan, M., Muscalu, A., Voicea, I. and Pruteanu, A. General aspects of the extreme meteorological phenomenon: hail. *Annals Fac. Engng Hunedoara: Int. J. Engng*, 2019, 17, 2 (May), 81–88.

A4.388. Kovačević, N. Hail suppression effectiveness for varying solubility of natural aerosols in water. *Meteorol. Atmos. Phys*, 2019, 131, 3 (Jun.), 585–599.

A4.389. Bozic, V. S. Development of antihail rocket with reaction liftoff. *Int. J. Energetic Materials & Chemical Propulsion*, 2019, 18, 4, 355–366.

A4.390. Rivera, J. A., Otero, F., Tamayo, E. N. and Silva, M. Sixty years of hail suppression activities in Mendoza, Argentina: uncertainties, gaps in knowledge and future perspectives. *Frontiers Environ. Sci.*, 2020, 8, 45, May, 6 pp.

A4.391. Auf der Maur, A. and Germann, U. A re-evaluation of the Swiss hail suppression experiment using permutation techniques shows enhancement of hail energies when seeding. *Atmosphere*, 2021, 12, 1623, 21 pp.

A5. HAILSTONE IMPACT: SIMPLIFIED FLEXURAL ANALYSIS OF LAMINATED GLASS BEAM-STRIP

A5.1. Introduction

It is a curious and interesting fact – perhaps not entirely obvious to the casual observer – that in some temperate or sub-tropical regions of our planet, including parts of the east coast of Australia during the spring and summer seasons, atmospheric conditions occasionally can lead to precipitation in the form of large hailstones. In a 19th-century account by Russell [A5.1], for example:

> '*It appears that severe thunder and hailstorms are not very uncommon in New South Wales, especially in the first year after a drought, and in the summer season. In wet years they are neither severe nor numerous. In February 1847, in New South Wales, hail fell of various forms, which were compared with eggs and oranges, and some of the stones were said to be 14 inches in circumference.*'

A century later, according to Newman [A5.2] and others, a supercell thunderstorm traversed the Sydney region during the afternoon on New Year's Day in 1947, when wind-gust speeds of up to 74 km/h (46 mph) were recorded, and hailstones with a diameter greater than 8 cm were observed. The collateral structural damage throughout the city was huge, with widespread damage to property and crops, together with numerous personal injuries from hailstones and flying glass during the afternoon of the public holiday (maximum temperature 27°C, relative humidity 67%); whilst not forgetting the desperate plight of countless sunbathers frolicking on Bondi beach and elsewhere under such precipitous heavenly bombardment.

In particular, there was extensive fragmentation of the historic glass roof at the Central railway station, thereby endangering the public below; a possible scenario that could be envisaged within the concourse areas beneath the north-facing glass walls of the Opera House, had not laminated safety glass been deployed. Indeed, the unfinished glass walls endured a substantial hailstorm in August 1971 without damage, although it appears from Bahr *et al.* [A5.3] and others that the hailstones seldom exceeded 2 cm in diameter.

Earlier hailstorms occurred in the Sydney region during October 1919 and July 1931, while possibly more severe subsequent hailstorms struck in April 1999 and December 2018.

Among many documents describing various aspects of such events are those by Yeo *et al.* [A5.4], Lee and Collings [A5.5], Buckley *et al.* [A5.6] and Schuster *et al.* [A5.7]–[A5.9], together with more general revues by Leslie *et al.* [A5.10], Crompton and McAneney [A5.11], Rasuly *et al.* [A5.12], Allen and Allen [A5.13], Warren *et al.* [A5.14] and Dowdy *et al.* [A5.15]. The photographic images in Figure A5.1 illustrate various portentous hailstorm cloud formations within the central region of the city during more recent times. It is therefore appropriate in the present context to assess the likely effect of hailstone impact upon the structural integrity of the glass walls of the Sydney Opera House.

Whereas full dynamic analyses in civil engineering practice are normally limited to investigating the wind excitation of structures such as tall buildings and suspension bridges, or the vibration characteristics of large machine foundations and similar installations, it is usually sufficient for design purposes to carry out an approximate structural analysis on the basis of quasi-static loading. A common example of this approach occurs when designing entire buildings or their external cladding components to withstand a specified maximum wind loading deemed likely to occur over a given return period, generally specified in terms of a number of years, largely depending on the usage and estimated life-span of the building.

When assessing the structural response of architectural glazing to wind pressure, it is normal practice to base the analysis on the maximum expected three-second gust pressure, treated as uniform quasi-static loading. This key assumption usually enables maximum glass panel deflections and bending stresses to readily be determined, upon reference to published theoretical solutions for the flexure of thin elastic plates under prescribed edge-support conditions. If the duration of applied loading is very much shorter, as in hailstone impact, then some form of dynamic analysis becomes necessary; and because rigorous theoretical solutions to such problems are highly complex, even on the basis of linear elasticity, simplified methods commonly are adopted in structural design.

Accordingly, attention herein is focussed on the comparatively vulnerable lower cone region of glass wall A4, where the near-horizonal laminated glass panels are essentially simply supported along two opposite edges, resting on adjacent glazing bars that diverge with radial distance from the apex on the geometrical mullion origin, with a maximum 'free' span of 2.1 m; as partially depicted in Figures 18 and 21 in the main text. Because these trapezoidal panels are butt-jointed along almost the entire length of each partial sector, the present structural calculations assume a continuous panel of constant width (2.1 m) and infinite length subjected to a uniform line load (*i.e.* the dimensions of the hailstone contact area are small compared to the 'free' span) acting on a representative beam-strip at mid-span, utilising a lumped mass and a linear elastic spring in conjunction with a single degree-of-freedom structural model (*i.e.* movements restricted to only one direction).

In what follows, brief overviews are given covering the physical characteristics and basic dynamics of natural hailstones, with an emphasis on diverse historical developments. Some of the original design calculations have been regenerated and expanded to give a more complete account herein, including the structural beam-strip analysis and summary numerical results.

HAILSTONE IMPACT: FLEXURAL ANALYSIS

(a)

(b)

FIGURE A5.1. Anonymous photographic images (1999–2018) of hailstorm cloud formations in Sydney region: (a); (b)

(c)

(d)

FIGURE A5.1. Anonymous photographic images (1999–2018) of hailstorm cloud formations in Sydney region: (c); (d)

A5.2. Hailstone characteristics

The physical processes associated with the evolutionary formation of hailstones are complex, and have been studied extensively over many decades. Reference can be made to the early work of Schumann [A5.16] and Ludlam [A5.17], and to monographs by Chisholm and English [A5.18], Gokhale [A5.19], Foote and Knight [A5.20], Pruppacher and Klett [A5.21], Wang [A5.22] and Kumar [A5.23], which cover a multiplicity of scientific subject matter. The wide variety in both external form and internal structure observed in large natural hailstones is indicative of the many factors that influence their ultimate size, shape and density, while most hailstorm observations are surrounded by an immense penumbra of theory. Some scientific aspects have yet to be resolved, and numerous investigations continue to the present day.

In general terms, hailstones tend to be formed within cumulonimbus clouds of immense height in the presence of intense updrafts and high liquid water content. This process is illustrated in the splendid composite graphics prepared for Encyclopaedia Britannica, reproduced in Figure A5.2. The rapid upward movement of initially low-level water droplets into low-temperature cloud zones can lead to progressive accretion, at a rate that depends mainly on the concentration of supercooled water encountered upon their multifarious journeys through the turbulent atmosphere during a thunderstorm. Accordingly, large hailstones frequently are of variable density and acquire a layered structure of translucent and clear ice, the opacity of the former indicating the presence of minute bubbles of entrapped air; as described by Carte [A5.24], Kidder and Carte [A5.25], Browning *et al.* [A5.26] and others. Such hailstones tend to be approximately spherical in their final form as a result of random tumbling, although multiple collisions with smaller hailstones during descent can lead to irregular shapes and distinct variations in density. Measured values of the mean density of large hailstones are typically within the range 800–900 kg/m^3, in some cases only slightly lower than the density of pure ice (917 kg/m^3), while measurements of the compressive strength of natural and artificial hailstones have been reported by Giammanco *et al.* [A5.27].

In the era of pre-history, during a geological survey in Natal, South Africa, circular or slightly elliptical depressions (about 14 mm diameter, 5 mm deep) observed in freshly exposed deposits of sandy shale were tentatively attributed by Kent [A5.28] to the action of hailstones. More direct observations on the shape and internal structure of natural hailstones have been reported by Girardin [5.29], Waller [A5.30], [A5.31], Hopkins [A5.32], Dufour [A5.33], Sutcliffe [A5.34], Nash [A5.35], Anon. [A5.36]–[A5.39], Reinsch [A5.40], Allen [A5.41], Reynolds [A5.42]–[A5.44], Ernst [A5.45], Herdman [A5.46], Wilson [A5.47], Landis and Smith [A5.48], Gregory [A5.49], Blair [A5.50], Johnson [A5.51], Lonsdale and Owston [A5.52], List and Quervain [A5.53], [A5.54], List [A5.55], Schleusener [A5.56], Browning *et al.* [A5.57], Levi *et al.* [A5.58], Rogers [A5.59], Macklin *et al.* [A5.60] and Knight and Heymsfield [A5.61], among many others. Brief early accounts summarised by Abbe [A5.62], for example, refer to very large hailstones falling at Topeka, Kansas, U.S.A., in June 1897 (ambient temperature 31 °C). The measured diameter of a dozen large specimens varied from 75 to 150 mm (average 100 mm); some were aggregations of small hailstones, while others comprised solid ice formed in onion-type layers, with one weighing 0.9 kg.

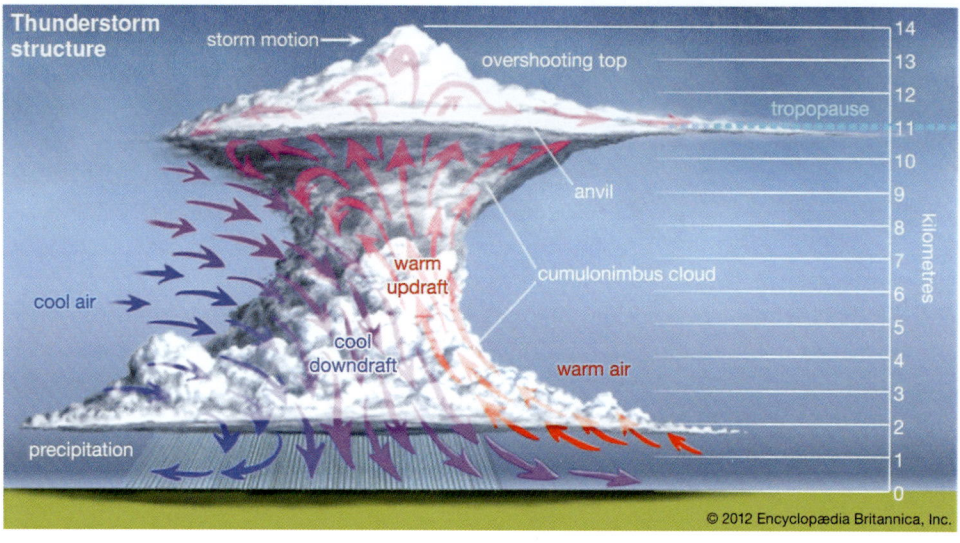

(a)

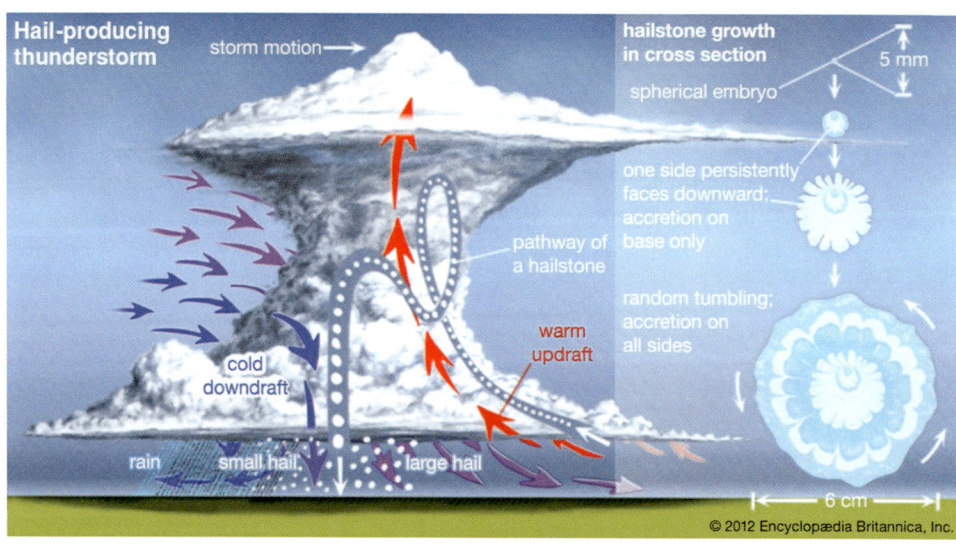

(b)

FIGURE A5.2. Characteristic atmospheric conditions for generation of large hailstones (by courtesy of *Encyclopaedia Britannica, Inc.*): (a) thunderstorm structure; (b) hail-producing thunderstorm

Lowe [A5.63] reported on large hailstones of roughly ellipsoidal shape that fell near Keewatin, Ontario, Canada, in June 1964 (ambient temperature 29 °C). A photograph of one bisected specimen is reproduced Figure A5.3, together with a sketch illustrating the internal layered structure (albeit uncommonly symmetrical), some melting of the outer skin having

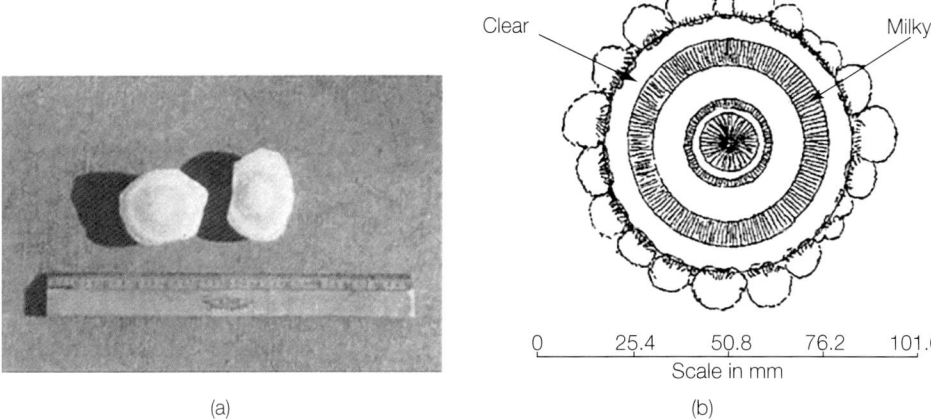

FIGURE A5.3. Large bisected hailstone from Keewatin, Ontario, June 1964, after Lowe [A5.63] (by courtesy of *Taylor & Francis Group*): (a) photograph; (b) sketch of internal structure

already occurred. Because of practical difficulties inherent in acquiring and handling large natural hailstones, much detailed laboratory work on simulating and examining their internal structure has been performed on artificial hailstones, with extensive test results for cylindrical specimens reported by Pflaum [A5.64].

To illustrate the internal lobe structure exhibited by some types of natural hailstone, photographs of thin sections taken from several large hailstones that fell near Oklahoma City, U.S.A., in May 1963 were presented by Browning and Landry [A5.65] and Browning [A5.66]. Figure A5.4 illustrates the convoluted structure of one such specimen of spheroidal shape, sectioned to contain its major axis and to pass close to its growth centre. Figure A5.4(a) shows the bubble structure visible under reflected white light, with regions of clear and milky ice appearing black and white, respectively. The grey-scale image in Figure A5.4(b) shows the general crystalline structure visible under transmitted plane polarised light, on account of variations in the relative retardation of orthogonal rays passing through crystals having different molecular forms. Detailed interpretation of these and similar images can help to elucidate the evolutionary process of forming the accreted layers of ice upon the initial small embryo. Further observations on the shape and internal structure of large natural hailstones were reported by Browning and Beimers [A5.67], in particular to explain the marked oblateness of large spheroidal specimens often encountered in severe hailstorms.

Investigations along similar lines were described in several papers by Knight [A5.68] and Knight and Knight [A5.69]–[A5.75], and included some notable coloured photographic images illustrating the crystalline texture of natural hailstones. Figure A5.5 shows thin sections from various specimens viewed under transmitted white light in a 'crossed' or 'dark-field' plane polariscope. Figure A5.5(a) refers to a hailstone collected at Boulder, Colorado, U.S.A., on 10 June 1969, with a longest cross-section dimension of 48 mm. A slightly smaller hailstone that fell at Friend, Kansas, U.S.A., on 21 June 1969 is depicted in Figure A5.5(b), where the broad array of smaller crystals was considered to be indicative

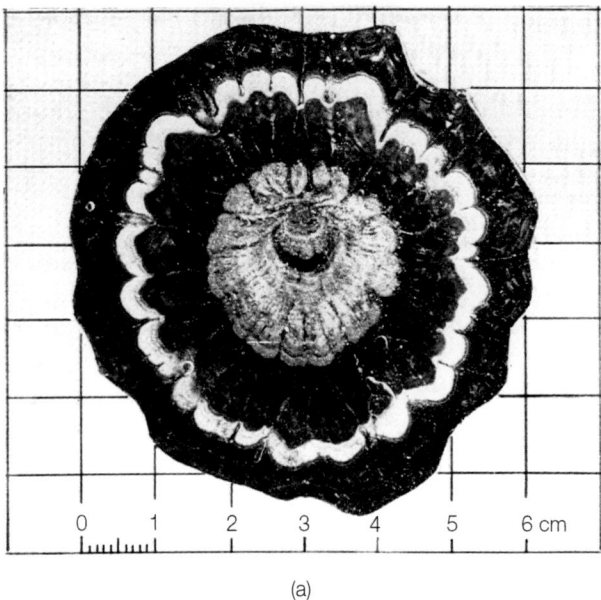

(a)

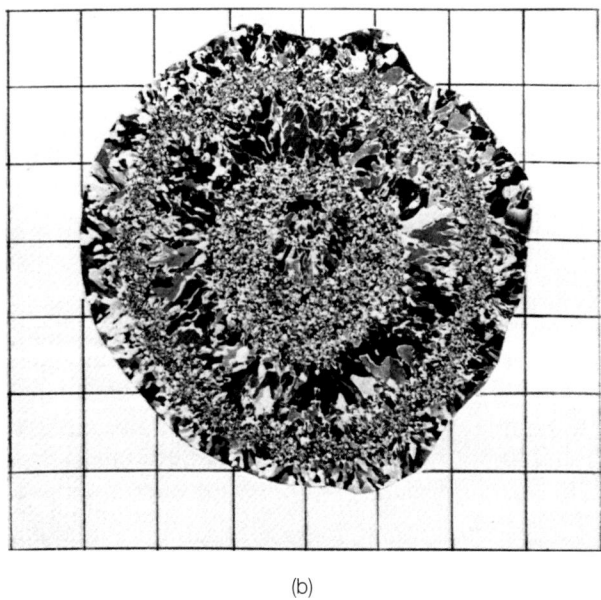

(b)

FIGURE A5.4. Photographs of thin section of large hailstone collected near Oklahoma City, U.S.A., in May 1963, after Browning [A5.66] (by courtesy of *Royal Meteorological Society*): (a) viewed under reflected white light; (b) viewed under transmitted plane polarised light

of a colder environment during formation, compared to the preceding (Boulder) specimen. Figure A5.5(c) represents a large spheroidal hailstone collected at Iowa City, U.S.A., reckoned to have tumbled symmetrically during its descent, while Figure A5.5(d) shows a thin section taken from a giant hailstone that fell at Coffeyville, Kansas, U.S.A., on 3 September 1970 – one of the largest authenticated specimens ever recorded – having a diameter of about 150 mm and an approximate weight of 0.75 kg.

Measurements of ice density within thin slices taken from natural and artificial hailstones were undertaken by Prodi [A5.76], based on photometric observations of X-ray micrographs. Steps were taken during the preparation of these planar test specimens to avoid experimental errors due to melting and re-freezing, and to variations in thickness. Measured values of the mean density of several large natural hailstones collected in the U.S.A. were in the range 820–870 kg/m^3. Figure A5.6(a) summarises the results for a small symmetric hailstone that fell in New Raymer, Colorado, U.S.A., on 10 June 1969, where the low density of the central core was ascribed to the presence of air bubbles. The plain reflected light image is shown on the left, with the corresponding X-ray micrograph on the right. Figure A5.6(b) shows the reflected light image for an exceptionally large asymmetric hailstone collected in Kansas City, U.S.A., on 21 June 1969, illustrating a sequence of opaque and translucent layers together with their measured densities. More recent field data on natural hailstone shapes are presented by Witt *et al.* [A5.77], Soderholm *et al.* [A5.78] Shedd *et al.* [A5.79].

The vast range in geographical locations prone to severe hailstorms is evident from the cornucopia of field data reported by Roth [A5.80], Frisby [A5.81] and Changnon [A5.82] in North America, Vittori and Caporiacco [A5.83], Giaiotti *et al.* [A5.84], [A5.85], Palencia *et al.* [A5.86] and Baldi *et al.* [A5.87] in Italy, Macklin *et al.* [A5.88] in England, Mossop and Kidder [A5.89] and Carte and Held [A5.90] in South Africa, Frisby and Sansom [A5.91] in the Tropics, List *et al.* [A5.92], Goyer [A5.93] and Paul [A5.94] in Canada, Federer and Waldvogel [A5.95], Waldvogel *et al.* [A5.96] and Trefalt *et al.* [A5.97] in Switzerland, Dessens [A5.98], [A5.99], Vinet [A5.100], Berthet *et al.* [A5.101] and Merino *et al.* [A5.102] in France, Steiner [A5.103] in New Zealand, Kotinis-Zambakas [A5.104] and Sioutas *et al.* [A5.105] in Greece, Michaelides *et al.* [A5.106] in Cyprus, Počakal *et al.* [A5.107] in Croatia, Webb *et al.* [A5.108] in Britain and Ireland, Sánchez *et al.* [A5.109] in France, Spain and Argentina, Tuovinen *et al.* [A5.110] and Saltikoff *et al.* [A5.111] in Finland, Xie *et al.* [A5.112] and Li *et al.* [A5.113], [A5.114] in China, Mezher *et al.* [A5.115] and Kumjian *et al.* [A5.116] in Argentina, Pilorz [A5.117] in Poland, Kahraman *et al.* [A5.118] in Turkey, Puskeiler *et al.* [A5.119] in Germany, Brázdil *et al.* [A5.120] in the Czech Republic, Martins *et al.* [A5.121] in Brazil, Jin *et al.* [A5.122] in South Korea, Zou *et al.* [A5.123] in Tibet, Wouters *et al.* [A5.124] in the Netherlands, Tani and Paulitsch [A5.125] in Austria, and Marcos *et al.* [A5.126] and Farnell *et al.* [A5.127] in northeastern Spain. Abundant records covering the worldwide distribution and size of hail are summarised by Hull [A5.128] and Prein and Holland [A5.129], while more data relating to Europe have been compiled by Suwała and Bednorz [A5.130], Punge and Kunz [A5.131] and Púčik *et al.* [A5.132]. Other related studies include those by Cheng *et al.* [A5.133], Cecil and Blankenship [A5.134], Gower *et al.* [A5.135], Martius *et al.* [A5.136] and Murillo and Homeyer [A5.137].

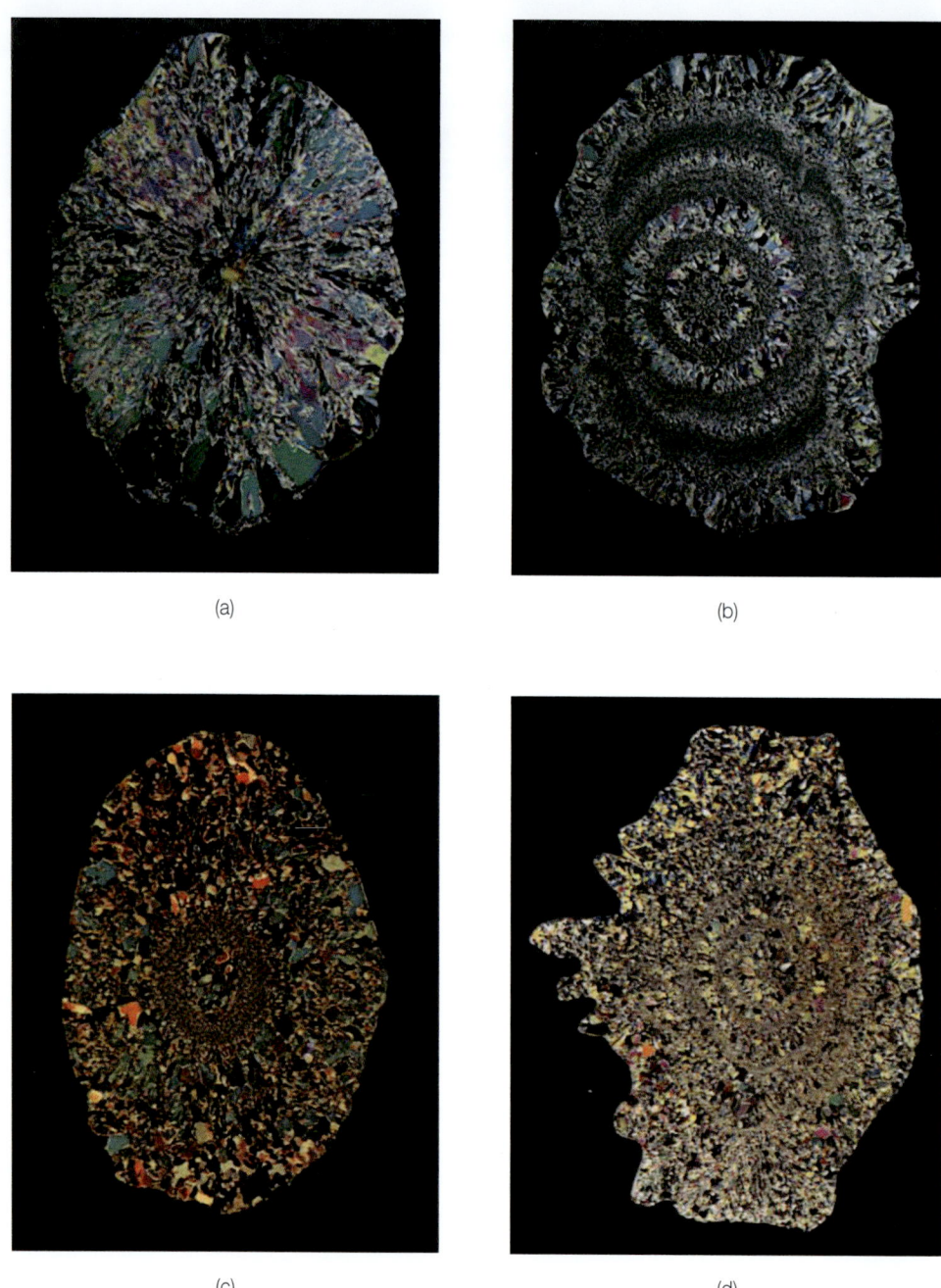

FIGURE A5.5. Photographs of thin sections from natural hailstones viewed under transmitted white light in 'crossed' plane polariscope, after Knight and Knight [A5.74] (by courtesy of *Scientific American, Inc.*): (a) from Boulder, Colorado, U.S.A., June 1969 (longest dimension 48 mm); (b) from Friend, Kansas, U.S.A., June 1969; (c) from Iowa City, U.S.A; (d) from Coffeyville, Kansas, U.S.A., September 1970 (diameter 150 mm, mass 0.75 kg)

HAILSTONE IMPACT: FLEXURAL ANALYSIS

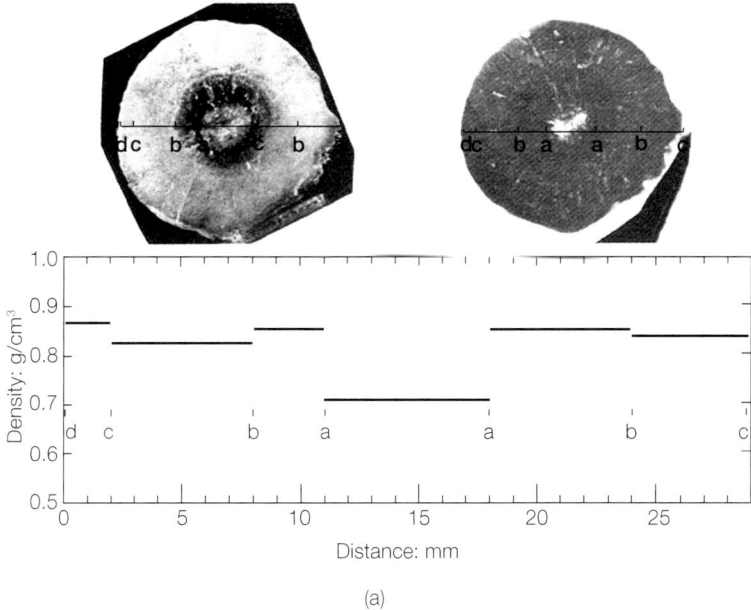

(a)

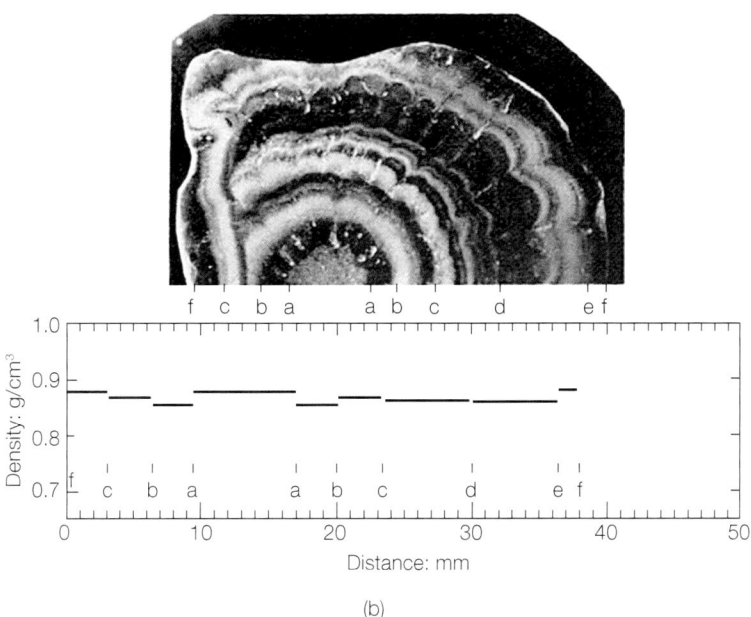

(b)

FIGURE A5.6. Measured density profiles within thin slices taken from natural hailstones in June 1969, after Prodi [A5.76] (by courtesy of *American Meteorological Society*): (a) reflected light image and X-ray micrograph, small symmetric hailstone from Raymer, Colorado, U.S.A; (b) reflected light image, large asymmetric hailstone from Kansas City, U.S.A.

A5.3. Basic hailstone dynamics

Of specific interest in the context of generalised structural design are the likely size and density of natural hailstones, and their velocity close to ground level. In the following analysis, hailstones are taken to be spherical in shape, with diameters in the range 10–100 mm; for although larger hailstones have been recorded, their occurrence is exceptionally rare. A uniform hailstone density $\rho_1 = 900$ kg/m³ is assumed for all sizes (*i.e.* close to that for pure ice, namely 917 kg/m³), while noting that non-spherical hailstones of a given weight would generally be subjected to increased drag resistance, resulting in a lower impact force than for a spherical specimen. Conversely, the effect of surface roughness might be to reduce air resistance, leading to higher terminal velocities and impact forces. Here it is deemed preferable to simulate severe yet realistic field conditions, without becoming excessively extreme in terms of theoretical probabilities.

The analysis considered herein relates to the non-rotationary motion of a sphere through an atmosphere in which the density (ρ_a) and kinematic viscosity (η_k) are constant and equal to the values at ground level; while also noting that $\eta_k = \eta_d/\rho_a$, where η_d is the dynamic viscosity. Assumed values of these quantities (corresponding to mean sea-level and an ambient temperature of 15°C) are $\rho_a = 1.225$ kg/m³, $\eta_d = 17.89$ µPa s, and $\eta_k = 14.60$ µm²/s. The sphere is taken to be rigid in the sense that it does not distort under the action of forces imposed during the descent, and the effects of air compressibility are considered to be negligible as maximum fall velocities are low compared to the speed of sound (*i.e.* Mach 1, approximately 340 m/s at 15°C at mean sea-level).

Notwithstanding the earlier work of Galileo Galilei (1564–1642), as well as that of the English mathematician and theologian John Wallis (1616–1703) and possibly others, limited historical records suggest that it was Isaac Newton (1643–1727) who first investigated the detailed nature of fluid motion around an infinitely stiff (rigid) solid object; which topic was partly covered in the original Latin version of *Philosophiae Naturalis Principia Mathematica* in 1687, and augmented in two subsequent editions [A5.138] published in 1713 and 1726. Upon referring to book 2 in the first English language translation by Andrew Motte in 1729, Newton deduced from basic reasoning that in the case of a sphere or 'globe':

> '*The resistances of globes in infinite compressed mediums are in a ratio compounded of the duplicate ratio of the velocity, and the duplicate ratio of the diameter, and the ratio of the density of the mediums.*'

To test this hypothesis, 'free-fall' experiments in air were undertaken by Newton and his collaborators inside the newly completed St Paul's Cathedral (the preceding cathedral having been destroyed during the Great Fire of London in 1666):

> '*From the top of St Paul's Church in London in June 1710, there were let fall together two glass globes, one full of quicksilver, the other of air; and in their fall they described a height of 220 English feet. A wooden table was suspended upon iron hinges on one side, and the other side of the same was supported by a wooden pin. The two globes lying upon this table were let fall together by pulling out the pin by means of an iron wire reaching from thence quite down to*

the ground; so that, the pin being removed, the table, which had then no support but the iron hinges, fell downwards; and turning round upon the hinges, gave leave to the globes to drop off from it. At the same instant, with the same pull of the iron wire that took out the pin, a pendulum oscillating to seconds was let go, and began to oscillate. The diameters and weights of the globes, and their times of falling, are exhibited in the following table.'

Further details of these spectacular experiments were reported by Hauksbee [A5.139] in 1710, and by Desaguliers [A5.140] in 1719, the latter using improved equipment and based on a greater specimen release height. Figure A5.7 illustrates the mechanism (or '*contrivance*') deployed by Desaguliers to ensure the simultaneous release of a pair of test specimens – in this case, a small solid lead ball and an air-filled hog's bladder (roughly spherical) – from the upper gallery within the dome of the cathedral:

'A,A,A,A is the hole through which the bodies fell: 1,2, is a board laid over the hole. C,D,D is another board fixed to the first board by the two woodscrews D,D, with a pulley G at the other end of it, over the hole. W is a two pound ball of lead fastened to a strong thread, which going over the pulley is stretched horizontally from G to the nails F: to which it is fastened, so as to be about a quarter of an inch above the board.

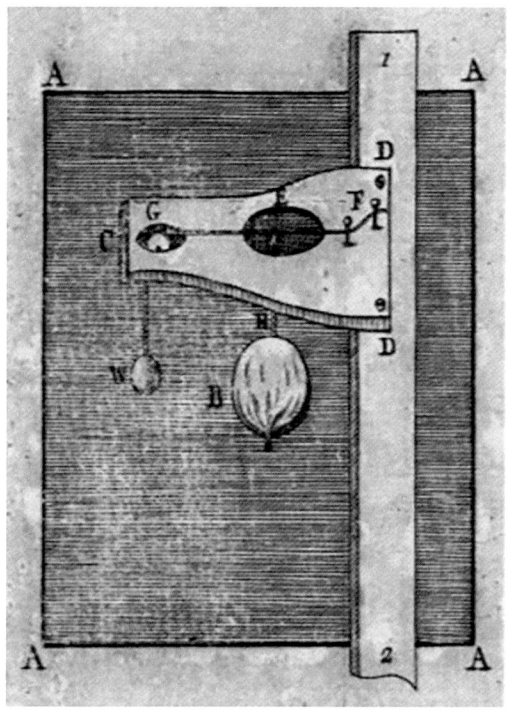

FIGURE A5.7. Mechanism deployed by Desaguliers [A5.140] in 1719 to release pairs of 'free-fall' test specimens simultaneously from upper gallery of St Paul's Cathedral (by courtesy of *Royal Society, London*)

> B is one of the bladders, hanging with the neck or heaviest part downwards, by means of a loop of fine thread as E,H, which goes over the thread G,E,F. Now when with a pair of scissors the thread of the lead (which in all is but one foot long) is cut just at E, before the loop of the bladder, the lead pulling away the string, the loop of the bladder flips off the remaining thread F,E, and begins to fall exactly in the same instant as the lead: but if the thread should be cut between E and F, as the lead falls its thread might give the bladder an oblique direction.
>
> He that observes the time either with a pendulum or chronometer may take it very exactly, by feeling the motion of the scissors as they cut the thread.'

With regard to the crucial step of timekeeping, Desaguliers states:

> 'The times of the falls were taken two ways above, viz: with a wheel-chronometer, which measures a small part of time accurately, nearer than to a quarter of a second (made and contrived by Mr George Graham, an ingenious clock-maker) and with a half second pendulum; and the differences of time between the fall of the leaden balls and the other balls were taken below, by the President, Martin Folkes Esq., F.R.S. and another person, who all agreed in their observations of the time, which they made each with a half second pendulum.'

On the basis of these experimental data, Newton was able to verify his theoretical proposition. And in considering the general approach to any such analysis, it is well to recall the wise words of Wallis [A5.141] written in 1687:

> 'I am aware of some objections to be made, whether to some points of the process, or to some of the suppositions. But I saw not well how to wave it, without making the computation much more perplexed. And in a matter so nice, and which must depend upon physical observations, t'will be hard to attain such accuracy as not to stand in need of some allowances.'

However, the original deployment of primitive equipment and antiquated units of measurement – prior to standardisation – can pose difficulties in achieving reliable results in any current retrospective analysis.

These early experiments demonstrated that, in modern terminology, the drag force (P_d) acting upon the moving object is directly proportional to the fluid density (ρ_a), the cross-sectional area (A_1) of the object normal to the flow (i.e. frontal surface area), and the square of its velocity relative to that of the fluid. Accordingly, for a spherical hailstone (diameter D_1) travelling vertically at its *terminal* velocity (v_T), the total force resisting downward vertical motion can be expressed as

$$P_d = \frac{\rho_a v_T^2 A_1 c_d}{2} = \frac{\rho_a v_T^2 \pi D_1^2 c_d}{8} \tag{A5.1}$$

where c_d denotes a dimensionless drag coefficient that usually is determined experimentally. Now assume that any such hailstone (mass m_1, density ρ_1) in 'free' vertical descent reaches a constant terminal velocity – relative to the adjacent airstream – when the vertical force (P_1) exerted by its mass is exactly balanced by the vertical drag force (P_d) generated in passing through the air. But

$$P_1 = m_1 g = \frac{\rho_1 g \pi D_1^3}{6} \tag{A5.2}$$

where 'g' denotes the standard acceleration due to gravity (taken herein as 9.807 m/s^2). Hence, with $P_1 = P_d$, the terminal velocity in air at a given temperature and atmospheric pressure is

$$v_T = \left(\frac{4\rho_1 g D_1}{3\rho_a c_d}\right)^{1/2} \tag{A5.3}$$

Within the cherished bailiwick of elementary Newtonian mechanics in a uniform gravitational field, the 'free-fall' motion of a sphere in terms of its downward velocity v at time t is governed by the elementary differential equation

$$m_1 \frac{dv}{dt} = m_1 g - \left(\frac{\pi \rho_a D_1^2 c_d}{8}\right) v^2 \tag{A5.4}$$

In experiments where the sphere is initially at rest, $v = 0$ at $t = 0$, and the theoretical (asymptotic) solution is

$$v(t) = v_T \tanh\left(\frac{gt}{v_T}\right) \tag{A5.5}$$

where v_T is the terminal velocity from equation (A5.3), and where $v \to v_T$ as $t \to \infty$. It is further noted that the time to reach any given velocity is

$$t = \frac{v_T}{g} \tanh^{-1}\left(\frac{v}{v_T}\right) \tag{A5.6}$$

Integration of equation (A5.5) gives the vertical distance y (upward positive) travelled at time t from an initial stationary position $y(0)$, namely

$$y(t) = y(0) - \frac{v_T^2}{g} \ln\left[\cosh\left(\frac{gt}{v_T}\right)\right] \tag{A5.7}$$

The general characteristics of this last equation are shown in Figure A5.8, giving the elapsed time (t_h) corresponding to a 'free-fall' height (h_1) for selected values of terminal velocity, where $h_1 = y(0) - y(t_h)$.

Largely as a result of pioneering investigations by the Irish-born engineer Osborne Reynolds (1842–1912), it became evident that the drag coefficient (c_d) depends mainly on the particular flow regime surrounding the sphere (a so-called 'bluff body') at its terminal velocity, usually characterised by the Reynolds number (Re) which, upon taking D_1 as the length parameter, is defined as

$$Re = \frac{v_1 D_1 \rho_a}{\eta_d} = \frac{v_1 D_1}{\eta_k} \tag{A5.8}$$

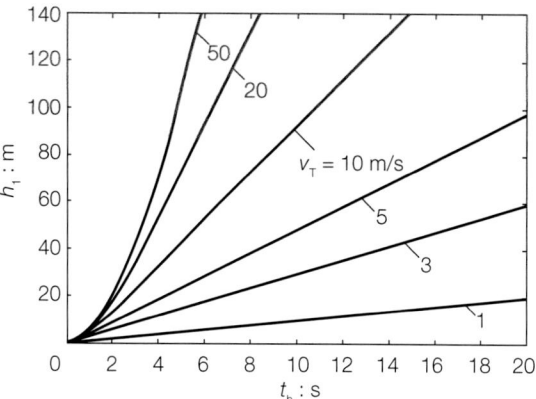

FIGURE A5.8. Idealised 'free-fall' characteristics of sphere in air at sea-level: elapsed time (t_h) for drop height (h_1) at given values of terminal velocity (v_T)

but which is not necessarily single-valued for any given diameter. More generally, the Reynolds number can also be regarded as the ratio of inertial to viscous forces acting upon the sphere. For many applications encountered in aeronautical science, the drag coefficient is sensibly constant over a wide range in the Reynolds number, but this is not always the case for a simple sphere.

It is instructive at this stage – not least in providing historical perspective – to examine Newton's original experimental results [A5.138], in order to deduce estimated values of the drag coefficient; while noting that the efficacy of such back-calculations presented by Loomis [A5.142] is inhibited by the inclusion of parameters with obscure physical dimensions. First consider the six tests using large thin air-filled glass globes (nominal diameter 130 mm) falling through a vertical height of approximately 67 m, carried out in July 1710. The calculated results are summarised in Table A5.1(a), together with the experimental data expressed in modern metric units (e.g. originally specimen weights were given in grains). The calculated mean c_d-value is 0.52, corresponding to a Reynolds number of 8.2×10^4.

The globes were spheroidal in shape, rather than precise spheres, and fall times were measured by means of a half-second pendulum device. Notwithstanding such practical difficulties, including those of calibrating the mechanical contrivance used to release the specimens, the tabulated drag coefficients are entirely plausible in the light of current knowledge, having regard to the related values of terminal velocity and Reynolds number. In companion tests using small mercury-filled glass globes (nominal diameter 20 mm), the estimated terminal velocities are higher by an order of magnitude, and the rudimentary nature of the timing procedure effectively precludes any reasonably accurate determination of the drag coefficient.

Similar tests were carried out in July 1719 using large yet much lighter air-filled hog's bladders, broadly spherical in shape (nominal diameter 132 mm), dropped from a height of approximately 83 m on to wooden boards covering the cathedral floor. Here the earlier timing equipment was enhanced by an auxiliary quarter-second timepiece, although some

TABLE A5.1. Data from 'free-fall' experiments within St Paul's Cathedral reported by Isaac Newton, for spheres of mass m_1, diameter D_1 falling from height h in time t_h, with deduced estimated values of drag coefficient c_d, terminal velocity v_1 and Reynolds number Re: (a) tests in June 1710, $h_1 \approx 67$ m, air-filled glass globes; (b) tests in July 1719, $h_1 \approx 83$ m, air-filled hog's bladders

m_1: g	D_1: mm	t_h: s	c_d	v_1: m/s	$Re \times 10^{-4}$
33.0	129.5	8.2	0.51	8.9	7.9
41.6	132.1	7.7	0.53	9.5	8.6
38.8	129.5	7.7	0.52	9.5	8.5
33.4	127.0	7.95	0.50	9.2	8.0
31.3	127.0	8.2	0.50	8.9	7.7
41.5	132.1	7.7	0.53	9.5	8.6

(a)

m_1: g	D_1: mm	t_h: s	c_d	v_1: m/s	$Re \times 10^{-4}$
8.3	134.1	19.0	0.48	4.4	4.1
10.1	131.8	17.0	0.48	5.0	4.5
8.9	134.6	18.5	0.48	4.6	4.2
6.3	133.6	22.0	0.50	3.8	3.5
6.4	127.0	21.125	0.51	4.0	3.5

(b)

specimens were observed to deviate slightly from the vertical during their descent, with one specimen being described as 'rough, with several wrinkles and inequalities'. The corresponding results are listed in Table A5.1(b), and again are broadly compatible with much later accepted values for the drag coefficient (mean tabulated value of $c_d = 0.49$ at $Re = 3.9 \times 10^4$): altogether a remarkable achievement considering that these tests were performed some three centuries ago.

The notable results presented by Bilham and Relf [A5.143] in 1937 were based on values of the drag coefficient deduced from observations on the behaviour of a sphere towed by an aeroplane, ostensibly to avoid wind tunnel side-effects, although no further details are given (e.g. size, material composition and surface roughness of the sphere). However, according to a detailed review by Heymsfield and Wright [A5.144], the original experimental data were obtained some years earlier by Millikan and Klein [A5.145]. The corresponding graphical results in the range $10^5 \leq Re \leq 10^6$ presented by Bilham and Relf are re-plotted in Figure A5.9, taking account of differences in notation and drag coefficient definition. Evidently, if the Reynolds number is lower than around 3×10^5, values of c_d vary from approximately 0.45 to 0.5, but their rapid decrease thereafter can lead to c_d-values of 0.1 or even lower when the Reynolds number exceeds about 4×10^5.

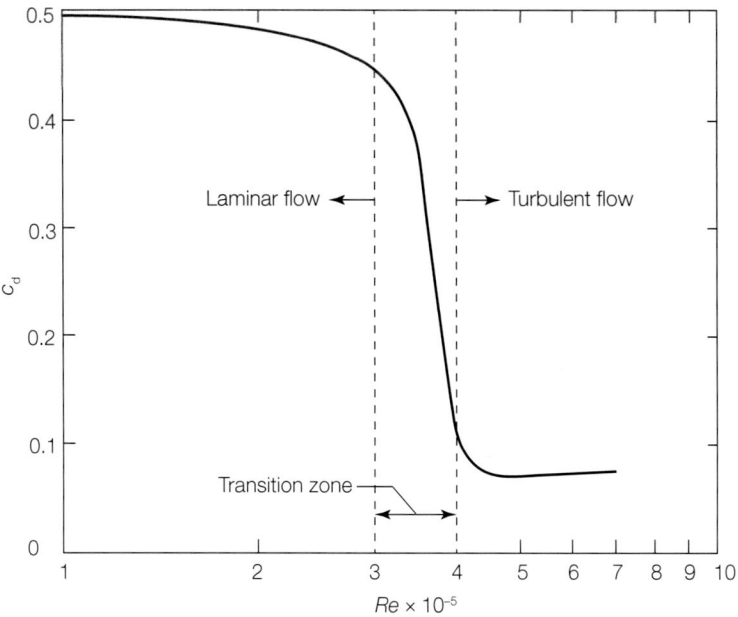

FIGURE A5.9. Experimental relationship between drag coefficient (c_d) and Reynolds number (Re) for sphere towed by aeroplane, after Bilham and Relf [A5.143], indicating zonal regions of boundary layer flow

Historical notes by Gorbushin and Volobuev [A5.146] and others indicate that this so-called 'drag crisis' was observed by the Russian scientist Nikolay Zhukovsky (1847–1921) – his surname frequently Romanised to Joukowsky – during wind-tunnel tests in around 1905, whilst postulating that the paradox could be explained by the detachment of streamlines at different points of the sphere at different velocities. Low drag coefficients at relatively high velocities ($Re \approx 6 \times 10^5$) can also be deduced from the results of tests described in 1901 by Zahm [A5.147], in which 100 mm diameter polished pinewood spheres were projected horizontally in still air. Similarly low c_d-values were obtained by the French civil engineer Gustave Eiffel (1832–1923) from wind tunnel tests on spheres [A5.148]; which findings were discussed briefly by Lord Rayleigh [A5.149] and augmented by further experimental results reported by Maurain [A5.150]. Many such early measurements on spheres obtained by various investigators were reviewed by Bacon and Reid [A5.151], who also gave their own wind-tunnel test results for large polished spheres of laminated wood at air flows corresponding to high values of Reynolds number.

A partial explanation of this puzzling phenomenon was given by the German aeronautical engineer and elastician Lugwig Prandtl (1875–1953) in 1914, following extensive theoretical and experimental investigations; see the original paper [A5.152], reprinted by Tollmien et al. [A5.153], and two later publications [A5.154], [A5.155], together with the contemporary observations reported by Carl Wieselsberger [A5.156], [A5.157]. From the 1927 paper by Prandtl [A5.154]:

'It was observed first of all in the case of spheres, and then of cylinders and other bodies; with large Reynolds numbers, 100,000 to 300,000, there is a somewhat sudden diminution in the resistance coefficients to about one-fourth of the value with lower Reynolds numbers. Observation shows that the point where the current leaves the boundary is pushed far back and that the pressure distribution approaches that of the theoretical frictionless current. The explanation was that the boundary layer at the body becomes turbulent. Owing to this, there appears a rapid intermixture of the frictional boundary layer and the outer flow which accelerates the frictional layer again and thus prevents the separation of the flow from the boundary or at least postpones it. That this is the correct explanation I was able to prove in the case of a sphere, by fixing a wire hoop on the sphere which rendered the boundary layer vortical.'

Whence the landmark general conclusion that predominantly laminar flow within a thin boundary layer enveloping part of the sphere at relatively low velocities changes to predominantly turbulent flow during a narrow unstable transitional stage of increasing velocity, leading to large reductions in drag resistance.

A qualitative perspective of this transitional effect is provided by a pair of photographs included by Prandtl in the original 1914 paper [A5.152], and reproduced in Figure A5.10. Spheres having diameters of 7 to 28 cm were mounted in a horizontal wind tunnel, and tested at air speeds in the range 5–23 m/s, with smoke added to aid flow visibility. Figure A5.10(a) illustrates the subcritical flow pattern around a given sphere, while Figure A5.10(b) shows the reduced wake that is characteristic of supercritical flow, generated by adding a thin wire ring on the leading face of the same sphere (*i.e.* to create a rudimentary form of surface roughness); a technique subsequently deployed by Maxworthy [A5.158] and Ikushima *et al.* [A5.159]. Prandtl [A5.154] also added:

'The turbulence is probably connected with the fact that the speed distribution in the boundary layer has a point of inflexion behind the point of maximum speed. When the Reynolds number is sufficiently high, these speed distributions are unstable.'

Detailed explanations for the substantial reduction in measured drag force when passing from subcritical to supercritical flow are not self-evident, especially as the necessary reasoning is to some degree counter-intuitive; as explained in the reference works of Goldstein [A5.160] and Schlichting [A5.161]. A century later, results of quantitative analyses of this difficult problem continue to be reported, based on some advanced form of computational fluid dynamics in which several *billion* degrees of freedom within the Newtonian space-time domain may be required in the numerical simulation of turbulent flow.

Bilham and Relf [A5.143] surmised from physical considerations that, in most practical circumstances, it is the laminar flow regime – with values of c_d in the range 0.45–0.5, as indicated in Figure A5.9 – that is applicable to the air flow around large hailstones; a most welcome slice of good fortune proffered by nature, as otherwise the impact forces would be much higher. They also concluded that, assuming $\rho_1 = 600$ kg/m^3, the largest possible spherical hailstone would have a diameter of about 130 mm and an approximate mass of 700 g, which is broadly compatible with observed sightings subsequently reported in the literature.

Wind tunnel measurements on hailstone models were discussed by List [A5.162], while experimental results obtained by Macklin and Ludlam [A5.163] for artificial hailstones

(a) (b)

FIGURE A5.10. Observed air flow (left to right, with smoke) past sphere in horizontal wind tunnel reported by Ludwig Prandtl in original 1914 paper [A5.152] (by courtesy of *Göttinger Archiv des Deutschen Zentrums für Luft- und Raumfahrt*): (a) sphere only; (b) same sphere with thin wire ring attached to leading face

in 'free fall' gave values of the drag coefficient ranging from 0.45 (for a sphere) to about 0.8, depending on the shape and surface roughness of the test specimen. Further tests on artificial hailstones grown within a vertical wind tunnel were carried out by Bailey and Macklin [A5.164], where lobe-like growth similar to that found in some natural hailstones was reproduced. None of the measured c_d-values exceeded 0.66, but some of the moderately large hailstones (40–60 mm diameter) entered the critical flow regime where their drag coefficients were considerably reduced. However, this transition did not occur for large (100 mm diameter) irregular hailstones.

In tests using high-precision tracking radar, Willis *et al*. [A5.165] measured the fall speeds of cast ice spheres having diameters in excess of 50 mm, and released from an altitude of 6100 m. Those spheres that remained dry fell according to supercritical Reynolds numbers, with drag coefficients of only 0.24 to 0.30; values well below those normally associated with large hailstones in 'free fall', and ascribed to the effect of surface roughness. The one sphere (51 mm in diameter) that became wet during its fall led to a doubling of its drag coefficient, which might be viewed as more representative of natural hailstones.

Following a series of early wind tunnel tests conducted by Hoerner [A5.166], similar yet more refined experiments were carried out by Young and Browning [A5.167] on 100 mm diameter wooden spheres coated with sand (two grades) or lead shot to give specimens of different surface roughness, together with a smooth sphere to provide a reference datum. For the three rough spheres, there occurred a sharp drop in drag coefficient at each critical Reynolds number, indicating that the transition to a lower drag coefficient could apply to rough natural hailstones of diameter greater than about 40 mm, whereas the corresponding transition for a smooth hailstone is unlikely occur unless its diameter exceeds 100 mm. Broadly similar results were obtained by Landry and Hardy [A5.168] using radar tracking equipment to measure the fall speeds of plastic spheres. Moreover, investigations by Raithby and Eckert [A5.169] focussed on the surface region of a sphere in the range $4\times10^4 \le Re \le 2.5\times10^5$ using a flow visualisation technique, while Clamen and Gauvin [A5.170] used a particle tracing procedure to obtain the drag coefficients of smooth spheres in a supercritical flow regime.

Field observations by Roos [A5.171] on replicas of the giant Coffeyville hailstone – shown in Figure A5.4(d) – in which specimens were dropped from a helicopter at heights up to 900 m above ground level, suggested that the terminal velocity of the real hailstone (assumed density 900 kg/m^3) was approximately 47 m/s (≈105 mph). In companion horizontal wind tunnel tests to determine aerodynamic drag, deduced values of c_d varied from 0.46 to 0.65, depending on the orientation of the irregular shaped replica hailstone. It was concluded that although the critical Reynolds number was not quite reached by the replicas in 'free fall', the original hailstone probably did enter the transition zone shortly before impact; see also Roos and Carte [A5.172]. Further experimental results relating to non-spherical hailstones were reported by List et al. [A5.173] for wooden models of smooth oblate spheroids with different axis ratios, including values of the drag coefficient at various inclinations to the direction of flow, measured over a range in Reynolds number from 0.4×10^5 to 4×10^5. Their associated numerical modelling also yielded data on the possible oscillatory or tumbling motion of 'free-falling' hailstones.

High-speed photography of 48 small natural hailstones (5–10 mm maximum diameter) near ground level carried out in Canada by Lozowski and Beattie [A5.174] furnished an empirical relationship between vertical velocity and diameter, which inferred a mean c_d-value of 0.61. In similar field studies conducted by Matson and Huggins [A5.175] in the U.S.A., stroboscopic photography was deployed to record the kinematic behaviour of over 600 hailstones – overwhelmingly spheroidal in shape – having diameters in the range 5–25 mm. These results yielded a mean drag coefficient of 0.87 over a range in Reynolds number from 10^3 to 2×10^4. Correlations between terminal velocity and hailstone diameter were presented by Böhm [A5.176], while the measured physical properties of over two thousand hailstones collected in the Great Plains of America were subjected to statistical analysis by Heymsfield et al. [A5.177], although their mass was generally only one-half that of pure ice. The equivalent diameter of some 30% of these hailstones exceeded 25 mm (with only 1% exceeding 50 mm), and empirical relations were established for terminal velocity, based partly on the methods adopted by Laurie [A5.178]; see also the subsequent detailed investigation reported by Heymsfield et al. [A5.179].

Achenbach [A5.180], [A5.181] conducted a notable series of experiments on the flow past smooth and rough hollow aluminium spheres (20 cm diameter) at high Reynolds numbers in a horizontal wind tunnel (2 m diameter), with strain gauges attached to a rear support rod (20 mm diameter) used to measure drag forces. The deduced dependence of drag coefficient on Reynolds number for smooth spheres was similar to that reported by Millikan and Klein [A5.145] and Bilham and Relf [A5.143]. The corresponding results for rough spheres were markedly different, with the critical Reynolds number progressively decreasing with increasing surface roughness, accompanied by increases in the transcritical drag coefficient. Figure A5.11 compares the results for a smooth sphere with those for two similar rough spheres, produced by bonding glass *ballotini* (diameter $d_s = 0.5$ mm, 2.5 mm) on to a polished aluminium sphere ($D_1 = 20$ cm), giving roughness parameters (d_s/D_1) of 0.0025 and 0.0125. In Figure A5.11(b), the polar angle of boundary layer separation (ϕ_s) is measured clockwise positive to the circular separation line from the forward stagnation point ($\phi = 0$) on the horizontal axis.

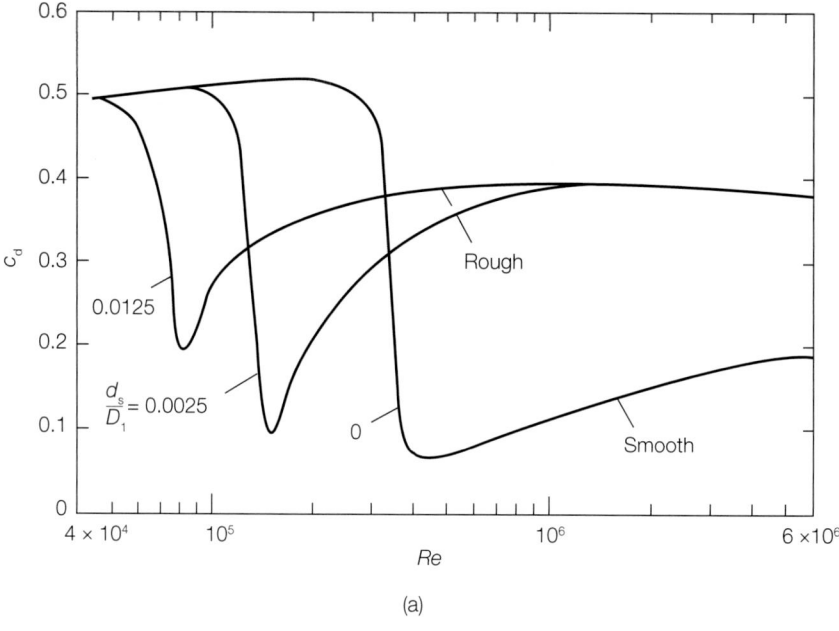

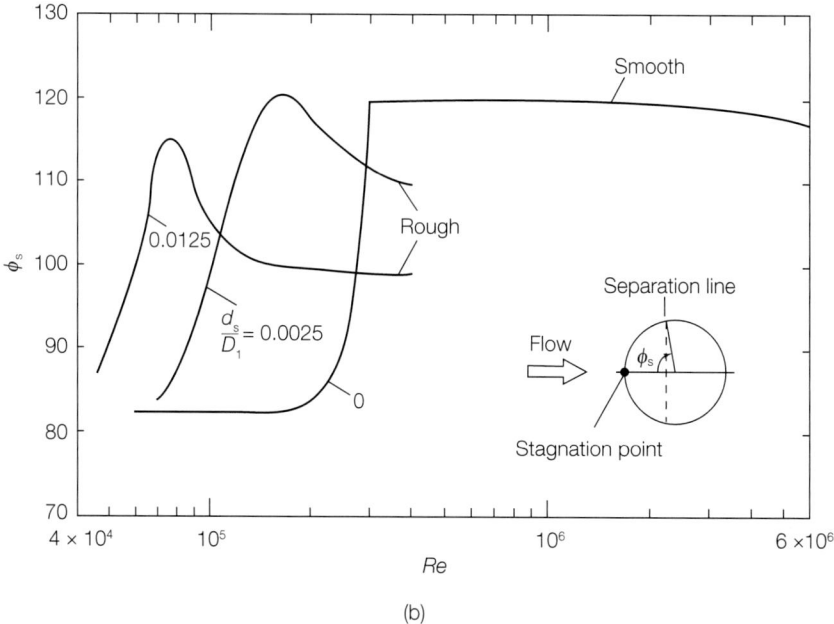

FIGURE A5.11. Measured effects of surface roughness produced by bonded glass *ballotini* (diameter d_s) on flow past hollow aluminium spheres (diameter $D_1 = 20$ cm) in horizontal wind tunnel at high Reynolds numbers (Re), after Achenbach [A5.181]: (a) drag coefficient (c_d); (b) polar angle (degrees) of boundary layer separation (ϕ_s).

It is the transition from laminar to turbulent flow close to the spherical surface and the downstream shift of the separation line that modify the viscous boundary layer resistance and lead to substantial pressure changes within a narrower wake, resulting in a lower net drag resistance; which conclusion may be verified by integrating the measured tangential shear forces and normal pressures over the entire surface of the test spheres at a given value of Reynolds number. Further aspects of the complex flow patterns around spheres at high flow rates, including vortex shedding and wake configuration, have been investigated experimentally by Achenbach [A5.182] and Taneda [A5.183], who also review the work of others. Intriguingly, it was noted that the wake is not axisymmetric, indicating that a sphere placed within a uniform flow field is subjected to a net resultant side force whose direction is random; see also the flow visualisation images obtained by Bakić and Perić [A5.184] and Bakić et al. [A5.185], and the force measurement results for spheres in the range $5\times10^4 \leq Re \leq 5\times10^5$ reported by Norman and McKeon [A5.186], [A5.187]. In closely related experiments, Venning et al. [A5.188] obtained high-resolution photographs illustrating the cavitation topology around a sphere mounted in a laboratory water tunnel ($1.25\times10^5 \leq Re \leq 1.5\times10^6$).

When investigating the relative motion between a sphere and a large expanse of viscous fluid, analytical solutions to the governing partial differential equations can only be obtained in special cases (*e.g.* at very low velocities). In general, therefore, theoretical solutions are obtained by some form of numerical analysis, ideally accompanied by experimental tests for verification purposes; as described by Jeong and Hussain [A5.189], Tomboulides and Orszag [A5.190], Constantinescu and Squires [A5.191], Yun *et al.* [A5.192], Hoffman [A5.193], Moradian *et al.* [A5.194], [A5.195], Wang *et al.* [A5.196], Linić *et al.* [A5.197], Deshpande *et al.* [A5.198], Terra *et al.* [A5.199], [A5.200], Pendar and Roohi [A5.201] and Nakhostin and Giljarhus [A5.202], among others. The complex issues associated with the physics of 'free-falling' spherical hailstones, including the dual aspects of heat and mass transfer, continue to attract the attention of scientists worldwide; see, for instance, a review by Böhm [A5.203], and the more recent developments outlined by List [A5.204], [A5.205], based on several decades of theoretical and experimental studies.

Cheng and Wang [A5.206] conducted a three-dimensional numerical analysis of the flow fields around smooth spherical hailstones falling vertically in air at their terminal velocity, corresponding to high Reynolds numbers and based on the governing Navier–Stokes equations for incompressible viscous flow within an infinite isotropic fluid medium at constant temperature. Some of their results are illustrated in Figure A5.12, where the various distributions relate to the central *y-z* plane, and where resultant numerical values are based somewhat arbitrarily on an empirical relationship between terminal velocity and hailstone diameter. The randomly selected streamlines in Figure A5.12(a) form a cocoon-shaped downstream region (recirculation bubble) at comparatively low air pressures within the turbulent wake. Figure A5.12(b) shows the pressure distribution (Pa, compression positive) and velocity vectors, emphasising the marked differences in maximum upstream and downstream air pressures. The distribution of vertical air velocity (m/s, upward positive) is shown in Figure A5.12(c), with low positive upstream values close to the stagnation point, and maximum (positive) values located slightly upstream of the equator. The large velocity gradients within this latter region are reflected in the corresponding intensity of vorticity (s^{-1}) shown in Figure A5.12(d). Other computed results include hailstone size effects and

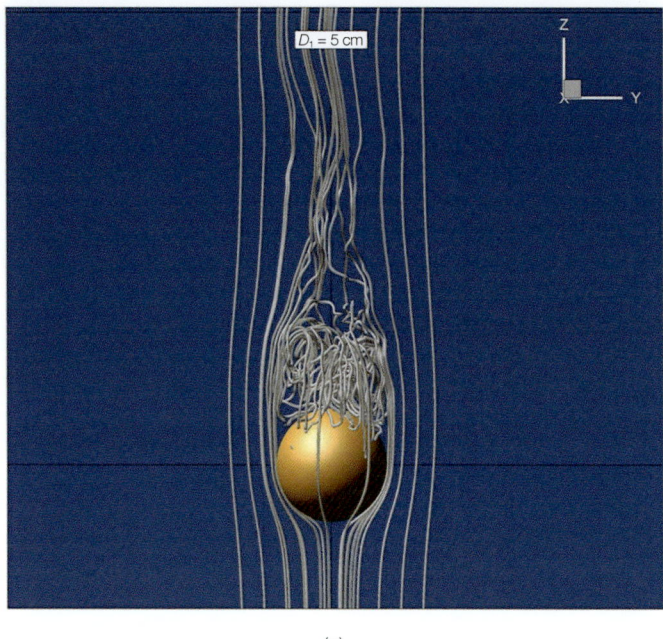

FIGURE A5.12. Computed flow fields (central *y-z* plane) around smooth spherical hailstone (5 cm diameter) falling vertically in air at terminal velocity, after Cheng and Wang [A5.206] (by courtesy of *Elsevier B.V.*): (a) random streamline pattern; (b) air pressure distribution (Pa) and vertical velocity vectors

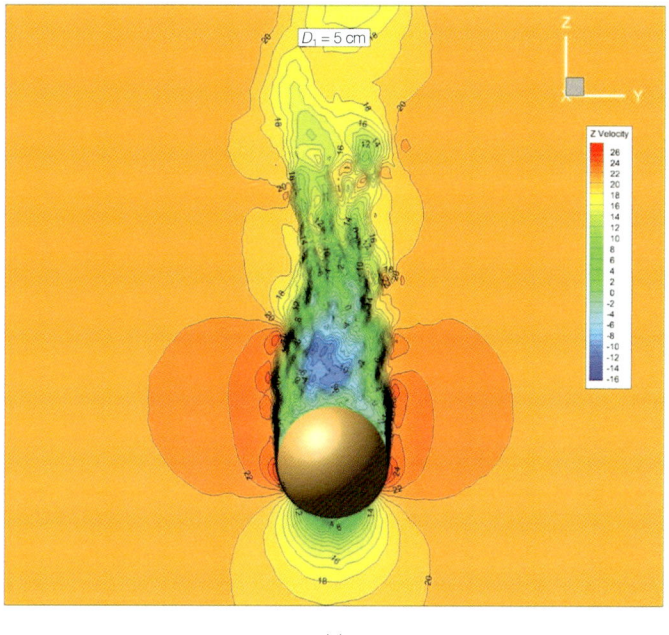

(c)

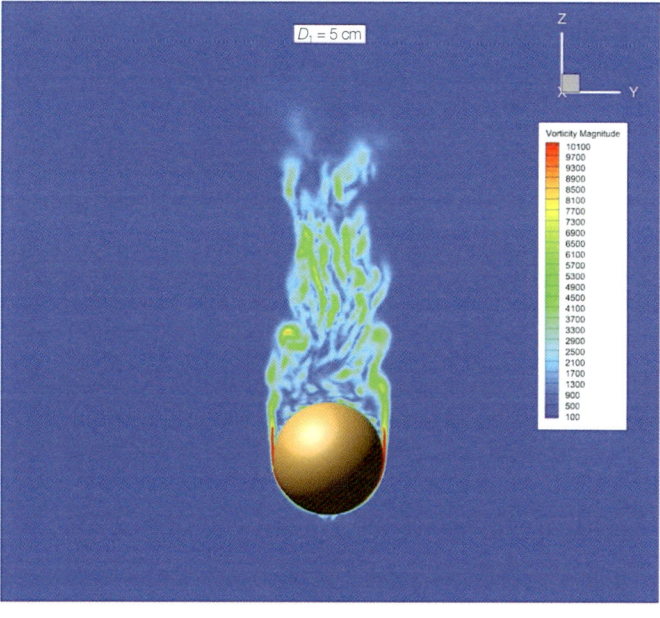

(d)

FIGURE A5.12. (continued) Computed flow fields (central y-z plane) around smooth spherical hailstone (5 cm diameter) falling vertically in air at terminal velocity, after Cheng and Wang [A5.206] (by courtesy of *Elsevier B.V.*): (c) vertical air velocity (m/s); (d) vorticity (s^{-1})

ring-shedding phenomena, together with the influence of convective water vapour diffusion discussed by Cheng et al. [A5.207], while subsequent companion studies of lobed hailstones are described by Wang et al. [A5.208] and Wang and Chueh [A5.209], augmented by wind tunnel test results reported by Theis et al. [A5.210].

Further computational progress has been made by Geier et al. [A5.211], [A5.212] in simulating the turbulent flow around a sphere during the 'drag crisis', obtaining detailed results within the range $10^4 < Re < 10^6$ for the case of a smooth stationary sphere located within a relatively large fluid medium and subjected to constant upstream horizontal flow. Figure A5.13 illustrates the essential features of the transitional flow regime, with a critical Reynolds number of around 2.4×10^5; below this number, the wake is divergent, while above this number, the wake converges to much narrower form. Moreover, the polar separation angle of the wake increases markedly during transition, leading to significant changes in pressure acting on the rear of the sphere and thereby reducing the total net axial drag force. These results relate to a medium grid resolution (a mere 74×10^6 nodes), while the incident far-field fluid velocity is taken as unity, and pressures are normalised by the maximum surface pressure. The analysis has successfully captured the essential features of the 'drag crisis' phenomenon, while again noting that the absence of axially symmetry in such results is a consequence of the statistical nature of the fluid dynamics.

Closely related experimental investigations of the flow around dimpled spheres are described by Bearman and Harvey [A5.213], Choi et al. [A5.214], Smith et al. [A5.215], Alam et al. [A5.216], Aoki et al. [A5.217] and Bogdanović-Jovanović et al. [A5.218], while numerical analyses have been carried out by Li et al. [A5.219], Spálenský and Rozehnal [A5.220], Crabill et al. [A5.221] and Beratlis et al. [A5.222], among several others. Here the degree of difficulty increases substantially if the rotational effects of a spinning sphere are also taken into account. More generally, the ultimate practical objective of surface modification is to increase the potential range of a driven golf ball by reducing drag resistance in 'free flight'; in marked contrast to the converse attributes sought in a falling hailstone.

These and other investigations have confirmed that the influence of surface roughness on the flow around a rigid sphere can be substantial, primarily by shifting the onset of unsteady flow to a lower Reynolds number. Figure A5.14 displays a basic summary of test results prepared by the National Aeronautics and Space Administration, U.S.A., whilst recognising that the curve relating to smooth spheres is similar to those given by Zahm [A5.223], Clift et al. [A5.224] and others. Experimental and reference details are not given, although copious data are included in the aforementioned independent study by Heymsfield and Wright [A5.144], and in the reviews by Mitchell [A5.225], Almedeij [A5.226], Cheng [A5.227], Tiwari et al. [A5.228], [A5.229] and Dieling et al. [A5.230]. In some cases, therefore, a rough sphere will experience a far lower drag force than the corresponding smooth sphere, resulting in a much higher terminal velocity. Indeed, this may occur in nature with smaller spherical hailstones, but because most large natural hailstones are found to be non-spherical, it is likely that the effect of surface roughness is more than offset by the increased drag coefficients applicable to such bodies, especially if they undergo wandering downward motion. Herein, calculated terminal velocities of hailstones are based on smooth spheres within a laminar boundary layer flow regime.

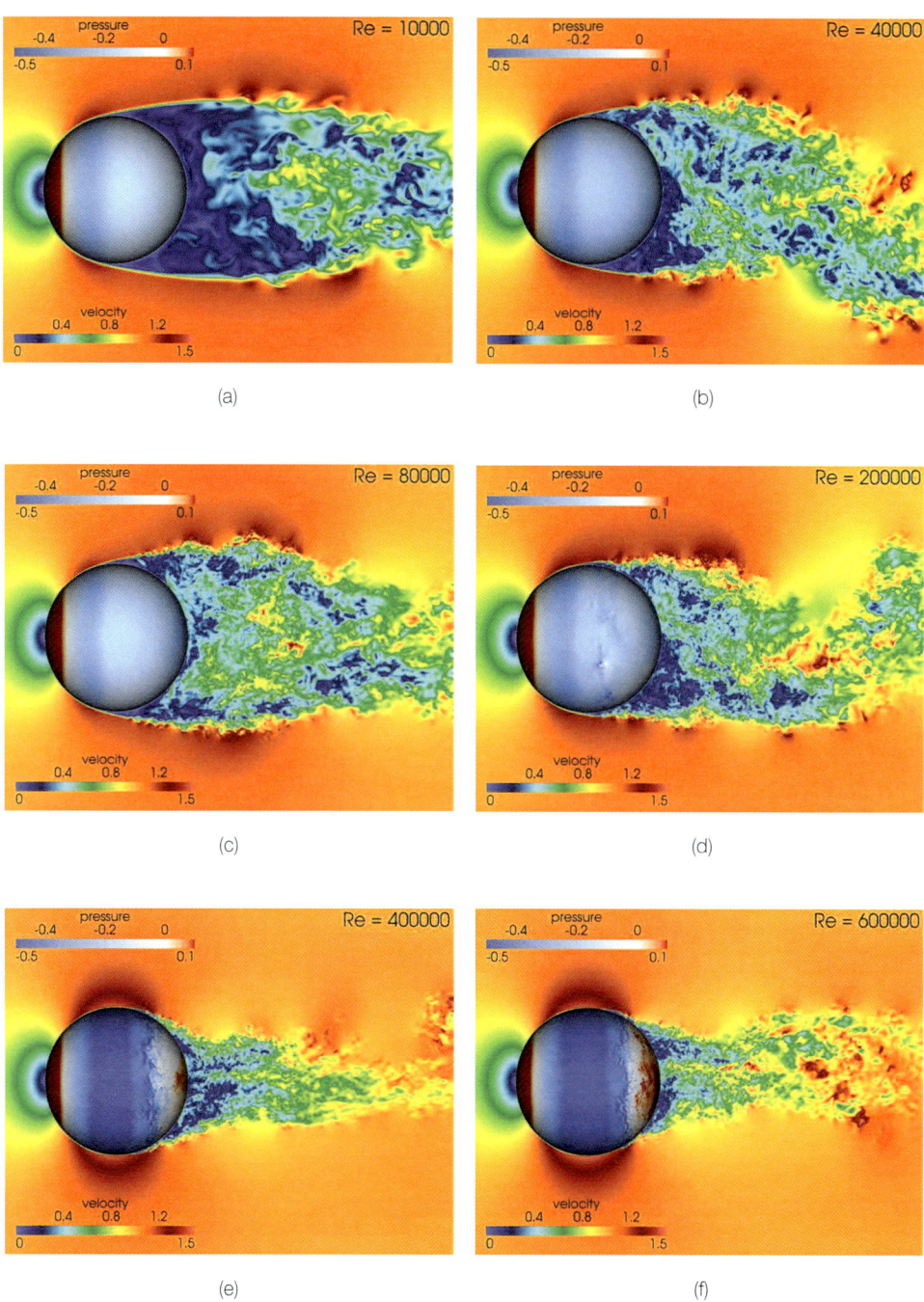

FIGURE A5.13. Development of wake behind smooth sphere under horizontal flow at increasing values of Reynolds number (Re), showing normalised fluid velocity and surface pressure, after Geier *et al.* [A5.212] (by courtesy of *Elsevier B.V.*): (a) $Re = 10^4$; (b) $Re = 4 \times 10^4$; (c) $Re = 8 \times 10^4$; (d) $Re = 2 \times 10^5$; (e) $Re = 4 \times 10^5$; (f) $Re = 6 \times 10^5$

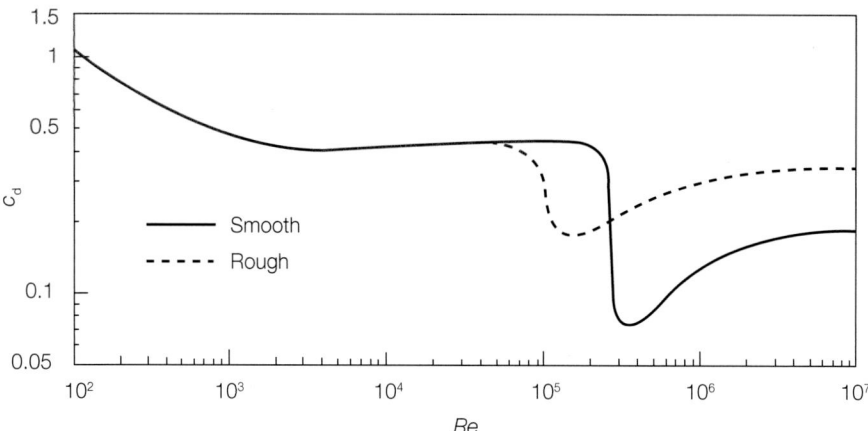

FIGURE A5.14. General effect of surface roughness on drag coefficient (c_d) for sphere over wide range in Reynolds number (Re), from wind tunnel tests and other experiments (by courtesy of *National Aeronautics and Space Administration, U.S.A.*)

As a general guide, the estimated dependence of terminal velocity on hailstone diameter based on equation A5.3 is shown in Figure A5.15, assuming that a drag coefficient of 0.47 applies to all spherical hailstones, irrespective of size. Also plotted are the corresponding values of hailstone momentum (U_1) and kinetic energy (K_1), where

$$U_1 = m_1 v_T, \qquad K_1 = m_1 v_T^2 / 2 \tag{A5.9}$$

Structural design calculations for laminated glass outlined in the following section are carried out on this basis.

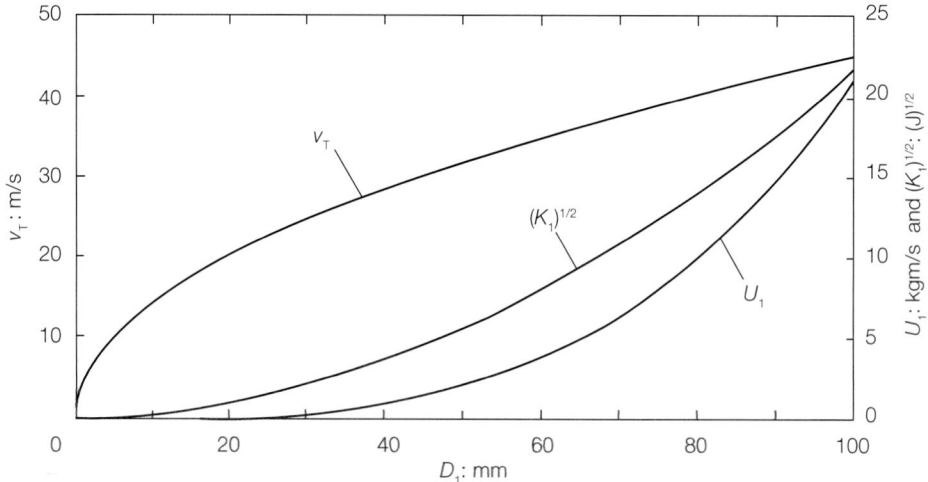

FIGURE A5.15. Calculated relationship between terminal velocity (v_T) and diameter (D_1) of idealised spherical hailstone ($\rho_1 = 900$ kg/m³, $\rho_a = 1.225$ kg/m³, g = 9.807 m/s², $c_d = 0.47$), with corresponding values of momentum (U_1) and kinetic energy (K_1)

A5.4. Elastic beam-strip modelling

Because the laminated glass panels that form the glass walls are simply supported only along two opposite edges, with the other two edges generally butt-jointed with silicone rubber to adjacent panels, it is reasonable to greatly simplify the structural analysis by examining the flexure of a representative *beam-strip* within an infinitely long panel of constant 'free' span, subjected to concentrated impact loading and based on one-degree-of-freedom modelling. It is also reckoned that the impact force acts normal to the glass surface, and that inclined impact can be vectored accordingly if necessary. Such methodology is likely to be adequate for most design purposes, as was assumed originally in the present application.

With reference to Figure A5.16, consider an initially straight *monolithic* glass beam (strictly a beam-strip) of length a, simply supported at both ends. The beam is of constant thickness T and effective width B_E, and has a flexural stiffness $E_g I$, where E_g denotes the Young's modulus of the linear elastic material. The second moment of area $I = B_E T^3/12$. An initially stationary mass m_1 is dropped from a vertical height h_1 above the centre of the horizontal beam, attaining a velocity v_1 upon impact. Hence the vertical force imparted to the beam (taken as a uniform line load) at the instant of contact is $P_1 = m_1 g$, where the physical constant 'g' denotes the acceleration of the mass due to gravity.

Now assume that a slow application of this force would generate a *static* vertical deflection w_S^* at the beam centre; whereupon it is convenient to represent the beam as a vertical massless linear spring, fixed at one end, having a stiffness $k_S = P_1/w_S^*$ (*i.e.* the force required to produce unit compression or extension of the spring) impacted by a lumped mass. It is further assumed that the deflected *shape* of the beam under dynamic loading is identical to that under static loading (*i.e.* a single-mode shape).

The most rudimentary approach to examining the dynamic response of the beam essentially follows from innovative studies contained in the monumental work [A5.231] of the English polymath Thomas Young (1773–1829) published in 1807, while also noting the subsequent investigations reported by Hodgkinson [A5.232]. Here the beam is assumed to be

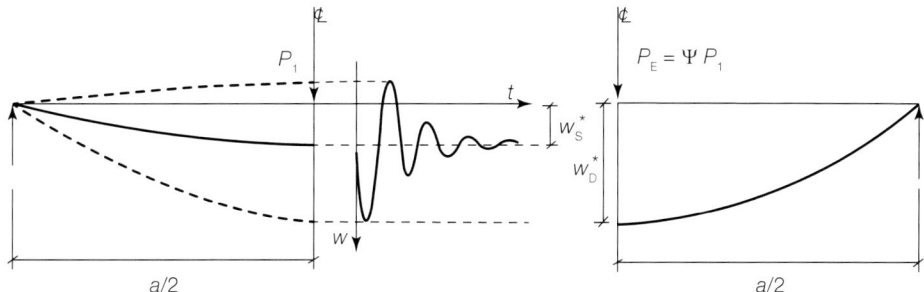

FIGURE A5.16. Schematic representation of dynamic load factor (Ψ) for impact of vertical line load (P_1) at centre of simply supported horizontal elastic beam-strip (span a) to give equivalent static line load (P_E)

weightless, and the energy conservation principle is invoked in the absence of any damping or dissipation losses. Relative to the final equilibrium position of the beam at mid-span (w_S^*), the entire kinetic energy of the impacting mass is equated to the total change in potential energy of the vibratory system at the instant of the maximum *dynamic* vertical deflection, w_D^*; namely the change in potential energy of the mass (proportional to $w_D^* - w_S^*$), together with the strain energy stored in the beam (*i.e.* the work done to deflect the beam). From the resulting differential equation governing the so-called 'free' or 'natural' vertical vibrations, which are shown to exhibit simple harmonic motion, it follows that

$$w_D^* = \frac{m_1 g}{k_S}\left[1 + \left(1 + \frac{k_S v_1^2}{m_1 g^2}\right)^{1/2}\right] \quad (A5.10)$$

as measured from the unloaded beam position. Accordingly, with $w_S^* = m_1 g / k_S$,

$$\Psi = \frac{w_D^*}{w_S^*} = 1 + \left(1 + \frac{k_S v_1^2}{m_1 g^2}\right)^{1/2} \approx 1 + \left(1 + \frac{2h_1}{w_S^*}\right)^{1/2} \quad (A5.11)$$

where the dynamic load factor Ψ also defines the equivalent applied static force $P_E = \Psi P_1$ required to give the calculated maximum dynamic deflection. It is also noted from idealised energy conservation considerations in elementary mechanics (*e.g.* zero drag, rigid beam) that the corresponding vertical velocity of the mass is $v_1 \approx (2gh_1)^{1/2}$ to a first approximation. This elementary theory implies that even if the impacting mass is released instantaneously from a vanishingly small height above the beam ($h_1 \to 0$), then $\Psi = 2$ and the dynamic effect is to double the maximum beam deflection, relative to a slow application of the same mass.

An improved simplified theory, communicated to the Cambridge Philosophical Society by the English lawyer and mathematician Homersham Cox (1821–1897) in 1849 and published in 1856, takes account of the inertial effect of the beam [A5.233]; see also a related contemporary paper [A5.234], [A5.235], and the wider investigations [A5.236] of Lord Rayleigh (1842–1919). Suppose the beam has a mass $m_2 = \rho_2 A_2 a$, where ρ_2 is the mass density and A_2 is the cross-sectional area, and that static beam deflections generated by this self-weight loading are neglected temporarily (*i.e.* these latter deflections are calculated separately). Further assume that, under conditions of completely 'inelastic' contact (*e.g.* no rebound or multiple impact), the attached impactor mass m_1 and a specified proportion of the beam mass (yet to be determined) together attain the same initial vertical velocity $v(0)$ at mid-span. Then based upon the conservation of momentum together with the energy balance relations, it is readily shown that the equivalent beam mass is $(17/35)m_2$. Whence

$$v(0) = \frac{m_1 v_1}{m_1 + (17/35)m_2} \quad (A5.12)$$

and the maximum dynamic deflection becomes

$$w_D^* = \frac{m_1 g}{k_S}\left(1 + \left\{1 + \frac{k_S v_1^2}{[m_1 + (17/35)m_2]g^2}\right\}^{1/2}\right) \quad (A5.13)$$

which expression immediately reduces to equation (A5.10) in the limiting case of a weightless beam ($m_2 = 0$). The elegance and originality of this approximate solution was recognised by Todhunter and Pearson [A5.237], who highlighted the close match between numerical results given by equation (A5.13) and those obtained from a more refined solution obtained some years later by the French elastician Barré de Saint-Venant (1797–1886), whose *hypothèse plausible* had been pre-empted on this occasion by the veritable Homersham Cox.

More generally, the theoretical vertical motion of the beam at mid-span at time t after initial impact is given by

$$w(t) = \frac{m_1 g}{k_S}(1 - \cos \zeta t) + \frac{\zeta m_1 v_1}{k_S} \sin \zeta t \tag{A5.14}$$

as measured from the final equilibrium position (w_S^*), where

$$\zeta = \left[\frac{k_S}{m_1 + (17/35)m_2} \right]^{1/2} \tag{A5.15}$$

This analysis essentially retains only the first mode of vibration, whereas in reality, beam deflections will have the characteristics of a damped oscillator, as sketched in Figure A5.16. The maximum deflection, occurring at time t^*, is

$$w(t^*) = \frac{m_1 g}{k_S}\left[1 + \left(1 + \frac{\zeta^2 v_1^2}{g^2}\right)^{1/2} \right] \tag{A5.16}$$

in agreement with equation (A5.13), where

$$t^* = \frac{1}{\zeta} \tan^{-1}\left(\frac{\zeta v_1}{g}\right) \tag{A5.17}$$

Hence the modified dynamic load factor now becomes

$$\Psi = 1 + \left(1 + \frac{\zeta^2 v_1^2}{g^2}\right)^{1/2} \tag{A5.18}$$

For a *monolithic* glass beam, the equivalent spring stiffness (k_S) and maximum longitudinal bending stresses under *dynamic* central line loading are

$$k_S = \frac{48EI}{a^3}, \quad \sigma_D^* = \pm \frac{3\Psi P_1 a}{2 B_E T^2} \tag{A5.19}$$

based on conventional small-deflection theory. In the case of a *laminated* glass beam, with the cross-sectional layout depicted in Figure A5.17, two glass layers of thickness t_1 and t_2 ($t_1 \leq t_2$) enclose a plastic interlayer of thickness c and shear modulus G_p. First define the dimensionless parameters

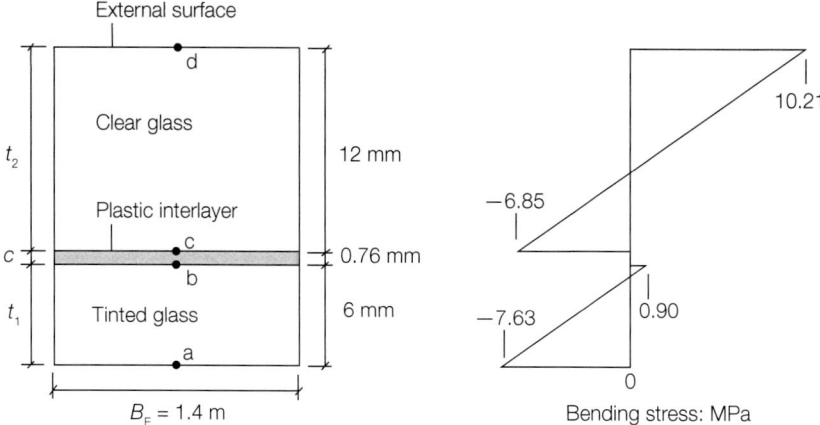

FIGURE A5.17. Calculated distributions of elastic bending stress (MPa, tension negative) from simplified dynamic beam-strip analysis of simply supported laminated glass panel (lower cone, A4 glass wall, span 2.1 m) subjected to central normal impact of 10 cm diameter hailstone at terminal velocity (45.2 m/s)

$$\mu = 1 + \frac{(t_1 + t_2)(t_1^3 + t_2^3)}{12 t_1 t_2 h^2}, \qquad \Omega = ah\left[\frac{3\mu G_p}{cE_g(t_1^3 + t_2^3)}\right]^{1/2} \tag{A5.20}$$

where $h = c + (t_1 + t_2)/2$. Then reference to appendix A1 (section A1.3.2) gives the mid-span deflection of the beam under a central *static* vertical line load P_1 as

$$w_S^* = \frac{P_1 a^3}{4 B_E \mu E_g (t_1^3 + t_2^3)}\left[\mu - 1 + \frac{3}{\Omega^3}(\Omega - \tanh\Omega)\right] \tag{A5.21}$$

so that the elastic spring stiffness becomes

$$k_S = \frac{4 B_E \mu E_g (t_1^3 + t_2^3)}{a^3\left[\mu - 1 + (3/\Omega^3)(\Omega - \tanh\Omega)\right]} \tag{A5.22}$$

Upon assuming as a first approximation that the earlier energy balance procedure for a monolithic beam directly carries over to a laminated beam of mass m_2 (e.g. linear elastic deformation of the plastic interlayer under transient loading), the maximum dynamic deflection is $w_D^* = \Psi w_S^*$, with Ψ given by equation (A5.18).

To check the structural integrity of the most vulnerable region of the glass walls, consider the largest laminated glass panels in the lower cone segment (wall A4), which have a 'free' span $a = 2.1$ m, subjected to the normal impact of a 10 cm diameter hailstone ($m_1 = 0.471$ kg) at mid-span. As a reasonable approximation for calculation purposes, such a panel is represented by a horizontal rectangular plate of *infinite* length, simply supported along the two edges, with the effective width of a beam-strip taken as $B_E = 2a/3$. The thicker glass layer is positioned *uppermost*, as sketched in Figure A5.17, and the cross-section dimensions are

$t_1 = 6$ mm, $t_2 = 12$ mm, $c = 0.76$ mm ($T = 18.76$ mm) and $B_E = 1.4$ m. The assumed material properties are $\rho_g = 2500$ kg/m^3 and $E_g = 72.4$ GPa for the density and Young's modulus of the glass, respectively, and $G_p = 0.5$ MPa for the shear modulus of the thin plastic interlayer (self-weight neglected herein).

Under a vertical *static* load $P_1 = m_1 g$, with the dimensionless parameters $\mu = 1.425$ and $\Omega = 2.90$, the small central deflection of the beam from equation (A5.21) is $w_S^* = 0.025$ mm, giving the equivalent spring stiffness $k_S = 184$ kN/m. Now assume that the hailstone strikes the centre of the beam at a vertical terminal velocity $v_T = 45.2$ m/s, as given by equation (A5.3). With $v_1 = v_T$ and $m_2 = 137.9$ kg, equation (A5.12) implies that the combined beam and attached hailstone move vertically downward at an initial velocity $v(0) = 0.329$ m/s. The dynamic load factor is $\Psi = 247$, giving a maximum *dynamic* beam deflection $w_D(t^*) = 6.2$ mm occurring at time $t^* = 0.03$ s after impact, from equations (A5.16) and (A5.17). The corresponding bending stresses (tension negative) at mid-span at the surface of each glass layer under *dynamic* central line loading are

$$\begin{Bmatrix} (\sigma_a^*)_D \\ (\sigma_b^*)_D \end{Bmatrix} = \frac{3\Psi P_1 a}{2B_E \mu (t_1^3 + t_2^3)} \left\{ \mp \mu t_1 \pm \left[t_1 \mp \frac{2t_2 h(\mu-1)}{t_1 + t_2} \right] \left(1 - \frac{\tanh \Omega}{\Omega} \right) \right\} \quad (A5.23)$$

$$\begin{Bmatrix} (\sigma_c^*)_D \\ (\sigma_d^*)_D \end{Bmatrix} = \frac{3\Psi P_1 a}{2B_E \mu (t_1^3 + t_2^3)} \left\{ \mp \mu t_2 \pm \left[t_2 \pm \frac{2t_1 h(\mu-1)}{t_1 + t_2} \right] \left(1 - \frac{\tanh \Omega}{\Omega} \right) \right\} \quad (A5.24)$$

and calculated values are given in Figure A5.17, with a maximum tension of -7.63 MPa (-1110 psi). For comparison, the corresponding surface bending stresses for a *monolithic* glass beam of the same overall thickness ($T = 18.76$ mm) are $\sigma_D^* = \pm 8.94$ MPa (± 1300 psi), giving a somewhat higher maximum tensile stress than that in the laminate. In contrast, the same laminated panel positioned with the thinner glass layer uppermost would incur a maximum tensile stress of -10.21 MPa (-1480 psi) under the same loading, thereby emphasising a tangible structural benefit of the *in-situ* laminate orientation. The calculated maximum tensile bending stress in the laminated panel can readily be accommodated by the glass, in addition to those induced by wind and self-weight loading (as given in Figure 44, section 5.7, main text), noting that the estimated impact duration is two orders of magnitude lower than for the design wind-gust loading, thereby permitting a much higher allowable design stress.

Of particular interest is that the *combined* effect of hailstone impact and self-weight loading is to increase the maximum tensile stress to -15.47 MPa (-2240 psi) at the internal surface of the thicker glass layer, as indicated in Figure A5.18(c). In the unlikely event of hailstone impact occurring simultaneously with extreme wind pressure (1.44 kPa), the superposition of stress distributions in Figures A5.18(c) and (d) increase the maximum tensile stress to -25.97 MPa (-3770 psi) in the thinner glass layer, which is only marginally higher than the maximum tensile stress in the thicker glass layer, as shown in Figure A5.18(e).

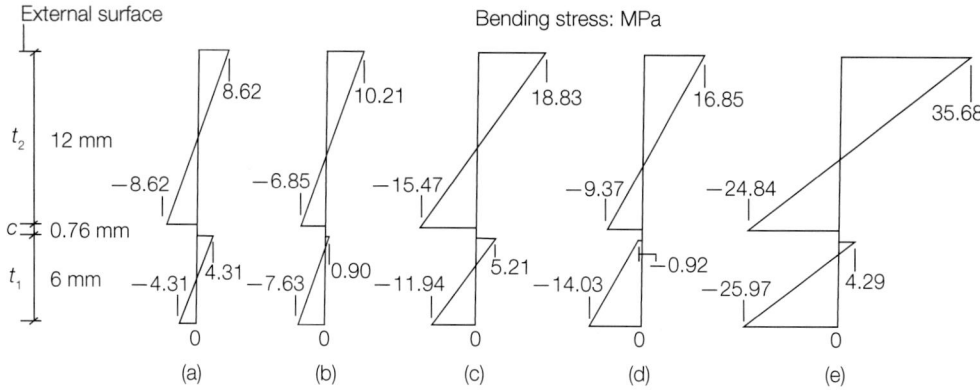

FIGURE A5.18. Combined distributions of calculated elastic bending stress (MPa, tension negative) from simplified beam-strip analysis of simply supported laminated glass panel (lower cone, A4 glass wall, span 2.1 m) subjected to hailstone impact, self-weight and wind loading (±1.44 kPa): (a) self-weight and hailstone impact; (b) self-weight and wind pressure; (c) self-weight, wind pressure and hailstone impact; (d) self-weight and wind suction; (e) self-weight, wind suction and hailstone impact

In some respects, this calculated flexure might be regarded as too severe because, in reality, other forms of energy will be dissipated during the process of 'soft' impact. Gross distortions of the hailstone, and possible partial rebound, will reduce the incident kinetic energy imparted to a laminated glass panel. Localised (non-Hertzian) deformations within the impact surface region of the glass also will absorb energy, as will the propagation of stress waves throughout the laminate. Further losses will occur at the glazing supports (*i.e.* glass panel edges typically resting on neoprene rubber shims) and within the plastic interlayer, largely due to frictional effects and nonlinear material hysteresis. Conversely, higher maximum bending stresses would be expected if the hailstone impact occurred at a butt-joint connecting two adjacent laminated panels, or if the shear modulus of the plastic interlayer is effectively much higher under such short-duration loading.

It is also noted that the bending stresses calculated using elementary beam theory will be modified within a localised region beneath the surface contact area, resulting from the superposition of a two-dimensional stress field within an elastic continuum; as revealed by the early experimental results obtained by Carus Wilson [A5.238] from photoelastic examinations of monolithic glass beams under three-point loading, with some typical graphical results reproduced by Hemsley [A5.239], together with further landmark studies by Filon [A5.240], [A5.241], Mindlin [A5.242] and others.

However, the cumulative effect of these modifying factors is difficult to quantify theoretically, and further progress in this regard would require substantial experimentation, not least to determine the nonlinear time-dependent stiffness properties of the plastic interlayer over a wide temperature range, together with details of the instantaneous physical breakdown of artificial hailstones upon impact. More rigorous solutions utilising such data could then be computed using advanced numerical methods, and compared with the observed behaviour. A brief reconnoitre of the substantial corpus of recent technical literature reveals encouraging progress towards understanding the complex dynamic response of architectural laminated glass to soft impact loading. Quite how far removed from physical reality are the core results of the present simplified analysis will doubtless be enthusiastically revealed by others in the fullness of time.

A5.5. References

A5.1. RUSSELL, F. A. R. *On hail*. Edward Stanford, London, 1893, 61–62.

A5.2. NEWMAN, B. W. *Phenomenal hailstorm with thunderstorm, Sydney 1 January 1947.* Australian Government, Bureau of Meteorology, Melbourne, 1947, Weather R&D Bull. 8, 23–41.

A5.3. BAHR, V. J., KEMP, R. L. and KURZEME, D. *Report on a study of the severe thunderstorms over Sydney Metropolitan Area on 21 August, 1971.* Australian Government, Bureau of Meteorology, Melbourne, Feb. 1973, 28 pp.

A5.4. YEO, S., LEIGH, R. and KUHNEL, I. The April 1999 Sydney hailstorm. *Natural Hazards Quarterly (Macquarie Univ.)*, 1999, 5, 2 (Jun.), 1–4; *Australian J. Emergency Management*, 1999, 14, 4, 23–25.

A5.5. LEE, L. A. and COLLINGS, A. Sydney hailstorms: the health role in the recovery process. *Medical J. Australia*, 2000, 173, 11–12 (Dec.), 579–582.

A5.6. BUCKLEY, B. W., LESLIE, L. M. and WANG, Y. The Sydney hailstorm of April 14, 1999: synoptic description and numerical simulation. *Meteorology & Atmos. Phys*, 2001, 76, 3–4 (Apr.), 167–182.

A5.7. SCHUSTER, S. S., BLONG, R. J., LEIGH, R. J. and MCANENEY, K. J. Characteristics of the 14 April 1999 Sydney hailstorm based on ground observations, weather radar, insurance data and emergency calls. *Natural Hazards & Earth System Sci.*, 2005, 5, 613–620.

A5.8. SCHUSTER, S. S., BLONG, R. J. and SPEER, M. S. A hail climatology of the Greater Sydney area and New South Wales, Australia. *Int. J. Climatology*, 2005, 25, 12, Oct., 1633–1650.

A5.9. SCHUSTER, S. S., BLONG, R. J. and MCANENEY, K. J. Relationship between radar-derived hail kinetic energy and damage to insured buildings for severe hailstorms in Eastern Australia. *Atmospheric Res.*, 2006, 81, 3 (Sept.), 215–235.

A5.10. LESLIE, L. M., LEPLASTRIER, M. and BUCKLEY, B. W. Estimating future trends in severe hailstorms over the Sydney Basin: a climate modelling study. *Atmospheric Res.*, 2008, 87, 1 (Jan.), 37–51.

A5.11. CROMPTON, R. P. and MCANENEY, K. J. Normalised Australian insured losses from meteorological hazards: 1967–2006. *Environ. Sci. & Policy*, 2008, 11, 5 (Aug.), 371–378.

A5.12. RASULY, A. A., CHEUNG, K. K. W. and MCBURNEY, B. Hail events across the Greater Metropolitan Severe Thunderstorm Warning Area. *Natural Hazards & Earth System Sci.*, 2015, 15, 5 (May), 973–984.

A5.13. ALLEN, J. T. and ALLEN, E. R. A review of severe thunderstorms in Australia. *Atmospheric Res.*, 2016, 178–179, Sept., 347–366.

A5.14. WARREN, R. A., RAMSAY, H. A., SIEMS, S. T., MANTON, M. J., PETER, J. R., PROTAT, A. and PILLALAMARRI, A. Radar-based climatology of damaging hailstorms in Brisbane and Sydney, Australia. *Quart. J. Royal Meteorological Soc.*, 2020, 146, 726 (Jan.), 505–530.

A5.15. DOWDY, A. J., SODERHOLM, J., BROOK, J., BROWN, A. and McGOWAN, H. Quantifying hail and lightning risk factors using long-term observations around Australia. *J. Geophysical Res: Atmospheres*, 2020, 125, 21, Nov., 14 pp.

A5.16. SCHUMANN, T. E. W. The theory of hailstone formation. *Quart. J. Royal Meteorological Soc.*, 1938, 64, 273 (Jan.), 3–17; Discussion: 17–21.

A5.17. LUDLAM, F. H. The composition of coagulation-elements in cumulonimbus. *Quart. J. Royal Meteorological Soc.*, 1950, 76, 327 (Jan.), 52–58.

A5.18. CHISHOLM, A. J. and ENGLISH, M. *Alberta hailstorms*. American Meteorological Society, Boston, Massachusetts, U.S.A., 1973. (Series: *Meteorological Monographs*, Vol. 14).

A5.19. GOKHALE, N. R. *Hailstorms and hailstone growth*. State Univ. New York Press, Albany, 1975.

A5.20. FOOTE, G. B. and KNIGHT, C. A. (Eds). *Hail: a review of hail science and hail suppression*. American Meteorological Society, Boston, Massachusetts, U.S.A., 1977. (Series: *Meteorological Monographs*, Vol. 16).

A5.21. PRUPPACHER, H. R. and KLETT, J. D. *Microphysics of clouds and precipitation*. Kluwer Academic Publ., Dordrecht, 1997, 2nd edn.

A5.22. WANG, P. K. *Physics and dynamics of clouds and precipitation*. Cambridge University Press, Cambridge, 2013.

A5.23. KUMAR, P. *Hailstorms: prediction, control and damage assessment*. CRC Press, Boca Raton, 2017, 2nd edn.

A5.24. CARTE, A. E. Air bubbles in ice. *Proc. Physical Soc.*, 1961, 77, 3 (Mar.), 757–768.

A5.25. KIDDER, R. E. and CARTE, A. E. Structures of artificial hailstones. *Journal de Recherches Atmosphériques*, 1964, 4, 169–181.

A5.26. BROWNING, K. A., LUDLAM, F. H. and MACKLIN, W. C. The density and structure of hailstones. *Quart. J. Royal Meteorological Soc.*, 1963, 89, 379 (Jan.), 75–84. Discussion: 1964, 90, 384 (Apr.), 213–214.

A5.27. GIAMMANCO, I. M., BROWN, T. M., GRANT, R. G., DEWEY, D. L., HODEL, J. D. and STUMPF, R. A. Evaluating the hardness characteristics of hail through compressive strength measurements. *J. Atmospheric & Oceanic Technol.*, 2015, 32, 11 (Nov.), 2100–2113.

A5.28. KENT, L. E. Probable fossil hail impressions in Natal Coal Measures (Middle Ecca series). *Nature*, 1938, 141, 3575, May, 835.

A5.29. GIRARDIN, M. Analysis of hailstones. *Phil. Mag.*, 1839, Ser. 3, 15, 95 (Sept.), 252–253.

A5.30. WALLER, A. Microscopic observations on hail. *Phil. Mag.*, 1846, Ser. 3, 29, 192 (Aug.), 103–111.

A5.31. WALLER, A. Additional observations on hail, and on the organic bodies contained in hailstones, &c. *Phil. Mag.*, 1847, Ser. 3, 30, 200 (Mar.), 159–171 (& Plate).

A5.32. HOPKINS, W. On the motion of glaciers. *Trans Camb. Phil. Soc.*, 1849, 8, 2, 159–179.

A5.33. DUFOUR, M. L. On the freezing of water and the formation of hail. *Phil. Mag.*, 1861, Ser. 4, 21, 143 (Suppl.), 543–544.

A5.34. SUTCLIFFE, T. Notice of remarkable hailstones which fell at Headingley, near Leeds, on the 7th of May, 1862. *Proc. Royal Soc. (Lond.)*, 1862, 12, 239–242.

A5.35. NASH, W. C. Notes on the unusual heat and great dryness of the atmosphere on May 19, 1868; and on the remarkable thunderstorm of May 29, 1868, as observed at Greenwich. *Proc. British Meteorological Soc.*, 1869, 4, Jun., 196–198 (& Plate).

A5.36. ANONYMOUS. Nocturnal hailstorms: hailstones of singular form. *Scientific American*, 1869, 21, 5, Jul., 72.

A5.37. ANONYMOUS. Remarkable hailstorm. *Scientific American*, 1871, 24, 21, May, 322–323.

A5.38. ANONYMOUS. A remarkable hailstorm in Arkansas. *Scientific American*, 1881, 44, 19, May, 296.

A5.39. ANONYMOUS. Severe hailstorms. *Scientific American*, 1911, 105, 25, Dec., 551.

A5.40. REINSCH, P. On the microscopic structure of hail. *Phil. Mag.*, 1871, Ser. 4, 42, 277 (Jul.), 79–80.

A5.41. ALLEN, G. W. A hailstorm at St. Louis, Mo. *Scientific American*, 1873, 28, 25, Jun., 389.

A5.42. REYNOLDS, O. On the manner in which raindrops and hailstones are formed. *Mem. Lit. & Phil. Soc. Manchester*, 18/9, Ser. 3, 6, 48–60.

A5.43. REYNOLDS, O. On the formation of hailstones, raindrops, and snowflakes. *Mem. Lit. & Phil. Soc. Manchester*, 1879, Ser. 3, 6, 161–170.

A5.44. REYNOLDS, O. *Papers on mechanical and physical subjects: reprinted from various transactions and journals.* Univ. Press, Cambridge, 1900, Vol. 1 (1869–1882).

A5.45. ERNST, A. A remarkable hailstorm. *Nature*, 1886, 34, 867, Jun., 122.

A5.46. HERDMAN, W. A. Unusually large hail. *Nature*, 1889, 40, 1023, Jun., 126.

A5.47. WILSON, A. W. G. A peculiar hailstorm. *Science*, New Ser., 1902, 16, 414, 5 Dec., 909–910.

A5.48. LANDIS, D. S. and SMITH, W. The structure of hailstones. *Monthly Weather Revue*, 1906, 34, 6 (Jun.), 277–278.

A5.49. GREGORY, W. K. The hailstorm of June 23. *Science*, New Ser., 1906, 24, 604, 27 Jul., 115–116.

A5.50. BLAIR, T. A. Hailstones of great size at Potter, Nebr. *Monthly Weather Revue*, 1928, 56, 8 (Aug.), 313 (including added notes by J. J. Norcross and A. Anderson, with Plate).

A5.51. JOHNSON, R. P. Structure of hailstones. *Nature*, 1938, 142, 3586, Jul., 172.

A5.52. LONSDALE, K. and OWSTON, P. G. X-ray examination of hail. *Nature*, 1946, 157, 3989, Apr., 479.

A5.53. LIST, R. and DE QUERVAIN, M. Zur Struktur von Hagelkörnern. *Zeit. angew. Math. Phys (ZAMP)*, 1953, 4, 6 (Nov.), 492–496.

A5.54. LIST, R. and DE QUERVAIN, M. *On the structure of hailstones.* National Research Council, Canada, 1955, N.R.C. Tech. Transl. TT-551, 7 pp.

A5.55. LIST, R. Kennzeichen atmosphärischer Eispartikeln. Teil 2: Hagelkörner. *Zeit. angew. Math. Phys (ZAMP)*, 1958, 9, 3 (Sept.), 217–234.

A5.56. SCHLEUSENER, R. A. Hailstorm characterisation and the crystal structure of hail. In: *Meteorological Monographs, American Meteorological Society, Boston*, Sept. 1963, Vol. 5 (*Severe local storms*), 27, 173–176.

A5.57. BROWNING, K. A., HALLETT, J., HARROLD, T. W. and JOHNSON, D. The collection and analysis of freshly fallen hailstones. *J. Appl. Meteorology*, 1968, 7, 4 (Aug.), 603–612.

A5.58. LEVI, L., ACHAVAL, E. and AUFDERMAUR, A. N. Crystal orientation in a wet growth hailstone. *J. Atmospheric Sci.*, 1970, 27, 3 (May), 512–513.

A5.59. ROGERS, L. N. Characteristics of a large number of hailstones from a single Alberta hailstorm. *Proc. Conf. Cloud Phys, Fort Collins, Colorado, U.S.A., Aug. 1970.* Amer. Meteorological Soc., Washington, D.C., 1970, 89–90.

A5.60. MACKLIN, W. C., CARRAS, J. N. and RYE, P. J. The interpretation of the crystalline and air bubble structures of hailstones. *Quart. J. Royal Meteorological Soc.*, 1976, 102, 431 (Jan.), 25–44.

A5.61. KNIGHT, N. C. and HEYMSFIELD, A. J. Measurement and interpretation of hailstone density and terminal velocity. *J. Atmospheric Sci.*, 1983, 40, 6 (Jun.), 1510–1516.

A5.62. Abbe, C. (Ed.). Depth of hail fall. *Monthly Weather Rev.*, 1897, 25, 9 (Sept.), 399.

A5.63. Lowe, A. B. Some unusual hailstones. *Weatherwise*, 1965, 18, 2 (Apr.), 87.

A5.64. Pflaum, J. C. New clues for decoding hailstone structure. *Bull. Amer. Meteorological Soc.*, 1984, 65, 6 (Jun.), 583–593.

A5.65. Browning, K. A. and Landry, C. R. The structure and growth of the giant hailstones. In: *A family outbreak of severe local storms: a comprehensive study of the storms in Oklahoma on 26 May 1963, Part I.* Air Force Cambridge Research Laboratories, Bedford, Massachusetts, U.S.A., Sept. 1965 (eds K. A. Browning and T. T. Fujita), Rep. AFCRL-65-695(I), Ch. 14, 283–317.

A5.66. Browning, K. A. The lobe structure of giant hailstones. *Quart. J. Royal Meteorological Soc.*, 1966, 92, 391 (Jan.), 1–14.

A5.67. Browning, K. A. and Beimers, J. G. D. The oblateness of large hailstones. *J. Appl. Meteorology*, 1967, 6, 6 (Dec.), 1075–1081.

A5.68. Knight, C. A. On the mechanism of spongy hailstone growth. *J. Atmospheric Sci.*, 1968, 25, 3 (May), 440–444.

A5.69. Knight, C. A. and Knight, N. C. Spongy hailstone growth criteria. I. Orientation fabrics. *J. Atmospheric Sci.*, 1968, 25, 3 (May), 445–452.

A5.70. Knight, C. A. and Knight, N. C. Spongy hailstone growth criteria. II. Microstructures. *J. Atmospheric Sci.*, 1968, 25, 3 (May), 453–459.

A5.71. Knight, C. A. and Knight, N. C. Hailstone embryos. *J. Atmospheric Sci.*, 1970, 27, 4 (Jul.), 659–666.

A5.72. Knight, C. A. and Knight, N. C. Lobe structures of hailstones. *J. Atmospheric Sci.*, 1970, 27, 4 (Jul.), 667–671.

A5.73. Knight, C. A. and Knight, N. C. The falling behaviour of hailstones. *J. Atmospheric Sci.*, 1970, 27, 4 (Jul.), 672–681.

A5.74. Knight, C. A. and Knight, N. C. Hailstones. *Scientific American*, 1971, 224, 4 (Apr.), 96–104.

A5.75. Knight, C. A. and Knight, N. C. Very large hailstones from Aurora, Nebraska. *Bull. Amer. Meteorological Soc.*, 2005, 86, 12 (Dec.), 1773–1781.

A5.76. Prodi, F. Measurements of local density in artificial and natural hailstones. *J. Appl. Meteorology*, 1970, 9, 6 (Dec.), 903–910.

A5.77. Witt, A., Burgess, D. W., Seimon, A., Allen, J. T., Snyder, J. C. and Bluestein, H. B. Rapid-scan radar observations of an Oklahoma tornadic hailstorm producing giant hail. *Weather & Forecasting*, 2018, 33, 5 (Oct.), 1263–1282.

A5.78. Soderholm, J. S., Kumjian, M. R., McCarthy, N., Maldonado, P. and Wang, M. Quantifying hail size distributions from the sky: application of drone aerial photogrammetry. *Atmospheric Meas. Tech.*, 2020, 13, 2, Feb., 747–754.

A5.79. Shedd, L., Kumjian, M. R., Giammanco, I., Brown-Giammanco, T. and Maiden, B. R. Hailstone shapes. *J. Atmospheric Sci.*, 2021, 78, 2 (Feb.), 639–652.

A5.80. Roth, R. J. Hailstones and hailstorms. *Weatherwise*, 1952, 5, 3 (Jun.), 51–54.

A5.81. Frisby, E. M. Hailstorms of the Upper Great Plains of the United States. *J. Appl. Meteorology*, 1963, 2, 6 (Dec.), 759–766.

A5.82. Changnon, S. A. The scales of hail. *J. Appl. Meteorology*, 1977, 16, 6 (Jun.), 626–648.

A5.83. Vittori, O. and Caporiacco, G. di. The density of hailstones. *Nubila*, 1959, 2, 1, 51–57.

A5.84. GIAIOTTI, D., GIANESINI, E. and STEL, F. Heuristic considerations pertaining to hailstone size distributions in the plain of Friuli-Venezia Giulia. *Atmospheric Res.*, 2001, 57, 4 (May), 269–288.

A5.85. GIAIOTTI, D., NORDIO, S. and STEL, F. The climatology of hail in the plain of Friuli Venezia Giulia. *Atmospheric Res.*, 2003, 67–68 (Jul.–Sept.), 247–259.

A5.86. PALENCIA, C., GIAIOTTI, D., STEL, F., CASTRO, A. and FRAILE, R. Maximum hailstone size: relationship with meteorological variables. *Atmospheric Res.*, 2010, 96, 2–3 (May), 256–265.

A5.87. BALDI, M., CIARDINI, V., DALU, J. D., DE FILIPPIS, T., MARACCHI, G. and DALU, G. Hail occurrence in Italy: towards a national database and climatology. *Atmospheric Res.*, 2014, 138, Mar., 268–277.

A5.88. MACKLIN, W. C., STRAUCH, E. and LUDLAM, F. H. The density of hailstones collected from a summer storm. *Nubila*, 1960, 3, 1, 12–17.

A5.89. MOSSOP, S. C. and KIDDER, R. E. Hailstorm at Johannesburg on 9th November 1959. Part II: structure of hailstones. *Nubila*, 1961, 4, 74–86.

A5.90. CARTE, A. E. and HELD, G. Variability of hailstorms on the South African Plateau. *J. Appl. Meteorology*, 1978, 17, 3 (Mar.), 365–373.

A5.91. FRISBY, E. M. and SANSOM, H. W. Hail incidence in the Tropics. *J. Appl. Meteorology*, 1967, 6, 2 (Apr.), 339–354.

A5.92. LIST, R., CANTIN, J. G. and FERLAND, M. G. Structural properties of two hailstone samples. *J. Atmospheric Sci.*, 1970, 27, 7 (Oct.), 1080–1090.

A5.93. GOYER, G. G. The hailstorm giant. *Bull. Amer. Meteorological Soc.*, 1978, 59, 4 (Apr.), 429.

A5.94. PAUL, A. H. Hailstorms in southern Saskatchewan. *J. Appl. Meteorology*, 1980, 19, 3 (Mar.), 305–314.

A5.95. FEDERER, B. and WALDVOGEL, A. Hail and raindrop size distributions from a Swiss multicell storm. *J. Appl. Meteorology*, 1975, 14, 1 (Feb.), 91–97.

A5.96. WALDVOGEL, A., KLEIN, L., MUSIL, D. J. and SMITH, P. L. Characteristics of radar-identified big drop zones in Swiss hailstorms. *J. Climate & Appl. Meteorology*, 1987, 26, 8 (Aug.), 861–877.

A5.97. TREFALT, S., MARTYNOV, A., BARRAS, H., BESIC, N., HERING, A. M., LENGGENHAGER, S., NOTI, P., RÖTHLISBERGER, M., SCHEMM, S., GERMANN, U. and MARTIUS, O. A severe hail storm in complex topography in Switzerland – Observations and processes. *Atmospheric Res.*, 2018, 209, Sept., 76–94.

A5.98. DESSENS, J. Hail in southwestern France. I: Hailfall characteristics and hailstorm environment. *J. Climate & Appl. Meteorology*, 1986, 25, 1 (Jan.), 35–47.

A5.99. DESSENS, J. Hail in southwestern France. II: Results of a 30-year hail prevention project with silver iodide seeding from the ground. *J. Climate & Appl. Meteorology*, 1986, 25, 1 (Jan.), 48–58.

A5.100. VINET, F. Climatology of hail in France. *Atmospheric Res.*, 2001, 56, 1–4 (Jan.), 309–323.

A5.101. BERTHET, C., DESSENS, J. and SANCHEZ, J. L. Regional and yearly variations of hail frequency and intensity in France. *Atmospheric Res.*, 2011, 100, 4 (Jun.), 391–400.

A5.102. MERINO, A., WU, X., GASCÓN, E., BERTHET, C., GARCÍA-ORTEGA, E. and DESSENS, J. Hailstorms in southwestern France: incidence and atmospheric characterisation. *Atmospheric Res.*, 2014, 140–141 (Apr.–May), 61–75.

A5.103. STEINER, J. T. New Zealand hailstorms. *New Zealand J. Geol. & Geophys*, 1989, 32, 2, 279–291.

A5.104. KOTINIS-ZAMBAKAS, S. R. Average spatial patterns of hail days in Greece. *J. Climate*, 1989, 2, 5 (May), 508–511.

A5.105. SIOUTAS, M., MEADEN, G. T. and WEBB, J. D. C. Hail frequency, distribution and intensity in Northern Greece. *Atmospheric Res.*, 2009, 93, 1–3 (Jul.), 526–533.

A5.106. MICHAELIDES, S. C., SAVVIDOU, K., NICOLAIDES, K. A., ORPHANOU, A., PHOTIOU, G. and KANNAOUROS, C. Synoptic, thermodynamic and agroeconomic aspects of severe hail events in Cyprus. *Nat. Hazards Earth Syst. Sci.*, 2008, 8, 3, May, 461–471.

A5.107. POČAKAL, D., VEČENAJ, Ž. and ŠTALEC, J. Hail characteristics of different regions in continental part of Croatia based on influence of orography. *Atmospheric Res.*, 2009, 93, 1–3 (Jul.), 516–525.

A5.108. WEBB, J. C. D., ELSOM, D. M. and MEADEN, G. T. Severe hailstorms in Britain and Ireland, a climatological survey and hazard assessment. *Atmospheric Res.*, 2009, 93, 1–3 (Jul.), 587–606.

A5.109. SÁNCHEZ, J. L., GIL-ROBLES, B., DESSENS, J., MARTIN, E., LOPEZ, L., MARCOS, J. L., BERTHET, C., FERNÁNDEZ, J. T. and GARCÍA-ORTEGA, E. Characterisation of hailstone size spectra in hailpad networks in France, Spain, and Argentina. *Atmospheric Res.*, 2009, 93, 1–3 (Jul.), 641–654.

A5.110. TUOVINEN, J-P., PUNKKA, A-J., RAUHALA, J., HOHTI, H. and SCHULTZ, D. M. Climatology of severe hail in Finland: 1930–2006. *Monthly Weather Revue*, 2009, 137, 7 (Jul.), 2238–2249.

A5.111. SALTIKOFF, E., TUOVINEN, J-P., KOTRO, J., KUITUNEN, T. and HOHTI, H. A climatological comparison of radar and ground observations of hail in Finland. *J. Appl. Meteorology & Climatology*, 2010, 49, 1 (Jan.), 101–114.

A5.112. XIE, B., ZHANG, Q. and WANG, Y. Observed characteristics of hail size in four regions in China during 1980–2005. *J. Climate*, 2010, 23, 18, Sept., 4973–4982.

A5.113. LI, X., ZHANG, Q., ZOU, T., LIN, J., KONG, H. and REN, Z. Climatology of hail frequency and size in China, 1980–2015. *J. Appl. Meteorology & Climatology*, 2018, 57, 4 (Apr.), 875–887.

A5.114. LI, M., ZHANG, D-L., SUN, J. and ZHANG, Q. A statistical analysis of hail events and their environmental conditions in China during 2008–15. *J. Appl. Meteorology & Climatology*, 2018, 57, 12 (Dec.), 2817–2833.

A5.115. MEZHER, R. N., DOYLE, M. and BARROS, V. Climatology of hail in Argentina. *Atmospheric Res.*, 2012, 114–115 (Oct.), 70–82.

A5.116. KUMJIAN, M. R., GUTIERREZ, R., SODERHOLM, J. S., NESBITT, S. W., MALDONADO, P., LUNA, L. M., MARQUIS, J., BOWLEY, K. A., IMAZ, M. A. and SALIO, P. Gargantuan hail in Argentina. *Bull. Amer. Meteorological Soc.*, 2020, 101, 8 (Aug.), E1241–E1258.

A5.117. PILORZ, W. Very large hail occurrence in Poland from 2007 to 2015. *Contemp. Trends Geosci.*, 2015, 4, 1, Oct., 46–56.

A5.118. KAHRAMAN, A., TILEV-TANRIOVER, Ş., KADIOGLU, M., SCHULTZ, D. M. and MARKOWSKI, P. M. Severe hail climatology of Turkey. *Monthly Weather Rev.*, 2016, 144, 1 (Jan.), 337–346.

A5.119. PUSKEILER, M., KUNZ, M. and SCHMIDBERGER, M. Hail statistics for Germany derived from single-polarisation radar data. *Atmospheric Res.*, 2016, 178–179, Sept., 459–470.

A5.120. BRÁZDIL, R., CHROMÁ, K., VALÁŠEK, H., DOLÁK, L., ŘEZNÍČKOVÁ, L., ZAHRADNÍČEK, P. and DOBROVOLNÝ, P. A long-term chronology of summer half-year hailstorms for South Moravia, Czech Republic. *Climate Research*, 2016, 71, 2, Dec., 91–109.

A5.121. MARTINS, J. A., BRAND, V. S., CAPUCIM, M. N., FELIX, R. R., MARTINS, L. D., FREITAS, E. D., GONÇALVES, F. L. T., HALLAK, R., SILVA DIAS, M. A. F. and CECIL, D. J. Climatology of destructive hailstorms in Brazil. *Atmospheric Res.*, 2017, 184, Feb., 126–138.

A5.122. JIN, H-G., LEE, H., LKHAMJAV, J. and BAIK, J-J. A hail climatology in South Korea. *Atmospheric Res.*, 2017, 188, May, 90–99.

A5.123. Zou, T., Zhang, Q., Li, W. and Li, J. Responses of hail and storm days to climate change in the Tibetan Plateau. *Geophysical Res. Lett.*, 2018, 45, 9, May, 4485–4493.

A5.124. Wouters, L., Boon, M., van Putten, D., van't Veen, B., Koks, E. and de Moel, H. *A hail climatology of the Netherlands*. Institute for Environmental Studies, Vrije Universiteit, Amsterdam, Dec. 2019, 37 pp.

A5.125. Tani, S. and Paulitsch, H. A case study on severe hailstorm on 27 July 2019 in the province of Styria, Austria. *European Geophysical Society General Assembly, May 2020*, EGU2020-18626 (online).

A5.126. Marcos, J. L., Sánchez, J. L., Merino, A., Melcón, P., Mérida, E. and García-Ortega, E. Spatial and temporal variability of hail falls and estimation of maximum diameter from meteorological variables. *Atmospheric Res.*, 2021, 247, Jan., 105142.

A5.127. Farnell, C., Rigo, T. and Heymsfield, A. Shape of hail and its thermodynamic characteristics related to records in Catalonia. *Atmospheric Res.*, 2022, 271, Jun., 106098.

A5.128. Hull, B. B. *Hail size and distribution*. Environmental Protection Research Division, Quartermaster Research and Engineering Centre, U.S. Army, Natick, Massachusetts, U.S.A., Feb. 1958, Tech. Rep. EP-83, 101 pp.

A5.129. Prein, A. F. and Holland, G. J. Global estimates of damaging hail hazard. *Weather & Climate Extremes*, 2018, 22, Dec., 10–23.

A5.130. Suwała, K. and Bednorz, E. Climatology of hail in central Europe. *Quaestiones Geographicae*, 2013, 32, 3, 99–110.

A5.131. Punge, H. J. and Kunz, M. Hail observations and hailstorm characteristics in Europe: a review. *Atmospheric Res.*, 2016, 176–177 (Jul.–Aug.), 159–184.

A5.132. Púčik, T., Castellano, C., Groenemeijer, P., Kühne, T., Rädler, A. T., Antonescu, B. and Faust, E. Large hail incidence and its economic and societal impacts across Europe. *Monthly Weather Revue*, 2019, 147,11 (Nov.), 3901–3916.

A5.133. Cheng, L., English, M. and Wong, R. Hailstone size distributions and their relationship to storm thermodynamics. *J. Appl. Meteorology & Climatology*, 1985, 24, 10 (Oct.), 1059–1067.

A5.134. Cecil, D. J. and Blankenship, C. B. Toward a global climatology of severe hailstorms as estimated by satellite passive microwave imagers. *J. Climate*, 2012, 25, 2, Jan., 687–703.

A5.135. Gower, K., Fontaine, J. B., Birnbaum, C. and Enright, N. J. Sequential disturbance effects of hailstorm and fire on vegetation in a Mediterranean-type ecosystem. *Ecosystems*, 2015, 18, 7 (Nov.), 1121–1134.

A5.136. Martius, O., Hering, A., Kunz, M., Manzato, A., Mohr, S., Nisi, L. and Trefalt, S. Challenges and recent advances in hail research. *Bull. Amer. Meteorological Soc.*, 2018, 99, 3 (Mar.), 51–54.

A5.137. Murillo, E. M. and Homeyer, C. R. Severe hail fall and hailstorm detection using remote sensing observations. *J. Appl. Meteorology & Climatology*, 2019, 58, 5 (May), 947–970.

A5.138. Newton, I. *Philosophiae naturalis principia mathematica*. Joseph Streater for the Royal Society, London, 1687, 1st edn; Cambridge, 1713, 2nd edn; William and John Innys, London, 1726, 3rd edn: *The mathematical principles of natural philosophy*. Benjamin Motte, London, 1729, 2 vols (transl. Andrew Motte).

A5.139. Hauksbee, F. Experiments concerning the time required in the descent of different bodies, of different magnitudes and weights, in common air, from a certain height. *Phil. Trans Royal Soc. (Lond.)*, 1710, 27, 328 (Oct.–Dec.), 196–198.

A5.140. DESAGULIERS, J. T. An account of some experiments made on the 27th day of April, 1719, to find how much the resistance of the air retards falling bodies. *Phil. Trans Royal Soc. (Lond.)*, 1719, 30, 362 (Sept.–Oct.), 1071–1075; A further account of experiments made for the same purpose, upon the 27th day of July last. By the same. 1075–1078 (& Plate).

A5.141. WALLIS, J. A discourse concerning the measure of the airs resistance to bodies moved in it. *Phil. Trans Royal Soc. (Lond.)*, 1687, 16, 186 (Jan.–Mar.), 269–280.

A5.142. LOOMIS, E. On the resistance experienced by bodies falling through the atmosphere. *Amer. J. Sci. & Arts*, 1854, Ser. 2, 18, 52, Jul., 67–70.

A5.143. BILHAM, E. G. and RELF, E. F. The dynamics of large hailstones. *Quart. J. Royal Meteorological Soc.*, 1937, 63, 269 (Apr.), 149–160. Discussion: 160–162.

A5.144. HEYMSFIELD, A. J. and WRIGHT, R. L. Graupel and hail terminal velocities: does a 'supercritical' Reynolds number apply? *J. Atmospheric Sci.*, 2014, 71, 9, Sept., 3392–3403.

A5.145. MILLIKAN, C. B. and KLEIN, A. L. The effect of turbulence: an investigation of maximum lift coefficient and turbulence in wind tunnels and in flight. *Aircraft Engng & Aerospace Technol.*, 1933, 5, 8 (Aug.), 169–174.

A5.146. GORBUSHIN, A and VOLOBUEV, V. S. The first aerodynamic balances in Russia. *Proc. 9th Int. Symp. Strain-Gauge Balances, Seattle, U.S.A., May 2014*, 14 pp.

A5.147. ZAHM, A. F. Resistance of the air at speeds below one thousand feet a second. *Phil. Mag.*, 1901, Ser. 6, 1, 5 (May), 530–535.

A5.148. EIFFEL, G. Sur la résistance des sphères dans l'air en mouvement. *Compt. Rend. Acad. Sci., Paris*, 1912, 155, 27, Dec., 1597–1599.

A5.149. STRUTT, J. W. (3RD BARON RAYLEIGH). Sur la résistance des sphères dans l'air en mouvement. *Compt. Rend. Acad. Sci., Paris*, 1913, 156, 2, Jan., 109–110.

A5.150. MAURAIN, C. Action d'un courant d'air sur des sphères. *Bull. de l'Inst. Aérotechnique de l'Univ. de Paris*, 1913, 3, 76–85.

A5.151. BACON, D. L. and REID, E. G. *The resistance of spheres in wind tunnels and in air*. National Advisory Committee for Aeronautics, Washington, D.C., U.S.A., Jan. 1924, N.A.C.A. Tech. Rep. 185, 24 pp.

A5.152. PRANDTL, L. Der Luftwiderstand von Kugeln. *Nachrichten von der Königlichen Gesellschaft der Wissenschaften zu Göttingen, Mathematisch-physikalische Klasse*, 1914, 177–190.

A5.153. TOLLMIEN, W., SCHLICHTING, H., GÖRTLER, H. and RIEGELS, F. W. (Eds). *Ludwig Prandtl Gesammelte Abhandlungen zur angewandten Mechanik, Hydro- und Aerodynamik*. Springer-Verlag, Berlin, 1961, 597–608.

A5.154. PRANDTL, L. The generation of vortices in fluids of small viscosity. *J. Royal Aeronautical Soc.*, 1927, 31, 200 (Aug.), 718–741 (*15th Wilbur Wright Memorial Lecture*, 16 May 1927).

A5.155. PRANDTL, L. *Führer durch die Strömungslehre*. Friedrich Vieweg & Sohn, Brunswick, 1942. *Essentials of fluid dynamics: with applications to hydraulics aeronautics, meteorology, and other subjects*. Blackie & Son, London, 1952 (transl. W. M. Deans).

A5.156. WIESELSBERGER, C. Der Luftwiderstand von Kugeln. *Zeitschrift für Flugtechnik und Motorluftschiffahrt*, 1914, 5, 9, May, 140–145.

A5.157. WIESELSBERGER, C. Weitere Feststellung über die Gesetze des Flüssigkeits-und Luftwiderstandes. *Physikalische Zeitschrift*, 1922, 23, 10, May, 219–224. *Further information on the laws of fluid resistance*. National Advisory Committee for Aeronautics, Washington, D.C., U.S.A., Dec. 1922, N.A.C.A. Tech. Note 121, 12 pp.

A5.158. MAXWORTHY, T. Experiments on the flow around a sphere at high Reynolds numbers. *J. Appl. Mech., Trans Amer. Soc. Mech. Engrs*, 1969, 36, 3 (Sept.), 598–607.

A5.159. IKUSHIMA, Y., GUO, W. W. and OHJI, M. Measurements of surface pressure distributions on a freely falling sphere with a tripping wire. *JSME International Journal, Ser. B, Fluids & Thermal Engng*, 1993, 36, 2 (May), 272–278.

A5.160. GOLDSTEIN, S. (Ed.). *Modern developments in fluid dynamics. An account of theory and experiment relating to boundary layers, turbulent motion and wakes*. Clarendon Press, Oxford, 1957, 2 vols; Dover Publ., New York, 1965, 2 vols.

A5.161. SCHLICHTING, H. *Boundary-layer theory*. McGraw-Hill, New York, 1979 (transl. J. Kestin), 7th edn.

A5.162. LIST, R. Zur Aerodynamik von Hagelkörnern. *Zeit. angew. Math. Phys (ZAMP)*, 1959, 10, 2 (Mar.), 143–159.

A5.163. MACKLIN, W. C. and LUDLAM, F. H. The fallspeeds of hailstones. *Quart. J. Royal Meteorological Soc.*, 1961, 87, 371 (Jan.), 72–81.

A5.164. BAILEY, I. H. and MACKLIN, W. C. The surface configuration and internal structure of artificial hailstones. *Quart. J. Royal Meteorological Soc.*, 1968, 94, 399 (Jan.), 1–11.

A5.165. WILLIS, J. T., BROWNING, K. A. and ATLAS, D. Radar observations of ice spheres in free fall. *J. Atmospheric Sci.*, 1964, 21, 1 (Jan.), 103–108.

A5.166. HOERNER, S. Versuche mit Kugeln betreffend Kennzahl, Turbulenz und Oberflächenbeschaffenheit. *Luftfahrtforschung*, 1935, 12, 1 (Mar.), 42–54. *Tests of spheres with reference to Reynolds number, turbulence, and surface roughness*. National Advisory Committee for Aeronautics, Washington, D.C., U.S.A., Oct. 1935, N.A.C.A. Tech. Memo. 777 (transl. J. Vanier), 36 pp.

A5.167. YOUNG, R. G. E. and BROWNING, K. A. Wind tunnel tests of simulated spherical hailstones with variable roughness. *J. Atmospheric Sci.*, 1967, 24, 1 (Jan.), 58–62.

A5.168. LANDRY, C. R. and HARDY, K. R. Fall speed characteristics of simulated ice spheres: a radar experiment. *Proc. 14th Conf. Radar Meteorology, Tucson, Arizona, U.S.A., Nov. 1970*. American Meteorological Society, Boston, Massachusetts, U.S.A., 1970, 27–30.

A5.169. RAITHBY, G. D. and ECKERT, E. R. G. The effect of support position and turbulence intensity on the flow near the surface of a sphere. *Wärme- und Stoffübertragung*, 1968, 1, 2 (Jun.), 87–94.

A5.170. CLAMEN, A. and GAUVIN, W. H. Effects of turbulence on the drag coefficients of spheres in a supercritical flow regime. *AIChE J. (Amer. Inst. Chem. Engrs)*, 1969, 15, 2 (Mar.), 184–189.

A5.171. ROOS, D. VAN DER S. A giant hailstone from Kansas in free fall. *J. Appl. Meteorology*, 1972, 11, 6 (Sept.), 1008–1011.

A5.172. ROOS, D. VAN DER S. and CARTE, A. E. The falling behaviour of oblate and spiky hailstones. *J. de Recherches Atmosphériques*, 1973, 7, 39–52.

A5.173. LIST, R., RENTSCH, U. W., BYRAM, A. C. and LOZOWSKI, E. P. On the aerodynamics of spheroidal hailstone models. *J. Atmospheric Sci.*, 1973, 30, 4 (May), 653–661.

A5.174. LOZOWSKI, E. P. and BEATTIE, A. G. Measurements of the kinematics of natural hailstones near the ground. *Quart. J. Royal Meteorological Soc.*, 1979, 105, 444 (Apr.), 453–459.

A5.175. MATSON, R. J. and HUGGINS, A. W. The direct measurement of the sizes, shapes and kinematics of falling hailstones. *J. Atmospheric Sci.*, 1980, 37, 5 (May), 1107–1125.

A5.176. BÖHM, H. P. A general equation for the terminal fall speed of solid hydrometeors. *J. Atmospheric Sci.*, 1989, 46, 15, Aug., 2419–2427.

A5.177. HEYMSFIELD, A. J., GIAMMANCO, I. M. and WRIGHT, R. L. Terminal velocities and kinetic energies of natural hailstones. *Geophys. Res. Lett.*, 2014, 41, 23, Dec., 8666–8672.

A5.178. LAURIE, J. A. P. *Hail and its effects on buildings*. National Building Research Institute, Pretoria, South Africa, 1960, Bull. 21, Rep. 176, 12 pp. South African Council for Scientific and Industrial Research, 1960, Vol. 176.

A5.179. HEYMSFIELD, A. J., SZAKÁLL, M., JOST, A., GIAMMANCO, I. and WRIGHT, R. L. A comprehensive observational study of graupel and hail terminal velocity, mass flux, and kinetic energy. *J. Atmospheric Sci.*, 2018, 75, 11 (Nov.), 3861–3885.

A5.180. ACHENBACH, E. Experiments on the flow past spheres at very high Reynolds numbers. *J. Fluid Mech.*, 1972, 54, 3, Aug., 565–575.

A5.181. ACHENBACH, E. The effects of surface roughness and tunnel blockage on the flow past spheres. *J. Fluid Mech.*, 1974, 65, 1, Aug., 113–125.

A5.182. ACHENBACH, E. Vortex shedding from spheres. *J. Fluid Mech.*, 1974, 62, 2, Jan., 209–221.

A5.183. TANEDA, S. Visual observations of the flow past a sphere at Reynolds numbers between 10^4 and 10^6. *J. Fluid Mech.*, 1978, 85, 1, Mar., 187–192.

A5.184. BAKIĆ, V. and PERIĆ, M. Visualisation of flow around sphere for Reynolds numbers between 22000 and 400000. *Thermophys & Aeromech.*, 2005, 12, 3, 307–315.

A5.185. BAKIĆ, V., SCHMID, M. and STANKOVIĆ, B. Experimental investigation of turbulent structures of flow around a sphere. *Thermal Sci.*, 2006, 10, 2, 97–112.

A5.186. NORMAN, A. K. and MCKEON, B. J. The effect of a small isolated roughness element on the forces on a sphere in uniform flow. *Exper. Fluids*, 2011, 51, 4 (Oct.), 1031–1045.

A5.187. NORMAN, A. K. and MCKEON, B. J. Unsteady force measurements in sphere flow from subcritical to supercritical Reynolds numbers. *Exper. Fluids*, 2011, 51, 5 (Nov.), 1439–1453.

A5.188. VENNING, J. A., PEARCE, B. W. and BRANDNER, P. A. Scale effects on cavitation about a sphere. *Proc. 22nd Australasian Fluid Mech. Conf. (AFMC2020), Brisbane, Australia, Dec. 2020*, 4 pp.

A5.189. JEONG, J. and HUSSAIN, F. On the identification of a vortex. *J. Fluid Mech.*, 1995, 285, Feb., 69–94.

A5.190. TOMBOULIDES, A. G. and ORSZAG, S. A. Numerical investigation of transitional and weak turbulent flow past a sphere. *J. Fluid Mech.*, 2000, 416, Aug., 45–73.

A5.191. CONSTANTINESCU, G. S. and SQUIRES, K. D. Numerical investigations of flow over a sphere in the subcritical and supercritical regimes. *Phys Fluids*, 2004, 16, 5 (May), 1449–1466.

A5.192. YUN, G., KIM, D. and CHOI, H. Vortical structures behind a sphere at subcritical Reynolds numbers. *Phys Fluids*, 2006, 18, 1 (Jan.), 015102.

A5.193. HOFFMAN, J. Simulating drag crisis for a sphere using skin friction boundary conditions. *Proc. Euro. Conf. Computational Fluid Dynamics (ECCOMAS CFD 2006), Egmond aan Zee, Netherlands, Sept. 2006*, 7 pp.

A5.194. MORADIAN, N., TING, D. S-K. and CHENG, S. The effects of freestream turbulence on the drag coefficient of a sphere. *Exper. Thermal & Fluid Sci.*, 2009, 33, 3 (Mar.), 460–471.

A5.195. MORADIAN, N., TING, D. S-K. and CHENG, S. Advancing drag crisis of a sphere via the manipulation of integral length scale. *Wind & Structures*, 2011, 14, 1 (Jan.) 35–53.

A5.196. WANG, Y. F., TAO, G. Q., LIU, D. X., HU, J. Z. and WU, Z. Numerical study of flow over sphere at supercritical Reynolds numbers. *Adv. Mater. Res.*, 2014, 1077, Dec., 207–214.

A5.197. LINIĆ, S., RISTIĆ, S., STEFANOVIĆ, Z., KOZIĆ, M. and OCOKOLJIĆ, G. Experimental and numerical study of super-critical flow around the rough sphere. *Sci. Tech. Rev.*, 2015, 65, 2, Jan., 11–19.

A5.198. DESHPANDE, R., KANTI, V., DESAI, A. and MITTAL, S. Intermittency of laminar separation bubble on a sphere during drag crisis. *J. Fluid Mech.*, 2017, 812, Feb., 815–840.

A5.199. Terra, W., Sciacchitano, A. and Scarano, F. Aerodynamic drag of a transiting sphere by large-scale tomographic-PIV. *Exper. Fluids*, 2017, 58, 7 (Jul.), 83.

A5.200. Terra, W., Sciacchitano, A., Scarano, F. and Oudheusden, B. W. van. Drag resolution of a PIV wake rake for transiting models. *Exper. Fluids*, 2018, 59, 7 (Jul.), 120.

A5.201. Pendar, M-R. and Roohi, E. Cavitation characteristics around a sphere: an LES investigation. *Int. J. Multiphase Flow*, 2018, 98, Jan., 1–23.

A5.202. Nakhostin, S. M. and Giljarhus, K. E. T. Investigation of transitional turbulence models for CFD simulation of the drag crisis for flow over a sphere. *IOP Conf. Ser: Mater. Sci. & Engng*, 2019, 700, 012007.

A5.203. Böhm, J. P. Review of flow characteristics and kinematics of hydrometeors in free fall. *Atmospheric Res.*, 1991, 26, 4 (Aug.), 285–302.

A5.204. List, R. New hailstone physics. Part I: heat and mass transfer (HMT) and growth. *J. Atmospheric Sci.*, 2014, 71, 4 (Apr.), 1508–1520.

A5.205. List, R. New hailstone physics. Part II: interaction of the variables. *J. Atmospheric Sci.*, 2014, 71, 6 (Jun.), 2114–2129.

A5.206. Cheng, K-Y. and Wang, P. K. A numerical study of the flow fields around falling hails. *Atmospheric Res.*, 2013, 132–133 (Oct.–Nov.), 243–263. Corrigendum: 2014, 135–136 (Jan.), 205.

A5.207. Cheng, K-Y., Wang, P. K. and Wang, C-K. A numerical study on the ventilation coefficients of falling hailstones. *J. Atmospheric Sci.*, 2014, 71, 7 (Jul.), 2625–2634.

A5.208. Wang, P. K., Chueh, C-C. and Wang, C-K. A numerical study of flow fields of lobed hailstones falling in air. *Atmospheric Res.*, 2015, 160, Jun., 1–14.

A5.209. Wang, P. K. and Chueh, C-C. A numerical study on the ventilation coefficients of falling lobed hailstones. *Atmospheric Res.*, 2020, 234, Apr., 104737.

A5.210. Theis, A., Borrmann, S., Mitra, S. K., Heymsfield, A. J. and Szakáll, M. A wind tunnel investigation into the aerodynamics of lobed hailstones. *Atmosphere*, 2020, 11, 5 (May), 494, 18 pp.

A5.211. Geier, M., Pasquali, A. and Schönherr, M. Parametrisation of the cumulant lattice Boltzmann method for fourth order accurate diffusion. Part I: derivation and validation. *J. Computational Phys*, 2017, 348, Nov., 862–888.

A5.212. Geier, M., Pasquali, A. and Schönherr, M. Parametrisation of the cumulant lattice Boltzmann method for fourth order accurate diffusion. Part II: application to flow around a sphere at drag crisis. *J. Computational Phys*, 2017, 348, Nov., 889–898.

A5.213. Bearman, P. W. and Harvey, J. K. Golf ball aerodynamics. *Aeronaut. Quart.*, 1976, 27, 2 (May), 112–122.

A5.214. Choi, J., Jeon, W-P. and Choi, H. Mechanism of drag reduction by dimples on a sphere. *Phys Fluids*, 2006, 18, 4 (Apr.), 0417021–0417024.

A5.215. Smith, C. E., Beratlis, N., Balaras, E., Squires, K. and Tsunoda, M. Numerical investigation of the flow over a golf ball in the subcritical and supercritical regimes. *Proc. 6th Int. Symp. Turbulence and Shear Flow Phenomena (TSFP6), Seoul, Korea, Jun. 2009*, 1013–1018. In: *Int. J. Heat & Fluid Flow*, 2010, 31, 3 (Jun.), 262–273.

A5.216. Alam, F., Steiner, T., Chowdhury, H., Moria, H., Khan, I., Aldawi, F. and Subic, A. A study of golf ball aerodynamic drag. *Procedia Engng*, 2011, 13, 226–231.

A5.217. Aoki, K., Muto, K. and Okanaga, H. Mechanism of drag reduction by dimple structures on a sphere. *J. Fluid Sci. & Technol.*, 2012, 7, 1, 1–10.

A5.218. Bogdanović-Jovanović, J. B., Stamenković, Ž. M. and Kocić, M. M. Experimental and numerical investigation of flow around a sphere with dimples for various flow regimes. *Thermal Sci.*, 2012, 16, 4, 1013–1026.

A5.219. Li, J., Tsubokura, M. and Tsunoda, M. Numerical investigation of the flow around a golf ball at around the critical Reynolds number and its comparison with a smooth sphere. *Flow, Turbulence & Combustion*, 2015, 95, 2–3 (Oct.), 415–436.

A5.220. Spálenský, V. and Rozehnal, D. CFD simulation of dimpled sphere and its wind tunnel verification. *Proc. Conf. Dynamics Civil Engng & Transport Struct. & Wind Engng (DYN-WIND 2017), Trstena, Slovakia, May 2017*. MATEC Web of Conferences Vol. 107, 2017 (eds J. Melcer and K. Kotrasova), EDP Sciences, Paris, 531–538.

A5.221. Crabill, J., Witherden, F. and Jameson, A. High-order computational fluid dynamics simulations of a spinning golf ball. *Sports Engng*, 2019, 22, 1 (Mar.), Art. 9.

A5.222. Beratlis, N., Balaras, E. and Squires, K. On the origin of the drag force on dimpled spheres. *J. Fluid Mech.*, 2019, 879, Nov., 147–167.

A5.223. Zahm, A. F. *Flow and drag formulas for simple quadrics*. National Advisory Committee for Aeronautics, Washington, D.C., U.S.A., Jan. 1927, N.A.C.A. Tech. Rep. 253, 29 pp.

A5.224. Clift, R., Grace, J. R. and Weber, M. E. *Bubbles, drops, and particles*. Academic Press, New York, 1978; Dover, New York, 2005 (with Errata).

A5.225. Mitchell, D. L. Use of mass- and area-dimensional power laws for determining precipitation particle terminal velocities. *J. Atmospheric Sci.*, 1996, 53, 12, Jun., 1710–1723.

A5.226. Almedeij, J. Drag coefficient of flow around a sphere: matching asymptotically the wide trend. *Powder Technol.*, 2008, 186, 3, Sept., 218–223.

A5.227. Cheng, N-S. Comparison of formulas for drag coefficient and settling velocity of spherical particles. *Powder Technol.*, 2009, 189, 3, Feb., 395–398.

A5.228. Tiwari, S. S., Pal, E., Bale, S., Minocha, N., Patwardhan, A. W., Nandakumar, K. and Joshi, J. B. Flow past a single stationary sphere, 1. Experimental and numerical techniques. *Powder Technol.*, 2020, 365, Apr., 115–148.

A5.229. Tiwari, S. S., Pal, E., Bale, S., Minocha, N., Patwardhan, A. W., Nandakumar, K. and Joshi, J. B. Flow past a single stationary sphere, 2. Regime mapping and effect of external disturbances. *Powder Technol.*, 2020, 365, Apr., 215–243.

A5.230. Dieling, C., Smith, M. and Beruvides, M. Review of impact factors of the velocity of large hailstones for laboratory hail impact testing consideration. *Geosciences*, 2020, 10, 12, 500, 16 pp.

A5.231. Young, T. *A course of lectures on natural philosophy and the mechanical arts*. Joseph Johnson, London, 1807, 2 vols.

A5.232. Hodgkinson, E. Impact upon beams. *Rep. Brit. Assoc. Adv. Sci.*, 1835, 5, 93–116. Reprinted by Richard Taylor, London, 1836.

A5.233. Cox, H. On impact on elastic beams. *Trans Camb. Phil. Soc.*, 1856, 9, 1, 73–78.

A5.234. Cox, H. The dynamical deflection and strain of railway girders. *The Civil Engineer and Architect's Journal; Scientific and Railway Gazette*, 1848, 11, Sept., 258–264.

A5.235. Cox, H. The dynamical deflection and strain of railway girders. *J. Franklin Inst.*, 1849, 3rd Ser., 17, 2 (Feb.), 73–81: 3 (Mar.), 145–153.

A5.236. Strutt, J. W. (3rd Baron Rayleigh). *The theory of sound*. Macmillan, London, 1877, Vol. 1.

A5.237. Todhunter, I. *A history of the theory of elasticity and of the strength of materials from Galilei to the present time. Edited and completed by K. Pearson*. University Press, Cambridge, 1893, Vol. 2 (*From Galilei to Lord Kelvin*), Pt 1; Dover Publ., New York, 1960.

A5.238. CARUS WILSON, C. A. The influence of surface-loading on the flexure of beams. *Phil. Mag.*, 1891, 5th Ser., 32, 199, Dec., 481–503.

A5.239. HEMSLEY, J. A. *Glass in engineering science.* Society of Glass Technology, Sheffield, 2015, Vol. 1 (*Optical birefringence in glass*). Ch. 3 (*Optical birefringence in glass: diverse applications*), 251–387.

A5.240. FILON, L. N. G. On an approximate solution for the bending of a beam of rectangular cross-section under any system of load, with special reference to points of concentrated or discontinuous loading. *Phil. Trans Royal Soc. (Lond.)*, 1903, A201, 63–155.

A5.241. FILON, L. N. G. The investigation of stresses in a rectangular bar by means of polarised light. *Phil. Mag.*, 1912, 6th Ser., 23, 133, Jan., 1–25.

A5.242. MINDLIN, R. D. A reflection polariscope for photoelastic analysis. *Rev. Sci. Instrum.*, 1934, 5, Jun., 224–228.

PRINCIPAL NOTATION

a	width of rectangular plate; cantilever length of beam
a	length of beam-strip (hailstone impact)
b	length of rectangular plate; distance between load and end of beam
c	interlayer thickness (glass laminate)
c_d	dimensionless drag coefficient (hailstone impact)
c_p	non-dimensional air pressure coefficient ($2p/\rho_a v^2$)
d_s	diameter of bonded glass *ballotini* (flow around sphere)
$f(\phi)$	general function
$f'(\phi)$	first derivative of $f(\phi)$
$\hat{f}(k)$	Laplace transform of $f(\phi)$
g	gravitational acceleration
h	distance between mid-planes of glass layers $[c + (t_1 + t_2)/2]$
h_1	vertical drop height of hailstone
i, j	non-negative integers
jnd	just noticeable difference (chromaticity)
k	positive integer; Laplace transform parameter
k_S	spring stiffness of beam-strip (hailstone impact)
l	half-length of loaded portion of beam
m	non-negative *even* integer
m_1	mass of hailstone
m_2	mass of beam-strip (hailstone impact)
n	positive or negative *odd* integer
p	wind pressure or suction
p	applied normal stress; colour purity (chromaticity)
p_s	maximum pressure of N-wave (sonic boom)
r, θ, z	cylindrical co-ordinates
s	positive integer
t	thickness of monolithic plate or laminated beam layer; elapsed time
t^*	elapsed time to maximum deflection of beam-strip (hailstone impact)
t_1, t_2	glass layer thicknesses in 3-ply laminate ($t_1 \leq t_2$)
v	velocity
v_T	terminal velocity of hailstone
v_1	vertical impact velocity of hailstone
w	normal displacement or deflection
w^*	maximum deflection

w_D^*	maximum dynamic vertical deflection (hailstone impact)
w_S^*	equivalent maximum static vertical deflection (hailstone impact)
x, y	chromaticity coefficients
x_s, y_s	chromaticity coefficients (colour standard)
x, y, z	Cartesian co-ordinates; normalised tristimulus values (chromaticity)
A_j, B_j	arbitrary constants (laminated beam)
A_n, A_n^*, A_n^+	Fourier series coefficients (rectangular plate)
A_1	cross-sectional area of hailstone
A_2	cross-sectional area of beam-strip (hailstone impact)
B	width of beam
B_n^*	Fourier series coefficients (rectangular plate)
B_E	effective width of beam-strip (hailstone impact)
C	colour (chromaticity)
$\mathbb{C}(\alpha_n, y)$	bespoke exponential function
D	flexural stiffness of plate $[Et^3/12(1 - v^2)]$
D_1	diameter of hailstone
D_{6500}	standard illuminant (chromaticity)
E	Young's modulus of elasticity
E	equal-energy white illuminant (chromaticity)
E_g	Young's modulus of glass
E_n	collective term in infinite series (hyperbolic functions)
$F_j(\xi, \lambda)$	influence factors (beam flexure, partial uniform loading)
$F_j^*(\xi, \lambda)$	maximum value of $F_j(\xi, \lambda)$
F_n	Fourier series coefficients
G	shear modulus $[E/2(1 + v)]$
G	Catalan's constant
$G_j(\xi, \eta, \delta)$	influence factors (beam flexure, with overhang, four-point loading)
$G_j^*(\xi, \eta, \delta)$	maximum value of $G_j(\xi, \eta, \delta)$
G_p	shear modulus of plastic interlayer
$H_j(\xi, \lambda)$	influence factors (beam flexure, no overhang, four-point loading)
$H_j^*(\xi, \lambda)$	maximum value of $H_j(\xi, \lambda)$
$H(\phi)$	Heaviside unit step function
I	second moment of area $(BT^3/12)$
I_h	second moment of area of glass layers $[B(t_1^3 + t_2^3)/12]$
K_1	kinetic energy of hailstone
L	half-span of beam
M	bending moment
$\bar{M}$	average moment sum
M^*	maximum bending moment
M_x	bending moment per unit length parallel to x-axis
M_{xy}	twisting moment per unit length in section perpendicular to x-axis
M_{xy}^*	maximum twisting moment

PRINCIPAL NOTATION

M_y	bending moment per unit length parallel to y-axis
M_1, M_2	major and minor principal bending moments
M_1^*	maximum principal bending moment
P	normal load
P_c	concentrated corner force (rectangular plate)
P_d	drag force acting on hailstone
P_E	equivalent applied static force (hailstone impact)
P_1	vertical force exerted by hailstone
R_x	normal edge reaction per unit length perpendicular to x-axis
R_y	normal edge reaction per unit length perpendicular to y-axis
R_x^T	total normal edge reaction
Re	Reynolds number (air flow)
S	interfacial shear force (laminated beam)
S_A, S_B, S_C	standard illuminants (chromaticity)
$\mathbb{S}(\alpha_n, y)$	bespoke exponential function
T	plate or beam thickness; total thickness of glass laminate ($t_1 + t_2 + c$)
T_s	duration of N-wave (sonic boom)
U_1	momentum of hailstone
V_x	normal shear force per unit length perpendicular to x-axis
V_y	normal shear force per unit length perpendicular to y-axis
V^T	total vertical shear force
W	white point (chromaticity, representing illuminant)
X, Y, Z	tristimulus values (chromaticity)
Y	percentage light transmission
α	reciprocal of characteristic length (glass laminate) $[h(\mu B G_p / c E_g I_h)^{1/2}]$
α_n	$n\pi/a$
γ_1	$t_1/(t_1 + t_2)$
γ_2	$t_2/(t_1 + t_2)$
δ	$(b - a)/L$ (beam)
ζ	time factor (hailstone impact)
$\zeta(s)$	Riemann Zeta function
η	a/L (beam)
η_d	dynamic viscosity (hailstone impact)
η_k	kinematic viscosity (hailstone impact)
θ	angle; airstream orientation (wind tunnel tests)
θ_1, θ_2	orthogonal directions of principal bending moments
λ	l/L (beam); wavelength
λ_c	Lamé elastic constant $[\nu E/(1 + \nu)(1 - 2\nu)]$
λ_d	dominant wavelength (chromaticity)
μ	Lamé elastic constant $[E/2(1 + \nu)]$
μ	dimensionless parameter (glass laminate) $\{1 + [(t_1 + t_2)I_h / Bt_1 t_2 h^2]\}$
ν	Poisson's ratio
ν_g	Poisson's ratio of glass

ξ	x/L (beam)
ρ	density
ρ_a	density of air
ρ_g	density of glass
ρ_1	density of hailstone
ρ_2	density of beam-strip (hailstone impact)
σ	normal stress
σ^*	maximum normal stress
σ_D^*	maximum bending stress (hailstone impact)
$\sigma_x, \sigma_y, \sigma_z$	normal stress components (Cartesian co-ordinates)
τ	interfacial shear stress (laminated beam)
τ^*	maximum shear stress
$\tau_{xy}, \tau_{xz}, \tau_{yz}$	shear stress components (Cartesian co-ordinates)
ϕ	angle; slope of beam; argument of function
ϕ^*	maximum slope
ϕ_s	polar angle of boundary layer separation (flow around sphere)
ϕ_x, ϕ_y	Cartesian components of plate slope
Δ_n, Δ_n^+	series coefficients (rectangular plate)
$\Theta(\xi,\eta,\delta)$	auxiliary function (laminated beam)
$\Lambda(s)$	infinite series (rectangular plate)
$\Upsilon(s)$	infinite series (rectangular plate)
$\Phi(\eta,\delta)$	auxiliary factor (laminated beam)
Ψ	dynamic load factor (hailstone impact)
Ω	dimensionless stiffness parameter αL (laminated beam)
∇^2	two-dimensional Laplacian operator
∇^4	two-dimensional biharmonic operator

SUBJECT INDEX

abraded metal   85
abrasive material   77
abrasive wheel   82
abseiler access   75
absolute temperature   68
absorption spectra   38
accelerated weatherometer test   31
acetic acid   89
acidic sealant   34, 89, 90
acoustic design   14, 108, 109
acoustic insulation   40
acoustic panel   14
acoustic performance   40, 108
adhesion   1, 30, 34, 39, 51, 70, 77, 90, 98, 155
adhesive contact   51, 115, 116
adjustable fixing bracket   31, 35–37, 41, 58, 74, 75, 77, 107
aerial view   7, 9, 99, 101
aerodynamic drag   413
aerodynamics   24, 107, 413, 434–437
aerodynamic similarity   24
aerosol   390
aesthetic quality   5, 30, 32, 34, 36, 38, 39, 41, 99
agriculture   256, 262–265, 317, 356, 366, 387, 389, 392
air bubble   34, 338, 397, 401, 428, 429
air-conditioning   77, 90
aircraft   61, 62, 66, 67, 84, 366, 390, 409
aircraft industry   36, 47, 56, 116
aircraft noise   40
aircraft propeller thrust   62
aircraft windscreen   51
air density   26–28, 404
airflow   25
air pressure   26, 27, 61, 67
airstream   26, 27
air velocity   24, 26, 27
Akkadian narrative   224–227
albumen photoprint   7
allowable tensile stress   65, 425

alternating infinite series   183, 184, 191
aluminium alloy   32
aluminium bronze   36
aluminium sphere   413
ambient lighting   39, 166
ambient temperature   34, 46, 47, 49, 51, 52, 54–62, 85, 154–159, 393, 397, 398, 404
American narrative   2, 226, 262, 267, 269, 270, 273, 305, 306, 319, 325, 330, 331, 334, 343, 352, 359, 377, 380, 383, 401, 413
amulet   356, 357, 387
analysis   1, 2, 5, 10, 11, 29, 30, 51, 52, 60, 63–66, 68, 90, 91, 112
analytical solution   1, 52, 60
Anglo-Saxon codex   273
Anglo-Saxon narrative   270–274, 375
annealed glass   1, 30, 33, 34, 36, 37, 45–49, 51, 53, 57, 59, 60, 65, 67, 75, 77, 98, 112
annealing lehr   75
annular disc   45, 48
anticlastic flexure   1, 48, 49, 176
antiplane core   115
Arabic narrative   269, 349, 374, 375
arch   11, 73
archaeology   269, 354
architect   1–3, 29–32, 35, 39–42, 45–48, 61, 99
architectural brief   41, 99
architectural consortium   2
architectural form   4, 11, 25, 28, 32, 40–42, 47, 61, 80
architectural glass   38, 47, 116, 158, 184, 394, 427
architectural laminated glass   1, 47, 51, 61, 112, 115, 116, 154, 155
architectural model   2, 3, 10, 15, 18–24
archive material   1, 2, 8, 15, 107, 223, 240, 272, 412
arrised edge   34, 48, 53, 82, 83
Arthurian legend   277, 280, 299, 335, 336
articulated model   23
arts centre   11

aryballos 253
aspect ratio 63, 175, 182–185, 188–194
astrology 267, 293, 363
astronomy 173, 256, 267, 372
asymmetric laminate 40, 63, 81, 154, 160–162
asymptotic solution 407
atmospheric aerosol 390
atmospheric conditions 34, 54, 63, 66, 255, 262, 294, 374, 382, 390, 393, 397, 398, 404, 407, 428–431, 434
atmospheric electric field 262
atmospheric moisture 63
Australian narrative 108, 111, 333, 427
autoclave 80
auxetic material 183
axial hole 48
axial load 48

Babylonian narrative 224, 226, 228, 370
ballotini 413, 414
bamboo engraving 250
beam deflection 52, 54–57, 60–62, 115–119, 123–126, 129, 132–137, 144, 146, 148, 421–425, 438
beam flexure 1, 51, 55, 115–149
beam influence factor 119–133, 137–147
beam slope 116–126, 129, 132, 134–137, 146, 148
beam vibration 422, 423
bending moment 48, 51, 117–119, 139, 141, 152, 156, 157, 171–221
bending resistance 51, 115
bending stiffness 28, 29, 52, 53, 61, 171
bending strain 51, 52, 56, 60
bending stress 52, 53, 62–64, 66, 124, 125, 132, 137, 148, 152, 153, 159–162
Bennelong Point 2, 5–9, 107
bespoke exponential function 176
Biblical narrative 228–230, 233–237, 240, 246, 247, 250, 268, 303, 305, 358, 370, 371
Bicheroux process 75
bifurcation 53, 162
biharmonic equation 172, 179
binocular vision 72
birefringent glass transducer 30, 48, 50, 112
blue-green glass 71
bluff body 24, 407
body force 172
body-tinted glass 37–39, 41, 67, 75, 78–80
bolted patch fixing 45, 48
bond 32, 33, 40, 52, 55, 56, 89, 413, 414

bond durability 32
bonded cover piece 22
bonded fibre pad 84, 88
bone carving 240, 243, 363, 364
boundary condition 28, 62, 118, 119, 142, 146, 171, 173, 179, 181
boundary element method 183
boundary layer 156, 410–415, 418, 435
boundary-wall interference 22
box-beam 41, 116
brittle material 65
bronze amulet 356, 357
bronze cover piece 22, 32, 33, 89, 90
bronze fixing bracket 36, 37, 74, 77
bronze glazing bar 15, 32, 33, 35, 36, 74, 77, 81, 85
bronze louvre wall 11, 29
bronze monorail 75
bronze nosing 41
bronze sheeting 15, 22
buckling 29
butt-joint 30–32, 34, 35, 42, 45, 58, 89–91, 151, 421, 426

calibration 51, 56, 70, 154, 408
cantilever 25, 41, 139, 148
Cartesian co-ordinates 173
cartouche 224, 225
carved bone 240, 243, 363, 364
cascade failure 61, 168, 169
Catalan's constant 184
cavitation topology 415
Celtic narrative 277, 280, 343
ceramic tiles 12, 14
chalk drawing 295
characteristic length 117
characteristic loading 57
chemical bond 32, 40
chemical compatibility 32, 89
chemical composition 77
chemical reaction 89, 90
chevron pattern 12, 14
Chinese narrative 247, 250, 270, 288, 290, 346, 353, 362, 371, 391, 401, 432
Christian narrative 228, 246, 272, 297, 327, 356, 359, 371, 375, 388
chromaticity chart 68–70, 72, 73
chromaticity co-ordinates 68–70
chromaticity space 68
C.I.E. coefficients 68, 71–73, 113
cine photography 166, 168

INDEX 447

circular clamp   48
circular glass disc   48
circular patch fixing   42, 47
circular platen   48
circular profile   7, 12
circular saw   81
circular table   71, 80
circular tube   35, 151
cladding   2, 12, 14, 108, 394
clay pot   75, 78
clay tablet   226–228, 370
clear glass   33, 38–41, 45, 46, 48, 49, 53, 54, 57, 59, 60, 63, 64, 67, 71, 75
clear plastic interlayer   39, 77
closed-form solution   52, 61, 62, 116, 174, 182
cloud formation   239, 294, 395, 396
cloud seeding   366, 370, 390–392
coating   34, 37, 38, 50, 82
codex   230, 245, 272, 273, 294, 371, 375–378
colorimetric analysis   68, 113
colorimetric data   67, 77
colour control   72, 73
colour-defective vision   68
colour difference   68, 70, 72, 113
colouring oxide   38, 67
colour matching   72
colour measurement   67, 68, 70, 77, 113, 330
colour mixture   72
colour perceptibility ellipse   72
colour purity   68, 72
colour quality   39, 68, 71
colour reference stimuli   68
colour rendering   68
colour saturation   71
colour standard   67, 71, 73
colour temperature   68
colour vision   68
Commission Internationale de l'Éclairage (C.I.E.)   68, 69
competition   2, 3, 7, 14, 15, 107, 173, 350
complementary colorant   38
composite beam   52
composite mullion   22
composite panel   37, 62, 77
compound fracture   30
compressive force   29, 33, 48, 66, 89, 111
compressive strength   66
compressive stress   63
computational fluid dynamics   411, 418, 438
concentrated loading   33, 45, 66, 99, 181, 421, 439

concert hall   2, 11, 13, 15, 18, 22, 40–44, 91, 101, 102, 108, 109, 160
concrete box-beam   41
concrete canopy   40, 93
concrete corbel   29, 74, 75, 77
concrete mullion   15
concrete pedestal   11, 12, 91
concrete rib   11, 12, 14, 16, 28, 29, 40, 41, 73, 75
concrete roof   11, 12, 73
concrete tile lid   12
conical surface   28, 29, 40, 41, 54, 63, 65, 81, 84, 90, 91, 97
conservation of energy   66, 422
conservation of momentum   422
constitutive model   159
construction tolerance   30, 34, 36, 63, 67, 73, 75, 81, 84, 151
contact region   33, 45, 48, 55, 66, 172
contact separation   172, 181
control specimen   48, 54, 60, 67, 70–73, 154
convergence   176, 182, 184, 191
copper oxide   38
copper plate engraving   230, 293, 380
corner patch support   30, 42, 45, 46, 48
corner reaction   181, 182
corona discharge   262
correlated colour temperature   68
corrosion   32, 36, 108
coupled shear wall   116, 149
cover piece   22, 32, 33, 89, 90
cover strip   35
crack bifurcation   53, 162
crack pattern   48, 53, 54, 59, 61, 162, 164
crack propagation   34, 82, 90, 162, 166
crack speed   162
creep deformation   32, 51, 52, 55, 116, 156
creep-loading test   54, 55
critical flaw   61
critical path   151
cross-wind mode   22
crystalline structure   338, 399, 429
cumulonimbus cloud   397
cuneiform text   226–228, 370
curing   32, 34, 85, 89, 90
curved glass panel   28, 30, 98
curvilinear triangle   10
cutting fluid   82
cutting table   82, 83
cyclone   28

cylindrical profile   176
cylindrical surface   28, 29, 81

damped oscillator   423
damping effect   40, 422
Danish narrative   2, 3, 111, 326, 331, 375
decolorisation   38
deflection   1, 29, 41, 51–57, 60–62, 115–149, 151–169, 171–221, 394, 421–425, 438
deformation   32, 34, 48, 51, 52
delamination   35, 56, 82, 90, 98, 99
demi-topaze glass   39, 41, 53, 54, 67, 75
density   26–28, 63, 294, 393–439
design competition   2, 3, 7, 14, 15, 107
design criterion   29, 65
design loading   425
diagonal bracing   29
dial gauge   51, 54, 60, 154
diametral loading   112
diamond-tipped saw   81, 82
differential equation   117, 173, 176, 407, 415, 422
differential temperature   90
differential thermal expansion   32, 39
diffusion   418, 437
dimensional tolerance   67, 81
dimpled sphere   418, 438
discoloration   39, 98
displacement   32, 34, 41, 56, 60, 115, 158, 172
distortion   30, 37, 98, 115
divalent cation   38
dominant wavelength   68, 70–72
drag coefficient   406–414, 418, 420, 435–438
drag crisis   410, 418, 436, 437
drag resistance   393–439
drawing   2, 10, 15, 20, 22, 47, 48, 230, 245, 277, 278, 294, 331, 334, 335, 343, 344, 354, 355, 376, 377
drill attachment   84, 89
durability   32, 39
Dutch narrative   148, 235, 236, 239, 305
dynamic air pressure   61
dynamic amplification   67
dynamic analysis   2, 65, 159, 394, 427
dynamic fatigue   55, 56, 58
dynamic interaction   151, 169
dynamic load factor   423
dynamic loading   62, 66
dynamic shear modulus   62
dynamic viscosity   404

edge condition   34, 35, 46, 48, 58, 65, 82, 85, 171–221
edge crack   48, 53, 54, 61, 90
edge deflection   60, 61
edge delamination   35, 90, 98, 99
edge reaction   171–221
edge restraint   32, 60, 158, 171, 182
edge strength   65
edge support   1, 31, 35, 39, 60, 62, 84, 112, 159, 169, 173, 180, 394, 421, 424, 426
effective width   421, 424
elapsed time   159, 407, 408
elastic analysis   1, 33, 60, 63, 91, 156, 158, 160
elastic flexure   1, 115–149, 152, 157, 171–221, 393–439
elastic property   34, 63
elastic spring   394
elliptical profile   7
elliptic cylinder   28, 81
empirical relationship   47, 159, 413, 415
energy absorption   426
energy conservation principle   66, 422, 424
energy dissipation   162, 422
engraving   230, 234–239, 250, 286, 293, 296, 304, 310–312, 320, 340, 341, 356, 360, 363, 372, 378, 383, 384
environmental condition   32, 51, 62, 66
equal-energy white   68
equilibrium   180, 181, 422, 423
ester plasticiser   155
Estonian narrative   332, 333
etching   235, 238, 239, 378
experimental method   4, 30, 45, 48, 52, 53, 62, 65, 112, 113, 115, 116, 149, 151–169
exponential function   176
extruded bronze   32, 35, 36
extruded neoprene   32

fable   223, 252, 333, 372
faceted glazing   28
failure load   48, 58, 60
failure mode   46–49, 53, 54, 61, 89, 98, 112, 154, 162, 166–169
failure probability   65
fall speed   404, 412, 435
fatigue   34, 36, 55
fatigue machine   56, 58
ferric oxide   38
ferrous oxide   38
fibre gasket   48

INDEX 449

fibreglass–epoxy disc 501
fibre optics 502
fibre pad 33, 84, 88
finite difference method 183
finite element analysis 30, 60, 91, 183
Finnish narrative 2, 331, 384, 401, 432
fire-polished surface 33, 75
fixing bracket 36, 37, 74, 77
flexural analysis 2, 64, 116, 171–221, 393–439
flexural stiffness 53, 152, 171–221, 421
flexure 1, 49, 51, 55, 171–221, 393–439
float glass 33, 39, 40, 45, 46, 48, 49, 51, 53, 54, 57, 59, 60, 67, 71, 75, 98, 154
flow field 415–417, 437
flow separation 25, 418
flow visualisation 411, 415
fluid dynamics 24, 404, 411, 418, 434–438
fluorescent light 72
foam spacer rod 32–35, 89
folklore 262, 306, 332, 333, 346
formwork 11, 75
four-edge support 31, 117, 157, 182
Fourier coefficient 174, 175
Fourier series 173, 174
four-point loading 51–58, 134–148, 156
foyer 14, 15, 37, 38, 91, 99, 101, 102
fracture origin 48, 53, 54, 61, 81, 162, 166
fracture pattern 30, 37, 46–49, 53, 61, 162–169
frameless glazing 42
free-fall motion 404–413, 435, 437
free-stream flow 25, 28
French narrative 39, 75, 89, 154, 173, 230, 231, 235, 238–240, 246, 247, 250, 255, 275, 281, 283, 286, 294, 296, 298–302, 317, 318, 334, 337, 345, 356, 357, 360, 365–368, 371, 373, 377, 379, 382, 387–391, 401, 410, 423, 431, 432
frictionless contact 115, 182
fringe pattern 51, 179
full-scale loading test 1, 30, 45–47, 57–59, 61, 62, 151–169
furnace 67, 75, 78, 79

geographical location 5–9
Georgian narrative 337, 339, 348, 385
German narrative 45, 173, 230, 232, 235, 262, 270, 290–293, 296–298, 303, 325, 328, 332, 333, 349, 363, 374, 375, 379, 384, 386, 392, 401, 410, 432
glass arrising 34, 48, 82, 83
glass assembly 154

glass *ballotini* 413, 414
glass beam 1, 51–58, 60–64, 66
glass coating 37, 38, 50
glass colour 12, 38, 39, 67–73, 113
glass crack 34, 53, 54, 59, 61, 82, 151–169
glass cutting 34, 36, 75, 77, 81–83
glass cylinder 48, 50
glass debris 166
glass decolorisation 38, 98
glass density 63, 425
glass dice 37
glass disc 48, 50
glass fatigue 34, 36, 55
glass fire-polished surface 33, 75
glass flaw 34
glass fracture 33, 37, 40, 46–48, 53, 54, 61, 67, 81, 90, 159
glass fragmentation 37, 393
glass furnace 67, 75, 78, 79
glass globe 404, 405, 408, 409
glass grinding 75, 77, 80
glass ground surface 33, 48, 67, 77
glass installation 38, 83–86
glass–interlayer bond 55
glass laminate
  annealed 1, 30, 33, 34, 36, 37, 45–49, 51, 53, 57, 59, 60, 65, 67, 75, 77, 98, 112
  beam 1, 51–58, 60–64, 66, 115–149
  beam-strip 393–439
  colour standard 67, 71, 73
  curved 28
  manufacture 75–83, 154, 155
  rectangular panel 58–60, 80, 84, 151–169
  sawing 80–83
  toughened 30, 37
glass layer 1, 32, 33, 37, 39–41, 48, 49, 51–65, 75, 77, 81, 98
glass melt 38
glass–metal contact 33, 45, 48
glass panel alignment 84
glass panel shaping 30, 80–83
glass polishing 33, 39, 48, 67, 75, 77, 80, 98
glass processing factory 73, 82, 83
glass replacement 67, 77, 98
glass ribbon 67, 75
glass safety 1, 30, 32, 36, 51, 61, 65, 393
glass sawing 80–83
glass shard 1, 36, 154
glass silanol 40
glass–silicone adhesion 34, 40, 90
glass spalling 34, 61, 82

glass–substrate bond   9, 10, 45
glass surface   3, 28, 30, 31, 33–35, 37, 41, 48, 51, 52, 61, 63, 64, 74, 75, 77, 81, 84, 89, 90, 98
glass temperature   35, 39, 41, 51, 54–57, 90, 91, 97
glass transducer   30, 48, 50, 112
glass walls   1–113
glazed assembly   46, 58, 61, 62, 151–169
glazing bar   31, 32, 35, 36, 58, 60, 61, 75, 77, 81, 84, 85, 151, 394
glazing butt-joint   30–32, 34, 35, 42, 45, 58, 89–91, 151
glazing substrate   32, 85, 89, 90, 98
glazing support system   1, 3, 5, 11, 15, 25, 28–35, 37, 39, 41, 42, 45–48, 58–66, 73–77, 81, 83–85, 88, 89, 99, 151, 169
globular aryballos   253
gold-leaf decoration   230
gouache drawing   277, 278
gravitational acceleration   407, 421
gravitational field   407
Greek manuscript   239, 246, 250, 256
Greek narrative   239, 246, 250–261, 267–269, 290, 323, 326, 331, 354–358, 361, 371–374, 387, 388, 401, 431, 432
green glass   38, 39, 71
grinding   75, 77, 80
ground glass   33, 48, 67, 77
gust pressure   28, 57, 61, 62, 66, 67, 112

hail cannon   363–369, 388, 389
hail damage   65, 223, 293, 301, 306, 308–310, 315, 318, 339–341, 348, 349, 353–358, 366–368, 381, 382, 385–393, 427, 428
hail prevention   353–392
hailstone density   397, 401–406, 413, 428–431
hailstone impact (analysis)   1, 65, 66, 393–439
hailstone impact (historical)   223–392
hailstone structure   307, 308, 339, 393–439
hailstorm   223–439
hand-coloured illustration   296
hard plastic interlayer   48, 51–53, 56–58, 154–158, 164–168
harmonic equation   179
hazard   1, 389, 427, 432, 433
heated roller   77
heat gain   38, 90
heat transfer   415, 437
heat treatment   36
Heaviside unit step function   141

Hebrew narrative   228–230, 282, 376
Hertzian contact   426
heterogeneous material   183
hieroglyph   224, 225, 269
high-energy fracture   48, 53, 162
high-speed photography   413
Hindu narrative   251
historical material   2, 5–7, 171, 173, 223–392, 394, 404, 408, 410
homogeneous material   115, 116, 183
honeycomb material   115
horizontal butt-joint   31–35, 42, 45, 58, 89–91
horizontal glazing   14, 32, 54
horticulture   339, 366, 369, 385, 392
hot-bonded sheeting   15, 22
hot-wire technique   82
humidity   85, 393
hurricane   313, 317
hydroxyl group   40

ice   255, 262, 268, 294, 297, 302–310, 317, 343, 349, 397, 399, 401, 404, 412, 413, 428, 435
Icelandic narrative   277
illuminant   68–72
illuminated manuscript   230–233, 239, 245–247, 250, 263, 264, 282–285, 288, 289, 371, 373, 376
impact loading   65, 66, 393–439
impact resistance   47, 66
imposed displacement   56
incident light beam   70, 179
inclined glazing   28, 33, 34, 37, 40, 63, 74, 75, 169
inclined view window   28–35, 40, 41, 54, 64, 72, 81, 83, 86, 90, 94, 98, 99, 102, 169
incompressible flow   415
Indian narrative   7, 251, 267, 275, 297, 323, 330, 333, 342–346, 353, 361, 362, 378, 385, 386
inelastic behaviour   116, 422
inertial force   408
infinite series   60, 118, 152, 174–184, 187, 191
influence factor   119–133, 137–147, 156
initial fracture   48, 53, 54, 166, 168
in-plane deformation   32, 34, 35, 62, 158, 172
in-plane force   32, 84
in-plane restraint   158, 172, 182
in-plane stress   35, 39, 41
in-plane support   99

INDEX 451

instrumentation   30, 51, 70, 151, 352, 382
integral transform   118, 141, 142, 144, 146
integration   144, 173, 407, 415
interfacial shear force   116–119, 134, 142
interfacial shear stress   33, 55, 118, 125, 134, 142, 144
interfacial slip   55, 115
interlayer adhesion   1, 30, 39, 51, 70, 77, 155
interlayer discoloration   39
interlayer moisture content   19, 31, 155
interlayer softening   54, 81
interlayer stiffness   51–63, 116, 125, 126, 132, 148, 154, 155, 160, 171
interlayer toughness   1, 39
ionised gas   261
Irish narrative   263, 275, 303, 314, 333, 341, 346, 384, 407
iron oxide   38, 39
irregular quadrilateral   81
I-section member   32, 151
isochromatic fringe   51, 179
isoclinic line   179
isopachic line   179
isotropic medium   1, 115, 152, 172, 183, 415
Italian narrative   173, 280–283, 290, 294–299, 340, 348, 349, 356, 376, 387

Japanese narrative   109, 281, 313, 314, 346, 360–362, 381, 382, 385
jnd   72
jnd ellipse   72
jnd ellipsoid   72
joint replacement   34, 98
joint sealant   30–35, 42, 45, 58, 60, 84, 85, 89–91, 98
just noticeable difference   72

Kew Palm House   38, 112
kinematic behaviour   413, 435, 437
kinematic viscosity   404
kinetic energy   420–436
Kirchhoff conditions   181

laboratory loading test   1, 30, 31, 34, 45–61, 65, 98, 151–169
Lamé elastic constants   174
laminar flow   410, 411, 415, 418
laminated glass
  annealed   1, 30, 33, 34, 36, 37, 45–49, 51, 53, 57, 59, 60, 65, 67, 75, 77, 98, 112
  beam   1, 51–58, 60–64, 66, 115–149

beam-strip   393–439
colour standard   67, 71, 73
curved   28
manufacture   75–83, 154, 155
rectangular panel   58–60, 80, 84, 117, 151–169
sawing   80–83
toughened   30, 37
Laplace transform   141, 142, 144, 146
Laplacian operator   173
large-deflection theory   116, 158, 159, 182
lateral buckling   29
lateral displacement   33, 75
Latin narrative   226, 230, 232, 240, 245–249, 261–268, 272, 273, 290, 299, 353, 356, 358, 372–378, 404
Latvian narrative   333, 384
lead ball   405, 406
lead iodide   366
lead shot   412
leeward suction   22
light absorption   38
light beam   70
lightning   226, 228, 254, 261, 266, 267, 281, 286, 287, 298, 302, 306, 309–311, 317, 321, 325, 326, 330, 334, 337, 346, 350, 358–363, 380, 381, 386, 428
light reflection   15, 40, 51
light transmission   68–72
lime   38, 67, 154, 184
limiting tensile stress   33, 34, 64
linear elasticity   52, 55, 60, 63, 115, 158, 394, 421, 424
linear spring   394, 421
loading duration   47, 60, 61, 65, 66, 154, 394, 425, 426
loading frequency   56
loading platen   33, 50
loading test   1, 30, 31, 34, 45–61, 65, 98, 151–169
load transducer   30, 48, 50, 112
load transfer   169
localised air pressure   24, 25
localised cracking   46
localised stress   30, 33, 51, 66
long-chain polymer   40
long-term effect   51, 65, 99, 154
louvre wall   11, 29
low-energy fracture   162
lower cone   28, 29, 40, 41, 54, 63, 65, 81, 90, 91, 97, 160, 394, 424, 426

Low German narrative   303
low-iron glass   38
luminous plasma   261
lumped mass   394, 421

MacAdam's perceptibility ellipse   72, 113
machine vibration   82, 394
Mach number   66, 404
magic   272, 290, 291, 299, 300, 303, 322, 331, 346, 351–356, 363, 364, 375, 377, 379, 387, 388
magnetophotoelasticity   179
main shell   7, 11, 12, 14, 29, 57
major principal moment   178, 198, 199, 208, 209, 216, 217
manganese bronze   32
manganese dioxide   39
manganese oxide   38
manometer   25
manufacturing cost   28, 32, 38
Maori narrative   362
maritime narrative   316, 317, 345
mass transfer   415
material properties   53, 115, 123, 156, 158, 425
Maya civilisation   269, 374
mechanical contrivance   75, 405, 408
mechanical jig   84, 85, 89
medieval narrative   230, 245, 246, 270, 272, 275, 276, 282, 290, 359, 374, 375, 379
membrane action   11, 60, 61, 115, 118, 172
mercury   404, 408
metal clamp   37, 45, 47, 48, 50, 56, 84, 85, 89
metallic oxide   39
meteorite   226, 370, 386
meteorological records   28, 65, 371
meteorology   28, 65, 112, 250, 255, 262, 268, 294, 331, 359, 360, 371–374, 382, 386–392, 400, 403, 427–435
methyl ethyl ketone   82
mezzotint engraving   239, 320
Middle Welsh narrative   277
mild steel   28, 48
military narrative   2, 7, 261, 286, 290, 306, 325, 355
miniature carving   240, 243
miniature illustration   230–233, 240, 246, 247, 250, 282–285, 371, 376
minor principal moment   178, 198, 199, 208, 209, 216, 217
mobile crane   83
model   2, 3, 10, 15, 18–28, 58, 107

moisture   63, 89
moisture content   51, 70
molten glass   38, 67, 75, 79
molten tin   75
moment–curvature relation   116, 144
moment sum   179
momentum   420
Mongolian narrative   280, 284, 323, 376
monolithic glass   30, 36, 40, 42, 45, 46, 48, 52–55, 60, 63, 64, 71
monolithic plate   1, 36, 42, 45, 48, 64, 154, 157, 160, 171–221
monorail   75
mosaic   243, 244
mullion   15, 17, 22, 23, 28, 29, 31, 35, 37, 40, 41, 58, 73–75, 77, 81, 169
mullion chord   29
mullion origin   28, 29, 394
mullion plane   29, 75
mullion tie   29, 31, 41
mullion web   29, 99
multi-layered laminate   116
multiple fracture   37
multiple impact   66, 422
mythology   223, 225, 254, 255, 258, 259, 261, 263, 266, 269, 273, 283, 303, 323, 333, 343, 346, 374, 388

natural frequency   67
natural sand   38
natural ventilation   90
nature   15, 258, 315–320, 323, 327, 334, 385, 388, 411, 418
naval aircraft   61, 62
Navier–Stokes equations   415
near-infrared spectroscopy   51, 68
neoprene rubber   32, 33, 60, 89, 426
Newtonian mechanics   407, 411
nickel oxide   39
nodal pattern   173
nonhomogeneous equation   172
nonlinear analysis   158, 159
nonlinear deformation   158
nonlinear stiffness   426
Nordic narrative   270, 331, 358, 359
normal contact pressure   415
normal displacement   115, 172, 182
Norwegian narrative   277–279
numerical analysis   5, 30, 51, 90, 116, 158, 159, 182, 413, 415, 418, 421, 427
numerical convergence   176, 182, 184, 191

INDEX 453

oblate spheroid   413
ogival arch   11
oil painting   239, 243, 244, 258, 260, 275, 276, 315, 320
Old English narrative   273, 336, 375
Old French narrative   230, 275
Old Norse narrative   276
opaque glazing   14, 38
opaque interlayer   155
opera hall   11–13, 24, 73, 160
optical distortion   37, 98
optical glass   38
optical quality   39, 98
optical reflection   40
optical retardation   179
ordinary differential equation   117, 176, 407
orthogonal angles   179
orthogonal bending moment   179
orthogonal ray   399
orthogonal shear force   180
oscillatory motion   413
oscilloscope   56, 58
outdoor testing   31, 61
out-of-plane distortion   30, 41, 98, 182
overhead glazing   14

Palm House   38, 112
papyrus   224, 256, 257, 372
papyrus parapegma   256, 257
parabolic profile   7
parapegma   256–258, 372
parchment   228–230, 239, 245–249, 270, 271, 371, 376
partial adhesion   77
partial combustion   82
partial differential equation   173, 415
partial separation   172, 181
partial shading   35, 41, 90, 91, 97
patch fixing   30, 42, 45–49
pen and ink drawing   245, 334, 335
pendulum device   405–408
perceptibility ellipse   72, 73
permanent load   33, 84, 160
permanent stress   63
permanent support pin   88, 89
Persian narrative   230, 273, 275, 328, 343, 375, 385
perturbation   24, 184
Peruvian narrative   306, 362, 380, 383
photoelastic analyser   51
photoelasticity   179, 426

photoelastic load gauge   48, 50, 51
photoelastic transducer   30, 48, 50
photometric analysis   401
phylactery   356, 357, 387
piloti   29
Pitot-static tube   62
Planckian locus   68, 69
Planckian radiator   68
plane elasticity   52
plaster of Paris   82
plastic interlayer   1, 32, 37–40, 47–63, 70, 77, 80–82, 90, 98, 115–117, 123, 154–169, 171, 423–426
plasticised polyvinyl butyral   39, 40, 51, 56, 77, 155
plasticiser content   47, 51, 155
plastic sheeting   89
plate aspect ratio   63, 175, 182–185, 188–194
plate flexure   171–221
plate glass   33, 39, 53, 54, 67, 71, 75, 77, 80, 98
platen   33, 50
plate resonance   173
plate vibration   55, 173
plywood mullion   15, 17, 22
plywood strip   116
podium substructure   1, 2, 11, 15, 22, 38, 40, 64, 83, 84, 91, 98, 99
Poisson's ratio   60, 152, 156, 172–175, 183–189, 190–221
polariscope   399, 401, 439
polarised light   51, 179, 399–402, 439
polished glass edge   48
polished plate glass   33, 39, 67, 77, 98
polyethylene foam rod   32, 34
polymer   40, 155
polyvinyl butyral   39, 40, 51, 56, 77, 155
Portuguese narrative   297, 378
postage stamp illustration   258, 259, 303, 304
post-failure behaviour   6, 61, 159–169
post-failure condition   151, 159–169
post-tensioned concrete   12
pot-casting process   67, 75, 78, 79
potential energy   182, 183, 422
precast concrete   12
pressure coefficient   25–28
primary crack   53, 54
primary design criteria   65
primary loading   62
primary seal   89
primary structure   22
primer   85

principal bending moment   178, 179, 193, 198–200, 208–210, 216–218
principal moment orientation   193, 200, 210, 218
principal moment trajectory   200, 210, 218
principal stress   179
probability   65, 404
propeller thrust   62
prototype adjustable bracket   58, 151
prototype glazing   25–27, 46, 59, 61, 62, 151–169
prototype patch fixing   45, 48, 49
prototype pin support   84
psalter   239, 245
pure bending   48
pvb resin   60

quicksilver   404, 408

radar tracking   366, 369, 391, 392, 412, 427, 430–432, 435
radiation   34, 35, 38, 39, 41
raindrop   429, 431
rainfall   31, 112, 391
rectangular glass panel   15, 45, 58, 60, 75, 77, 80, 83, 151–169
rectangular plate   1, 60, 64, 112, 171–221
reflected light   51, 321, 399–403
reflective coating   38
refractory clay pot   67, 75, 78
reinforced concrete   11, 29, 73
reinforced plywood   15
reinforcing cable   73
relative humidity   85, 393
relative retardation   399
replacement glass   77
research   22, 30, 31, 37, 47, 48
residual strength   30, 40
residual stress   82
resin   155
restaurant structure   11, 12, 15, 21, 29, 41–43, 93, 94, 99
retrospective analysis   11, 156, 182, 406, 408
return period   57, 394
Reynolds number   24, 407–415, 418–420, 434–436, 438
ribbon speed   75
ridge beam   11
Riemann Zeta function   183
rigid body   182, 268, 404, 418, 422
rigid support   60, 115, 152, 172, 182

risk assessment   32, 34
roller   51, 67, 75, 76
roller casting machine   67, 75, 79
Roman narrative   254, 258–269, 353–361, 374, 385–388
roof geometry   7, 10–16, 22
roof shell   1, 2, 10, 11, 14, 28, 40
roof structure   7, 10–12, 90, 93
rotary sawing   82
rotating disc   77
rotating table   77
rotational restraint   62
rotation invariance   179
rouge   77, 80
rough-cast glass   77, 80
roughness parameter   413
Royal Botanic Gardens   38, 112
rubber extrusion   32
rubber membrane   60
rubber pad   48
rubber strip   32, 60, 89
rune   271, 331, 332
runic alphabet   270, 271
Russian narrative   220, 325, 340, 362, 366, 369, 383, 385, 389, 410, 434

safety glass   1, 36, 61
sand   38, 67, 173
sandwich panel   115, 116
Sanskrit narrative   251, 371
satellite orbit   353, 386, 433
saw-cut edge   34, 35, 53, 65, 82, 85
sawing machine   82, 83
scale model   2, 3, 20, 26, 27, 58
Scottish narrative   258, 297, 305, 316, 322, 343, 349, 350, 358, 383
screw mechanism   84
sealant   31, 32, 34, 84, 85, 89, 90, 98, 154
sealant manufacture   32, 34, 89
secondary crack   54
secondary loading   65, 67
secondary seal   32
second moment of area   125, 153, 174, 421
security glazing   116
selenium   38
self-weight loading   1, 32, 33, 47, 54, 62–65, 88, 98, 116, 160–162, 422, 425, 426
series convergence   176, 182, 184, 187, 191
series summation   174, 175
service life   22, 31, 32
setting-out table   82

INDEX

shading blind   38, 90
shading effect   35, 39, 41, 90, 91
shear coupling   52, 62
shear displacement   158
shear force   116–119, 134, 142, 180, 171–221
shear modulus   52, 55, 56, 61, 62, 116, 117, 136, 137, 153–157, 160, 174, 423–426
shear strain   115, 159
shear strength   32, 55
shear stress   33, 55, 118, 125, 134, 142, 144
shear wall   116, 149
sheet glass   38
shell action   11
shell curvature   11
shell geometry   7, 10–14, 22
shell rib   14, 16, 28, 29, 40, 41, 44, 73, 75
shell roof   1–3, 7, 10–15, 28, 29, 38–41, 57, 58, 73, 75, 86, 93, 99, 112
shell tile   12, 14, 38, 108
shock wave   66, 67, 366, 390
short-term effect   90
short-term loading   52–56, 65, 66, 156, 160
side shell   12, 14, 15, 41, 44, 73, 94–98
silica   154
silicone adhesion   32, 34, 40, 89, 90, 98
silicone–bronze adhesion   32, 89, 90
silicone curing   32, 34, 85, 89, 90, 154
silicone–glass adhesion   34, 40, 90
silicone rubber   30–35, 42, 45, 50, 58, 60, 72, 82, 84, 85, 89–91, 98, 151, 154, 421
silver iodide   366, 391, 431
simple harmonic motion   422
simply supported beam   51, 54, 61, 62, 66, 152, 153, 157, 160–162, 421
simply supported panel   60–64, 152, 157, 158, 171–221, 394, 421, 424, 426
sketch   11, 66, 72, 85, 245, 278, 321, 350, 398, 399, 423, 424
slate amulet   356
slip   55, 115
slope   28, 116, 115–149, 171–221
small-deflection theory   60, 62, 152, 158, 172
snow loading   65
soda   67
soda–lime glass   38, 154, 184
soft impact   426, 427
soft plastic interlayer   48–61, 116, 154–157, 162, 163, 168
solar blind   38, 90, 97
solar energy   41
solar glare   38

solar heat gain   38, 90, 98
solar radiation   38–41, 54, 90
solar shading   35, 38–41, 90, 91, 97
solvent   82, 85
sonic boom   65–67, 112, 113
sonic wave   365, 366
sorcery   272, 290, 291, 299, 356, 361–364
sound attenuation   40
South African narrative   154, 340, 341, 363, 364, 397, 401, 431, 436
South American narrative   325, 330, 362
spacer rod   33–35, 89
spalling   34, 61, 82
Spanish narrative   269, 270, 302–306, 326, 373
spectral absorption   38
spectral colour   68, 113
spectral transmission   70–72
spectrophotometry   68, 70, 113
spectroscopy   51
spectrum locus   68, 69
spell   223, 358, 361
sphere   10, 12, 110, 404, 407–415, 418–420, 434–438
spherical geometry   11, 14, 16, 22
spherical seating   51
spherical segment   11
spherical triangle   12
spinning sphere   178
spring stiffness   394, 421–425
square plate   48, 49, 67
stagnation point   413–415
stagnation pressure   28
stainless steel platen   48, 50
stainless steel support pin   32, 33, 84, 88
standard deviation   72
standard glass specimen   67, 70–73
standard illuminant   68, 70
statically determinate structure   29
static fatigue   34
static loading   25, 45, 46, 51–59, 65, 66, 116, 151–169, 171–221, 394, 421–425
statistical analysis   65, 383, 391, 413, 418, 432
steel alloy   32
steel-body load gauge   51
steel disc   48
steel mullion   35, 41, 58, 74–77, 81
steel platen   48, 50
steel reinforcement   73
steel structure   3, 4, 25, 28, 29, 36, 42
steel tie   75
steel truss   29

steel tube   45, 151
steel web   29, 32, 99
stele   354, 355
stepped glass edge   82
stiffness   28, 29, 34, 52, 53, 61, 116, 137, 142, 144, 148, 152, 156, 171–176, 182, 185–189, 194–196, 202–206, 212–214, 421–426
stiffness parameter   119, 123–126, 132
stone parapegma   256–258, 372
strain energy   48, 183, 422
strain gauge   51, 52, 60, 156, 159
streamline   410, 415, 416
stress analysis   5, 30, 51, 90
stress concentration   30, 33, 45, 51
stress-corrosion   36
stress distribution   53, 63–66, 118, 125, 142, 159–162
stress-optical coefficient   51
stress reversal   56
stress wave   54, 169, 426
stroboscopic photography   413
structural analysis   10, 29, 65
structural behaviour   1, 47
structural design   62, 169, 394, 404, 420
structural glazing   30, 36
structural integrity   5, 66, 70, 151, 394, 424
structural safety   30, 51
structural sandwich   115, 116
structural steelwork   74, 75
structural stiffness   28, 29
structural support   28–31, 35
subcritical flow   411, 436, 437
substrate   98
suction box   45, 46, 58–60, 151, 154, 155, 162, 164, 166
suction force   58
suction lifting   83
suction pad   82
Sumerian narrative   224–228, 370, 371
sunlight   38, 39, 112
supercell thunderstorm   393
supercooled water   397
supercritical flow   411–413, 434–437
superposition principle   62, 118, 134, 160, 425, 426
supersonic flight   66, 67
superstition   272
surface abrasion   77, 82, 85
surface coating   34, 37, 38, 82
surface distortion   30, 37, 98
surface flatness   37, 98

surface flaw   34, 61
surface geometry   10
surface roughness   77, 80, 404, 409, 411–414, 418, 420, 435, 436
surface strain   51, 52, 56, 60
sustained loading   48, 54–57, 116, 123, 154, 156
Swedish narrative   12, 358, 359
synthetic polymer   40, 155

tangential shear force   415
tapered beam   56
tapestry   246, 247, 277, 279
tempera painting   250
temperature effect   34, 39, 54–57, 61, 62, 68, 90
temperature gradient   35, 41, 90, 91
tempered glass   36
tempest   254–256, 261–268, 275, 286, 291, 299, 303, 308, 317, 322, 328, 332, 335, 337, 373, 377, 381
tensile strength   42, 66
tensile stress   30, 33, 34, 41, 53, 61, 63–66, 124, 157, 160, 425
tensor matrices   10
terminal velocity   66, 162, 393–439
terracotta aryballos   253
test specimen   31, 38, 39, 47–54, 56, 60, 65, 67, 70–73, 151, 154–159, 162, 193, 401, 405, 408, 409, 412, 413
texture   12, 38, 399
theological narrative   235, 246, 301, 316, 359, 374, 388
theoretical analysis   1, 30, 47, 52, 53, 60, 61, 171–221
thermal distortion   37
thermal expansion   32, 39, 158
thermal fracture   90
thermal gradient   41, 90
thermal loading   90
thermally toughened glass   30, 36, 42
thermal movement   34, 35, 39, 158
thermal stress   35, 39, 41, 91
thermocouple   41, 90
thermodynamics   432, 433
thermoelastic analysis   90
thermoplastic material   47, 116, 159
thick plate   171, 172, 181, 182
thin film   37
thin-plate theory   181
three-dimensional analysis   415
three-dimensional form   10

three-ply laminate 1, 115
three-point loading 426
three-second gust 28, 57, 62, 156, 394
thunder 226, 229, 252, 263, 265, 267, 270, 272, 275, 277, 280, 286, 293, 296, 302–306, 309, 311, 317, 321, 323, 328, 333, 345, 346, 360–363, 375, 380, 381, 393
thunderstorm 254, 270, 283, 330, 348, 363, 382, 390, 393, 397, 398, 427, 428
Tibetan narrative 346, 360–364, 385, 401, 433
tile colour 12, 14, 38
time-dependent flow 34
time-dependent stiffness 426
tinted glass 33, 37–41, 63, 67, 70, 71, 78–80, 90
tinted interlayer 37
topaze normale glass 39
tornado 28
torsion 171
toughened glass 30, 36, 37, 42
transducer 30, 48, 50, 112
transient loading 32, 62, 67, 116, 424
transitional flow 410–415, 418, 436, 437
transitional surface 12, 28
translucent material 32–34, 397, 401
translucent silicone rubber 32–34
transmittance 71, 72
transverse hole 30, 45, 47, 48, 51
transverse loading 65, 116, 118, 123, 157, 159
trichromatic system 67
tristimulus value 68, 70
trivalent cation 38
tumbling motion 397, 401, 413
turbulent flow 62, 294, 410, 411, 415, 418, 434–438
Turkish narrative 246, 255–257, 349, 354, 355, 401, 432
twisting moment 171–221
two-dimensional analysis 72, 173, 426
two-dimensional photoelasticity 179
two-edge support 60, 62

ultraviolet radiation 34, 35, 90
updraft 397
uncoupled laminate 123
upper cone 28, 29, 81, 84

vacuum cup rig 82
vacuum sputtering 37
vault structure 11
V-column 29

vellum 230, 263, 288, 296, 373
velocity gradient 415
velocity vector 415, 416
vertex 10
vertical butt-joint 31, 89
vertical glazing 15, 16, 32, 33, 40, 46, 58, 84, 86
vibration 55, 82, 173, 394, 422, 423
view window 28–35, 40, 41, 54, 64, 72, 81, 83, 86, 90, 94, 98, 99, 102, 169
viscoelastic analysis 159
viscous boundary layer 415
viscous flow 415
viscous fluid 415
viscous force 408
Visigothic artifact 356, 387
visual discrimination 68, 72
visual inspection 91, 98
visual perception 12, 14, 29–31, 37, 39, 51, 68, 72, 90, 91, 98–100, 113
vortex shedding 415, 436
vorticity 294, 415, 417
vulcanised fibre 33

wake configuration 411, 415, 418, 419, 435, 437
wandering motion 418
warped surface 12, 14, 28
watercolour painting 5–7
water content 397
water ingress 98
water spray 61, 62, 82
wavelength 51, 68–72
weather lore 256, 262, 268, 306, 354, 358, 374
weatherometer test 31
weather protection 14, 22, 32, 35, 89, 151
weather resistance 31–34, 61, 98
Welsh narrative 277, 303, 309, 347, 350, 376
wet rouge 77, 80
white light 399, 400, 402
white point 68
wind buffeting 55
wind excitation 394
wind funnelling 24
wind gusting 28, 112, 156, 393
wind gust pressure 28, 57, 61, 62, 66, 67, 112, 425
wind loading 1, 22, 29, 57, 62–67, 116, 158, 394

wind pressure   25, 47, 63–65, 107, 160, 394
windshield   47, 51, 155, 156
wind speed   24, 25, 28, 393
wind suction   32, 63–65, 160–162, 426
wind tunnel test   22, 58, 409–438
windward pressure   22
witchcraft   223, 272, 290–293, 299–301, 354, 363, 377–379
woodblock print   346, 347
woodcut illustration   230, 233–235, 238, 290–292, 296–300
working platform   75, 83, 84

World Heritage Site   3, 111
Wright spectrophotometer   70, 113

X-ray micrograph   401, 403

yellow-green glass   38
Yemeni narrative   281
Young's modulus   52, 60, 117, 152, 156, 172, 421, 425

zig-zag pattern   12, 15, 16
Zulu narrative   340, 341, 363, 364